T0178464

Lecture Notes in Computer Science 12702

Founding Editors

Gerhard Goos
  Karlsruhe Institute of Technology, Karlsruhe, Germany

Juris Hartmanis
  Cornell University, Ithaca, NY, USA

Editorial Board Members

Elisa Bertino
  Purdue University, West Lafayette, IN, USA

Wen Gao
  Peking University, Beijing, China

Bernhard Steffen
  TU Dortmund University, Dortmund, Germany

Gerhard Woeginger
  RWTH Aachen, Aachen, Germany

Moti Yung
  Columbia University, New York, NY, USA

More information about this subseries at https://link.springer.com/bookseries/7412

João Manuel R. S. Tavares · João Paulo Papa ·
Manuel González Hidalgo (Eds.)

Progress in Pattern Recognition, Image Analysis, Computer Vision, and Applications

25th Iberoamerican Congress, CIARP 2021
Porto, Portugal, May 10–13, 2021
Revised Selected Papers

 Springer

Editors
João Manuel R. S. Tavares (ID)
Universidade do Porto
Porto, Portugal

João Paulo Papa (ID)
Universidade Estadual Paulista
São Paulo, Brazil

Manuel González Hidalgo (ID)
University of the Balearic Islands
Palma de Mallorca, Spain

ISSN 0302-9743   ISSN 1611-3349 (electronic)
Lecture Notes in Computer Science
ISBN 978-3-030-93419-4   ISBN 978-3-030-93420-0 (eBook)
https://doi.org/10.1007/978-3-030-93420-0

LNCS Sublibrary: SL6 – Image Processing, Computer Vision, Pattern Recognition, and Graphics

© Springer Nature Switzerland AG 2021
This work is subject to copyright. All rights are reserved by the Publisher, whether the whole or part of the material is concerned, specifically the rights of translation, reprinting, reuse of illustrations, recitation, broadcasting, reproduction on microfilms or in any other physical way, and transmission or information storage and retrieval, electronic adaptation, computer software, or by similar or dissimilar methodology now known or hereafter developed.
The use of general descriptive names, registered names, trademarks, service marks, etc. in this publication does not imply, even in the absence of a specific statement, that such names are exempt from the relevant protective laws and regulations and therefore free for general use.
The publisher, the authors and the editors are safe to assume that the advice and information in this book are believed to be true and accurate at the date of publication. Neither the publisher nor the authors or the editors give a warranty, expressed or implied, with respect to the material contained herein or for any errors or omissions that may have been made. The publisher remains neutral with regard to jurisdictional claims in published maps and institutional affiliations.

This Springer imprint is published by the registered company Springer Nature Switzerland AG
The registered company address is: Gewerbestrasse 11, 6330 Cham, Switzerland

Preface

The 25th Iberoamerican Congress on Pattern Recognition (CIARP 2021) is an annual international conference that publishes original, high-quality papers related to pattern recognition, artificial intelligence, and related fields, welcoming contributions on any aspect of theory as well as applications.

CIARP has become a key research event and one of the most important in Pattern Recognition for the Iberoamerican community. As has been the case for previous editions of the conference, CIARP 2021 hosted worldwide participants to promote and disseminate ongoing research on mathematical methods and computing techniques for artificial intelligence and pattern recognition, in particular in bioinformatics, cognitive and humanoid vision, computer vision, image analysis and intelligent data analysis, as well as their application in many diverse areas such as industry, health, robotics, data mining, opinion mining, sentiment analysis, telecommunications, document analysis, and natural language processing and recognition. Moreover, CIARP 2021 provided a forum for the scientific community to exchange research experience, share new knowledge, and increase cooperation among research groups in artificial intelligence, pattern recognition, and related areas.

CIARP has always been an open international event, and this edition received 82 contributions from 266 authors based in 21 countries. The most significant presence was from Brazil, Portugal, and Colombia. Also, we received contributions from South Africa, Argentina, Austria, Canada, France, Italy, India, the Netherlands, Spain, Sweden, Iraq, Switzerland, Mexico, Chile, Belgium, Uruguay, Cuba, and the Czech Republic.

After a rigorous blind reviewing process, where up to three highly qualified reviewers reviewed each submission (145 reviews from 63 reviewers who spent significant time and effort in reviewing the papers), 49 papers were accepted, which is an acceptance rate of 69.38%. All the accepted papers had scientific quality scores above the overall mean rating. The reviewers were chosen based on their expertise, ensuring that they came from different countries and institutions worldwide.

We want to thank all the Program Committee members for their work, which we are sure contributed to improving the selected papers' quality. The conference was held during May 10–13, 2021, in a virtual format and consisted of four days of papers, tutorials, and keynotes.

As has been the case for the most recent editions of the conference, CIARP 2021 was a single-track conference. The program comprised 12 sessions on the following topics: Natural Language Processing; Metaheuristics; Image Segmentation; Databases; Deep Learning (two sessions); Explainable Artificial Intelligence; Image Processing; Machine Learning (two sessions); and Computer Vision (two sessions).

CIARP 2021 was endorsed by IAPR, the International Association for Pattern Recognition. For this reason, the conference conferred the CIARP-IAPR Best Paper Award. CIARP 2021 also featured the Aurora Pons-Porrata Medal, which is awarded to an Iberoamerican woman in recognition of a significant contribution to pattern recognition and related fields. The authors of the best paper award and the winner of the Aurora

Pons-Porrata Medal were invited to submit their papers for publication in the Pattern Recognition Letters journal. In addition, a special issue of the Computer Methods in Biomechanics and Biomedical Engineering: Imaging & Visualization journal devoted to the event will be organized with extended versions of the best full papers relating to bio-imaging or visualization.

CIARP 2021 was organized by the Faculdade de Engenharia da Universidade do Porto (U.Porto) and the Universidade Estadual Paulista, Brazil (UNESP). We recognize and appreciate their valuable contributions to the success of CIARP 2021. We gratefully acknowledge the help of all members of the Organizing Committee and of the Local Committee for their unflagging work in the organization of CIARP 2021 that has allowed an excellent conference and proceedings.

We are especially grateful to the LNCS team at Springer for their support and advice during the preparation of this volume.

Special thanks are due to all authors who submitted their work to CIARP 2021, including those of papers that could not be accepted. Finally, we hope that these proceedings will result in a fruitful reference volume for the pattern recognition research community.

May 2021   João Manuel R. S. Tavares
                                                              João Paulo Papa
                                                     Manuel González-Hidalgo

Organization

General Chairs

João Manuel R. S.   Tavares, Universidade do Porto, Portugal
João Paulo Papa   Universidade Estadual Paulista, Brazil

Program Chair

Manuel González-Hidalgo   University of the Balearic Islands, Spain

Aurora Pons-Porrata Award Committee Chair

Leila Maria Garcia Fonseca   National Institute for Space Research, Brazil

Steering Committee

Luis Teixeira	APRP, Portugal
Eduardo Bayro-Corrochano	MACVNR, Mexico
César Beltrán-Castañón	APeRP, Peru
Julian Fierrez	AERFAI, Spain
José Ruiz Shulcloper	ACRP, Cuba
Marta Mejail	SARP, Argentina
Marcelo Mendoza	AChiRP, Chile
Joao Paulo Papa	SIGPR-BR, Brazil
Álvaro Pardo	APRU, Uruguay

Local Committee

André Pilastri	Universidade do Porto, Portugal
Gonçalo Almeida	Universidade do Porto, Portugal
Hugo Oliveira	Universidade do Porto, Portugal
Jessica Demoral	Universidade do Porto, Portugal
Zhen Ma	Universidade do Porto, Portugal
Vahid Hajihashemi	Universidade do Porto, Portugal

Aurora Pons-Porrata Award Committee

Leila Maria Garcia Fonseca	National Institute for Space Research, Brazil
Marta Mejail	Universidad de Buenos Aires, Argentina

Bernardete Ribeiro   University of Coimbra, Portugal
Reynier Ortega Bueno   Universidad de Oriente, Cuba

Program Committee

Alejandro Rosales-Perez   Centro de Investigación en Matemáticas, Mexico
Alberto Taboada-Crispi   Universidad Central "Marta Abreu" de Las Villas,
                               Cuba
Amadeo José Argüelles-Cruz   Instituto Politécnico Nacional, Mexico
Armando J. Pinho   University of Aveiro, Portugal
Alceu Britto   Pontifical Catholic University of Paraná, Brazil
Alejandro Frery   Victoria University of Wellington, New Zealand
Alessandro B. Oliveira   Federal University of Pampa, Brazil
Alexandre Levada   Federal University of São Carlos, Brazil
Alexei M. C. Machado   Pontifical Catholic University of Minas Gerais,
                               Brazil
Alicia Fernández   Universidad de la República, Uruguay
Arnaldo de Albuquerque Araujo Federal University of Minas Gerais, Brazil
Barbara C. Benato   University of Campinas, Brazil
Carlos Morimoto   University of São Paulo, Brazil
Daniel G. Acevedo   University of Buenos Aires, Argentina
Douglas Rodrigues   São Paulo State University, Brazil
Eanes Torres Pereira   Federal University of Campina Grande, Brazil
Fabricio Martins Lopes   Federal University of Technology – Paraná, Brazil
Felipe Belém   University of Campinas, Brazil
Gustavo H. de Rosa   São Paulo State University, Brazil
Gustavo Fernandez Dominguez   Austrian Institute of Technology, Austria
Hemerson Pistori   Dom Bosco Catholic University, Brazil
Humberto Sossa   National Polytechnic Institute, Mexico
Ílis de Paula   Federal University of Ceará, Brazil
Jesús Ariel Carrasco Ochoa   National Institute for Astrophysics, Optics and
                               Electronics, Mexico
Jacques Facon   Federal University of Espírito Santo, Brazil
Jefersson Alex dos Santos   Federal University of Minas Gerais, Brazil
Juan Vorobioff   National Atomic Energy Commission, Argentina
Jurandy G. A. Almeida   Federal University of São Paulo, Brazil
Luis Déniz Gómez   University of Las Palmas de Gran Canaria, Spain
Luis Enrique Sucar   Instituto Nacional de Astrofísica, Óptica y
                               Electrónica, Mexico
Leandro A. Passos   São Paulo State University, Brazil
Léo Pini Magalhaes   University of Campinas, Brazil
Leopoldo Altamirano Robles   Instituto Nacional de Astrofísica, Óptica y
                               Electrónica, Mexico

Luciano Silva	Federal University of Paraná, Brazil
Luciano Pereira Soares	Institute of Education and Research, Brazil
Luis Antonio de Souza Júnior	Federal University of São Carlos, Brazil
María del Pilar Gómez Gil	Instituto Nacional de Astrofísica, Óptica y Electrónica, Mexico
Manuel Montes-y-Gomes	National Institute of Astrophysics, Optics and Electronics, Mexico
Mara Bonates	Federal University of Ceará, Brazil
Marcos C. Silva Santana	São Paulo State University, Brazil
Maria Elena Buemi	Universidad de Buenos Aires, Argentina
Martin Kampel	Vienna University of Technology, Austria
Mateus Roder	São Paulo State University, Brazil
Michal Haindl	Institute of Information Theory and Automation, Czech Republic
Murilo Varges da Silva	Federal Institute of Education, Science and Technology of São Paulo, Brazil
Nelson Mascarenhas	Campo Limpo Paulista University Center, Brazil
Olga Bellon	Federal University of Paraná, Brazil
Pedro Real Jurado	University of Seville, Spain
Pablo Cancela	University of the Republic, Uruguay
Paulo E. Ambrosio	Santa Catarina State University, Brazil
Pedro H. Bugatti	Federal University of Technology – Paraná, Brazil
Priscila T. M. Saito	Federal University of Technology – Paraná, Brazil
Rafael Berlanga Llavori	Jaume I University, Spain
Rafael Gonçalves Pires	São Paulo State University, Brazil
Rafael Santos	National Institute for Space Research, Brazil
Sang-Woon Kim	Carleton University, Canada
Virgina L. Ballarin	National University of Mar del Plata, Argentina
Vitaly Kober	Centro de Investigacion Cientifica y de Educacion Superiorde Ensenada, Mexico
Volodymyr Ponomaryov	National Polytechnic Institute of Mexico, Mexico
Walter G. Kropatsch	Vienna University of Technology, Austria
Xiaoyi Jiang	University of Münster, Germany

Contents

Deep Learning

Explainable Artificial Intelligence

Image Processing

Machine Learning

Computer Vision

Medical Applications

Predicting the Use of Invasive Mechanical Ventilation in ICU COVID-19 Patients

Diana Serrano[1]([✉) (iD), Celeste Dias[2] (iD), Bruno Cardoso[1] (iD), and Inês Domingues[3] (iD)

[1] B-Simple, Porto, Portugal
{dserrano,bcardoso}@b-simple.pt
[2] Faculty of Medicine, University of Porto, Porto, Portugal
celeste.dias@med.up.pt
[3] Medical Physics, Radiobiology and Radiation Protection Group,
IPO Porto Research Centre (CI-IPOP), Porto, Portugal
inesdomingues@gmail.com

Abstract. Critically ill patients often need Invasive Mechanical Ventilation (IMV) when treated at intensive care units (ICU). However, it is a complex treatment that most medical doctors avoid when possible. This technique demands appropriate equipment such as ventilators and specialized personal to operate it. Patients with Coronavirus Disease (COVID-19) may need IMV, usually for an extensive period. Due to the pandemic, IMV resources became scarce, and the decision to institute mechanical ventilation based on medical judgement should be avoided unless it is absolutely necessary. This study proposes the use of clinical and laboratory data from the 24 h preceding and succeeding the ICU admission and Machine Learning classifiers such as Random Forest (RF) to predict the probability of a patient requiring IMV. The proposed methodology is split into pre-processing, modelling, and feature selection. A wide range of different classifiers with a diverse set of variables were tested. The final model is an RF model with sixteen features and a 91.88% out of sample accuracy. It can predict if a patient needs IMV, and produce an explanation for the model using Local Interpretable Model-agnostic Explanation in seven seconds. We believe this to be an advantageous tool for supporting clinical decisions, minimize ventilator-associated complications and optimize resources allocation.

Keywords: Machine learning · COVID-19 · Invasive mechanical ventilation · Intensive care unit

1 Introduction

An outbreak of Severe Acute Respiratory Syndrome Coronavirus 2 (SARS-CoV-2), causing the Coronavirus Disease (COVID-19) marked the beginning of 2020 [25]. As there is no treatment with proven efficacy, the World Health Organization (WHO) advice to delay the COVID-19 spread is self-isolation, social

© Springer Nature Switzerland AG 2021
J. M. R. S. Tavares et al. (Eds.): CIARP 2021, LNCS 12702, pp. 3–12, 2021.
https://doi.org/10.1007/978-3-030-93420-0_1

distance, and washing hands several times a day [25]. On March 11, 2020, WHO declared COVID-19 as a pandemic, the first caused by a coronavirus. The WHO has recorded more than 79 million confirmed cases of COVID-19 and more than 1.7 million deaths directly caused by this pandemic [25] on December 28, 2020. Gam-COVID-Vac known as Sputnik V became the first vaccine to have governmental approval for distribution in Russia, in August 2020 [4]. However, this vaccine was approved prior to the publication of the early results and the start of phase 3 trials [4]. The United Kingdom became the first country to authorize the use of a vaccine in Europe and the United States. The Tozinameran also know as Comirnaty received emergency authorizations from the United Kingdom, European Union and United States in December 2020 [7,9,11].

A person with a SARS-CoV-2 infection can be completely asymptomatic or present anosmia, fever, cough, or fatigue. Although these symptoms mimic the "common cold" [17], it is more severe than other influenza viruses diseases. COVID-19 can affect anyone, but is particularly severe for the elderly or patients with other comorbidities [8]. Since it affects primary the pulmonary and cardiac systems, 16.3% of COVID-19 inpatients need treatment in an Intensive Care Unit (ICU) and 11.3% require Invasive Mechanical Ventilation (IMV) [2,5]. Between 20% and 30% of COVID-19 cases develop a moderate or severe infection resulting in need of IMV, extended periods of hospitalization and eventually death with multiple organ failure. *Richardson et al.* [21] and *Myers et al.* [16] studied the statics factors of hospitalizations in New York City Area and California respectively. The first have considered 5700 and the second 377 hospitalizations due to infection by SARS-CoV-2 and reported that 14.2% and 29.97% of these inpatients needed ICU treatment. During hospitalization, 12.2% in New York City Area and 29.2% in California required IMV. Hence, health systems and hospitals had an enormous pressure in the inpatients' wards and in the ICU departments, especially with the IMV equipment and specialized personal needed.

This work focuses on predicting whether an ICU patient needs IMV. The model uses data from the 24 h preceding and succeeding the ICU admission. It uses clinical data of the COVID-19 patient recorded and available in that period, such as blood tests, or vital signs. As far as we know, there is no other study developed using as many hospitalizations and types of data as this. These features are used to train Machine Learning (ML) classifiers that will be able to predict the probability of a patient needing IMV. The model explanation will help physicians interpreter the most important features for the model. In the future, the selected model will be incorporated into Patient. CARE ® and it will be available for doctors and nurses use in real-time. There is no public knowledge of other software capable of this type of classification using ML methodologies.

This study is split into six parts. In Sect. 2, the work previously developed in this field is presented. In Sect. 3, one can find the description of the data set. Section 4 presents the methodology used in this study from data pre-processing to the modelling phase. The results of the application of the previous methodology and the out of sampling tests are discussed in Sect. 5. Finally, the main conclusions and future work of this study are given in Sect. 6.

2 State of the Art

Although the use of ML in medicine is relatively new, its potential has been recognized for helping to make wiser, more precise, and personalized medical decisions. Big Data and ML approaches in medicine can be divided into Predictive and Prescriptive analytics [19]. While the first focus on predicting whether the patient is going to experience an event, the second concentrates on what can or should be done for the patient.

Several studies use ML to help predict mortality rates, readmission, or the outcome of sepsis. *Fialho et al.* [10] used physiologic variables before discharging to predict patient readmission. In 2019, *Junqueira et al.* [12] used ML classification algorithms such as Random Forest Classifier (RF) and Support Vector Machine (SVM) to identify most relevant risk factors for ICU readmission. *Sacramento et al.* [22] use data from B-Simple [1] to predict the likelihood of a patient to be readmitted after surgery with an accuracy of 91%. Similar work has been developed to predict ICU mortality. *Krajnak et al.* [14] and *Bhattacharya et al.* [3] use ML techniques to achieve it. Both found an efficient and accurate model, however, the first reported some problems with excessive time to create a solution. The second focus on a solution for a typical problem when using health data: imbalanced data [6]. A recent study in Denmark examined the use of ML methods and time-series data to enhance the mortality prognostication of ICU patients. *Thorsen-Meyer et al.* [23] use a Recurrent Neural Network (RNN) with one-hour sampling interval for each feature to deliver real-time predictions of 90-day mortality. This study investigated the possibility and benefit of interpretable models. The RNN proved to be explainable and the most important features, for each patient, can be visualised. Sepsis is an usual diagnosis for critically hill patients so, it is necessary a tool to help treat or prevent it. *Komorowski et al.* [13] developed the Artificial Intelligence Clinician capable of analysing patient data and select a reliable course of treatment for patients with sepsis. It showed lower mortality when the clinical decision matched the proposed treatment by the Artificial Intelligence Clinician. The development of predictive tools for the use of IMV is still slim. *Fabregat et al.* developed a ML tool capable of predicting the outcome of programmed extubations. They use three classification models: Logistic Discriminant Analysis, Gradient Boosting Method and SVM. The best one presented a 94.6% accuracy.

3 Data Set

The anonymized clinical data of all patients hospitalized in ICU's using the critical care management software B-ICU.CARE® [1] was compiled and organized in a database. The included data represent most of the ICU's in continental Portugal and in the Community of Madrid in Spain. This database does not save any personal information, however, it records and encrypts all clinical data regarding the patient ICU staying, namely Heart Rate (HR), Respiratory Rate (RR) or if the patient needed IMV, among others. Only adult patients that had COVID-19 were considered.

The database contains data from patients admitted at the ICU's after the 24th of February 2020 in Spain and the 1st of March 2020 in Portugal and all of them have been medical discharged from the ICU's before the 16th of August 2020.

In Portugal, 66.93% of ICU patients were male (33.07% female), with an Average (AVG) age of 64.29 and a standard deviation (std) of 13.01. The AVG Lenght of Stay (LOS) in ICU was 13.08 days and 57.57% of these patients needed IMV. In Spain, 72.37% of ICU patients were male (27.63% female), with an AVG age of 60.59 and a std of 11.71. The AVG LOS in ICU was 17.86 days and 76.08% of these patients needed IMV. In this sample, the AVG duration of IMV was higher for the patients who survived, see Table 1. Spain presented longer duration of IMV and in AVG, patients that survived were ventilated for 19.69 days.

Table 1. AVG IMV duration (in days) at the end of the ICU stay.

	Total	Portugal	Spain
Alive	17.31	15.68	19.69
Deceased	16.13	14.25	17.31

The mortality rate in the Portuguese ICU was 17.00% while in Spain was 35.05%. We have no information about the clinical condition and follow-up after ICU discharge for patients who survived. We included information about drug therapy in our database, but we had to split it into two subsets, since this information was not available for all hospitals included. The first subset has all ICU hospitalizations but does not consider any drug-related features. The second includes the drugs, but excludes every patient treated at hospitals for which this information was not available. All analysis, modelling and statistics are performed in both sets to access the difference between models that use drug information and the ones that do not use it.

4   Methods

The methodology adopted for this study is explained in Fig. 1. It has six phases: data preparation, processing, modelling, accuracy evaluation, feature selection, and the evaluation of the final model.

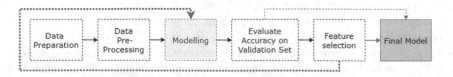

Fig. 1. Flowchart for the overall methodology.

4.1 Data Preparation

The B-ICU.CARE ® is a customizable ICU product. In this way, the enormous amount of clinical data recorded in different hospitals requires normalization in a central database. Techniques were thus developed to organize the information. Diagnosis were analysed by a medical doctor, specialist in intensive care medicine and divided into 65 categories. For instance, the category "Acute Respiratory Failure" includes all patients with acute respiratory failure with or without previous pulmonary diseases.

A similar process is applied to a list of drugs since each hospital uses a different nomenclature, units and concentrations. In this case, the same physician chooses the drugs and the drug classes that could have beeen prescribed for COVID-19 treatment. The drug classes considered in this study are as follow: Corticosteroid, Antimalarial, Antiviral, Anticoagulant, Vasopressor, Antibiotic, Bronchodilator.

4.2 Data Pre-processing

The initial variables set includes age, gender, Glasgow Coma Scale (GCS), Simplified Acute Physiology Score II (SAPS II), Acute Physiology And Chronic Health Evaluation II (APACHE II), vital signs, and results for blood and urine analysis such as bilirubin or urea. Most variables are Boolean, however, the vital signs, blood and urine analysis are time series that were replaced by their minimum, AVG and maximum value within the 24 h preceding and succeeding the ICU admission. Moreover, the diagnosis and drugs classes are also taken into account. For these variables, the number of times a drug of each class is prescribed was considered. For the diagnosis, the number of times a diagnosis of each class is associated with the hospitalization was counted. The type of diagnosis was not taken into consideration.

The percentage of missing data in each variable is first evaluated. There are some problems with missing data in the vital signs, blood and urine analysis due to the nature of these variables. The vital signs were obtained through the monitor and, if missing, the values inserted by the nurse were considered. All variables with more than 45% of missing data were not included. Furthermore, two more stages of missing data were considered: 25% and 40%. Second, the correlation between each variable and the IMV was computed. All variables with a correlation between −0.1 and 0.1 were excluded.

4.3 Modelling

This investigation is based on supervised learning, Fig. 2, that allows checking which set of variables best describes the data for classification. The cycle consists of five main steps: divide data, data normalization and imputation, balance the data with Synthetic Minority Over-sampling Technique for Nominal and Continuous (SMOTENC), train and validate the models using accuracy.

This work is developed in Python 3.8.6 [24]. The focus is on classification models given the nature of the primary objective. In total, 149 different classification models were tested by using different parameters within each model. To test different classification techniques, six types of classification models were used:

- RF
- XGBoost Classifier
- Stochastic Gradient Descent Classifier

- K-Neighbors Classifier
- Decision Tree Classifier
- Support Vector Classification

The normalization step applies Equation $x_i = \frac{X_i - \bar{X}}{\sigma_X}$ where x_i represents the normalized value of the variable X for the hospitalization i, $\bar{X}$ represents the AVG and σ_X is the std of the variable X in the training set. The normalization is previous to the SMOTENC function [15] and to the imputation of a constant value of 0 (note that imputation with 0 after normalizing corresponds to use the average value). Categorical features such as drugs, GCS, or prone position, are not normalized.

As indicated in Fig. 2, this process is repeated 50 times, which guarantees the use of different random training and validation sets. It reduces the possibility of overfitting since, for each model, the AVG and std of accuracy, precision and recall are evaluated. As mention in Sect. 3, there are two collections of data in the study: data with drug information ("Drugs") and without ("No Drugs"). For each collection, as discussed in Sect. 4.2, there are three sets of variables regarding the percentage of missing data. Note that no hospitalization is excluded, only variables.

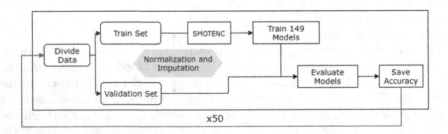

Fig. 2. Flowchart for modelling phase.

The previous analysis will find the best type of classification model. Next, the feature selection is imperative to ensure only the use of significant features. Figure 3 describes the algorithm applied for feature selection. The Recursive Feature Elimination (RFE) tool tests the possibility of using fewer variables [18]. It can be changed to choose only one variable or all of them (N). In this way, the algorithm tests all possibilities in 30 different training sets. The number of training sets is inferior to the initial since this process takes a significant amount of time when compared to the first. It ensures the best choice for minimizing the overfitting problem.

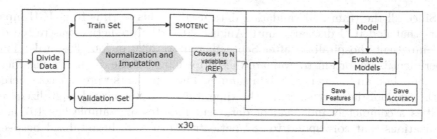

Fig. 3. Flowchart for the Choosing Features phase using RFE. N represents the maximum number of features.

5   Results

Following the application on Fig. 2, it is possible to determine the best initial set of variables, and the classification model with best accuracy. In this case, the set of features that presents the best accuracy was the initial set with all variables with less than 40% missing data. The best classification model was a RF. This model was used for the remaining processes. At this point, the model was trained using 50 different variables, for the data with the drugs, and 55 variables for the other case. Since the main objective for this tool is to be easily used within ICU context, the possibility of using fewer variables was evaluated.

Next, the algorithm presented in Fig. 3 was used to evaluate the behaviour of the three best classification models using between 1 and N features. To evaluate if the mean accuracy of other combination of variables is significantly different from the best, one can use the t-Test Paired Two Sample for Means. This tool allows the identification of all the models whose mean accuracy does not significantly differ from the best model. For the "No Drugs", case the minimum number of variables is 23, and the "Drugs" is 16. In Table 2, it is possible to observe the final results for these validations processes. The best accuracy results for booth "No Drugs" and "Drugs" possibilities are associated with a large number of variables, 54 and 49 respectively. The last has the best Accuracy, Precision, and Recall so, it was the chosen for the test with unseen data.

Table 2. AVG and std values for Accuracy, Precision, and Recall for the best set of variables and for the set with fewer variables and no significant difference.

	Models	Variables	Accuracy		Precision		Recall	
			AVG	std	AVG	std	AVG	std
No drugs	RF300	54	0.8648	0.0214	0.8893	0.0223	0.9291	0.0222
	RF300	23	0.8604	0.0227	0.8922	0.0239	0.9182	0.0208
Drugs	RF350	49	0.8852	0.0222	0.9011	0.0228	0.9489	0.0176
	RF300	16	0.8826	0.0217	0.9022	0.0213	0.9435	0.0219

Since all the train and validation data used in this work is from ICU inpatients that had ICU discharge until August 2020, the predictive was tested on ICU inpatients hospitalized after September first, 2020 in four hospitals. This experiment is important to evaluate the performance of the discussed classification model on an independent data sample. These patients were not considered on the original data set since, at the time, they had not been hospitalized yet so, it is a completely independent data set. This testing sample has 197 hospitalizations that correspond to 187 patients which were hospitalized between September 2020 and November 2020.

In Table 3, it is possible to observe the confusion matrix for the 197 predictions. The classification model made 16 wrong predictions, nine of them are false positive. The accuracy measures in this experiment are:

Accuracy = 91.88%   Recall/Sensitivity = 93.02%
Precision = 94.49%   Specificity = 89.71%

Table 3. Confusion matrix for the results obtained when testing the predictive on a new data sample. The "TRUE" represents the use of IMV by the patient.

		Actual class	
		TRUE	FALSE
Predicted class	TRUE	120	9
	FALSE	7	61

The algorithm is prepared to be incorporated into the B-Simple clinical software. It will output not only the probability that a patient will need IMV, but also which features had the most weight in the final decision. This can be achieved by using Local Interpretable Model-agnostic Explanation (LIME) [20]. LIME is capable of explaining the importance of each feature on the final decision as shown in Fig. 4. This explanation can be vital on a clinical support decision tool since it gives extra confidence to the medical doctor using it. Additionally, it may alert the physician for clinical factors that, at first sight, could be ruled out as unimportant.

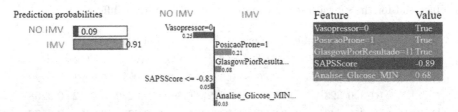

Fig. 4. Example of an explanation produced using LIME when predicting the use of IMV for a test hospitalization.

6   Conclusions

The main objective for this project was to develop a classification tool capable of forecasting whether a patient will need IMV during the ICU stay. Although this work is focused on COVID-19 patients, it can be extended to all ICU patients. This tool may be helpful when allocating ICU resources since it is capable of predicting how many ICU patients will need ventilators.

The algorithms developed for this project allowed to test the performance of numerous classification models and feature selection. The set of features and the RF classification model reached an accuracy of 91.88% in an out of sample set. The algorithm can handle missing data for numerical features.

The next step for this work is to extend the classification model to all ICU patients. This can be accomplished by two different approaches: training a general model or training several models for each type of ICU patient. Both approaches will result in a clinical support tool that has the potential to help medical doctors decide to use IMV or not.

Acknowledgements. This work is a result of the project NORTE-01-02B7-FEDER-048416 supported by Norte Portugal Regional Operational Programme (NORTE 2020), under the PORTUGAL 2020 Partnership Agreement, through the European Regional Development Fund (ERDF).

References

1. B-Simple. https://www.b-simple.pt/. Accessed 28 Dec 2020
2. Bennett, S., et al.: Clinical features and outcomes of adults with coronavirus disease 2019: a systematic review and pooled analysis of the literature. Int. J. Clin. Pract. **75**, e13725 (2020)
3. Bhattacharya, S., Rajan, V., Shrivastava, H.: ICU mortality prediction: a classification algorithm for imbalanced datasets. In: AAAI Conference on Artificial Intelligence, pp. 1288–1294 (2017)
4. Burki, T.K.: The Russian vaccine for COVID-19. Lancet Respir. Med. **8**(11), e85–e86 (2020)
5. Cheng, F.Y., et al.: Using machine learning to predict ICU transfer in hospitalized COVID-19 patients. J. Clin. Med. **9**, 1668 (2020)
6. Domingues, I., Amorim, J.P., Abreu, P.H., Duarte, H., Santos, J.J.: Evaluation of oversampling data balancing techniques in the context of ordinal classification. In: IEEE International Joint Conference on Neural Networks, pp. 1–8 (2018)
7. European Medicines Agency: Treatments and vaccines for COVID-19. https://tinyurl.com/y8t5jkv8. Accessed 28 Dec 2020
8. Fauci, A.S., Lane, H.C., Redfield, R.R.: Covid-19 - navigating the uncharted. New England J. Med. **382**(13), 1268–1269 (2020)
9. FDA: Pfizer-BioNTech COVID-19 Vaccine. https://tinyurl.com/yd3gwlcn. Accessed 28 Dec 2020
10. Fialho, A., Cismondi, F., Vieira, S., Reti, S., Sousa, J., Finkelstein, S.: Data mining using clinical physiology at discharge to predict ICU readmissions. Expert Syst. Appl. **39**(18), 13158–13165 (2012)
11. GOV.UK: Information for Healthcare Professionals on Pfizer/BioNTech COVID-19 vaccine (2020). https://tinyurl.com/yyhmeu6u. Accessed 28 Dec 2020

12. Junqueira, A.R.B., Mirza, F., Baig, M.M.: A machine learning model for predicting ICU readmissions and key risk factors: analysis from a longitudinal health records. Heal. Technol. **9**(3), 297–309 (2019)

13. Komorowski, M., Celi, L.A., Badawi, O., Gordon, A.C., Faisal, A.A.: The artificial intelligence clinician learns optimal treatment strategies for sepsis in intensive care. Nat. Med. **24**(11), 1716–1720 (2018)

14. Krajnak, M., Xue, J., Kaiser, W., Balloni, W.: Combining machine learning and clinical rules to build an algorithm for predicting ICU mortality risk. Comput. Cardiol. **39**, 401–404 (2012)

15. Lemaître, G., Nogueira, F., Aridas, C.K.: Imbalanced-learn: a Python toolbox to tackle the curse of imbalanced datasets in machine learning. J. Mach. Learn. Res. **18**(17), 1–5 (2017)

16. Myers, L.C., Parodi, S.M., Escobar, G.J., Liu, V.X.: Characteristics of hospitalized adults with COVID-19 in an integrated health care system in California. JAMA **323**(21), 2195–2198 (2020)

17. Paules, C.I., Marston, H.D., Fauci, A.S.: Coronavirus infections-more than just the common cold. JAMA **323**(8), 707–708 (2020)

18. Pedregosa, F., et al.: Scikit-learn: machine learning in Python. J. Mach. Learn. Res. **12**, 2825–2830 (2011)

19. Pirracchio, R., et al.: Big data and targeted machine learning in action to assist medical decision in the ICU. Anaesth. Crit. Care Pain Med. **38**(4), 377–384 (2019)

20. Ribeiro, M.T., Singh, S., Guestrin, C.: "Why should I trust you?" Explaining the predictions of any classifier. In: ACM SIGKDD International Conference on Knowledge Discovery and Data Mining, 13–17-August, pp. 1135–1144 (2016)

21. Richardson, S., Hirsch, J.S., Narasimhan, M., Crawford, J.M., McGinn, T., Davidson, K.W.: Northwell COVID-19 research consortium: presenting characteristics, comorbidities, and outcomes among 5700 patients hospitalized with COVID-19 in the New York City area. JAMA **323**(20), 2052–2059 (2020)

22. Sacramento, R., Silva, R., Domingues, I.: Artificial intelligence in the operating room: evaluating traditional classifiers to predict patient readmission. In: Portuguese Conference on Pattern Recognition, pp. 17–18 (2020)

23. Thorsen-Meyer, H.C., et al.: Dynamic and explainable machine learning prediction of mortality in patients in the intensive care unit: a retrospective study of high-frequency data in electronic patient records. Lancet Digit. Health **2**(4), e179–e191 (2020)

24. Van Rossum, G., Drake, F.L.: Python Software Foundation (2019). https://www.python.org. Accessed 28 Sept 2020

25. World Health Organization: Coronavirus Disease (COVID-19) (2020). https://www.who.int/emergencies/diseases/novel-coronavirus-2019. Accessed 16 Nov 2020

A Coarse to Fine Corneal Ulcer Segmentation Approach Using U-net and DexiNed in Chain

Helano Miguel B. F. Portela[1]([✉]) [iD], Rodrigo de M. S. Veras[1] [iD],
Luis Henrique S. Vogado[1] [iD], Daniel Leite[2] [iD], Jefferson A. de Sousa[3] [iD],
Anselmo C. de Paiva[3] [iD], and João Manuel R. S. Tavares[4] [iD]

[1] Departamento de Computação, Universidade Federal do Piauí,
Teresina, Brazil
rveras@ufpi.edu.br
[2] Departamento de Medicina Especializada, Universidade Federal
do Piauí, Teresina, Brazil
[3] Núcleo de Computação Aplicada,
Universidade Federal do Maranhão, São Luiz, Brazil
paiva@nca.ufma.br
[4] Instituto de Ciência e Inovação em Engenharia Mecânica
e Engenharia Industrial, Departamento de Engenharia Mecânica,
Faculdade de Engenharia, Universidade do Porto, Porto, Portugal
tavares@fe.up.pt

Abstract. A corneal ulcer is one of the most frequently appearing diseases that may affect eye health. The proper measurement of corneal ulcer lesions enables the physician to evaluate the treatment effectiveness and assist in decision-making. This article presents the solution for ulcer segmentation as a pixel-wise classification task, and proposes a novel coarse-to-fine method to extract corneal ulcers from ocular staining images. This study combines two classical convolutional neural networks (CNNs), known as U-net and DexiNed, following Morphological Geodesic Active Contour as a post-processing operation. We trained the CNNs using 358 point-flaky corneal ulcer images and evaluated its performance in 91 flaky corneal ulcer images. Our approach achieved 70.50% of Dice Coefficient on average, 87.4% of Recall, and 99.0% of Specificity, and True Dice Coefficient of 63.7%. These results corroborate our approach's efficacy and efficiency.

Keywords: Computer-aided diagnosis · Image segmentation · Deep learning · Eye health

1 Introduction

Many corneal diseases may affect eye health, such as Pytherigium, Infection, Conjunctival nevus and Corneal Ulcer. The corneal ulcer is one of the most frequently appearing of these, and it is defined as an inflammatory or even

© Springer Nature Switzerland AG 2021
J. M. R. S. Tavares et al. (Eds.): CIARP 2021, LNCS 12702, pp. 13–23, 2021.
https://doi.org/10.1007/978-3-030-93420-0_2

more severe condition. It may lead in some cases to epithelial layer disruption or corneal stroma disruption. There are some potential causes of corneal ulcers, such as topical steroid usage, contact lens usage, trauma and ocular disorders, leading to perforation, scarring and vision loss [3]. Corneal ulcers can be classified into three general types, considering their shape and distribution: point-like, point-flaky mixed and flaky. Figure 1 shows samples of these three types.

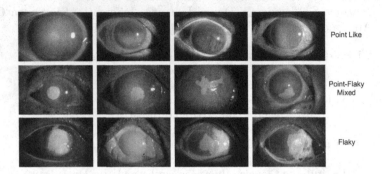

Fig. 1. Image samples from the SYSUTech-SYSU dataset, with the top row depicting point like corneal ulcers, the middle row depicting point-flaky mixed, and the bottom row depicting flaky corneal ulcers.

Usually, the point-like type appears at an early stage, when there are most chances of success in its treatment. This type of corneal ulcer has numerous ulcer dot distribution patterns that can appear anywhere within the corneal tissue. Therefore, it is not reasonably possible to segment it manually. A flaky corneal ulcer usually indicates a much more severe corneal disease. It has a uniform shape with clear boundaries, and may lead to scars and even vision loss. A point-flaky mixed corneal ulcer is a combination of point-like and flaky corneal ulcers. It indicates corneal disease with a severity degree, which lies between the aforementioned types. Measuring corneal ulcer lesion extension plays a crucial role in the treatment, as such a measurement may assist the specialist in the treatment follow-up.

The present study consisted of developing a computational method for corneal ulcer segmentation in ocular staining images. Hence, we evaluated and compared different CNN architectures and post-processing techniques found in the state-of-the-art. During the research process, we defined the following specific objectives: evaluate the U-net, DexiNed and LinkNet CNN architectures applied to this problem of image segmentation; estimate different pre- and post-processing image operations; and train the method using only point-flaky mixed corneal ulcer images, and validate it using flaky corneal ulcer images.

The remainder of this article has the following structure. Section 2 presents related works as to corneal lesion image segmentation; Sect. 3 details the evaluated CNN architectures, used image dataset, applied data preparing operations, and the adopted evaluation metrics. Section 4 presents the proposed approach.

Section 5 presents the results and discussion. Finally, Sect. 6 concludes the article and indicates future directions for research.

2 Related Works

We carried out a literature review looking for state-of-the-art articles related to computer-aided diagnosis solutions to segment corneal lesions. The survey aimed to identify and classify the works available in the literature based on the techniques employed, image dataset, year of publication and application domain. In this context, we can highlight the articles of Sun et al. [15], Deng et al. [5], Patel et al. [12], Deng et al. [4], Lima et al. [10], and Liu et al. [11].

We noticed that only the work of Sun et al. [15] uses a CNN-based approach to segment corneal ulcer images. In contrast, the other methods are based either on classical clustering or classification algorithms. In this context, our work contributes to exploring the limits of applying CNNs to segment corneal lesions, specifically corneal ulcer lesions.

Table 1 summarises the works found in the reviewed literature in terms of the year of publication, used technique(s), number of used images, and the application domain. In all of these works, the images used for training and test purposes were from the same dataset, and none of them used publicly available datasets, except for the CLID dataset used in Lima et al. [10].

Table 1. Summary of the works found in the reviewed literature in terms of the year of publication, used technique(s), number of used images and application domain.

Work	Year	Technique(s)	N. of images	Domain
Sun et al. [15]	2017	Path-based CNN	48	Corneal Ulcer
Deng et al. [5]	2018	SVM with Superpixel	150	Corneal Ulcer
Patel et al. [12]	2018	Random Forest and Active Contour	50	Corneal Ulcer
Deng et al. [4]	2018	Iterative k-means, Morphological Operations Region growing	48	Corneal Ulcer
Liu et al. [11]	2019	Gaussian mixture modeling and Otsu method	150	Corneal Ulcer
Lima et al. [10]	2020	Random Forest Classifier	30	Infection, Pterygium and Conjunctival nevus

3 Materials and Methods

This study aimed to propose an automatic method for corneal ulcer segmentation. We performed experiments using different combinations of the U-net, LinkNet and DexiNed CNNs architectures applied to the SUSTech-SYSU [6] dataset. Combined with that, we used other post-processing techniques such as Binary Threshold, Otsu Threshold, Geodesic Active Contour, Fill holes and Morphological Operations. We evaluated the models' performance using four different metrics to identify the proposed method's best settings.

3.1 Evaluated CNN Architectures

U-net [14] is a convolutional neural network proposed for biomedical image segmentation. The general U-net architecture consists of two paths: a contracting path capable of capturing the image's context, and an expansive path capable of building the segmented image. The primary strategy that differentiates the U-net architecture from the other fully connected ones is combining the feature maps from the contraction layers with their symmetric correlated feature maps from the expansion layers. This characteristic allows the propagation of context information to high-resolution feature maps.

Chaurasia et al. [2] proposed **LinkNet** aiming to provide a semantic segmentation approach using less computational complexity comparing to other CNN architectures. LinkNet is based on encoders and decoders concepts, and is designed to perform a convolutional operation followed by a max-pooling operation on its output data; after that, there are four encoders blocks, followed by four decoders blocks. Finally, the architecture applies a sequence of full convolution, followed by a simple convolution and another full convolution as output.

DexiNed (Dense Extreme Inception Network for Edge Detection) [13] is a convolutional neural network for edge detection. It is built using a stack of filters that predict an edge map based on an input image. DexiNed comprises two sub-network architectures: Dense Extreme Inception Network (Dexi) and an up-sampling block (UB). Whereas the Dexi architecture has an image as input, the up-sampling block gets a feature map from the Dexi architecture block. The resulting architecture generates thin edge maps avoiding edge losses in the deep layers. DexiNed provides two outputs: Pred-a and Pred-f. The upsampling block returns six edge map outputs; by calculating the average from these six edge maps, one gets the Pred-a output, and by fusing these six edge maps, one gets the Pred-f output. In this work, we use DN-a and DexiNed-a to refer to the DexiNed model using the Pred-a output, and DN-f to refer to the DexiNed model using the Pred-f output.

3.2 Image Dataset

SUSTech-SYSU [6] is a dataset for automatically segmenting and classifying corneal ulcers from ocular fluorescein staining images. It was prepared to supply the lack of high-quality datasets to develop segmentation and classification algorithms for corneal diseases. The dataset contains 712 ocular staining images, and the segmentation ground truth of flaky corneal ulcers: 263, 358 and 91 images for point-like, point-flaky and flaky general types, respectively. The dataset also provides the three-fold class labels for each image: 1) labels in terms of general ulcer pattern, 2) labels in terms of its specific ulcer pattern, and 3) the corresponding ulcer severity degree.

The work of Gross et al. [8] is the only official article published using the SUSTech-SYSU dataset. It describes a CNN based image classification approach to identify different types of Corneal Ulcers from fluorescein staining images.

3.3 Evaluation Metrics

In this work, we use the term 'positive' to designate ulcer areas and 'negative' to define non-ulcer areas. We calculated the confusion matrix to obtain the segmentation Recall (R), Specificity (S) and Dice Coefficient (DC).

In order to evaluate the segmentation quality, we calculated two more metrics: Average Dice Coefficient (ADC) and TDC (True Dice Coefficient). We can define the Average Dice Coefficient (ADC) as the mean value of all DC_i divided by the number of the images in a given dataset (n), i.e.:

$$ADC = \frac{\sum_{i=1}^{n} DC_i}{n}. \tag{1}$$

We consider that a "good" image segmentation result should have a DC value over a threshold (t). The True Dice Coefficient (TDC) metric for a dataset d is the number of automatic segmentation executions that achieved $DC_i > t$ divided by the total number of images. To compute the TDC metric, a score for each image (i) is calculated based on the Dice Coefficient as:

$$\{score_i = 0, \text{if } DC < 0.7, score_i = 1, \text{otherwise.} \tag{2}$$

Given a dataset with n images, the final TDC value is defined as the mean of all per-image scores:

$$TDC = \frac{\sum_{i=1}^{n} score_i}{n}. \tag{3}$$

According to Genctav et al. [7], Dice scores greater than 0.7 indicate a remarkable similarity between the segmented and ground truth regions. So, we considered the threshold $t = 0.7$, to calculate the TDC.

4 Proposed Method

This work proposes an automatic segmentation method that combines two CNNs Architectures: U-net for image segmentation and DexiNed, which was initially proposed for edge detection, but it is here combined with the U-net output for image segmentation. We then apply 300 operations of the Morphological Geodesic Active Contour (MorphGAC) algorithm [1]. Figure 2 illustrates the proposed method's process.

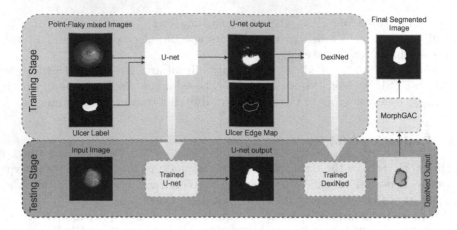

Fig. 2. Flowchart of the proposed method.

One should note that we do not propose any corneal segmentation method in this study. Instead, we focus exclusively on the corneal ulcer lesion segmentation, assuming that an automatic [4,9] or even manual [15] process had previously segmented the cornea. Therefore, the first performed preprocessing step is the corneal area segmentation using the corneal ground truth provided by the SUSTech-SYSU.

Once the training sets were prepared, we submitted the 358 point-flaky images and their corneal ulcer labels to the U-net. The U-net model was trained during 70 epochs using its classical architecture [14]. The number of 70 epochs was empirically defined, varying the number of epochs in increments of 10 until the model converges. The U-net output is a grayscale image that may contain undesirable segmented areas. Besides, a grayscale image is not the ideal final result for an image segmentation method. Thus, we then use the DexiNed to refine it.

To train the DexiNed model, we generated an edge map for each of the ulcer cornea labels. We did this by applying two iterations of erosion in the original ulcer label using a 3×3 structuring element. Then, we subtracted the resulting image from the erosion operation to the original ulcer label resulting in the ulcer edge map.

We submitted the 91 corneal flaky images (Fig. 3(a) to the trained U-net model. The U-net model (Fig. 3(b)) was then connected to the DexiNed input. Finally, on the DexiNed-a output (Fig. 3(c)), 300 iterations of the MorphGAC were executed to get the final result (Fig. 3(d)) using the Otsu threshold method to find the threshold value for the MorphGAC parameter. After that, we calculated the quality metrics using the final segmentation achieved by the proposed method (Fig. 3(d)) against the corresponding ground truth (Fig. 3(e)).

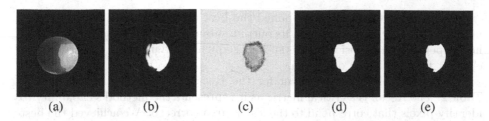

Fig. 3. Input image (a), U-net output (b), DexiNed-a output (c), Final result after submitted to MorphGAC (d), and corresponding ground truth provided by the SUSTech-SYSU dataset (e).

5   Results and Discussion

To find the best possible approach, we tested the combination of the U-net, DexiNed and LinkNet models, and each of these CNN models separately with different post-processing methods.

In the experiments, we used the following settings. **Dataset:** 358 Point-flaky images as training data and 91 Flaky images as testing data. **Architectures:** U-net, LinkNet, and DexiNed arranged according to the following settings: LinkNet connected to U-net, LinkNet itself, U-net connected to LinkNet, DexiNed connected to U-net, DexiNed itself, U-net connected to DexiNed and U-net itself. LinkNet and U-net were set with 1e-5 of learning rate and executed for 70 epochs, and DexiNed with a learning rate of 1e-4 and 1000 iterations at most.

Post-processing: For each of the architectures previously mentioned, we tested several combinations of Otsu threshold, MorphGAC and morphological operations.

Table 2 indicates the best results found for each combination. It is essential to point out that we did not use the point-like images for testing, because the used dataset does not provide the point-like corneal ulcer ground truth.

Table 2. Results obtained using combinations of the DexiNed (DN), LinkNet (LN) and U-net (UN) architectures, and MorphGAC, Otsu thresholding, Binary thresholding and Morphological operations. (Best values found in bold.)

Experimental settings	ADC (%)	R (%)	S (%)	TDC (%)
LN → UN	06.50 ± 0.069	00.00 ± 00.00	00.00 ± 00.00	00.00
LN → Otsu → fill holes	15.70 ± 17.00	40.50 ± 37.00	84.30 ± 30.10	00.00
UN → LN → Otsu	23.00 ± 20.10	38.10 ± 21.50	91.70 ± 19.90	00.00
DN-a → UN	26.90 ± 25.50	23.60 ± 29.10	95.50 ± 26.80	08.70
DN-a → Otsu → fill holes → DN-f	66.70 ± 25.50	91.20 ± 18.30	98.80 ± 01.10	49.40
DN-a → MorphGAC → Otsu value	68.30 ± 25.20	89.20 ± 19.30	**99.00 ± 01.00**	53.80
UN → binary threshold	70.30 ± 28.40	**96.40 ± 11.80**	98.30 ± 02.10	60.40
UN → Otsu → morph. op.	**74.10 ± 27.10**	92.10 ± 31.90	98.70 ± 01.90	62.60
UN → DN-a + MorphGAC	70.50 ± 25.10	87.40 ± 21.50	**99.00 ± 01.10**	**63.70**

As to the ADC metric, we obtained the best result (74.10%) with the U-net model trained for 70 epochs, with its output submitted to the Otsu threshold, fill holes operation and erosion, in this specific sequence and using a 3×3 structuring element for the erosion operation.

The results of each experiment for the Recall (R) metric are indicated in Table 2. The Recall is a critical metric as it represents the method's capability to identify pixels that correspond to the lesion area correctly. We achieved the best result (96.40%) using the U-net model combined with a classical binary threshold operation to get an utterly binary image from the U-net output directly.

The S column of Table 2 indicates the specificity metric results. This metric represents the method's capability to identify pixels that correspond to non-lesion areas correctly. We achieved a value of 99.00% for this metric as the best result by using the U-net model with its output connected to the DexiNed model. Using the DexiNed-a output, we applied MorphGAC using the Otsu threshold function to set its threshold value.

The same settings previously mentioned also achieved the TDC metric's best results with a value of 63.70%. This metric indicates the percentage of the testing data that the method could correctly segment, considering the criteria defined in Sect. 3.1.

As one can notice from the third row of Table 2, the results using the DexiNed-a output combined with the MorphGAC using the Otsu threshold parameter are promising. These results suggest that the DexiNed CNN architecture may be a reasonable option for image segmentation problems, although it was primarily designed for edge detection.

Figure 4 depicts the results of the proposed segmentation method (in red) overlapped with the correspondent ground truths (in white). The figure shows the worst (Figs. 4(a)), median (Figs. 4(b)) and best (Fig. 4(c)) cases obtained as to the Dice Coefficient.

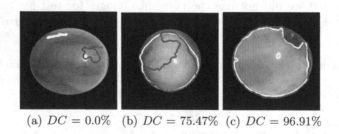

(a) $DC = 0.0\%$ (b) $DC = 75.47\%$ (c) $DC = 96.91\%$

Fig. 4. Examples of corneal ulcer images from the test dataset of 91 flaky corneal ulcers: The results of the proposed segmentation method (in red) are overlapped with the correspondent ground truths (in white). (Color figure online)

Although there are several approaches for corneal lesion segmentation, there was no benchmark image dataset in the literature. For a long time, this fact prevented a direct comparison between the existing methods over the years. Only in 2020, two corneal lesion image datasets were publicly released: SUSTech-SYSU [6] (used in this work) and the CLID dataset [10]. Therefore, to the best of our knowledge, this is the first method applied to corneal ulcer segmentation using the SUSTech-SYSU dataset.

From Table 3, one can verify that the number of images used in this work is higher than the number of images included in the datasets used by Sun et al. [15], Deng et al. [4] and Lima et al. [10], and only lower than the number used in the works of Deng et al. [5] and Liu et al. [11]. It is essential to point out that our method uses different corneal ulcer lesions for training and testing stages using 449 images. Additionally, we evaluated it with all 91 flaky images available on the dataset, which are the only ones that have clear ground truth for validation.

Table 3. Comparison of the results obtained by the proposed method with the ones of the state-of-the-art.

Work	Dataset size	ADC (%)	R (%)	S (%)	TDC (%)
Sun et al. [15]	48	86.00 ± 07.30	82.00 ± 11.20	–	–
Deng et al. [5]	150	–	–	–	–
Deng et al. [4]	48	87.90	–	–	–
Lima et al. [10]	30	87.82	98.05	98.20	82.00
Liu et al. [11]	150	88.05 ± 06.11	–	–	–
Proposed Method	449	70.50 ± 25.10	87.40 ± 21.50	$99.00 \pm \pm 01.10$	63.70

Although the methods presented in the state-of-the-art had achieved better results, we believe that our method is relevant because it could generalize the features from the point-flaky images to segment the flaky corneal images. Based on that, we can train our method to segment point-like corneal ulcers, bringing up the possibility of using it to assist physicians in measuring point-flaky corneal ulcers. Not to mention that some of those methods are not entirely automatic as the one proposed by Lima et al. [10].

6   Conclusion

We proposed an automatic segmentation method for corneal lesion images that was applied to the SUSTech-SYSU image dataset. The new method uses two different CNNs: U-net, originally proposed for image segmentation, and Dex-iNed, initially designed for edge detection. We also tested various combined post-processing techniques (Binary Threshold, Otsu threshold, Fill holes and Geodesic Active Contour) to improve the CNN model outputs.

We found that the combination of the U-net output connected to the Dex-iNed model achieved better overall results when using the DexiNed-a prediction output with MorphGAC using the Otsu threshold value as parameter.

It is essential to point out that we only used point-flaky corneal images with not accurate ground truth to train our model. Considering that the used dataset does not provide the point-like ulcer ground truth, we used only the flaky corneal ulcer images to test the proposed models. However, we achieved encouraging results. Thus, we think that our model could generalize the training data (358 point-flaky corneal images) to segment the test data (91 flaky corneal ulcer images).

In the future, we intend to apply the proposed method on point-like corneal ulcer images and perform a manual validation by ophthalmologists. We believe that our method would be able to successfully segment point-like corneal ulcer images.

References

1. Caselles, V., Kimmel, R., Sapiro, G.: Geodesic active contours. Int. J. Comput. Vision **22**(1), 61–79 (1997)
2. Chaurasia, A., Culurciello, E.: LinkNet: exploiting encoder representations for efficient semantic segmentation. In: 2017 IEEE Visual Communications and Image Processing (VCIP), pp. 1–4. IEEE (2017)
3. Cohen, E.J., Laibson, P.R., Arentsen, J.J., Clemons, C.S.: Corneal ulcers associated with cosmetic extended wear soft contact lenses. Ophthalmology **94**(2), 109–114 (1987)
4. Deng, L., Huang, H., Yuan, J., Tang, X.: Automatic segmentation of corneal ulcer area based on ocular staining images. In: Medical Imaging 2018: Biomedical Applications in Molecular, Structural, and Functional Imaging, vol. 10578, p. 105781D. International Society for Optics and Photonics (2018). https://doi.org/10.1117/12.2293270
5. Deng, L., Huang, H., Yuan, J., Tang, X.: Superpixel based automatic segmentation of corneal ulcers from ocular staining images. In: 2018 IEEE 23rd International Conference on Digital Signal Processing (DSP), pp. 1–5. IEEE (2018)
6. Deng, L., Lyu, J., Huang, H., Deng, Y., Yuan, J., Tang, X.: The SUSTech-SYSU dataset for automatically segmenting and classifying corneal ulcers. Sci. Data **7**(1), 1–7 (2020)
7. Gençtav, A., Aksoy, S., Onder, S.: Unsupervised segmentation and classification of cervical cell images. Pattern Recogn. **45**(12), 4151–4168 (2012). https://doi.org/10.1016/j.patcog.2012.05.006
8. Gross, J., Breitenbach, J., Baumgartl, H., Buettner, R.: High-performance detection of corneal ulceration using image classification with convolutional neural networks. In: Proceedings of the 54th Hawaii International Conference on System Sciences, p. 3416 (2021)
9. Hezekiah, J.D., Chacko, S.: A review on cornea imaging and processing techniques. Current Med. Imaging **16**(3), 181–192 (2020)
10. Lima, P.V., et al.: A semiautomatic segmentation approach to corneal lesions. Comput. Electr. Eng. **84**, 106625 (2020)
11. Liu, Z., Shi, Y., Zhan, P., Zhang, Y., Gong, Y., Tang, X.: Automatic corneal ulcer segmentation combining Gaussian mixture modeling and Otsu method. In: 2019 41st Annual International Conference of the IEEE Engineering in Medicine and Biology Society (EMBC), pp. 6298–6301. IEEE (2019)

12. Patel, T.P., et al.: Novel image-based analysis for reduction of clinician-dependent variability in measurement of the corneal ulcer size. Cornea **37**(3), 331–339 (2018). https://doi.org/10.1097/ICO.0000000000001488
13. Poma, X.S., Riba, E., Sappa, A.: Dense extreme inception network: towards a robust CNN model for edge detection. In: The IEEE Winter Conference on Applications of Computer Vision, pp. 1923–1932 (2020)
14. Ronneberger, O., Fischer, P., Brox, T.: U-net: convolutional networks for biomedical image segmentation. In: Navab, N., Hornegger, J., Wells, W.M., Frangi, A.F. (eds.) MICCAI 2015. LNCS, vol. 9351, pp. 234–241. Springer, Cham (2015). https://doi.org/10.1007/978-3-319-24574-4_28
15. Sun, Q., Deng, L., Liu, J., Huang, H., Yuan, J., Tang, X.: Patch-based deep convolutional neural network for corneal ulcer area segmentation. In: Cardoso, M.J., et al. (eds.) FIFI/OMIA -2017. LNCS, vol. 10554, pp. 101–108. Springer, Cham (2017). https://doi.org/10.1007/978-3-319-67561-9_11

Replacing Data Augmentation with Rotation-Equivariant CNNs in Image-Based Classification of Oral Cancer

Karl Bengtsson Bernander[(✉)], Joakim Lindblad[(✉)], Robin Strand[(✉)], and Ingela Nyström[(✉)]

Centre for Image Analysis, Department of Information Technology,
Uppsala University, Uppsala, Sweden
{karl.bengtsson_bernander,robin.strand,ingela.nystrom}@it.uu.se,
joakim.lindblad@cb.uu.se

Abstract. We present how replacing convolutional neural networks with a rotation-equivariant counterpart can be used to reduce the amount of training images needed for classification of whether a cell is cancerous or not. Our hypothesis is that data augmentation schemes by rotation can be replaced, thereby increasing weight sharing and reducing overfitting. The dataset at hand consists of single cell images. We have balanced a subset of almost 9.000 images from healthy patients and patients diagnosed with cancer. Results show that classification accuracy is improved and overfitting reduced if compared to an ordinary convolutional neural network. The results are encouraging and thereby an advancing step towards making screening of patients widely used for the application of oral cancer.

Keywords: Machine learning · Accuracy · Cell image analysis · Malignancy

1 Introduction

In the last years, convolutional neural networks (CNNs) have been increasingly applied to classification of biomedical data. Our application concerns cell samples collected from patients' oral cavity used to efficiently screen for oral cancer [4]. Performing analysis of the cell images by experienced pathologists is time-consuming and thereby costly. In addition, the expertise may not always be readily available to classify whether a cell is malignant or not. In order to make the screenings widely used, it is therefore desirable to perform these screenings automatically [8]. In this study, we make an advancement towards feasible automated classification of cell images.

It is well-known that despite being popular, the CNNs need a large amount of varied training data to avoid overfitting. One approach to increase the data

© Springer Nature Switzerland AG 2021
J. M. R. S. Tavares et al. (Eds.): CIARP 2021, LNCS 12702, pp. 24–33, 2021.
https://doi.org/10.1007/978-3-030-93420-0_3

during training and thereby enrich the data is to use data augmentation [6]. For image analysis tasks, the input data is typically altered with different transformations such as rotations, scalings, and reflections [10]. One example is when the amount of data is increased four-fold by rotating the images by 90°. By using such a scheme, even though being unvaried, new data is retrieved for training of the network while still keeping the ground truth classification labels intact. However, this comes with the cost of an extra step in the data preparation. In addition, interpolation artifacts may be introduced in the images.

CNNs are, with some exceptions such as edge effects, equivariant to translations [9], which means that if an object is shifted in the image, the resulting features after the convolution operation are shifted as well. In more mathematical terms, the translation operator T commutes with the convolution operator C:

$$T(C(x)) = C(T(x)) \tag{1}$$

However, this relation does not hold for other transformations such as rotations. In order to achieve this handy property, several methods have been proposed and tested. One of the most common ways to achieve rotation-equivariant CNNs is to generalize the convolution operator [3,15,16]. Most of these approaches are limited to specific discrete rotations. Other advances in the field extend to arbitrary angular resolutions using atomic steerable filters [13]. Another alternative is to use conic convolutions and discrete Fourier transforms for rotation-invariant classification [2]. Recently, a framework was released to aid in the development of E(2)-equivariant steerable networks (E2CNN) [12].

One domain where CNNs are heavily used is cytopathology, where cell samples from patients are collected, imaged and diagnosed for diseases. When these samples are collected from, for example, the oral cavity, they can serve as biomarkers for cancer development. For these images, we make the assumption that the cells are rotationally invariant on a global scale, implying that regardless how the sample is rotated during the microscopy imaging process, the classification output should be identical. However, on a smaller scale, the rotations of parts of the cells (for example, organelles and chromosomes) could be an indicative of cancer, which is what we expect the CNNs to reveal.

Our hypothesis is that by replacing ordinary CNNs with equivariant CNNs in our application, the data augmentation step could be skipped since the networks in effect perform the corresponding transformations at each layer of the network. That is, instead of performing data augmentations by rotating the data when training the CNNs, we can use rotation-equivariant networks with no data augmentations during training and testing. This should improve the expressive capacity of the network while reducing the number of trainable weights.

To verify our hypothesis, we compare the accuracy of a rotation-equivariant CNN with that of a standard CNN on a large oral cancer dataset which we use as a reference. Previous works indicate it would be interesting to investigate these networks further on this type of dataset, due to the importance of texture-based features in cell images [14]. In this study, we also vary the size of the training datasets from a few hundred up to several thousands of images, to verify that the accuracy is comparable to the results for the baseline network.

Section 2 introduces the methodology of the experiments. Section 3 details the oral cancer dataset, while Sect. 4 outlines the architectures and training procedures for classifiers on this dataset. The results are then presented in Sect. 5 and interpreted in Sect. 6.

2   Methodology

In this work, we focus on how data augmentation, the amount of training data, and equivariance properties of the architecture affect the ability of the network to generalize. We also inspect and compare the models in terms of sensitivity and specificity. These statistical measures are calculated in the following way:

$$\text{Sensitivity} = \frac{TP}{TP + FN}, \tag{2}$$

$$\text{Specificity} = \frac{TN}{TN + FP} \tag{3}$$

where TN is the sum of true negatives, FN is the sum of false negatives, TP is the sum of true positives, and FP is the sum of false positives.

The core idea is however to investigate overfitting. According to the classic bias-variance trade-off, an optimally trained network strikes a good balance between underfitting and overfitting [6]. In this region, the network has captured the characteristics of the underlying distribution and is able to generalize to unseen data from similar distributions. To measure overfitting, we compare the empirical risk with the testing accuracy.

The empirical risk for a classifier $R_{emp}(h)$ [11] is defined as:

$$R_{emp}(h) = \frac{1}{n} \sum_{i=1}^{n} L(h(x_i), y_i) \tag{4}$$

where h is the hypothesis (i.e., our network), L is the loss function, and x_i and y_i are n independent input and output samples, respectively.

The accuracy is defined as the percentage of examples classified with the correct label. Since a low empirical risk is directly proportional to a high classification accuracy on the training set, we can use the training accuracy as a proxy for the empirical risk. Based on this, we form the following metric of overfitting:

$$Overfitting\ Ratio = \frac{Training\ Accuracy}{Testing\ Accuracy} \tag{5}$$

A high number will then indicate high overfitting, while a low ratio will indicate low overfitting.

3   Oral Dataset

Screening for malignant cancer is a highly specialized task, requiring good sample quality, experience, and patience. The procedure for oral cancer diagnosis is similar to the one for cervical cancer [1]. One can expect issues of very different origin in the procedure, for example, staining artifacts, bacterial and fungal infections, ruptured cells, or a combination thereof. Indicators of cancer include atypical ratios of the nucleus to cytoplasm, the presence of several nuclei, or asymmetrical shapes of the cell. Recently, a dataset of images has been prepared to automate this diagnostic process [8] (we use images from the liquid-based prepared Dataset 3). A number of colored cell samples from this dataset are exemplified in a mosaic setting in Fig. 1a and one of the cells is selected for larger illustration in Fig. 1b. The corresponding gray-scale image is shown in Fig. 1c. The single cell images are 80×80 pixels, centered on cell nuclei. We use the same dataset in this study, with a few modifications.

We opt to remove the color information the three bands RGB carry, in order to simplify the image processing and analysis for our prototype system for malignant cancer classification. While the gray-scale conversion eliminates hue and saturation, our aim is to keep the luminance intact. The conversion is formed by a weighted sum of the red, green, and blue channels:

$$0.289R + 0.587G + 0.114B. \tag{6}$$

This way, the green channel is given the highest impact, which is reasonable since a wavelength of around 530 nm (i.e., green) is well-known in the literature to produce the best contrast for this type of biological material [5,7]. This conversion is achieved by the Matlab standard function rgb2gray.

In phase one, we partition the dataset into training data and test data as follows, in two phases. For phase one, the training data consist of 4.254 images balanced evenly between healthy and malignant samples. The test data consist of 9.942 images, divided into roughly one third healthy and two thirds malignant samples. For phase two, we investigate the effects of varying the amount of training data. Here, the number of cells in the training set is increased from 266 to 8.508 in steps. We use the same test set as in phase one.

The data set contains samples from 12 patients, 6 healthy and 6 with confirmed cancer diagnosis. The data are split on a patient level, so that the test set and the training set include cells from three healthy and three sick patients, respectively. No cells from an individual patient appear in both training and test sets, to avoid information leakage. Despite having annotations only on patient level, we aim to classify each cell nucleus as originating either from a sick or a healthy patient. That is, if the patient was diagnosed with cancer, then we associate all the individual cell images from that patient with the cancer label. However, it is not necessarily true that all the individual cells are malignant cells, or even indicative of cancer, and therefore it is not expected to near a 100% classification accuracy.

4 Experiments

The experiments were performed in two phases following the methodology described in Sect. 2. We investigated the effects on empirical risk and accuracy

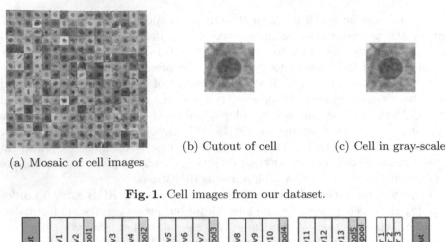

(a) Mosaic of cell images   (b) Cutout of cell   (c) Cell in gray-scale

Fig. 1. Cell images from our dataset.

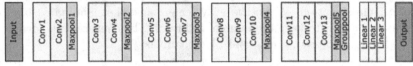

Fig. 2. The modified VGG16 architecture.

by varying data augmentation, the equivariant properties of the network, and the amount of data. For all experiments, we modified the VGG16 architecture, seen in Fig. 2 to work with greyscale images of size 80×80 pixels.

The training hyper-parameters are listed in Table 1 along with additional details of the network. Note that the standard number of channels was reduced to speed up the experiments. We also adapted the network to be equivariant to 90-degree rotations, by replacing the ordinary convolutions with group-equivariant convolutions on the C4 group. We call this architecture GCNN. The architecture with ordinary convolutions is named CNN. For the GCNN, we also introduced an additional group-pooling layer before the fully connected layer, which was not present in the ordinary CNN architecture. This functioned as a max-pooling layer over the rotations, and ensured a compatible input to the fully connected layer. This resulted in an architecture invariant to 90-degree rotations. We implemented our networks in the E2CNN framework, which is based on `pytorch` [12].

In the first phase, we investigated how the ordinary CNN performed compared to our new rotation-equivariant network. Here, we also investigated the effect of the use of data augmentation. We rotated the input images by 90, 180, and 270°, increasing the amount of data fourfold, before starting the training. For each experimental setting, we performed five runs in order to account for possible outlier results.

In phase two, we investigated how the amount of training data affected accuracy. The amount of training data was increased from 266 to 4254 in steps, following a law of $266 * 2^x$, where x is step. The final step diverged from this law and used 8508 samples, which was the full amount of data available. This ensured use of all the samples as well as intermediary multiples of the smallest number of samples. Data augmentation by multiples of 90° was also used.

Table 1. Settings for the CNN architecture and training procedure.

Parameter	Setting
Loss function	*Cross entropy*
Weight initialization	*He*
Optimizer	*Adam*
Learning rate	*0.00001*
Batch normalization	*Batches of size 128*
No. of epochs	*200*
Validation frequency	*1/5 epochs*
Activation functions	*ReLu*
Dropout	*0.5 between linear layers*
No. of channels	*16-16-32-32-64-64-64-128-128-128-128-128-128*
Convolution layers (size, stride, padding)	*Layers 1–12: (3, 1, 1)* *Layer 13: (4, 0, 0)*
Maxpooling layers (size, stride, padding)	*Layers 1–4: (2, 2, 0)* *Layer 5: (2, 1, 0)*
Linear layer parameters (input, output sizes)	*(128, 4096) – (4096, 4096) – (4096, 2)*

5 Results

The results from the first phase are presented in Figs. 3, 4, and 5. Figure 3 shows the training results for the experiments with the CNN. The training accuracy quickly converged to 100%. This was true for the GCNN as well.

Figures 4 and 5 show the accuracy on the test set for the CNN and GCNN, respectively. It can be seen that the accuracy for the CNN on the test set stabilizes at around 55–56%. This is true regardless of whether data augmentation is used or not. Using the GCNN increases the test accuracy to around 59–60%. Again, this holds regardless the use of using data augmentation. In Table 2 the sensitivity and the specificity for the last epoch for one instance of the experiments are reported. For both architectures, data augmentation by 90-degree rotations is performed. It is again evident that the GCNN outperforms the CNN.

The results from the second phase are presented in Fig. 6. For both architectures, the accuracy on the test set quickly reaches 100%. It converges quicker for increasing amounts of data. On the test set, for both architectures, the accuracy increases when more data is added. The GCNN again outperforms the CNN.

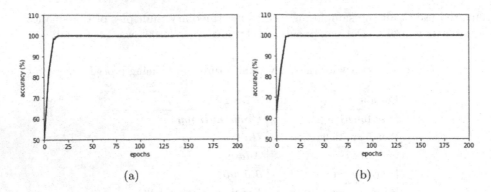

(a)   (b)

Fig. 3. Classification accuracy during training for an ordinary CNN. The figures show the experiments with a) no data augmentation b) with data augmentation.

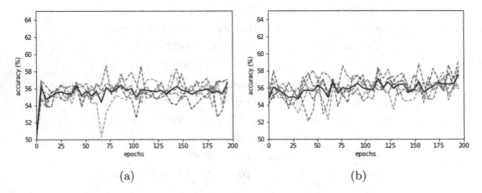

(a)   (b)

Fig. 4. Classification testing accuracy for an ordinary CNN. The plots show the experiments with a) no data augmentation b) with data augmentation. Dashed lines are runs with identical settings, the solid line is the mean of these runs.

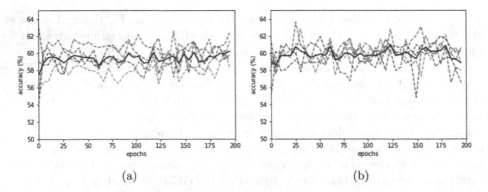

(a)   (b)

Fig. 5. Classification testing accuracy for a GCNN. The plots show the experiments with a) no data augmentation b) with data augmentation. Dashed lines are runs with identical settings, the solid line is the mean of these runs.

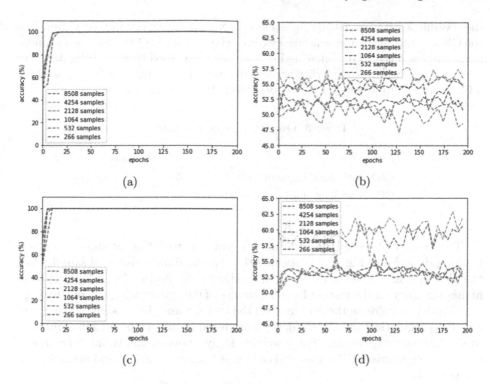

Fig. 6. Training and testing accuracy when varying the amount of training data. Figures a) and b) show the training and testing accuracy respectively for a CNN, while figures c) and d) show the training and testing accuracy respectively for a GCNN.

Table 2. CNN confusion matrix

Network	TN	FN	TP	FP	Sensitivity	Specificity
CNN	2980	1991	2630	2341	0.569	0.560
GCNN	3249	1722	2756	2215	0.615	0.595

6   Conclusions and Future Work

In this study, we investigated classifiers on a biomedical dataset. Our hypothesis was that replacing CNNs using data augmentation by rotations with architectures equivariant to those same rotations would decrease overfitting. Specifically, we experimented with 90-degree rotations.

By looking at Fig. 4, we see that performing data augmentation by multiples of 90-degree rotations does not improve test accuracy, for an ordinary CNN. However, designing the network to be equivariant to those same rotations by the use of the GCNN increases the test accuracy. By inspecting the values in Figs. 4b and 5a, we estimate the mean test accuracies for the CNN with data augmentation versus the GCNN with no data augmentation. For the last epoch, these values are 0.55 and 0.59 respectively. From Fig. 3b, the training accuracy for the CNN is 1.0. The training accuracy for the GCNN is also 1.0. This corresponds to an empirical risk of 0. Following Eq. 5 we calculate overfitting in Table 3. Since a lower value means less overfitting, we conclude that using a GCNN instead of a CNN with data augmentation reduces overfitting in this setting.

Table 3. Overfitting measurements

Network	Overfitting ratio
CNN with data augmentation	**1.82**
GCNN without data augmentation	**1.69**

The second phase of the experiments saw the variation of the amount of training data. From Fig. 6, we conclude that the minimal number of samples for the highest possible test accuracy is around 4000. Secondly, the GCNN yields a higher accuracy on the test set for all amounts of data, indicating less overfitting.

In order to increase the accuracy on the test set and decrease overfitting, we added L_2 regularization. Weight decay was varied between 0.5 and 256, doubling in size for each experiment. For a weight decay above 8, the training accuracy decreased significantly. However, this came at the cost of decreased accuracy on the test set.

A promising way forward is to repeat the experiments using other types of neural network architectures, particularly those who already have shown to yield high test accuracy on the dataset. Future work could also compare other rotations besides the current multiples of 90°, as well as other types of transformations, such as reflections. Finally, considering other, highly realistic datasets is of great interest for determining if equivariant networks can improve automatic diagnostics in more general biomedical settings.

Acknowledgment. We thank Gabriele Cesa *et al.* for their contributions during our tailoring of the framework to the specific application of cancer classification for oral images. We are grateful for Professor Emeritus Ewert Bengtsson's valuable input to this project on cell image analysis in general and color conversions in particular. We especially thank Professor Nataša Sladoje for initiating the project in conjunction with the Wallenberg AI, Autonomous Systems and Software Program.

References

1. Bengtsson, E., Malm, P.: Computational and mathematical methods in medicine. J. Big Data (2014)
2. Chidester, B., Zhou, T., Do, M.N., Ma, J.: Rotation equivariant and invariant neural networks for microscopy image analysis. Bioinformatics **35**(14), 530–537 (2019)
3. Cohen, T., Welling, M.: Group equivariant convolutional networks. In: International Conference on Machine Learning (ICML) (2016)

4. Forslid, G., et al.: Deep convolutional neural networks for detecting cellular changes due to malignancy. In: International Conference on Computer Vision Workshops (ICCVW) (2017)
5. Galbraith, W., Marshall, P., Lee, E., Bacus, J.: Studies on Papanicolaou staining. I. Visible-light spectra of stained cervical cells. Anal. Quant. Cytol. **1**(3), 160–168 (1979)
6. Goodfellow, I., Bengio, Y., Courville, A.: Deep Learning, p. 127. MIT Press, Cambridge (2016)
7. Holmquist, J., Imasoto, Y., Bengtsson, E., Olsen, B., Stenkvist, B.: A microspectrophotometric study of Papanicolaou-stained cervical cells as an aid in computerized image processing. J. Histochem. Cytochem. **24**(12), 1218–1224 (1976)
8. Lu, J., Sladoje, N., Stark Runow, C., Darai Ramqvist, E., Hirsch, J.M., Lindblad, J.: A deep learning based pipeline for efficient oral cancer screening on whole slide images. In: International Conference on Image Analysis and Recognition (ICIAR) (2020)
9. Semih Kayhan, O., van Gemert, J.C.: On translation invariance in CNNs: convolutional layers can exploit absolute spatial location. In: Conference on Computer Vision and Pattern Recognition (CVPR) (2020)
10. Shorten, C., Khoshgoftaar, T.: A survey on image data augmentation for deep learning. J. Big Data **6**, 60 (2019)
11. Vapnik, V.: Principles of risk minimization for learning theory. In: Moody, J.E., Hanson, S.J., Lippmann, R.P. (eds.) Advances in Neural Information Processing Systems, vol. 4, pp. 831–838. Morgan-Kaufmann (1992)
12. Weiler, M., Cesa, G.: General E(2)-Equivariant Steerable CNNs. In: Conference on Neural Information Processing Systems (NeurIPS) (2019)
13. Weiler, M., Hamprecht, F., Storath, M.: Learning steerable filters for rotation equivariant CNNs. In: Conference on Computer Vision and Pattern Recognition (CVPR) (2018)
14. Wetzer, E., Gay, J., Harlin, H., Lindblad, J., Sladoje, N.: When texture matters: texture-focused CNNs outperform general data augmentation and pretraining in oral cancer detection. In: International Symposium on Biomedical Imaging (ISBI) (2020)
15. Winkels, M., Cohen, T.: 3D G-CNNs for pulmonary nodule detection. In: International Conference on Medical Imaging with Deep Learning (2018)
16. Worrall, D., Brostow, G.: CubeNet: equivariance to 3D rotation and translation. In: Ferrari, V., Hebert, M., Sminchisescu, C., Weiss, Y. (eds.) ECCV 2018. LNCS, vol. 11209, pp. 585–602. Springer, Cham (2018). https://doi.org/10.1007/978-3-030-01228-1_35

A Multitasking Learning Framework for Dermoscopic Image Analysis

Lidia Talavera-Martínez[1,2]([✉]) [iD], Pedro Bibiloni[1,2] [iD],
and Manuel González-Hidalgo[1,2] [iD]

[1] SCOPIA Research Group, University of the Balearic Islands, Palma, Spain
l.talavera@uib.es
[2] Health Research Institute of the Balearic Islands (IdISBa), 07010 Palma, Spain

Abstract. Multitasking learning improves a model's ability to generalize by learning multiple tasks in parallel. However, it is difficult to know how each task influences the others' learning. In this work, we study in-depth the behavior of the tasks of skin lesion segmentation, hair mask segmentation, and the inpainting of those hairs, in a multitasking framework to discover how they influence each other. The experiments are performed using an encoder-decoder convolutional neural network and images from five public databases: PH2, dermquest, dermis, EDRA2002, and the ISIC Data Archive. To evaluate the tasks' performance, we use a series of metrics on which we apply a statistical test to check the superiority of each task in a multitasking model with respect to their individual performance. We also check, in a three-task model, whether there is a task that dominates the learning stage. Finally, we conclude that while the inpainting task does not benefit from this type of learning, the rest of the tasks improve their performance when compared to that obtained by their corresponding single-task model.

Keywords: Multitasking · Deep neural networks · Dermoscopy · Skin lesion segmentation · Hair removal · Inpainting

1 Introduction

In recent years, the development of Computer Aided Diagnosis (CAD) systems based on deep learning has proven to be a powerful tool for a large number of tasks in computer vision and image analysis. More specifically, they have achieved higher performance over traditional approaches for most applications within the medical field [7] being a support tool to help dermatologists and general practitioners, to provide an early, objective, and reproducible diagnosis of skin lesions.

Recently multitasking learning has become one of the most interesting approaches in many computer vision applications [10]. This technique intends to solve related tasks simultaneously and to improve their generalization ability, by sharing some of the hidden layers of the model, so that they learn a joint representation for the multiple tasks, leveraging both their commonalities and differences [3]. The reduction of the computational time, an improvement in the

© Springer Nature Switzerland AG 2021
J. M. R. S. Tavares et al. (Eds.): CIARP 2021, LNCS 12702, pp. 34–44, 2021.
https://doi.org/10.1007/978-3-030-93420-0_4

robustness of the model against overfitting, and a possible improvement in prediction accuracy are some of its main advantages. Among the extensive literature that we can find for the analysis of dermoscopic images with computer vision techniques, only a few works have explored the benefits of multitasking learning using CNNs for this purpose. However, multitask models have shown potential in this field [4,6,11,15].

In this work, we carry out a deeper analysis of the model presented in [13], where we introduced a multitasking model based on CNNs to simultaneously 1) segment the skin lesion, 2) segment the hairs that may occlude the lesion, and perform the 3) inpainting of these hairs. Thus, we explore in a deeper way the behavior of the different tasks combinations to discover how they influence each other. We believe that task 2 can be an auxiliary task of 1, and that 3 could be an auxiliary task of 2.

The rest of the document is structured as follows. First, in Sect. 2, we describe the architecture of our approach and some aspects of its learning process. In Sect. 3, we establish the experimental set up in which we describe the database, the implementation details of the network, as well as presenting the results obtained. Finally, in Sect. 4, we discuss the previous results.

2   Network Architecture and Learning Details

In this section, we describe the architecture and learning details of our proposed multitasking model for dermoscopic image analysis.

As can be seen in Fig. 1, the architecture of our model is based on a convolutional encoder-decoder network with a low-resolution module for a more complete view of the context of the images. Our model is fed with a set of dermoscopic images and three ground-truth (GT) images that correspond to each of the tasks to be learned. First of all, the images' size is reduced to a fixed size of $512 \times 512 \times 3$, to lower the computational cost. On the one hand, the encoder part consists of two blocks composed by one 3×3 convolution, of 128 filters in the first block and 256 filters in the second one, followed in both cases by a down-sampling operation. In the low-resolution module, we employ an average pooling layer to down-sample the input image to a size of $128 \times 128 \times 3$ and an encoder-decoder architecture where the encoder's blocks consist of a 3×3 convolution, of 8, 32, and 64 filters, respectively, followed each one by a down-sampling operation to reduce the spatial resolution, which is applied by a two-stride 3×3 convolution. The decoder performs the opposite operations by means of deconvolutions and up-sampling layers, until it reaches a feature map of $128 \times 128 \times 3$. Then, the hidden representations of the images' high-level features, obtained with a different resolution by both the low-resolution module and the encoder part, are merged and fed to the decoder. On the other hand, the decoder part also consists of two blocks composed by an up-sampling of the feature map by a deconvolution of 3×3 with strides of two, followed by its concatenation with the corresponding feature map from the layer of equal resolution of the encoder (skip connection), and a 3×3 convolution. Finally, we stack three convolutional layers and one sigmoid activation layer to obtain each of the two segmentation outputs, and a single 3×3 convolution to obtain the inpainting's task output with the number of output channels.

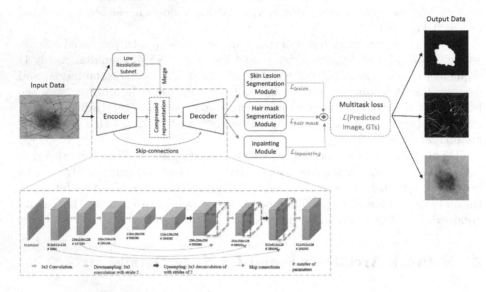

Fig. 1. Architecture of our proposed network. The input of our model consists of a dermoscopic image and three ground-truth (GT) images, one for each task that the model intends to learn. The input flows in parallel through the low-resolution module and the encoder. Their two hidden representations are merged into one by concatenating their features, which are fed to the decoder, which also leverages skip connections to provide useful features for the output modules. Our model optimizes during training a weighted average of three loss functions.

2.1 Learning Details

During training, the proposed model optimizes a weighted average of three loss functions: the dice coefficient loss, the weighted binary cross-entropy loss and the one presented in [12], which correspond to the tasks of skin lesion segmentation, the hair mask segmentation and the inpainting, respectively. We also performed data augmentation in the training phase to improve the model's ability to generalize (random zoom and rotation, shifts and flips in both horizontal and vertical directions). The performance of the model depends on the relative weighting between the loss of each task. Thus, we use a gradient normalization (GradNorm) strategy to automatically balance training by dynamically tuning each task's weight on the loss so that their contributions to the gradient in a certain shared layer are similar. Our model is trained following an early stopping policy based on monitoring the validation loss [5].

3   Experimental Design and Results

In this section, we perform a comparative study of the performance of different configurations of multitasking models based on the tasks of lesion segmentation, and hair segmentation and inpainting. First, we establish the experimental framework by describing the database, the performance measures, and the details of its implementation. Finally, we present and analyze the results obtained, comparing them from a qualitative and quantitative point of view.

3.1   Dataset and Implementation Details

Ideally, given a dermoscopic image, we want our model to be able to segment the skin lesion, extract any hairs that may exist, and finally, recover the underlying texture of these hair regions by means of inpainting. However, as far as we know, there is no dataset with such expert information, so we learned on a set of collected samples from different datasets. There are many public databases that provide the lesion segmentation as GT, as it is one of the most studied topics in this field. On the contrary, finding images with hair, along with their corresponding hair mask, and their "clean" version, – the same image without hair–, was a challenging task. Of course, the same dermoscopic image cannot be captured with and without hair. To address this problem, we selected hairless images from five publicly available data sets, *i.e.* PH2 [8], dermquest[1], dermis[2], EDRA2002 [1] and the ISIC Data Archive[3], and simulated the presence of skin hair on them. We have tried to avoid selecting images with other artifacts (e.g. ruler, bandages, etc.). We have used three different hair simulation methods, one based on generative adversarial networks [2], another is the "HairSim" software [9], and the last one consisted in extracting hair masks by an automated method, proposed by Xie *et al.* [14], and superimpose them on hairless images.

To train and evaluate the models, we constructed a dataset with 1060 images. Hair simulation has been performed on a limited number of them, so not all samples have GTs for all three tasks. In Fig. 2, we can see an example of a test image and its three corresponding GTs. Of the 1060 images, 556 images have the GT for the lesion segmentation task, and 557 images have the GTs for the hair segmentation and inpainting tasks. During the experimentation we divide the whole dataset into 80% for training and 20% for the test set.

The experiments were carried out with the Windows 10 Pro 64-bit operating system with a 64.0 GB RAM, a single NVIDIA GeForce GTX 1070 and an Intel® Xeon® CPU E5-2620 V4 @ 2.10 GHz. We implemented the proposed architecture using Keras and trained it from scratch with a batch size of 2, randomly initialized weights, and using the Adam optimizer with a learning rate experimentally set to 10^{-4}.

---

[1] Was deactivated on December 31, 2019.
[2] www.dermis.net.
[3] www.isic-archive.com.

3.2   Experiments and Analysis

We carried out a qualitative and quantitative analysis of the results obtained by training a single task model for each of the three tasks considered, as well as for each multitasking model formed by the pairs of tasks and the combination of all of them.

In Fig. 3 we present a visual comparison of the results obtained for each possible combination of tasks on Fig. 2a. Regarding the lesion's segmentation, we can see how a more accurate segmentation is obtained when the task is learned in combination with auxiliary tasks. Mainly with the inpainting task, see Fig. 3d, if we compare the results with the pair conformed by lesion and hair mask segmentation, in Fig. 3e. As for the hair mask segmentation task, although the results appear to be visually similar, when a multitasking model is trained, its results appear to be more complete and a crisper better defined segmentation of the hairs is achieved. It is worth mentioning that there is a great relationship between this task and the inpainting one, since the hair's areas that are not detected are not inpainted properly. This is easily seen in those hairs sections that are inside the lesion. Finally, concerning the inpainting task, when we add some auxiliary tasks during training, its results improve in terms of the reconstruction of the hair areas within the lesion, although in some cases leave some traces in the skin regions.

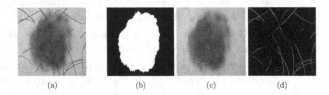

(a)   (b)   (c)   (d)

Fig. 2. Example of a test image (a) and its three GTs for skin lesion segmentation (b), inpainting (c), and hair mask segmentation (d).

A qualitative evaluation is not enough to assess the performance of the different tasks when considered independently or in a combination with the others. Therefore, in the following, we perform a quantitative evaluation which provides an objective and comparable evaluation of the performance of each task independently. Unlike the results presented in [13], in this case we have decided to use the Dice Coefficient metric instead of the Balanced Accuracy for the lesion segmentation task and the Structural Similarity Index (SSIM) instead of the Multiscale Structural Similarity Index (MSSSIM) for the inpainting task. In the first case, this metric seems to be more suitable for this task as it is one of the most used metrics for image segmentation. In the second case, it allows us to complement the results obtained based on a simpler performance metric for the quality of the restoration. Finally, for the hair mask segmentation task, where there is a clear imbalance between the background class pixels compared to hair pixels, we rely again on the Balanced Accuracy.

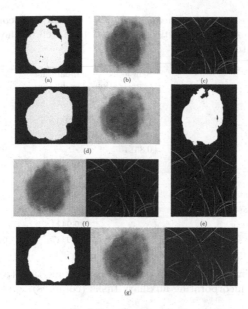

Fig. 3. Results obtained on the example of Fig. 2 for training a single task model (a) lesion segmentation, (b) hair removal, (c) hair mask segmentation. Multitask model for each of the pairs between (d) lesion segmentation and hair removal, (e) lesion and hair mask segmentation, (f) hair removal and hair mask segmentation; and a combination of all three tasks (g) lesion and hair mask segmentation, and hair removal.

The results obtained on the 214 images of the test set, for the performance measures corresponding to each task, are shown in Table 1. In addition, we have computed the gains and losses on the performance measures of single tasks models when incorporating auxiliary tasks, which is reflected in Table 2. As can be seen, in 5 out of 9 cases the results improve in terms of these performance measures. In particular, the lesion segmentation performance is boosted by the introduction of any other task. However, its inclusion only improves the hair mask segmentation task, while when combined with the inpainting task, the performance of the segmentation is increased by 2.7%, and the inpainting is reduced by 0.97%, thus contributing the latter more to the former task. We can also observe that the hair mask segmentation task does not improve the inpainting task with which we initially believed there was a great influence. We see that the performance measure of the inpainting task is not benefited by the rest of tasks. Finally, we notice that when training the three tasks together, both segmentation tasks increase their balanced accuracy by 1.80% and 0.98%, respectively.

The next step in our work is to perform a bilateral statistical test between the base models –single task– and the ones resulting from its combination with other tasks, to determine whether and under what conditions, these last models outperform significantly the former ones. We have used the t-test if the samples pass the Shapiro-Wilk normality test, or the Wilcoxon signed-rank test other-

Table 1. Performance measures' mean obtained for each task considered either for the single-task model or for the multitasking model, as well as the number of epochs needed for the model to learn.

Tasks	Skin lesion segmentation evaluation	Inpainting evaluation	Hair mask segmentation evaluation	Epochs
	Dice coefficient	SSIM	Balanced Acc.	
Skin lesion seg.	0.8514	–	–	100
Inpainting	–	0.9476	–	49
Hair mask seg.	–	–	0.7954	10
Skin lesion seg. and Inpainting	0.8791	0.9379	–	68
Skin lesion and Hair mask seg.	0.8594	–	0.7978	43
Inpainting and Hair mask seg.	–	0.9424	0.7937	29
Skin lesion and Hair mask seg. and Inpainting	0.8694	0.9347	0.8052	42

Table 2. Relative gains and losses (in percent) over single-task (row) model's performance measure when incorporating auxiliary tasks (columns).

	With skin lesion segmentation	With inpainting	With hair mask segmentation	With the rest of the tasks
Skin lesion segmentation Dice coefficient	–	2.77%	0.80%	1.80%
Inpainting SSIM	−0.97%	–	−0.53%	−1.29%
Hair mask segmentation Balanced Acc	0.24%	−0.17%	–	0.98%

wise, setting a confidence level of 95%. Table 3 summarizes the results obtained when applying the statistic test to the performance metrics of each task. As we can see, when we combine the three tasks, the performance metrics for the hair mask and the lesion segmentation tasks have a statistically better indicator when compared to their individual tasks' performance, although the difference is not significant. Regarding the pairwise combinations of the tasks, in the case of the hair mask segmentation, according to the Balanced Accuracy, both multitasking models obtain statistically comparable results to the base hair mask segmentation model. However, the multitasking model that has a better indicator is the one resulting from its combination with the task of lesion segmentation. In the same way, based on the SSIM measure, the inpainting task in a multitasking framework is statistically lower than its solo performance. Finally, the segmentation task benefits from incorporating either of the other two tasks, as its multitasking models' performance measure statistically outperforms that of the single-task lesion segmentation model.

Table 3. Results of the statistical test on the performance of each base-task model compared to its task' performance in a multitasking environment.

	Multitasking models	p-value	Statistically significant	Model with best performance
Hair mask Segmentation base model (Balanced Acc.)	Skin lesion and Hair mask seg.	0.38	✗	Multitask
	Inpainting and Hair mask seg.	0.51	✗	Single-task
	Skin lesion and Hair mask seg. and Inpainting	6.59e−05	✓	Multitask
Inpainting base model (SSIM)	Skin lesion and Hair mask seg.	4.01e−17	✓	Single-task
	Inpainting and Hair mask seg.	1.82e−19	✓	Single-task
	Skin lesion and Hair mask seg. and Inpainting	1.22e−19	✓	Single-task
Skin lesion Segmentation base model (Dice Coefficient)	Skin lesion and Hair mask seg.	0.91	✗	Multitask
	Inpainting and Hair mask seg.	8.68e−05	✓	Multitask
	Skin lesion and Hair mask seg. and Inpainting	0.05	✗	Multitask

To complete the study of how different tasks influence multitasking learning, we analyze how the validation loss change depending on whether it is a single task or a multitasking model. We can see in Figs. 4a, 4b and 4c, that the training of only the lesion segmentation task is slightly unstable at first and hinders to find an optimal point from which to substantially improve the loss function. However, when we add the inpainting task, the hair mask segmentation, or both, the learning seems to flow in a suitable way. Regarding the number of epochs required to train the models, we note that the individual model for the lesion segmentation task requires a greater number of epochs compared to the other models, while this is considerably reduced when this task is introduced in a multitasking model. On the contrary, the task that needs less epochs to be learned is the one for the hair mask segmentation. Finally, it should be mentioned that although in Fig. 4d the loss function of the inpainting task coincides with the loss of the multitasking model, and therefore it seems that this is the dominant task, in Fig. 5 we can see that the loss of each task has a high correlation with the combined one. Even for the correlation with respect to the inpainting task, where we obtain the lowest correlation index of 0.86, we can see that is due to the influenced by the presence of outliers, corresponding to the first epochs. For the losses of the hair mask and lesion segmentation tasks, we obtain a correlation index of 0.94 and 0.99, respectively.

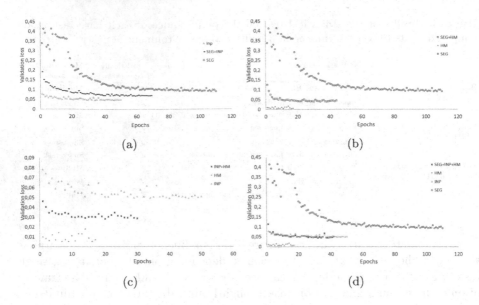

Fig. 4. Validation loss functions evolution depending on whether it is a single task model or a multitasking one.

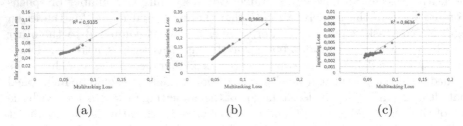

Fig. 5. Correlation between each single-task model and the multitasking model for the three tasks.

4   Conclusions

In this work, we have studied how different tasks on dermoscopic imaging are affected by multitasking learning. Specifically, our research focuses on the tasks of lesion segmentation, hair mask segmentation and the inpainting of these. Thus, exploring the behavior of the different tasks combinations we aim to discover how they influence each other. From the experimentation results, we can conclude that the lesion segmentation task is benefited from incorporating either of the other two tasks, as the multitasking models' performance measure statistically outperform the individual segmentation task's performance. On the contrary, the inpainting task in a multitasking framework is statistically lower to its solo performance. We also show that when we combine the three tasks, the performance metrics for the hair mask and the lesion segmentation tasks are statistically better than their corresponding base model's performance. Finally, regarding the contribution of each task, we have been able to see that each of the three tasks has a great correlation, individually, with the performance of the multitask model of the three tasks studied.

Acknowledgments. This work was partially supported by the Spanish Grant FEDER/Ministerio de Economía, Industria y Competitividad - AEI/TIN2016-75404-P. Lidia Talavera-Martínez also benefited from the fellowship BES-2017-081264 conceded by the *Ministry of Economy, Industry and Competitiveness* under a program co-financed by the *European Social Fund*. We thank M. Attia from the Deakin University, Australia, for providing the GAN-based simulated hair images.

References

1. Argenziano, G., Soyer, H., De Giorgi, V., Piccolo, D., Carli, P., Delfino, M.: Interactive atlas of dermoscopy (Book and CD-ROM). EDRA Medical Publishing & New media (2000)
2. Attia, M., Hossny, M., Zhou, H., Nahavandi, S., Asadi, H., Yazdabadi, A.: Realistic hair simulator for skin lesion images: a novel benchemarking tool. Artif. Intell. Med. **108**, 101933 (2020)
3. Caruana, R.: Multitask learning. Mach. Learn. **28**(1), 41–75 (1997)
4. Chen, E.Z., Dong, X., Li, X., Jiang, H., Rong, R., Wu, J.: Lesion attributes segmentation for melanoma detection with multi-task u-net. In: IEEE 16th International Symposium on Biomedical Imaging (ISBI 2019), pp. 485–488. IEEE (2019)
5. Chen, Z., Badrinarayanan, V., Lee, C.Y., Rabinovich, A.: Gradnorm: gradient normalization for adaptive loss balancing in deep multitask networks. In: International Conference on Machine Learning, pp. 794–803. PMLR (2018)
6. Kawahara, J., Daneshvar, S., Argenziano, G., Hamarneh, G.: Seven-point checklist and skin lesion classification using multitask multimodal neural nets. IEEE J. Biomed. Health Inform. **23**(2), 538–546 (2018)
7. Litjens, G., et al.: A survey on deep learning in medical image analysis. Med. Image Anal. **42**, 60–88 (2017)
8. Mendonça, T., Celebi, M., Mendonça, T., Marques, J.: PH^2: a public database for the analysis of dermoscopic images. In: Dermoscopy Image Analysis (2015)
9. Mirzaalian, H.: Hair Sim Software. http://www2.cs.sfu.ca/~hamarneh/software/hairsim/Welcome.html. Accessed 7 Mar 2019
10. Ruder, S.: An overview of multi-task learning in deep neural networks. arXiv preprint arXiv:1706.05098 (2017)
11. Song, L., Lin, J.P., Wang, Z.J., Wang, H.: An end-to-end multi-task deep learning framework for skin lesion analysis. IEEE J. Biomed. Health Inform. **24**(10), 2912–2921 (2020)
12. Talavera-Martínez, L., Bibiloni, P., González-Hidalgo, M.: Hair segmentation and removal in dermoscopic images using deep learning. IEEE Access **9**, 2694–2704 (2021). https://doi.org/10.1109/ACCESS.2020.3047258
13. Talavera-Martínez, L., Bibiloni, P., González-Hidalgo, M.: How do different tasks influence a multitasking environment? Case study in dermoscopic images. In: To be Published at IEEE 18th International Symposium on Biomedical Imaging (ISBI 2021) (2021, in press)

14. Xie, F.Y., Qin, S.Y., Jiang, Z.G., Meng, R.S.: PDE-based unsupervised repair of hair-occluded information in dermoscopy images of melanoma. Comput. Med. Imaging Graph. **33**(4), 275–282 (2009)
15. Yang, X., Zeng, Z., Yeo, S.Y., Tan, C., Tey, H.L., Su, Y.: A novel multi-task deep learning model for skin lesion segmentation and classification. arXiv preprint arXiv:1703.01025 (2017)

An Evaluation of Segmentation Techniques for Covid-19 Identification in Chest X-Ray

Arthur Rodrigues Batista[1]([✉]), Diego Bertolini[2], Yandre M. G. Costa[1],
Luiz Fellipe Machi Pereira[1], Rodolfo Miranda Pereira[3],
and Lucas O. Teixeira[1]

[1] State University of Maringá (UEM), Maringá, PR, Brazil
[2] Federal Institute of Paraná (IFPR), Pinhais, PR, Brazil
[3] Federal University of Technology Paraná (UTFPR), Campo Mourão, PR, Brazil

Abstract. COVID-19 is a highly contagious disease caused by the SARS-CoV-2 virus. Due to its high impact on society, several efforts have been made to design practical ways to support COVID-19 diagnosis. In this context, automated solutions based on chest x-rays (CXR) images and deep learning are among the popular ones. Although these techniques achieved exciting results in the literature, the use of regions that do not support pneumonia diagnosis, i.e., regions outside the lung area, may bias the recognition model. A strategy to avoid this issue is to use segmentation techniques to isolate the lung area before the classification process. In this work, we investigate the impact of three CNN segmentation architectures on COVID-19 identification: U-Net, MultiResUnet, and BCDU-NET. We also investigate which portions of the CXR most influence each model's predictions, using Explainable Artificial Intelligence. The BCDU-NET architecture achieved a Jaccard Index of 0.91 and a Dice Coefficient of 0.95. In the best scenario, lung segmentation improved the COVID-19 identification F1-Score by about 6.6%.

Keywords: COVID-19 · Semantic segmentation · Explainable artificial intelligence

1 Introduction

Since December 2019, a new virus (SARS-CoV-2) has spread from Wuhan to other parts of China and worldwide. As of March 23, 2021, over 124 million cases have been confirmed, and around 2.74 million deaths have been reported worldwide[1]. Due to the many uncertainties regarding the vaccine's availability, particularly considering less economically developed regions, there are many proposed methods for faster and automatic diagnoses, evaluation of disease evolution and severity.

As a result, many specialists consider chest X-rays (CXR) and computed tomography (CT) scan a vital ally to provide additional data, reinforcing evidence

---

[1] https://www.worldometers.info/coronavirus/.

© Springer Nature Switzerland AG 2021
J. M. R. S. Tavares et al. (Eds.): CIARP 2021, LNCS 12702, pp. 45–54, 2021.
https://doi.org/10.1007/978-3-030-93420-0_5

to reach a diagnosis, mainly when the patient has already developed signs of pneumonia. In these cases, they are considered essential tools to help doctors assess the disease evolution and optimize prevention and control measures. The CT scan is considered the gold standard since it generates very detailed images. However, CXR is still useful because they are cheap, produces fast results and, uses less radiation [30].

Given the high prevalence of X-ray machines in most of the health centers, one of the significant advantages of this type of exam lies in the portability of the equipment needed to perform it, making the CXR an excellent complement to other tests such as RT-PCR, that require an early infection stage to be efficient in detecting COVID-19 [37]. It is especially useful since CXR is a routine exam for patients with breathing difficulties [22].

One of the biggest bottlenecks faced when using CXR is the difficulty of radiologists interpreting X-ray images due to factors such as type of X-ray machine, amount of X radiation applied to the patient, particular lung characteristics, and different types of pathogens. It is tough to visually identify what caused pneumonia, especially on lungs that have not been drastically affected. Hence, computer-aided diagnostic systems, which can help radiologists interpret CXR more quickly and accurately, are highly desirable, especially for detecting emerging diseases like COVID-19.

In an ideal condition, the typical steps in a CXR analysis system should include: (1) identification of the region of interest (ROI) (for example, pulmonary lobes); (2) ROI extraction; and (3) application of machine learning to detect and diagnose the abnormality [21,26,35]. The process of finding and extracting the ROI is commonly called segmentation. Various works aiming to detect COVID-19 using CXR images did not perform the lung segmentation as the first step [1,5,24].

In the literature, various works applied segmentation to identify different lung diseases using CXR, such as lung cancer [14], tuberculosis [34], among others [7,38]. Neglecting this step might affect the performance of the subsequent steps and the overall system performance as a whole. Mainly because some features used might be outside the ROI, leading to questionable performance accuracies. Therefore, it is an important pre-processing step in an abnormality validation process [10].

In this work, our primary aim is to compare three deep learning inspired segmentation models: U-Net [28], MultiResUNet [19] and BCDU-NET [6]. The motivation is to evaluate how the use of these different models impacts the overall COVID-19 identification rates. In order to achieve that, we followed a method with three stages: i) lung segmentation, ii) COVID-19 identification, and iii) model inspection using Explainable Artificial Intelligence (XAI) techniques. The main reason for using XAI is to visually inspect what regions of the input image the machine learning model uses. In our context, we are interested in models that primarily focus on the lung area.

The XAI motivation is to develop approaches that help to explain the individual predictions of modern solutions based on Machine Learning (ML). We applied a technique called Local Interpretable Model-Agnostic Explanations (LIME) [27]. LIME works by locating specific regions of the image that increase the probability of the model predicting a specific class. These regions are considered important, given that the model actively uses them to make predictions.

The remaining of this paper is organized as follows: Sect. 2 introduces our purposed methodology. Section 3 shows details about our experimental setup, such as the trained models parameters, data augmentation techniques and evaluation metrics. Furthermore, Sect. 4 presents the results and discussions. Finally, Sect. 5 presents our conclusions and future works.

2   Proposed Method

This section presents the method used to evaluate the impact of lung segmentation in COVID-19 diagnosis, illustrated in Fig. 1.

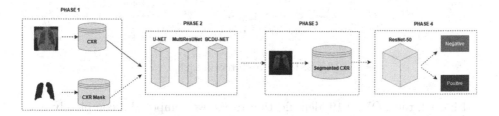

Fig. 1. COVID-19 identification using lung segmentation.

In phase 1, we collected CXR images used for the segmentation models. In this case, four data sets were used: Montgomery County X-ray Set, Shenzhen Hospital X-ray Set [11], Japanese Society of Radiological Technology (JSRT) [33] and Cohen v7labs[2], containing 138, 566, 247 and 489 CXR images respectively, totaling 1.440 images. We also created 205 binary lung masks manually using CXR images from Cohen data set[3] [12]. Finally, distributed as follows: ≈ 90% for training, ≈ 5% for testing, and the remaining ≈ 5% for validation.

Phase 2, the segmentation step, aims to identify and cut the ROI from a CXR image. In this case, the ROI is the lung region. This step aims to reduce the impact of noise that could interfere in the classification stage. Therefore, for the detection of lung diseases such as COVID-19, the identification system must not use information outside of ROI, given that it can lead to erroneous or not completely reliable results (further discussed in Sect. 4.3).

In order to perform CXR segmentation, we used three Convolutional Neural Network (CNN) inspired architectures: U-Net [28], MultiResUNet [19] and BCDU-NET [6]. U-Net's architecture is symmetrical (as seen in Fig. 2), with

---

[2] https://github.com/v7labs/covid-19-xray-dataset.
[3] https://github.com/ieee8023/covid-chestxray-dataset.

an encoder that extracts the spatial characteristics of the image and a decoder that builds the segmentation map from the coded characteristics. Both MultiResUNet and BCDU-NET follow a similar construction, and their difference with U-Net resides in how they introduce new convolution operations mostly to avoid learning redundant features.

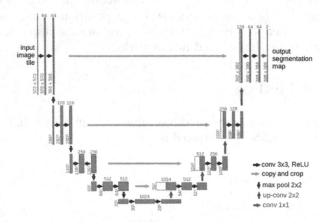

Fig. 2. U-Net architecture [28]

Phase 3, the COVID-19 identification tasks, we composed a completely new data set containing COVID-19 cases from two others: Covid Chestxray Dataset (CCD) [12], and CheXPert [20]. From CDD data set, we used 512 COVID-19 infected CXR images, 225 infected by other lung diseases, and 400 healthy lung images. From CheXPert, we used 3,375 CXR images from patients not infected by SARS-CoV-2. Table 1 presents the final distribution of images.

Table 1. Data distribution in training, testing and validation for COVID-19.

Step	COVID-19	Non COVID-19
Training	244	1,900
Validation	12	100
Testing	256	2,000
Total	512	4,000

Finally, in phase 4, we used COVID-19 identification as an external validation for each segmentation model. To do so, we applied each segmentation model to our COVID-19 data set and generated three data sets of segmented CXR images. Then, we used a ResNet-50 model to identify COVID-19 in each data set, plus the original data set without segmentation. In addition, we also applied an Explainable Artificial Intelligence method called LIME to inspect each trained model visually.

3   Experimental Setup

This section describes the overall experimental setup, including the parameters, data augmentation, and evaluation metrics.

3.1   Parameters

For the segmentation models, we used Adam optimizer, 100 epochs with a batch size of 16 and a learning rate of 0.5.

For the classification using ResNet-50, we applied transfer learning with pre-defined weights from ImageNet, and we followed a standard workflow used in the literature with two stages: warm-up and fine-tuning [16]. For the training, we used Adam optimizer, 100 epochs with a batch size of 24, a learning rate of 0.5 for training, and 50 epochs for fine-tuning with a batch size of 24 and a learning rate of 0.3.

3.2   Data Augmentation

In deep learning, the overall performance tends to increase as the training data increases [31]. Accordingly, we applied data augmentation techniques to increase our training set in Phase 1 and Phase 3. We applied the following transformations:

- Horizontal flip;
- Change of brightness and contrast;
- Change of scale, translation or random rotations;
- Unique elastic deformations, presented in [32].

3.3   Evaluation Metrics

To evaluate lung segmentation, we used two popular metrics: Jaccard Index [36], and Dice Coefficient [4]. For the performance evaluation of the COVID-19 binary classifier, we considered a popular metric named F1-score [15].

4   Results and Discussion

This section presents our results and discussions. To better conduct the discussion, we present the results in three subsections: i) segmentation performance; ii) COVID-19 identification rates using segmented CXR; and iii) visually inspection of the trained models using LIME.

Table 2. Segmentation results of each Segmentation Model

Model	Jaccard index	Dice coefficient
U-Net	0.90	0.94
MultiResUnet	0.45	0.63
BCDU-NET	**0.91**	**0.95**

4.1 Segmentation Performance

Table 2 presents the Jaccard Index and Dice coefficient for each evaluated segmentation model.

The U-Net and BCDU-NET models achieved satisfying performance, and MultiResUnet stood out negatively. Through our manual inspection (see Fig. 3), we found that MultiResUnet is performative in isolating only the pulmonary region, without the region superimposed by the heart. Such a result is not unexpected given that MultiResUnet can identify the curvatures of objects very well [19]. In contrast, the U-Net was able to identify the whole lung region, including the zones superimposed by the heart. We believe that many masks included the zones with superimposition, which benefited U-Net and BCDU-NET, and caused MultiResUnet to be outperformed.

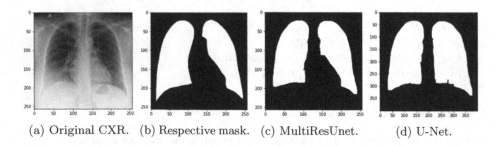

(a) Original CXR.   (b) Respective mask.   (c) MultiResUnet.   (d) U-Net.

Fig. 3. Segmentation example.

4.2 COVID-19 Identification Scores

To evaluate the impact of lung segmentation on COVID-19 identification, we generated a data set for each segmentation model and used ResNet-50 for classification. For completeness, we also applied ResNet-50 to the original images. Table 3 presents the F1-Score obtained in our classification experiments.

Overall segmented CXR images provide better COVID-19 identification rates. In BCDU-NET, the performance gain is approximately 6.6% points compared with non-segmented images. A decisive factor that deserves to be highlighted is that the amount of COVID-19 CXR images available at the time the experiments were conducted was scarce. Hence, we are dealing with an imbalanced data set, which led to low performance in all scenarios analyzed regarding

Table 3. F1-Score results.

Segmentation model	COVID-19	Non COVID-19	Macro-average
U-Net	0.59	0.97	0.78
MultiResUnet	0.62	0.97	0.80
BCDU-NET	0.63	0.98	**0.81**
No segmentation	0.55	0.96	**0.76**

detecting patients with COVID-19. Thus, in a real-world scenario, the use of class balancing techniques may be propitious.

4.3   Models Interpretability with LIME

Finally, to visually inspect the classification models, we applied an XAI technique called LIME. The experiments carried out using LIME aimed to evaluate which regions of the CXR images the ResNet-50 used for the classification. Figure 4 presents an example of LIME applied to a full and a segmented CXR image. In the full CXR example, although the model used some information of the pulmonary region (i.e., predominantly red zones), a considerable number of features were from regions outside the ROI (i.e., predominantly green zones). On the other hand, segmented CXR images only focused on features inside the pulmonary lobes. Hence, this provides evidence that segmentation tends to produce more reliable diagnoses in real scenarios, as only the ROI is taken into account.

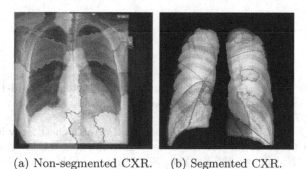

(a) Non-segmented CXR.   (b) Segmented CXR.

Fig. 4. LIME examples.

5   Conclusion

In real applications, especially in the medical field, automated techniques must be approached with extreme caution. An erroneous diagnosis may cause severe consequences for the health and treatment of patients. In a pandemic scenario, the impact is more aggravating since there is a possibility of affecting several other individuals if the outcome is not reliable.

The current COVID-19 pandemic attracted several researchers to investigate automated ways to identify COVID-19. However, there still has been a lack of critical papers analyzing a more trustworthy scenario at the time of this research. In this work, we demonstrated that segmentation techniques are vital to improving the reliability of the automatic detection of COVID-19 in CXR images. When using segmentation, classifiers focus only on the ROI, excluding non-relevant information, thus, providing more realistic and reliable diagnoses.

We compared different CNN segmentation architectures and, based on the results obtained, BCDU-NET stood out in segmenting the pulmonary lobes, achieving an Jaccard Index and Dice Coefficient of 0.91 and 0.95, respectively. Moreover, in our case study, we have achieved a 6.6% better F1-score rates with segmentation for COVID-19.

Although our primary objective was not to achieve state-of-the-art performances in lung segmentation or COVID-19 identification, we believe that the research fulfilled its purpose by providing some essential discussions regarding the automation of medical diagnoses.

For future work, we intend to increase the amount of COVID-19 CXR images. When we started to develop the experiments described here, COVID-19 CXR image sets were not abundant. However, it is important to observe that the availability of this content is growing rapidly. This way, we also intend to investigate the impact of the use of class balancing methods.

Acknowledgement. This research has been partly supported by the National Council for Scientific and Technological Development (CNPq) grant 312672/2020-9 for the financial support.

References

1. Aboughazala, L., Mohammed, K.K.: Automated detection of Covid-19 coronavirus cases using deep neural networks with X-ray images. Al-Azhar Univ. J. Virus Res. Stud. 2 (2020)
2. Acobi, A., Chung, M., Bernheim, A., Eber, C.: Portable chest x-ray in coronavirus disease-19 (COVID-19): a pictorial review. Clin. Imaging **64**, 35–42 (2020)
3. Alwarasneh, N., Chow, Y.S.S., Yan, S.T.M., Lim, C.H.: Bridging explainable machine vision in CAD systems for lung cancer detection. In: Chan, C.S., et al. (eds.) ICIRA 2020. LNCS (LNAI), vol. 12595, pp. 254–269. Springer, Cham (2020). https://doi.org/10.1007/978-3-030-66645-3_22
4. Yao, A.D., Cheng, D.L., Pan, I., Kitamura, F.: Deep learning in neuroradiology: a systematic review of current algorithms and approaches for the new wave of imaging technology. Radiol. Artif. Intell. **2**(2), e19002 (2020)
5. Asif, S., Wenhui, Y., Jin, H., Jinhai, S.: Classification of COVID-19 from Chest X-ray images using Deep Convolutional Neural Networks. Cold Spring Harbor Laboratory Press (2020)

6. Azad, R., Asadi-Aghbolaghi, M., Fathy, M., Escalera, S.: Bi-directional ConvLSTM U-Net with densley connected convolutions. In: Proceedings of the IEEE/CVF International Conference on Computer Vision (ICCV) Workshops (2019)
7. Baltruschat, I., et al.: When Does Bone Suppression And Lung Field Segmentation Improve Chest X-Ray Disease Classification?, pp. 1362–1366 (2019)
8. Beers, F.V.: Using intersection over union loss to improve binary image segmentation (2018)
9. Brunese, L., et al.: Explainable deep learning for pulmonary disease and coronavirus COVID-19 detection from X-rays. Comput. Methods Programs Biomed. **196**, 105608 (2020)
10. Candemir, S., Antani, S.: A review on lung boundary detection in chest X-rays. Int. J. Comput. Assist. Radiol. Surg. **14**(4), 563–576 (2019)
11. Candemirs, J.S., et al.: Lung segmentation in chest radiographs using anatomical atlases with nonrigid registration. IEEE Trans. Med. Imaging **33**(2), 577–590 (2014)
12. Cohen, J.P., et al.: COVID-19 Image Data Collection: Prospective Predictions are the Future. arXiv:2006.11988 [Cs, Eess, q-Bio], December 2020
13. Eelbode, T., et al.: Optimization for medical image segmentation: theory and practice when evaluating with dice score or jaccard index. IEEE Trans. Med. Imaging **39**(11), 3679–3690 (2020)
14. Gordienko, Y., et al.: Deep learning with lung segmentation and bone shadow exclusion techniques for chest X-ray analysis of lung cancer. In: Hu, Z., Petoukhov, S., Dychka, I., He, M. (eds.) ICCSEEA 2018. AISC, vol. 754, pp. 638–647. Springer, Cham (2019). https://doi.org/10.1007/978-3-319-91008-6_63
15. Goutte, C., Gaussier, E.: A probabilistic interpretation of precision, recall and F-score, with implication for evaluation. In: Losada, D.E., Fernández-Luna, J.M. (eds.) ECIR 2005. LNCS, vol. 3408, pp. 345–359. Springer, Heidelberg (2005). https://doi.org/10.1007/978-3-540-31865-1_25
16. Gulli, A., Pal, S.: Deep Learning with Keras. Packt Publishing, Birmingham (2017)
17. He, K., et al.: Deep Residual Learning for Image Recognition. arXiv:1512.03385 [Cs], December 2015
18. Holzinger, A., Biemann, C., Pattichis, C., Kell, D.: What do we need to build explainable AI systems for the medical domain? arXiv:1712.09923 (2017)
19. Ibtehaz, N., Rahamn, M.S.: Multiresunet: rethinking the u-net architecture for multimodal biomedical image segmentation. CoRR, abs/1902.04049 (2019)
20. Irvin, J., et al.: CheXpert: A Large Chest Radiograph Dataset with Uncertainty Labels and Expert Comparison (2019)
21. Jaeger, S., et al.: Automatic tuberculosis screening using chest radiographs. IEEE Trans. Med. Imaging **33**(2), 233–245 (2014)
22. Kong, W., Agarwal, P.P.: Chest imaging appearance of COVID-19 infection. Radiol. Cardiothorac. Imaging **2**(1), e200028 (2020)
23. Minaee, S., Kafieh, R., Sonka, M., Yazdani, S., Soufi, G.J.: Deep-covid: predicting covid-19 from chest x-ray images using deep transfer learning. Med. Image Anal. **65**, 101794 (2020). ISSN 1361-8415
24. Narin, A., Kaya, C., Pamuk, Z.: Automatic Detection of Coronavirus Disease (COVID-19) Using X-ray Images and Deep Convolutional Neural Networks (2020)
25. Joarder, R., Crundwell, N.: Chest X-Ray in Clinical Practice. Springer, London (2009). https://doi.org/10.1007/978-1-84882-099-9
26. Pereira, R.M., Bertolini, D., Teixeira, L.O., Silla, C.N., Jr., Costa, Y.M.G.: COVID-19 identification in chest X-ray images on flat and hierarchical classification scenarios. Comput. Methods Programs Biomed. **194**, 105532 (2020)

27. Ribeiro, M.T., Singh, S., Guestrin, C.: "Why should I trust you?": explaining the predictions of any classifier. In: Proceedings of the 22nd ACM SIGKDD International Conference on Knowledge Discovery and Data Mining. New York, NY, USA Computing Machinery, (KDD 2016), pp. 1135–1144 (2016). ISBN 9781450342322

28. Ronneberger, O., Fischer, P., Brox, T.: U-net: convolutional networks for biomedical image segmentation. CoRR, abs/1505.04597 (2015)

29. Ruder, S.: An overview of gradient descent optimization algorithms. CoRR, abs/1609.04747 (2016)

30. Self, W., Courtney, D., Mcnaughton, C., Wunderink, R., Kline, J.: High discordance of chest x-ray and computed tomography for detection of pulmonary opacities in ED patients: implications for diagnosing pneumonia. Am. J. Emerg. Med. **31**, 401–405 (2012)

31. Shorten, C., Khoshgoftaar, T.M.: A survey on image data augmentation for deep learning. J. Big Data **6**, 60 (2019)

32. Simard, P.Y., Steinkraus, D., Platt, J.C.: Best practices for convolutional neural networks applied to visual document analysis. In: Seventh International Conference on Document Analysis and Recognition, pp. 958–963 (2003)

33. Shiraishi, J., et al.: Development of a digital image data set for chest radiographs with and without a lung nodule: receiver operating characteristic analysis of radiologists' detection of pulmonary nodules. Am. J. Roentgenol. **174**, 71–74 (2000)

34. Stirenko, S., et al.: Chest X-Ray Analysis of Tuberculosis by Deep Learning with Segmentation and Augmentation (2018)

35. Teixeira, L.O., et al.: Impact of lung segmentation on the diagnosis and explanation of COVID-19 in chest X-ray images. arXiv:2009.09780 (2020)

36. Vorontsov, I.E., Kulakovskiy, I.V., Makeev, V.J.: Jaccard index based similarity measure to compare transcription factor binding site models. Algorithms Mol. Biol. **8**, 23 (2013)

37. Wang, L., Lin, Z.Q., Wong, A.: Covid-net: a tailored deep convolutional neural network design for detection of covid-19 cases from chest x-ray images. Sci. Rep. **10**(1), 1–12 (2020)

38. Gu, X., Pan, L., Liang, H., Yang, R.: Classification of bacterial and viral childhood pneumonia using deep learning in chest radiography. In: Proceedings of the 3rd International Conference on Multimedia and Image Processing (ICMIP 2018), pp. 88–93. ACM, New York (2018)

A Study on Annotation Efficient Learning Methods for Segmentation in Prostate Histopathological Images

Pedro Costa[1]([✉])[iD], Aurélio Campilho[1,2][iD], and Jaime Cardoso[1,2][iD]

[1] Faculty of Engineering, University of Porto, 4200-464 Porto, Portugal
{ei10011,campilho,jaime.cardoso}@fe.up.pt
[2] Institute for Systems and Computer Engineering, Technology and Science
INESC-TEC, 4200-465 Porto, Portugal

Abstract. Cancer is a leading cause of death worldwide. The detection and diagnosis of most cancers are confirmed by a tissue biopsy that is analyzed via the optic microscope. These samples are then scanned to giga-pixel sized images for further digital processing by pathologists. An automated method to segment the malignant regions of these images could be of great interest to detect cancer earlier and increase the agreement between specialists. However, annotating these giga-pixel images is very expensive, time-consuming and error-prone. We evaluate 4 existing annotation efficient methods, including transfer learning and self-supervised learning approaches. The best performing approach was to pretrain a model to colourize a grayscale histopathological image and then finetune that model on a dataset with manually annotated examples. This method was able to improve the Intersection over Union from 0.2702 to 0.3702.

Keywords: Histopathology · Transfer learning · Self-supervised learning · Segmentation · Computer vision · Deep learning

1 Introduction

Cancer is the second leading cause of death in the United States [14] and a major health concern worldwide. The diagnostic of most cancers is confirmed by tissue biopsy where a sample of the suspicious tissue is collected and stained, typically with hematoxylin and eosin (H&E) to better distinguish the cancerous cells, and then analyzed via optic microscope. Additionally, these samples can be scanned to giga-pixel sized images typically containing $100,000 \times 100,000$ pixels, referred to as whole-slide images (WSI), for posterior digital processing.

Pathologists analyze the histological properties of WSI in search for signs of cancer. The overall tissue architecture is analyzed as well as how nuclei are organized, their density and other morphological features [1]. Regions of tissue are identified as either non-malignant or malignant, where non-malignant regions

© Springer Nature Switzerland AG 2021
J. M. R. S. Tavares et al. (Eds.): CIARP 2021, LNCS 12702, pp. 55–64, 2021.
https://doi.org/10.1007/978-3-030-93420-0_6

can be further characterized as normal or benign and malignant regions as *in situ* carcinoma or invasive. This fine-grained distinction is clinically relevant as patients are treated differently depending on the diagnosis. Patients with malignant tissue may require chemotherapy or even surgery, while patients with benign tissue are usually only followed clinically [2].

A method capable of segmenting and classifying the malignant regions in WSIs could be of great interest to help pathologists analyze biopsies faster, with less missed malignant regions and improve the agreement between different pathologists. Deep Learning has achieved great success in computer vision, including in the medical imaging domain. Driving the success of Deep Learning are large manually annotated datasets which, in the case of medical imaging segmentation, are difficult to obtain.

In this work we evaluate existing popular annotation efficient methods, namely using transfer learning and self-supervised approaches, to improve the performance of an image segmentation model for the task of segmenting cancerous tissue in prostate histopathological images. We concluded that some transfer learning and self-supervised approaches can be useful in improving the performance of a segmentation model. However, the typical transfer learning approach where an encoder is pre-trained on ImageNet and then used as a backbone in a segmentation network did not improve the performance over the baseline. The best performing approach was to pretrain a model to colourize a grayscale histopathological image and then finetune that model on a dataset with manually annotated examples.

2   Related Work

2.1   Segmentation

Most of the state-of-the-art image segmentation methods are based on the Fully Convolutional Network (FCN) architecture [10]. FCN uses an image classifier network as the backbone, to classify regions of the input image, resulting in a coarse heatmap. This heatmap is then upsampled to the resolution of the input image by means of deconvolution layers.

Modern existing approaches that follow the FCN architecture propose changes to either keep spatial information from the input in order to predict a detailed output, or to increase the context information, for instance, by increasing the receptive field of the backbone. One of the most successful approaches to maintain the spatial information of the input is by replacing down-sampling operations of the network by dilated convolutions [3]. Some methods extend this idea to capture more context information, by employing convolution layers with different dilation rates and fusing that information before making the prediction [3].

Another approach to image segmentation that is particularly popular in the medical imaging domain is based on the U-Net [12]. The U-Net consists of an encoder-decoder architecture with skip-connections between the encoder and decoder. These skip connections allow the recovery of fine spatial information.

2.2   Cancer Detection in WSIs

However, the best performing method to detect cancer in WSI tiles in the Bach dataset [2] used an ensemble of CNNs pre-trained on Imagenet [4]. Regarding the segmentation of normal, benign, *in situ* carcinoma and invasive, the best performing approaches used patch classifiers. For this, CNNs were trained for the image classification task and then used to classify multiple patches from the WSI and create a heatmap [8]. One of the potential reasons for patch prediction methods obtaining better results than segmentation models is because of larger image classification datasets. The authors used the dataset for image classification in conjunction with the segmentation dataset, while segmentation methods were restricted to the much smaller dataset with segmentation annotations. Additionally, the most popular CNN architectures with pre-trained weights are suited for classification, as they are trained on Imagenet. Transfer learning can significantly boost the performance of image classification models, especially when data is scarce [9].

2.3   Unsupervised Representation Learning

One possibility to boost the performance of Deep Learning methods in segmenting WSIs would be to make use of the potentially large pool of unlabelled data to learn representations (features) that could be useful for the downstream task. More recently, self-supervised learning has shown great promise in using unlabelled examples to initialize the weights of a CNN to be used in a downstream task with limited data [7]. Self-supervised learning methods propose various pretext tasks for networks to solve in order to learn visual features. These pretext tasks can be automatically generated from the raw dataset, such as colourizing grayscale images [15], image inpainting [11], among others. However, most of existing self-supervised learning methods are mainly focused on learning representations for classification tasks and not for segmentation tasks.

One exception is a recent method capable of segmenting nuclei in a self-supervised way [13]. The authors propose to train a nuclei segmentation network and, then, use the predicted segmentation to classify the magnification of the input tile. However, the authors do not test if the resulting segmentation network is useful to the detection of cancerous tissue.

There is still a lack of unsupervised methods focused on image segmentation that produce good results. In fact, transfer learning of models pre-trained on Imagenet for the task of object detection in the COCO dataset has limited impact on results [5]. The same behaviour can be observed for the image segmentation task [16]. The authors propose a self-training framework, where a teacher model is trained on the labelled dataset, then the teacher is used to generate labels on an unlabelled dataset so that a larger student model can be trained to optimize for both the human and pseudo labels jointly. The authors show that this approach performs better than the standard transfer learning approach. These results indicate that existing pre-trained image classification models may not be suitable as initialization for image segmentation tasks.

3   Methodology

In this study we aimed to answer mainly 2 related questions: 1) Does Transfer Learning help improve segmentation performance of cancerous tissue in prostate WSIs? 2) Does Self-Supervised pre-training help improve segmentation performance of cancerous tissue in prostate WSIs?

In the following subsections, we will describe the segmentation model that we used and what transfer learning and self-supervised learning approaches we evaluated.

3.1   U-Net

The U-Net is the standard architecture in medical imaging applications, where small objects are important to identify and, therefore, we chose it as the base architecture in this study. As previously described, this network consists of an encoder-decoder architecture with skip connections between the encoder and the decoder. This design choice enables the network to retrieve high-frequency details from the input image, which is especially important when dealing with small objects, such as nuclei. The network was trained with cross-entropy loss to distinguish between cancerous and non-cancerous tissue. We used the original U-Net architecture [12] with Batch-Norm [6] after each convolution layer.

3.2   Transfer Learning

Transfer Learning approaches typically start with a model trained on a large dataset and then finetune that model in a smaller dataset. In this work we try two different approaches to transfer learning with U-Net: 1) pretrain the entire network on similar task; and 2) pretrain only the encoder.

Pretrain Entire Network. This is the transfer learning approach most similar to the ones typically followed in image classification tasks: the entire segmentation network is trained on a related task and, then, the last layer is replaced with a new one. The entire model is finetuned on the new task, reusing all the features learned from the first model.

The main advantage of this approach is that both the encoder and decoder are pretrained. However, to follow this approach we require a second segmentation dataset, which may not be easily available. Additionally, the decoder features may only be useful if the pretraining task is closely related to the downstream one.

Pretrain Encoder. Another typically used approach is to pretrain an image classification network and use it as the encoder in a segmentation network. To use the U-Net encoder as an image classification network, we append a Global Average Pooling layer to reduce the dimensionality of the features extracted by the encoder and, finally, we apply a single Fully Connected layer to classify the image. After training the image classifier, it is possible to transfer the layers of the encoder into the U-Net architecture and retrain the model for the segmentation task.

The main motivation for this approach is that image classification labels are more easily available than segmentation annotations. Furthermore, image labels tend to be less noisy than segmentation annotations. It is, then, possible to train the encoder on a larger dataset, extracting features that could potentially be useful for the image segmentation task.

3.3 Self-supervised Learning

For transfer learning approaches we require labelled datasets, either with image labels or segmentation annotations. However, there are many unlabelled WSIs that could potentially be used to learn feature representations in an unsupervised way. In this section, we describe two different self-supervised learning approaches that we evaluated: 1) Inpainting; and 2) Colourization.

Inpainting. For the inpainting task, we train a U-Net model to predict the pixel intensity of random patches from the input images. A random patch is erased from the input image $\hat{I} = I \times M$, where I is the input image, M is a binary mask with value 0 in the pixel locations to be erased, and $\hat{I}$ is the resulting image to be inpainted. The goal of the inpainting task is to retrieve I from $\hat{I}$. The model was trained to minimize a modified Mean Squared Error (MSE) between I and $\hat{I}$:

$$\mathcal{L} = \frac{1}{\sum(1-M)}\sum(1-M) \times (I - \hat{I})^2 + \frac{1}{\sum M}\sum M \times (I - \hat{I})^2. \qquad (1)$$

The main idea is to divide the loss into two tasks: the reconstruction of 1) erased regions and 2) non-erased regions. The second task is easier, as the network simply needs to copy the input pixels to the output. However, these two tasks are unbalanced, as there is fewer erased pixels than non-erased ones. By treating the two reconstruction tasks as separate, and providing a weight to each one inversely proportional to its area, the problem of the pixels in the erased region being underrepresented in the reconstruction loss is solved.

In this work, we randomly erase a rectangular area from the input image. The size of the largest side of rectangle randomly varies between 0.1 and 0.3 times the size of the input image, and the smaller side of the rectangle randomly varies between 0.1 and 1.0 times the size of the largest side of the rectangle.

Colourization. Another successful self-supervised approach proposes to colourize a grayscale input image. The main idea is that a model capable of predicting the RGB intensity from a grayscale image is required to learn features to distinguish the different objects in the image, such as nuclei.

However, a colourization model has a one-channel input and, after training, we want to finetune it on the RGB segmentation dataset. We repeat the grayscale filters learned at first layer, so that we can apply them to the input RGB images.

4   Evaluation

4.1   Dataset

We used a private dataset containing 694 WSIs, 126 of those with manual segmentation annotations. These WSIs were gathered from 2 different medical centers, 384 WSIs coming from Center1, 76 of those with annotations, and the remaining 310 coming from Center2, with 50 annotated slides. The WSIs from Center1 were used to training and testing, whereas WSIs from Center2 were only used for pretraining, either using Transfer Learning approaches or Self-Supervised ones.

We randomly selected 50 WSIs from Center1 as the *training* set and the 26 remaining WSIs to *validate* the model. To test the efficacy of the transfer learning and self-supervised approaches, we used 2 additional sets: 1) *Center2 segmentation*, with 50 WSIs; and 2) *Self-Supervised Pretraining*, with 308 unlabeled WSIs from Center1 and 260 WSIs from Center2. The dataset details can be seen in Table 1.

Tiles are extracted from the selected WSIs to train the segmentation models. We extracted 224×224 px tiles from the WSIs, without overlap. Additionally, we removed tiles containing background, meaning that all the used tiles only contain tissue. The number of resulting tiles can also be seen in Table 1.

4.2   Segmentation Results

To evaluate whether transfer learning approaches improve the performance of segmentation models in our dataset, we started by training a baseline on the entire train set. We also trained the U-Net encoder on the entire *SS pretrain* dataset, to classify tiles into having cancerous tissue or not. Additionally, a U-Net was trained on *Center2 segmentation* set.

Table 1. Dataset distribution. The *train* and *validation* sets only contain examples from Center1. The *Center2 segmentation* and *SS pretrain* sets are only used to pretrain the models, either with transfer learning or self-supervised approaches.

Set	Center1 (WSIs/Tiles)	Center2 (WSIs/Tiles)
Train	50/17408	0/0
Validation	26/9344	0/0
Center2 segmentation	0/0	50/23231
SS pretrain	308/122299	260/122962

The models were trained for 40000 iterations with a batch size of 10, optimized with Adam with an initial learning rate of 1×10^{-3}. The learning rate was decayed using a Cosine Annealing strategy every 1000 iterations. After training the U-Net encoder on the *SS pretrain* set and the U-Net on the *Center2*

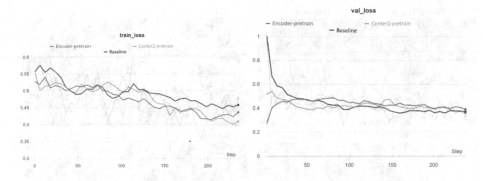

Fig. 1. The training and validation losses of the baseline and transfer learning approaches. The 3 approaches display comparable values regarding both training and validation losses.eps

segmentation set, we finetuned the models on the *train* set and compared the performance with the baseline. We can see in Fig. 1 that, in terms of training and validation losses, the baseline takes more time to converge. However, the differences are not significant, both in terms of training and validation losses.

The loss is not a good measure of performance in this task though due to the imbalance between positive and negative pixels. Only 16.8% of the pixels are cancerous, meaning that the majority of the WSIs' tissue is non-cancerous. In order to properly evaluate the difference in performance, we use the Intersection over Union (IoU), which is more robust in tasks where the objects of interest represent a small area of the entire images. As shown in Table 2, we can verify that the model pretrained on the *Center2 segmentation* set achieves a significantly higher IoU than both the baseline and the U-Net with the pretrained encoder. Even though the encoder was pretrained on a much larger dataset, the U-Net with the pretrained encoder performed significantly worse. These results seem to be aligned with the conclusions of [16] that pretrained image classification models do not boost the performance of either object detection or segmentation models and, in some cases, may even hurt performance. We show that

Table 2. Segmentation results of the different approaches on the *validation* set. *S* - Supervised; *TL* - Transfer Learning; *SSL* - Self-Supervised Learning

Method	Loss	IoU	Approach
U–Net Baseline	0.3859	0.2702	*S*
U–Net pretrained on Center2	**0.3641**	0.3421	***TL***
Pretrained U–Net Encoder	0.3918	0.2379	*TL*
U–Net Colourization pretrain	0.3692	**0.3702**	***SSL***
U–Net Inpaint pretrain	0.4161	0.2560	*SSL*

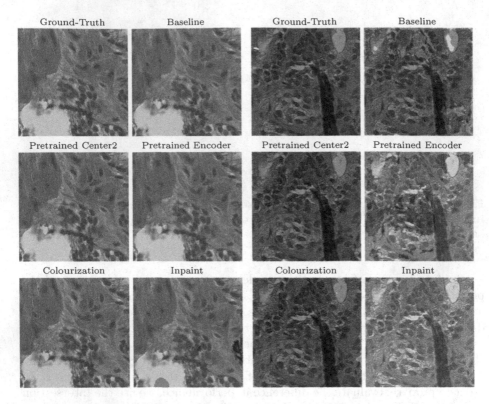

Fig. 2. Qualitative results of the different models. The model pretrained on Center2 and the colourization model are more consistent than the competing approaches, specially in images with mainly malignant tissue. The models are more accurate in images with mainly benign tissue.

pretraining the segmentation model on a closely related task, within the same domain, may help performance.

We also evaluated 2 different Self-Supervised approaches, as described in the previous section: inpainting and colourization. We trained the models on the *SS pretrain* set, using only the pixel intensity information, discarding the image labels. The models were trained for 100000 iterations with a batch size of 10, optimized with Adam, with an initial learning rate of 1×10^{-3} and a Cosine Annealing learning rate scheduler with a step every 2000 iterations.

As shown in Table 2, the colourization model obtained a better performance than all other evaluated methods. On the contrary the U-Net pretrained on the inpainting task performed worse than the baseline. One possible explanation for the poor performance of the inpainting method is that the model was still far from converging. This hypothesis is supported by the fact that the inpainting task converges slower than the colourization one in terms of training loss. The optimization is harder for the inpainting model probably due to the more challenging nature of the task when compared with the colourization task. Another explanation for the difficulty in optimizing the inpainting model is lack of capacity. However, we chose not to increase the number of parameters of the U–Net for the inpainting task in order to keep the comparison with the other approaches fair. Finally, the inpainting task may be extracting features that are not relevant for the task of segmenting malignant regions in WSIs.

After a visual inspection of the results, it was possible to confirm that the model pretrained on Center2 and the colourization model are more consistent, especially on examples with malignant tissue (Fig. 2).

5 Conclusion

In this work, we explored several approaches for annotation efficient prostate cancer grade assessment. We evaluated different transfer learning and self-supervised approaches for the segmentation task. Our results are aligned with recent works that state that typical transfer learning techniques, where an image classifier is used to extract features from an image before using a decoder to segment the input image, do not improve results. We verify that, for our dataset, it is better to pretrain the segmentation model on a large dataset to colourize grayscale images, before finetuning it on the manually annotated images.

We show that pre-training on large datasets with self-supervised or transfer learning approaches help the performance, even though most of the existing methods are designed for image classification tasks. In the future, we want to develop novel annotation efficient methods developed specifically for image segmentation models. Such methods could ease the development of robust imaging segmentation models, that could more easily adapt to WSIs of different organs and created with different staining techniques.

Acknowledgments. This work is financed by National Funds through the Portuguese funding agency, FCT - Fundação para a Ciência e a Tecnologia, within project UIDB/50014/2020.

References

1. Araújo, T., et al.: Classification of breast cancer based on histology images using convolutional neural networks. PLoS ONE **6**, 24680–24693 (2017). https://doi.org/10.1109/ACCESS.2018.2831280
2. Aresta, G., et al.: BACH: grand challenge on breast cancer histology images. Med. Image Anal. **56**, 122–139 (2019). https://doi.org/10.1016/j.media.2019.05.010
3. Chen, L.C., Papandreou, G., Kokkinos, I., Murphy, K., Yuille, A.L.: DeepLab: semantic image segmentation with deep convolutional nets, atrous convolution, and fully connected CRFs. IEEE Trans. Pattern Anal. Mach. Intell. **40**(4), 834–848 (2017)
4. Chennamsetty, S.S., Safwan, M., Alex, V.: Classification of breast cancer histology image using ensemble of pre-trained neural networks. In: Campilho, A., Karray, F., ter Haar Romeny, B. (eds.) ICIAR 2018. LNCS, vol. 10882, pp. 804–811. Springer, Cham (2018). https://doi.org/10.1007/978-3-319-93000-8_91

5. He, K., Girshick, R., Dollár, P.: Rethinking imagenet pre-training. In: Proceedings of the IEEE International Conference on Computer Vision, pp. 4918–4927 (2019)
6. Ioffe, S., Szegedy, C.: Batch normalization: accelerating deep network training by reducing internal covariate shift. In: International Conference on Machine Learning, pp. 448–456. PMLR (2015)
7. Jing, L., Tian, Y.: Self-supervised visual feature learning with deep neural networks: a survey. IEEE Trans. Pattern Anal. Mach. Intell. (2020)
8. Kwok, S.: Multiclass classification of breast cancer in whole-slide images. In: Campilho, A., Karray, F., ter Haar Romeny, B. (eds.) ICIAR 2018. LNCS, vol. 10882, pp. 931–940. Springer, Cham (2018). https://doi.org/10.1007/978-3-319-93000-8_106
9. Litjens, G., et al.: A survey on deep learning in medical image analysis. Med. Image Anal. **42**, 60–88 (2017). https://doi.org/10.1016/J.MEDIA.2017.07.005
10. Long, J., Shelhamer, E., Darrell, T.: Fully convolutional networks for semantic segmentation. In: Proceedings of the IEEE Conference on Computer Vision and Pattern Recognition, pp. 3431–3440 (2015)
11. Pathak, D., Krahenbuhl, P., Donahue, J., Darrell, T., Efros, A.A.: Context encoders: feature learning by inpainting. In: Proceedings of the IEEE Conference on Computer Vision and Pattern Recognition, pp. 2536–2544 (2016)
12. Ronneberger, O., Fischer, P., Brox, T.: U-Net: convolutional networks for biomedical image segmentation. In: Navab, N., Hornegger, J., Wells, W.M., Frangi, A.F. (eds.) MICCAI 2015. LNCS, vol. 9351, pp. 234–241. Springer, Cham (2015). https://doi.org/10.1007/978-3-319-24574-4_28
13. Sahasrabudhe, M., et al.: Self-supervised nuclei segmentation in histopathological images using attention. In: Martel, A.L., et al. (eds.) MICCAI 2020. LNCS, vol. 12265, pp. 393–402. Springer, Cham (2020). https://doi.org/10.1007/978-3-030-59722-1_38
14. Siegel, R.L., Miller, K.D., Jemal, A.: Cancer statistics, 2019. CA: Cancer J. Clin. **69**(1), 7–34 (2019). https://doi.org/10.3322/caac.21551
15. Zhang, R., Isola, P., Efros, A.A.: Colorful image colorization. In: Leibe, B., Matas, J., Sebe, N., Welling, M. (eds.) ECCV 2016. LNCS, vol. 9907, pp. 649–666. Springer, Cham (2016). https://doi.org/10.1007/978-3-319-46487-9_40
16. Zoph, B., et al.: Rethinking pre-training and self-training. arXiv preprint arXiv:2006.06882 (2020)

Natural Language Processing

Data-Augmented Emoji Approach to Sentiment Classification of Tweets

Tiago Martinho de Barros, Helio Pedrini[(✉)], and Zanoni Dias

University of Campinas, Institute of Computing, Campinas, SP, Brazil
helio@ic.unicamp.br

Abstract. The Natural Language Processing field has made great strides recently. As a result, many challenging tasks are being given better solutions. One of these tasks is Sentiment Analysis, which is the subject of this work. We propose a novel methodology to classify the sentiment of tweets, based on BERT and focusing on emoji. Our method also employs data augmentation to improve its generalization ability. Experiments on two Brazilian Portuguese datasets – TweetSentBR and 2000-tweets-BR – show that our methodology produces better results than BERT and outperforms the previously published results for TweetSentBR, with accuracy of 0.7726 (6.3 percentage points (p.p.) of improvement) and F_1 score of 0.7514 (9.5 p.p. of improvement), as well as for 2000-tweets-BR, with accuracy of 0.8247 (14.5 p.p. of improvement) and F_1 score of 0.8035 (23.4 p.p. of improvement).

Keywords: Natural language processing · Sentiment analysis · Emoji · Social media

1 Introduction and Background

Sentiment Analysis, also referred to as Opinion Mining, is a research field concerned with the computational study of sentiments. Since early 2000s, Sentiment Analysis has grown to be one of the most active research topics in the Natural Language Processing (NLP) field. It is also widely studied in Data Mining, Web Mining, Text Mining, and Information Retrieval.

With the widespread access to the Internet in the last decades, people have gained a new and powerful medium to voice their thoughts. With an unprecedented reach, it has never been easier to make our opinions visible worldwide. Entities, such as companies and government agencies, are frequently interested in knowing what people think in order to make informed decisions.

Usual sources of subjective texts (texts with an opinion) are social networking services, (micro)blogs, and websites featuring user reviews. In this work, we used two datasets of tweets from different domains. Since people express themselves in many different ways, processing user-generated content is not an easy task. Lack of punctuation and use of slang, for instance, are two of the challenges we face. On the other hand, this type of text also presents interesting opportunities, one of them being the use of emoji, which help to express sentiment.

© Springer Nature Switzerland AG 2021
J. M. R. S. Tavares et al. (Eds.): CIARP 2021, LNCS 12702, pp. 67–76, 2021.
https://doi.org/10.1007/978-3-030-93420-0_7

Fig. 1. Overview of our proposed method for sentiment classification of tweets.

In this work, we present a method to classify the sentiment of user-generated texts that makes use of emoji, when present, to enrich the text representation used to perform the classification. Our methodology is based on BERT [3] and features additional pre-training and data augmentation.

1.1 Bidirectional Encoder Representations from Transformers (BERT)

Since we used BERT [3] as a basis to build our proposed method for sentiment classification, we describe it in more detail here. One of the strong points of BERT is its bidirectionality. While previous models use left-to-right or independently trained left-to-right and right-to-left encoders to generate features for downstream tasks, BERT employs a bidirectional Transformer, producing representations that are jointly conditioned on both left and right context in all layers, which is reflected on the strong results obtained.

BERT is pre-trained using two unsupervised tasks: Masked Language Modeling (MLM) and Next Sentence Prediction (NSP). MLM is the solution that the authors found to train a language model, since standard conditional language models can only be trained left-to-right or right-to-left. In MLM, a percentage of the input tokens are masked at random and the model is trained to predict those masked tokens. The NSP task objective is to build understanding of the relationship between two consecutive sentences, to be used in downstream tasks such as Question Answering and Natural Language Inference. The model is trained to predict, given two sentences A and B, if B follows A or not.

2   Methodology

The core idea behind our proposed methodology is to extract the maximum information possible from emoji to have a richer representation of a piece of text and use that to improve the sentiment classification. The methodology comprises additional pre-training evaluation, data augmentation evaluation, emoji extraction, and fine-tuning. Figure 1 illustrates the process. To evaluate our method we adopted two datasets of tweets, which we briefly present.

2.1   Datasets

In this section, we present the datasets we used to conduct our experiments and assess our methodology: the TweetSentBR and the 2000-tweets-BR.

TweetSentBR. The TweetSentBR [1] is a sentiment corpus for Brazilian Portuguese, manually annotated, with 15000 tweets on TV show domain. The tweets were labeled in three classes: positive, neutral, and negative. The training set has the following distribution of tweets: 5741 (44.2%) positive, 3410 (26.3%) neutral, and 3839 (29.5%) negative. And the test set has the following distribution of tweets: 907 (45.1%) positive, 516 (25.7%) neutral, and 587 (29.2%) negative.

2000-tweets-BR. The 2000-tweets-BR [9] is a multi-domain Brazilian Portuguese corpus of tweets built to analyze the Brazilian and European varieties of the Portuguese language with respect to Sentiment Analysis. It was manually annotated and organized in four classes: positive, neutral, negative, and mixed. This last class refers to tweets having both positive and negative opinions. Following Vitório et al. [9], who introduced the dataset, we do not use the "mixed" class in the classification process. Additionally, since the dataset does not have predefined training and test sets, we split it using 15% of the samples, randomly selected, as test set. For the TweetSentBR dataset, the test set is 13.4% of the total, so we chose the nearest multiple of five here. Thus, the training set has the following distribution of tweets: 329 (20.0%) positive, 894 (54.2%) neutral, and 425 (25.8%) negative. And the test set has the following distribution of tweets: 61 (20.9%) positive, 146 (50.2%) neutral, and 84 (28.9%) negative.

2.2 Additional Pre-training

Since pre-training a Transformer-based model from scratch is very expensive time-wise and it also requires massive amounts of data, we make use of *BERTimbau* [7] – a BERT [3] model pre-trained on the *brWaC* corpus, which is composed of 2.7 billion tokens, from 120 thousand different websites. We can fine-tune this model on our downstream task – sentiment classification – using a labeled dataset and obtain a trained classifier. However, according to Gururangan et al. [4], performing a second phase of pre-training, this time using in-domain documents, tend to generate better results.

To pre-train our model, we prepared a corpus of user-generated texts from social media with 89458 entries, all of which contain at least one emoji. Some examples of entries are:

Linda a Jessica e tem senso de humor.[a]
Quando foi isso? A mulher não ganhou com um nhoque?[b]
Caramba, que nível.... circo de horrores[c]

---

[a] Jessica is beautiful and has a sense of humor.
[b] When was that? Didn't she win with a gnocchi recipe?
[c] That's terrible... it's like a horror freak show.

They were obtained from social media pages related to TV shows, so the domain should be similar to the one of the TweetSentBR [1] dataset.

The pre-training process is unsupervised – or, more accurately, semi-supervised, since the labels are obtained from the input samples – so we need nothing besides the corpus. Furthermore, using the same inputs, it is possible to generate different labels using different configurations (tasks). We conducted pre-training experiments using six different configurations for the methodology:

- **Masked Language Modeling (MLM)**: the same task used during pre-training of BERT. Random tokens are masked with a probability of 15% and the model is trained to predict those masked tokens. For more details, please refer to Sect. 1.1.
- **Masked Language Modeling 50% (MLM50)**: akin to the *Masked Language Modeling* configuration, but using probability of 50% to mask a token.
- **All Emoji (All)**: all emoji (and only emoji) are masked and the model is trained to predict those masked emoji.
- **First Emoji (First)**: the first occurring emoji of a text is masked and the model is trained to predict this masked emoji. If there is only one emoji in the text, it behaves like the *All Emoji* configuration.
- **Emoji Masked Language Modeling (EMLM)**: similar to the *Masked Language Modeling* configuration, but only emoji tokens are randomly masked, with a probability of 15%.
- **Emoji Masked Language Modeling 50% (EMLM50)**: similar to the *Emoji Masked Language Modeling* configuration, but using a probability of 50% to mask a token.

In addition to these six pre-training configurations, we also evaluated the scenario without additional pre-training.

2.3   Data Augmentation

We evaluated whether data augmentation could improve the results of sentiment classification of social media texts or not. The approach employed here is based on the work of Wei and Zou [10]. The central idea is to modify a piece of text using four operations:

- **Synonym Replacement:** Randomly choose n words from the sentence that are not stop words. Replace each of these words with one of its synonyms chosen at random.
- **Random Insertion:** Find a random synonym of a random word in the sentence that is not a stop word. Insert that synonym into a random position in the sentence. Do this n times.
- **Random Swap:** Randomly choose two words in the sentence and swap their positions. Do this n times.
- **Random Deletion:** For each word in the sentence, randomly remove it with probability p.

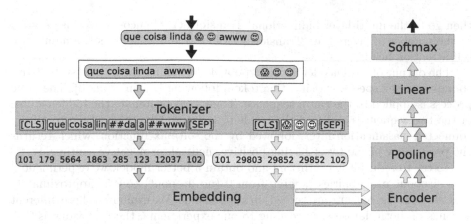

Fig. 2. Our proposed method for sentiment classification of tweets.

The number of words changed per sample, n, is based on the text length l and given by the formula $n = \alpha l$, where α is a parameter that indicates the percentage of words in a sample to be changed. For the random deletion operation, $p = \alpha$.

We experimented with the values of 0.1, 0.2, 0.3, 0.4, 0.5, and 0.6 for α to determine which one produces the best results for each dataset. In addition to the parameter α, another important one is the n_{aug}, which determines the number of augmented samples per original sample. We experimented with the values of 1, 2, 3, 4, and 5 for n_{aug}.

The original method of Wei and Zou [10] was built to work with English texts. We modified it to work with Portuguese text by employing the Open Multilingual WordNet to find synonyms of the words.

2.4 Emoji Extraction

For every tweet, we perform emoji extraction to separate them from the text, so that we end up with two sequences: one with words and another with emoji. We consider *emoticons* as emoji, because albeit not being the same thing, they essentially fulfill the same role[1]. All emoji found in the training sets are added to the vocabulary of the WordPiece embedding tokenizer.

2.5 Model Architecture

Figure 2 illustrates the model and the classification process, from the input text to the output sentiment probabilities. Each step is explained as follows.

First, the words sequence is processed by the tokenizer, which inserts a special classification token ([CLS]) as the first token of the sequence and translates all input tokens to WordPiece token IDs, according to the vocabulary. These token IDs are fed to the embedding layer, using hidden size $H = 768$. The embeddings

---

[1] Emoticons considered: S2 <3 ;D :D ;-) :-) ;) =) :) ;-(:-(;(=(:(.

then go to the multi-layer bidirectional Transformer [8] encoder. We use $L = 12$ as the number of layers (i.e., Transformer blocks) and $A = 12$ as the number of self-attention heads.

The output of the encoder is then "pooled" by taking the hidden state corresponding to the special classification token, following Devlin et al. [3]. The same process is applied to the emoji sequence. Then we concatenate the hidden states of the two sequences and feed the result to a dropout layer and then to a fully connected classification layer, followed by the softmax function, which returns the probability of the tweet having positive, neutral, or negative sentiment.

To try to reduce the overfitting and obtain a better model, we experimented with dropout probabilities ranging from 0 (no dropout) to 0.5 (approximately half the neurons' outputs are zeroed) in steps of 0.05. We evaluated these different settings on both datasets. According to our experiments, the best value is 0.35 for TweetSentBR, and 0.05 for 2000-tweets-BR.

Besides the aforementioned dropout layer, dropout is also used in the self-attention computation. We evaluated the same range of values, using the value of 0.35 for the general dropout probability with TweetSentBR and 0.05 for 2000-tweets-BR, as they performed the best. We found 0.05 to be the best value for TweetSentBR, and 0.15 for 2000-tweets-BR.

To summarize the results obtained in our dropout experiments, the best dropout settings for TweetSentBR are general dropout rate of 35% and self-attention dropout rate of 5%. As for 2000-tweets-BR, the best settings are general dropout rate of 5% and self-attention dropout rate of 15%.

2.6 Training Protocol

Starting with a pre-trained BERTimbau [7] model, we evaluated additional pre-training, according to Sect. 2.2. We then fine-tuned our model on the datasets using maximum sequence length of 128 tokens with padding.

We utilized Adam optimizer with initial learning rate of 1×10^{-5}, weight decay of 0.01, batch size of 32 and maximum number of epochs of 20 for Tweet-SentBR and 100 for 2000-tweets-BR, since it is smaller and the model was still learning in the 20th epoch in some cases. To determine the best values for the training parameters, we used a stratified 5-fold cross-validation schema.

To develop our methodology, we used PyTorch, the Transformers library from Hugging Face, BERTimbau, NumPy, Pandas, the Natural Language Toolkit, and scikit-learn. The hardware specifications of the computers used to conduct the experiments are: 2.2 GHz Intel Xeon processors, 12 GB of RAM memory, and NVidia Tesla P100 graphics cards, with 16 GB of memory HBM2, running Linux operating system.

3 Results

We evaluated our methodology and compared the obtained results with the published results for the TweetSentBR and 2000-tweets-BR datasets. Since our model is based on BERT, and we could not find published results for it on these datasets, we performed the evaluation of a standard BERT model using the same training protocol as our model. In all cases, the inputs are the full tweets, including emoji, if present.

3.1 Evaluation Metrics

The evaluation metrics traditionally used for the TweetSentBR and 2000-tweets-BR datasets are accuracy and F_1 score. We also present the results for precision and recall to offer a better representation of the effectiveness of the classifiers. All the metrics have their values in the interval $[0, 1]$, and the higher the better.

3.2 Pre-training Results

One of the first things we have to verify is whether additional pre-training improves the results of the final model or not. We experimented with six different pre-training configurations (tasks), presented in Sect. 2.2, in addition to no further pre-training at all. For the TweetSentBR dataset, the best results were obtained with the Masked Language Modeling 50% configuration (improvement of 1.2 p.p. in accuracy and 1.1 p.p. in F_1 score). For the 2000-tweets-BR dataset, the configuration with no additional pre-training yielded the best results.

3.3 Data Augmentation Results

Using an adaptation for the Portuguese language of the approach by Wei and Zou [10], we evaluated data augmentation for Sentiment Analysis. There are two important parameters in this method: α and n_{aug}, as explained in Sect. 2.3.

First, we executed experiments to determine the best value for α, using $n_{aug} = 1$. For the TweetSentBR dataset, we found $\alpha = 0.4$ to be the best value, while for the 2000-tweets-BR dataset, the best value was $\alpha = 0.2$.

Then, we evaluated the parameter n_{aug}. For the TweetSentBR dataset, we used $\alpha = 0.4$ and obtained $n_{aug} = 3$ as the best value. For the 2000-tweets-BR dataset, we used $\alpha = 0.2$ and found $n_{aug} = 3$ to be the best value as well.

To summarize the results obtained in our data augmentation experiments, the best data augmentation schema for the TweetSentBR dataset is 3 augmented samples per original sample, with 40% of the words changed. As for the 2000-tweets-BR dataset, the best schema is 3 augmented samples per original sample, with 20% of the words changed.

Having found the best settings for data augmentation, we compared these results with those obtained without data augmentation to determine whether it is beneficial to the method or not. Considering the TweetSentBR dataset, the accuracy is 0.7647 without augmentation and 0.7711 with it. For the 2000-tweets-BR dataset, the accuracy is 0.8213 without and 0.8247 with data augmentation. Based on these results, we consider that data augmentation is beneficial for both datasets.

Table 1. Experimental results for TweetSentBR.

Model	Accuracy	F$_1$ score	Precision	Recall
Brum and Nunes [1]	0.6462	0.5985	–	–
Brum and Nunes [2]	–	0.6214	–	–
Sakiyama et al. [6]	0.6840	0.6560	–	–
Nascimento [5]	0.7100	0.5000	–	–
BERT$_{\text{BASE}}$	0.7468	0.7292	0.7287	0.7297
Our model	**0.7726**	**0.7514**	**0.7529**	**0.7514**

Table 2. Experimental results for 2000-tweets-BR.

Model	Accuracy	F$_1$ score	Precision	Recall
Vitório et al. [9]	0.6451	–	–	–
Nascimento [5]	0.6800	0.5700	–	–
BERT$_{\text{BASE}}$	0.8110	0.7937	0.8135	0.7807
Our model	**0.8247**	**0.8035**	**0.8347**	**0.7849**

3.4 Fine-Tuning Results

Table 1 lists the experimental results for TweetSentBR. Note that the values for precision and recall are the arithmetic means of their per-class values. When building the model, we used the same size parameters as BERT$_{\text{BASE}}$ ($L = 12$, $H = 768$, and $A = 12$), enabling us to make a fair comparison. Our methodology produced better results than the ones previously published and achieved improvements of 2.6 percentage points (p.p.) in accuracy and 2.2 p.p. in F$_1$ score over BERT.

The results for the 2000-tweets-BR dataset are presented in Table 2. The results obtained represent an improvement of accuracy of 1.4 p.p. and an improvement of F$_1$ score of 1.0 p.p. over BERT.

Since our approach focuses on emoji, we tested on a subset of tweets of each dataset that have one or more emoji, which are about 20% of the TweetSentBR dataset and about 15% of the 2000-tweets-BR dataset. With this setting, we can compare the effectiveness of BERT$_{\text{BASE}}$ and our model. As can be seen in Table 3, in this scenario for TweetSentBR, our model excelled BERT$_{\text{BASE}}$ by 4.9 p.p. in accuracy and by 17.9 p.p. in F$_1$ score, indicating that our model can actually extract more information from emoji and use that to perform a better sentiment classification.

For the emoji subset of the 2000-tweets-BR, the results are listed in Table 4. We can see that our model outperforms BERT in the accuracy metric by 2.4 p.p., but is surpassed in the F$_1$ score by 1.3 p.p., since the lower recall value pulls the F$_1$ score down, even though the precision is high. It may be caused by having emoji which contradict the text in tweet or even by inconsistent labels in the dataset.

Table 3. Experimental results for TweetSentBR – emoji subset.

Classifier	Accuracy	F_1 score	Precision	Recall
BERT$_{BASE}$	0.7724	0.5631	**0.8342**	0.5747
Our model	**0.8208**	**0.7425**	0.7311	**0.7607**

Table 4. Experimental results for 2000-tweets-BR – emoji subset.

Classifier	Accuracy	F_1 score	Precision	Recall
BERT$_{BASE}$	0.7073	**0.6779**	0.7146	**0.6587**
Our model	**0.7317**	0.6652	**0.8323**	0.6270

4   Conclusions

In our competitive society, having a better understanding of what people think can represent a huge advantage for many companies, for instance evaluating user satisfaction. Deep Learning methods usually produce good results but often require copious amounts of labeled data, which, for many applications, are hard to acquire because it is very time-consuming to build a large manually-annotated dataset – the gold standard in Machine Learning. Unsupervised language representation learning methods alleviate this issue by means of Transfer Learning, pre-training a language representation model using a large amount of unlabeled data and then fine-tuning it on a labeled dataset, which does not need to be too large.

One prominent such model is the Bidirectional Encoder Representations from Transformers (BERT) [3], used as a basis to develop our method. We focused on extracting information not only from text but also from emoji. We extract emoji from the input text and process them through the Transformer [8] encoder independently, then concatenate the hidden states corresponding to the text and emoji before sending them to the classification layer of the model. Our model also employs data augmentation to improve its generalization ability.

We obtained good results in our experiments with two datasets of tweets in Brazilian Portuguese – TweetSentBR [1] and 2000-tweets-BR [9] – surpassing the previously published results for both datasets. For TweetSentBR, our method achieved accuracy of 0.7726 (6.3 p.p. of improvement) and F_1 score of 0.7514 (9.5 p.p. of improvement). For 2000-tweets-BR, accuracy of 0.8247 (14.5 p.p. of improvement) and F_1 score of 0.8035 (23.4 p.p. of improvement). It is possible to use a previously pre-trained BERT$_{BASE}$ model to warm start ours, greatly reducing the total training time.

As future work, we intend to apply our methodology to datasets in other languages, in particular English, since resources in this language are more plentiful. We also plan to investigate the application of other Transformer models other than BERT as the basis for our approach.

Acknowledgements. The authors would like to thank FAPESP (grants #2015/11937-9, #2017/12646-3, #2017/16246-0, #2017/12646-3 and #2019/20875-8), CNPq (grants #304380/2018-0 and #309330/2018-1) and CAPES for their financial support.

References

1. Brum, H.B., Nunes, M.G.V.: Building a sentiment corpus of tweets in Brazilian Portuguese. In: 11th International Conference on Language Resources and Evaluation (LREC), Miyazaki, Japan, pp. 4167–4172. European Language Resources Association (ELRA), ELRA (2018)
2. Brum, H.B., Nunes, M.G.V.: Semi-supervised sentiment annotation of large corpora. In: Villavicencio, A., et al. (eds.) PROPOR 2018. LNCS (LNAI), vol. 11122, pp. 385–395. Springer, Cham (2018). https://doi.org/10.1007/978-3-319-99722-3_39
3. Devlin, J., Chang, M.W., Lee, K., Toutanova, K.: BERT: pre-training of deep bidirectional transformers for language understanding. In: 20th Annual Conference of the North American Chapter of the Association for Computational Linguistics: Human Language Technologies (NAACL-HLT), Minneapolis, USA, pp. 4171–4186. Association for Computational Linguistics (ACL), ACL (2019)
4. Gururangan, S., et al.: Don't stop pretraining: adapt language models to domains and tasks. Computing Research Repository, pp. 1–19 (2020)
5. Nascimento, P.A.: Aplicando Ensemble para Classificação de Textos Curtos em Português do Brasil. Master's thesis, Universidade Federal de Pernambuco, Recife, Brazil (2019)
6. Sakiyama, K.M., Silva, A.Q.B., Matsubara, E.T.: Twitter breaking news detector in the 2018 Brazilian presidential election using word embeddings and convolutional neural networks. In: 37th International Joint Conference on Neural Networks (IJCNN), Budapest, Hungary, pp. 1–8. Institute of Electrical and Electronics Engineers (IEEE), IEEE (2019)
7. Souza, F., Nogueira, R., Lotufo, R.: BERTimbau: pretrained BERT models for Brazilian Portuguese. In: Cerri, R., Prati, R.C. (eds.) BRACIS 2020. LNCS (LNAI), vol. 12319, pp. 403–417. Springer, Cham (2020). https://doi.org/10.1007/978-3-030-61377-8_28
8. Vaswani, A., et al.: Attention is all you need. In: 31st Conference on Neural Information Processing Systems (NIPS), Long Beach, USA, pp. 5998–6008. Neural Information Processing Systems (NIPS) Foundation, Curran Associates Inc. (2017)
9. Vitório, D., Souza, E., Teles, I., Oliveira, A.L.: Investigating opinion mining through language varieties: a case study of Brazilian and European Portuguese tweets. In: 11th Brazilian Symposium in Information and Human Language Technology (STIL), pp. 43–52. Sociedade Brasileira de Computação (SBC), SBC, Uberlândia (2017)
10. Wei, J., Zou, K.: EDA: easy data augmentation techniques for boosting performance on text classification tasks. In: 24th Conference on Empirical Methods in Natural Language Processing and the 9th International Joint Conference on Natural Language Processing (EMNLP-IJCNLP), Hong Kong, pp. 6383–6389. Association for Computational Linguistics (ACL) (2019)

Detecting Hate Speech in Cross-Lingual and Multi-lingual Settings Using Language Agnostic Representations

Sebastián E. Rodríguez[1][✉], Héctor Allende-Cid[2], and Héctor Allende[1]

[1] Departamento de Informática, Universidad Técnica Federico Santa María,
Valparaíso, Chile
srodrigu@alumnos.inf.utfsm.cl, hector.allende@inf.utfsm.cl
[2] Escuela de Ingeniería Informática, Pontificia Universidad Católica de Valparaíso,
Valparaíso, Chile
hector.allende@pucv.cl

Abstract. The automatic detection of hate speech is a blooming field in the natural language processing community. In recent years there have been efforts in detecting hate speech in multiple languages, using models trained on multiple languages at the same time. Furthermore, there is special interest in the capabilities of language agnostic features to represent text in hate speech detection. This is because models can be trained in multiple languages, and then the capabilities of the model and representation can be tested on a unseen language.

In this work we focused on detecting hate speech in mono-lingual, multi-lingual and cross-lingual settings. For this we used a pre-trained language model called Language Agnostic *BERT* Sentence Embeddings (*LabSE*), both for feature extraction and as an end to end classification model. We tested different models such as Support Vector Machines and Tree-based models, and representations in particular bag of words, bag of characters, and sentence embeddings extracted from Multi-lingual *BERT*. The dataset used was the SemEval 2019 task 5 data set, which covers hate speech against immigrants and women in English and Spanish. The results show that the usage of *LabSE* as feature extraction improves the performance on both languages in a mono-lingual setting, and in a cross-lingual setting. Moreover, *LabSE* as an end to end classification model performs better than the reported by the authors of SemEval 2019 task 5 data set for the Spanish language.

Keywords: Hate speech detection · Natural Language Processing · Multi-lingual language models

ⓒ Springer Nature Switzerland AG 2021
J. M. R. S. Tavares et al. (Eds.): CIARP 2021, LNCS 12702, pp. 77–87, 2021.
https://doi.org/10.1007/978-3-030-93420-0_8

1   Introduction

Hate speech refers to the communicative action of promoting discriminatory actions which infringe upon the dignity of a group of people. For the most part, these actions are based in discrimination given the race, skin tone, ethnicity, gender, sexual orientation, nationality, religion and another characteristics of groups or individuals. Although hate speech is not a novel problem, it is still a relevant one as the growing use of social network platforms and the anonymity they provide present an ideal environment for the rise of such occurrences.

In order to tackle these discriminatory actions, automated hate speech detection tools can be employed. To this end, Machine learning based models using traditional algorithms (Naive Bayes, Support Vector Machines, Decision Trees, etc.) and Deep Neural network based algorithms have been used. Traditionally, the focus of hate speech detection has been in a mono-lingual setting (using only one target language at a time to detect hate-speech). This is problematic for other languages in which labeled data is scarce. Nonetheless, there have been recent efforts in providing multi-lingual datasets, in which hate speech can be detected in multiple languages across different domains [3,9,13].

With the new developments in Transformer-based Language Models [4,5] trained in multiple languages, these big models can be used to extract a sentence embedding. These representations should preserve the properties of linguistic characteristics, in which similar sentences should be in proximity given the embedding space. Moreover, with language models trained in multiple languages, they should preserve the similarity of sentences across all languages. Given this, we can leverage these representations and multi-lingual datasets to train models in a cross-lingual setting (training in A languages, and testing their performance on a held out language B).

In this work, we study the application of language agnostic feature representations with different Machine Learning models to detect Hate Speech detection in three settings: Mono-lingual, Multi-lingual and Cross-lingual. For these settings we use a benchmark dataset containing two languages, Spanish and English, and pre-process them with these feature representations. In order to measure the effectiveness of this language agnostic representation, we conduct a comparison with traditional approaches to vectorize natural text (for example, with a Bag of words approach), in addition to the use of two multi-lingual state of the art models (Multi-lingual *BERT* and *LabSE*).

This work is organized as follows. In Sect. 2, we present the related works in hate speech detection. Section 3 is where we present our proposal. The dataset, the different representations, models and metrics are detailed in Sect. 4. In Sect. 5 we present the results and a discussion on the experiments conducted in this work. Finally, in Sect. 6 we present the conclusions and future work.

2   Related Works

Automated hate speech detection has been in study as early as 2012, where Warner [21] employs a Bag of words representation and Support Vector Machines

(SVM), to detect messages with antisemitic connotation. In 2013, Kwok & Wang [10] propose the use of a Naive Bayes model to detect racist messages towards the Black community on Twitter. In 2015, Gitari et al. [8] created a rule-based algorithm that uses a lexicon created by the authors to detect hate speech in three domains: race, nationality, and religion.

Around the year 2017, word embedding approaches were being adopted for hate speech detection, using this representation along with Deep Neural Networks [2], specifically Long Short Term Memory (LSTM) Networks. In the same year, Convolutional Neural Networks (CNN) were being used to detect hate speech on Twitter [6]. In 2018 Zimmerman et al. [23] proposed the use of Deep Neural Network ensembles to improve the performance in automated hate speech detection. In 2019, with the release of the pre-trained Transformers language models, such as *BERT* [4], there was a widespread use of these architectures by fine tuning them on the hate speech classification domain, and in different languages [11,12,15].

In the domain of multi-lingual and cross-lingual hate speech detection, Ousidhoum et al. [13] explores the effectiveness of single task and mono-lingual models, compared to models trained on multiple tasks, and multiple languages. Sohn and Lee [18], use a multi-channel variant of BERT architecture, generating an ensemble of three versions of *BERT*: Multi-lingual, English and Chinese *BERT*. More recently, Aluru et al. [1] applied language agnostic sentence embeddings to detect hate speech in both multi-lingual, and cross-lingual settings. Furthermore, Stappen et al. [20], applied pre-trained language models in a zero-shot, and few-shot training scheme to carry out cross-lingual hate speech detection. In 2021, Ghosh Roy et al. [7] used pre-trained cross-lingual models, in addition to emoji embeddings and hashtags embeddings, to detect hate speech in three languages.

3 Proposal

For tackling the problem of cross-lingual and multi-lingual classification of hate speech, we propose the use of a pre-trained model named Language Agnostic BERT Sentence Embeddings (*LabSE*) [5]. This model is trained to produce language agnostic sentence embeddings for 109 languages, and uses a Dual *BERT* [4] architecture with shared parameters as the base model. Another difference with the multi-lingual version of *BERT* is the pre-training step. For multi-lingual *BERT* a fixed vocabulary of 110k WordPiece tokens is used, and the pre-training tasks are mono-lingual Masked Language Modeling (reconstructing a sentence given a corrupted input) and Next Sentence Prediction. For *LabSE*, the vocabulary consist of 500k WordPiece tokens, and is pre-trained on two tasks: Masked Language Modelling and Translation Language. The latter uses parallel sentences in different languages, which are concatenated and then corrupted by masking words in both sentences. Then, the model is trained to reconstruct both sentences in their respective language. This encourages the alignment of word embeddings for each parallel sentence. To obtain a sentence representation from multi-lingual *BERT*, we need to compute the mean pooling of each hidden state on the output of the last layer. On the other hand, for *LabSE* we obtain this sentence embedding by extracting the last hidden state of the *[CLS]* token.

To show the differences between *LabSE* sentence embeddings and multi-lingual *BERT* sentence embeddings, we encode the sentence *"The quick brown fox jumps over the lazy dog"* in 4 different languages (English, Spanish, German and French). Then, these sentence embeddings are projected to a 2 dimensional space using Principal Component Analysis. This is visualized on Fig. 1:

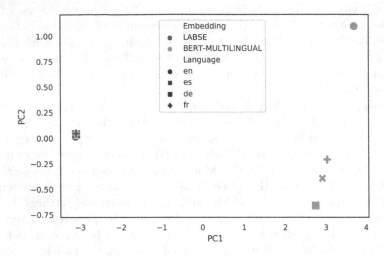

Fig. 1. 2*d* PCA projections for the sentence embeddings extracted with *LabSE* (left side of the scatter plot), and sentence embeddings extracted with multi-lingual *BERT*

We can see that for the different languages and *LabSE* sentence embeddings, a well defined cluster is located at the left side of the scatter plot, indicating that the embeddings for each language are highly similar. For multi-lingual *BERT* on the other hand, there is some proximity for the Spanish, French and German sentence embeddings, but there is no well defined cluster. The same analysis can be done computing the euclidean distance between sentence vectors for both pre-trained models.

In recent years, these pre-trained language models have been effective on tackling several Natural Language Processing Tasks by fine-tuning these models on a downstream task. For this reason, we propose fine-tuning *LabSE* to detect hate speech in a context of social media. Additionally, we can use *LabSE* as a language agnostic feature extractor, in which these dense representations can be used as inputs for traditional Machine Learning models.

4 Experiments

4.1 Dataset

The dataset used for this proposal belongs to the SemEval 2019 workshop - Task 5 [3]. This dataset contains messages from Twitter social network, and was used to tackle the detection of hate speech against immigrants and women in 2 languages: English and Spanish. In this case, we used the labels for the subtask A: binary classification of the messages of Hate Speech or Non-Hate Speech. Table 1 shows a summary of the number of documents for each language and each train/test sets. Also it is worth noting, that there is a small bias towards the Non-Hate Speech category. The best results in SemEval 2019 in terms of F_1-$score$ (defined in the next section) was 0.65 for English, and 0.73 for Spanish.

Table 1. Number of tweets for each language on the datasets

Language	Train set	Test Set	Non-HS/HS
English	10000	3000	≈58%/42%
Spanish	5000	1600	≈59%/41%

For this paper, we use this dataset in three settings:

1. **Mono-lingual:** training and testing in the same language, avoiding data from the other language corpus.
2. **Multi-lingual:** training and testing mixing both languages in one corpus.
3. **Cross-Lingual:** training in one language, and testing in the other language.

These settings permits us to benchmark our proposal, evaluating the generalization of models trained on language agnostic representations given these settings. There is minimal pre-processing applied to each message, mainly we remove emojis and URLs. Stopwords are not removed, nor a lemmatizing/stemming algorithm to each word. This is because, we try to avoid the sequentiality to be destroyed by the removal of stopwords, and aditionally, generating conflicts with the subword tokenization scheme applied by the pre-trained models.

4.2 Models and Evaluation Metrics

There are six baseline models implemented for each setting, three of them are Support Vector Machines (SVMs) with different kernels: Linear SVM (LSVM), SVM with RBF kernel (RSVM) and SVM with a Polynomial kernel (PSVM). The other three baseline models are tree-based classifiers, namely: Decision Trees (DT), Random Forest (RF) and Extremely Randomized Trees (ET). We conducted hyperparameter tuning for each of the baseline models, by doing 5-fold cross validation over the training set. For the SVM-based models, we tune the C parameter for all models, and γ and degree for RBF and Polynomial SVM respectively. For the tree-based models, the only parameter tuned is the maximum tree depth. Additionally, we implement two feature representations in order to compare the Transformer-based representations: a Bag of Words representation, and a Bag of Character representation (using a vocabulary of graphemes and symbols).

For the state of the art neural network models, we used both *LabSE* and Multi-lingual *BERT* as an end to end classification model. For both models we apply a 10% Dropout [19] on the last layer of the pre-trained model, followed by a dense layer with a sigmoid activation serving as the new output layer. For fine-tuning, we fixed the learning rate to 0.0001, used a batch size of 8, and used the Binary Cross Entropy Loss function (1). We conducted 5-fold cross validation to determine the number of epochs to fine-tune the architecture.

$$\mathcal{L} = -\frac{1}{N} \sum_{i=1}^{N} (y_i \cdot log(\hat{y}_i) + (1 - y_i) \cdot log(1 - \hat{y}_i)) \tag{1}$$

Two metrics were used to assess the quality of the classification for all the experiments. These metrics derive from the results of the confusion matrix: True Positives (TP), True Negatives (TN), False Positives (FP) and False Negatives (FN). The first metric was the Accuracy Score (A), which gave the ratio of correctly classified observations to the total of observations. The Accuracy Score was computed as follows (2):

$$Accuracy = \frac{TP + TN}{TP + TN + FP + FN} \tag{2}$$

The second metric was the F_1-*score*, which is the weighted harmonic mean of the Precision (P) and Recall (R):

$$P = \frac{TP}{TP + FP} \quad R = \frac{TP}{TP + FN} \tag{3}$$

$$F_1\text{-}score = 2 \cdot \frac{P \cdot R}{P + R} \tag{4}$$

The implementation of the baseline models was carried out using scikit-learn library [14], while the Deep Neural Network models were implemented with pytorch and Hugging Face pre-trained transformer library [22]. The machine used in the experiments had an Intel i7 processor, 32 Gb RAM and two Nvidia GTX 1080 TI graphics card. The code for reproducing these experiments will be made publicly available[1].

---

[1] https://github.com/capkuro/DetectingHatespeechLabse.

5   Results

In this section the results of the different tests carried out will be presented, as well as the different training approaches. For each table, an upper section detailing the performance of the baseline models using all representations is presented. The lower section of the table shows the performance of the Transformer-based models, in a end to end classification setting. Accordingly, these models only use the sentence embedding generated by their own architecture.

5.1   Mono-lingual

Table 2 shows the results obtained in the mono-lingual process with the traditional models and multiple input representations. The models with the best performance are SVM with Linear kernel for English with *LabSE* as the input vectors, obtaining 0.61 accuracy and 0.55 F_1-*score*. For Spanish, the best results are obtained by the SVM with RBF kernel, also with *LabSE* input vectors, with an accuracy of 0.74 and F_1-*score* of 0.73, matching the best results obtained at the competition [3] for the latter metric.

Table 2. Mono-lingual results for baseline algorithms and the fine tuned *LabSE* and *BERT*

Model	Word				Char				LabSE				BERT			
	EN		ES		EN		ES		EN		ES		EN		ES	
	Acc	F_1	Acc	F_1	Acc	F_1	Acc	F_1	Acc	F_1	Acc	F_1	Acc	F_1	Acc	F_1
LSVM	0.48	0.42	0.71	0.70	0.45	0.41	0.62	0.55	**0.61**	**0.57**	0.69	0.67	0.46	0.39	0.62	0.46
RSVM	0.48	0.42	0.71	0.71	0.47	0.44	0.63	0.60	0.53	0.50	0.74	0.73	0.53	0.50	0.67	0.63
PSVM	0.47	0.42	0.71	0.70	0.49	0.49	0.60	0.52	0.52	0.50	0.73	0.72	0.52	0.50	0.67	0.64
DT	0.47	0.43	0.68	0.68	0.48	0.47	0.62	0.55	0.55	0.55	0.60	0.59	0.55	0.54	0.62	0.52
RF	0.47	0.40	0.73	0.70	0.47	0.44	0.64	0.59	0.55	0.54	0.69	0.65	0.51	0.49	0.63	0.57
ET	0.54	0.51	0.71	0.70	0.47	0.44	0.63	0.59	0.54	0.54	0.68	0.63	0.51	0.50	0.63	0.56
LabSE									0.54	0.51	**0.76**	**0.75**				
BERT													0.53	0.49	0.75	0.74

Moreover, when we fine tune both Multi-lingual *BERT* and *LabSE* as an end to end classification architecture we get interesting results. For the English language, *LabSE* obtains the best results with 0.54 accuracy and 0.51 F_1-*score*. Yet the Linear SVM with *LabSE* embeddings obtains better results than fine-tuning the whole architecture. For the Spanish language on the other hand, both Multi-lingual *BERT* and *LabSE* architectures when fine tuned, obtain better performance than the SVM with RBF kernels and *LabSE* embeddings input. In this case, fine-tuning the *LabSE* model gives an 0.76 and 0.75 of accuracy and F_1-*score* respectively.

5.2   Multi-lingual

For the Multi-lingual task, Table 3 shows the results for all the models with different representations. In this case, the best results are obtained by the Random Forest with a Bag of Words representation. It is worth noting that tree-based models work on par with Spanish, and even better than SVM-based models for Spanish with Bag of Words representation. In this case, given that the Bag of Words contains both English and Spanish vocabulary, we theorize that tree-based models can generate sub-trees specialized on each language. Furthermore, the Bag of Characters representation gives smooth results across all the traditional models, obtaining similar results for SVM-based models and *LabSE* representation.

Table 3. Multi-lingual results for baseline algorithms and the fine tuned *LabSE* and *BERT*

Model	Word		Char		LabSE		BERT	
	Acc	F_1	Acc	F_1	Acc	F_1	Acc	F_1
LSVM	0.55	0.54	0.62	0.62	0.62	0.62	0.52	0.50
RSVM	0.56	0.55	0.63	0.63	0.63	0.63	0.58	0.58
PSVM	0.56	0.55	0.63	0.63	0.63	0.63	0.58	0.58
DT	0.68	0.68	0.58	0.56	0.57	0.55	0.56	0.55
RF	**0.73**	**0.71**	0.63	0.59	0.61	0.61	0.64	0.58
ET	0.70	0.69	0.64	0.59	0.61	0.60	0.64	0.58
LabSE					0.62	0.61		
BERT							0.59	0.57

The results of fine-tuning both Multi-lingual *BERT* and *LabSE* are presented next. In this setting, *LabSE* obtains better results than fine-tuning Multi-lingual *BERT*, yet it does not obtain comparable results with Random Forest and a Bag of Word representation.

5.3   Cross-Lingual

For the Cross-lingual task, Table 4 shows the results for all baseline models and representations. For this task, SVM-based models with *LabSE* input representation works best when compared with the other baseline models. For the $ES \rightarrow EN$ task, the SVM with RBF kernel obtains a 0.65 and 0.64 accuracy and F_1-*score* respectively. For the $EN \rightarrow ES$ task, the SVM with Polynomial kernel obtains a 0.66 and 0.65 of accuracy and F_1-*score* respectively. It is worth noting that a Bag of Words representation for this task is not suitable and it shows on the results that with this representation, the models tend to bias towards the Non Hate speech category.

For the Bag of Characters representation it can be observed that there is a partial transferring for the languages, given the increase in F_1-$score$. Nonetheless, it gives a worse performance when compared with the mono-lingual training. For Multi-lingual $BERT$ input representations, the results are slightly better than the mono-lingual training, indicating that these embeddings can indeed provide information across languages. It is worth mentioning that $LabSE$ input representation gives better results on the English language when compared with the mono-ligual English task, obtaining a performance close to the best model reported on [3].

Table 4. Cross-lingual results for baseline algorithms and the fine tuned $LabSE$ and $BERT$

Model	Word				Char				LabSE				BERT			
	$ES \to EN$		$EN \to ES$		$ES \to EN$		$EN \to ES$		$ES \to EN$		$EN \to ES$		$ES \to EN$		$EN \to ES$	
	Acc	F_1	Acc	F_1	Acc	F_1	Acc	F_1	Acc	F_1	Acc	F_1	Acc	F_1	Acc	F_1
LSVM	0.57	0.43	0.59	0.37	0.57	0.40	0.54	0.52	0.62	0.59	0.63	0.62	0.43	0.33	0.60	0.43
RSVM	0.58	0.53	0.59	0.37	0.57	0.40	0.58	0.44	**0.65**	**0.64**	0.65	0.64	0.57	0.54	0.61	0.49
PSVM	0.58	0.37	0.59	0.37	0.57	0.41	0.57	0.41	0.65	0.62	**0.66**	**0.65**	0.57	0.54	0.61	0.48
DT	0.58	0.37	0.59	0.38	0.56	0.42	0.55	0.46	0.59	0.56	0.56	0.56	0.54	0.49	0.58	0.54
RF	0.58	0.37	0.59	0.37	0.57	0.40	0.55	0.46	0.61	0.49	0.62	0.61	0.53	0.51	0.60	0.52
ET	0.58	0.37	0.59	0.37	0.58	0.41	0.55	0.48	0.60	0.46	0.63	0.61	0.53	0.50	0.59	0.52
LabSE									**0.68**	**0.65**	**0.69**	**0.68**				
BERT													0.66	0.64	0.66	0.61

When both Multi-lingual $BERT$ and $LabSE$ are used as an end to end model, the latter produces the best results for both cross-lingual tasks. For the $ES \to EN$ task, $LabSE$ model obtains a 0.68 and a 0.65 of accuracy and F_1-$score$ respectively, a better performance on the English mono-lingual task, and an equal F_1-$score$ reported by [3]. For the $EN \to ES$ task, $LabSE$ model delivers the best performance when compared with the baseline models. In this case, the cross-lingual setting for Spanish does not yield a better performance when compared with the mono-lingual setting, as opposed to the case for the English language.

6   Conclusions

In this paper, we tackle the problem of multi-lingual and cross-lingual hate speech detection using language agnostic representations. To this end, we proposed the use of $LabSE$ both as an feature extraction method, and as an end to end model for performing this classification. Experiments showed that $LabSE$ both as an embedding, and also as a model, outperforms other language agnostic representations. Additionally, $LabSE$ as an end to end model, outperforms the best results obtained in [3].

In future work we would like to explore different datasets with an increased number of both languages and samples, and include other language agnostic representations, such as Byte Pair Encoding [17]. Additionally, exploring data augmentation techniques, such as translating/back translating [16] the datasets, could provide a richer exploration of data intensive models such as Recurrent Neural Networks.

Acknowledgment. This work was supported in part by Basal Project AFB 1800082, in part by Project DGIIP-UTFSM PI-LIR-2020-17. Héctor Allende-Cid work is supported by PUCV VRIEA.

References

1. Aluru, S.S., Mathew, B., Saha, P., Mukherjee, A.: Deep learning models for multilingual hate speech detection. arXiv preprint arXiv:2004.06465 (2020)
2. Badjatiya, P., Gupta, S., Gupta, M., Varma, V.: Deep learning for hate speech detection in tweets. In: Proceedings of the 26th International Conference on World Wide Web Companion, pp. 759–760 (2017)
3. Basile, V., et al.: Semeval-2019 task 5: multilingual detection of hate speech against immigrants and women in twitter. In: 13th International Workshop on Semantic Evaluation, pp. 54–63. Association for Computational Linguistics (2019)
4. Devlin, J., Chang, M.W., Lee, K., Toutanova, K.: Bert: pre-training of deep bidirectional transformers for language understanding. arXiv preprint arXiv:1810.04805 (2018)
5. Feng, F., Yang, Y., Cer, D., Arivazhagan, N., Wang, W.: Language-agnostic bert sentence embedding. arXiv preprint arXiv:2007.01852 (2020)
6. Gambäck, B., Sikdar, U.K.: Using convolutional neural networks to classify hate-speech. In: Proceedings of the First Workshop on Abusive Language Online, pp. 85–90 (2017)
7. Ghosh Roy, S., Narayan, U., Raha, T., Abid, Z., Varma, V.: Leveraging multilingual transformers for hate speech detection. arXiv e-prints pp. arXiv-2101 (2021)
8. Gitari, N.D., Zuping, Z., Damien, H., Long, J.: A lexicon-based approach for hate speech detection. Int. J. Multimedia Ubiquitous Eng. **10**(4), 215–230 (2015)
9. Glavaš, G., Karan, M., Vulić, I.: Xhate-999: analyzing and detecting abusive language across domains and languages. In: Proceedings of the 28th International Conference on Computational Linguistics, pp. 6350–6365 (2020)
10. Kwok, I., Wang, Y.: Locate the hate: detecting tweets against blacks. In: Proceedings of the Twenty-Seventh AAAI Conference on Artificial Intelligence, pp. 1621–1622 (2013)
11. Mishra, S., Mishra, S.: 3idiots at hasoc 2019: fine-tuning transformer neural networks for hate speech identification in Indo-European languages. In: FIRE (Working Notes), pp. 208–213 (2019)
12. Mozafari, M., Farahbakhsh, R., Crespi, N.: A BERT-based transfer learning approach for hate speech detection in online social media. In: Cherifi, H., Gaito, S., Mendes, J.F., Moro, E., Rocha, L.M. (eds.) COMPLEX NETWORKS 2019. SCI, vol. 881, pp. 928–940. Springer, Cham (2020). https://doi.org/10.1007/978-3-030-36687-2_77
13. Ousidhoum, N., Lin, Z., Zhang, H., Song, Y., Yeung, D.Y.: Multilingual and multi-aspect hate speech analysis. arXiv preprint arXiv:1908.11049 (2019)

14. Pedregosa, F., et al.: Scikit-learn: machine learning in Python. J. Mach. Learn. Res. **12**, 2825–2830 (2011)
15. Polignano, M., Basile, P., De Gemmis, M., Semeraro, G.: Hate speech detection through Alberto Italian language understanding model. In: NL4AI@ AI* IA (2019)
16. Sennrich, R., Haddow, B., Birch, A.: Improving neural machine translation models with monolingual data. arXiv preprint arXiv:1511.06709 (2015)
17. Sennrich, R., Haddow, B., Birch, A.: Neural machine translation of rare words with subword units. arXiv preprint arXiv:1508.07909 (2015)
18. Sohn, H., Lee, H.: Mc-bert4hate: hate speech detection using multi-channel bert for different languages and translations. In: 2019 International Conference on Data Mining Workshops (ICDMW), pp. 551–559. IEEE (2019)
19. Srivastava, N., Hinton, G., Krizhevsky, A., Sutskever, I., Salakhutdinov, R.: Dropout: a simple way to prevent neural networks from overfitting. J. Mach. Learn. Res. **15**(1), 1929–1958 (2014)
20. Stappen, L., Brunn, F., Schuller, B.: Cross-lingual zero-and few-shot hate speech detection utilising frozen transformer language models and axel. arXiv preprint arXiv:2004.13850 (2020)
21. Warner, W., Hirschberg, J.: Detecting hate speech on the world wide web. In: Proceedings of the Second Workshop on Language in Social Media, pp. 19–26 (2012)
22. Wolf, T., et al.: Huggingface's transformers: state-of-the-art natural language processing. arXiv preprint arXiv:1910.03771 (2019)
23. Zimmerman, S., Kruschwitz, U., Fox, C.: Improving hate speech detection with deep learning ensembles. In: Proceedings of the Eleventh International Conference on Language Resources and Evaluation (LREC 2018) (2018)

Prediction of Perception of Security Using Social Media Content

Cristian Pulido, Luisa Fernanda Chaparro, Jorge Rudas[✉], Jorge Victorino,
Camilo Estrada, Luz Angela Narvaez, and Francisco Gómez

Department of Mathematics, Universidad Nacional de Colombia, Bogotá, Colombia
{cpulido,luchaparros,jerudasc,jevictorinog,cestradag,
lanarvaez,fagomezj}@unal.edu.co

Abstract. The Perception of Security (PoS) refers to the opinion that
persons have about security or insecurity in a place or situation. Real-
time monitoring and the capacity of anticipation of citizen's PoS are
highly relevant for citizen's security planning. Surveys represent the most
widely used strategy to quantify PoS. Nevertheless, this approach cannot be applied continuously to obtain real-time monitoring or to predict
future PoS. Recent evidence suggests that social network content may
provide valuable information to quantify PoS. However, the prediction
of these PoS quantifications remains poorly studied. We propose a novel
strategy to quantify and anticipate PoS in short time windows using
social network data. The model considers the external factors that may
contribute to the publication of posts related to PoS and the retweeting
phenomena. Results show that the proposed model may provide competitive predictive performances while keeps high levels of interpretability
about the factors influencing PoS.

Keywords: Perception of Security · Predictive security · Social
networks content · Natural languaje processing · Hawkes point process

1 Introduction

The Perception of Security (PoS) refers to the subjective opinion that persons
have about the feeling of security or insecurity in a place or situation [1]. The
PoS is closely related to the fear of crime and its consequences on the citizens [2].
Due to its subjective character, the PoS quantification is highly challenging [3].
The PoS may change due to several factors, such as the environmental quality,
information coming from third parties, previously suffered criminal experiences,
the occurrence of particular situations influencing crime rates, or the occurrence
of events that may affect safety, among others [3].

This work was funded by the project "Diseño y validación de modelos de analítica
predictiva de fenómenos de seguridad y convivencia para la toma de decisiones en
Bogotá", at Bank of National Investment Programs and Projects, National Planning
Department, Government of Colombia (BPIN: 2016000100036).

© Springer Nature Switzerland AG 2021
J. M. R. S. Tavares et al. (Eds.): CIARP 2021, LNCS 12702, pp. 88–96, 2021.
https://doi.org/10.1007/978-3-030-93420-0_9

Traditionally, the PoS is quantified using surveys applied to a small sample of people, commonly annually or twice per year [4]. This approximation is time expensive and also limited for performing real-time quantification of the PoS [4]. Besides, to our knowledge, there is no mechanism to predict or anticipate future PoS. These predictions may play an important role in citizen's security short planning.

Previous works have recently explored social media as an alternative to improve survey limitations on measuring PoS [5,6]. Mainly because of the kind of content commonly published in this network and the spreading information capacity of social networks. Nevertheless, these approaches focused on the quantification of PoS and did not consider PoS prediction's problem.

In this work, we propose a novel strategy to quantify and anticipate PoS in short time windows using social network data [5,6]. The quantification strategy focused on Twitter posts, which provide continuous observation of the citizen's PoS. These data are analyzed using natural language processing tools and supervised classification strategies to obtain PoS quantifications. Together with relevant covariates, these quantifications serve as input for a Hawkes point process model that anticipate PoS in short time windows [7]. We evaluated the strategy in predicting PoS for the city of Bogotá (Colombia). Results suggest that this strategy may successfully anticipate PoS for short time windows, providing valuable information to better understand the origin of PoS.

2   Materials and Methods

The predictions herein performed were based on Twitter's social network content. Specifically, geo-located data for the city of Bogotá (Colombia) between 2019 and 2020. This data resulted from a social media listening exercise performed by the City's Security Secretary. After acquiring and preprocessing the data a filtering process aimed to identify posts related to security was used. Later, a sentiment analysis model based on machine learning classifiers tagged the post text into classes (from 1 to 5), depending on the negativity (1) or positivity (5) level of PoS [8]. Additional tweet information, such as the number of followers of the account that performed the message and the posting time, were also considered input for the predictive model. Besides, these inputs included some previously known relevant event's in the city, which may alter the PoS. In particular, dates of citizen protests, dates of soccer matches in which the teams that historically their fans have participated in caused riots in the city, and other dates are known for their increase in criminal events or that affect security [1].

The model aimed to predict the number of tweets that can be published on the social network related to citizen security for future times [9,10]. A Hawkes point process provided this prediction [7]. This model aimed to estimate future events based on past events of near occurrence simultaneously considering the covariate's influences and the retweeting phenomena.

2.1 Proposed Model

Nowadays, people use social networks to share and discuss their opinions, particularly opinions related to their Perception of Security [6]. One of the most widely used social networks for this task is Twitter, where users can either post a message or share messages from other people [11]. This sharing mechanism results in the appearance of messages with similar content or retweeted from an original message, increasing the number of posts related to PoS [12]. For this reason, the proposed model considers both messages related to PoS and retweets of these messages. The model's ultimate goal is to predict the number of security-related tweets for a short time future window. Therefore, the total number of tweets related to PoS posted per hour was computed, resulting in a time series.

A Hawkes point process provided the prediction of this time series. This self-excitatory model was used to explain the tweets posting rate by simultaneously considering the effects on the PoS of particular covariates and the retweeting time. Specifically, the proposed model assumed a function for the intensity of events of tweeting posts of the form:

$$\lambda(t) = \mu(t) + \sum_{i:t_i<t} g(t - t_i)$$

where $\mu(t)$ captures the behavior of the posting events that arise naturally (background), in this case, the original tweets. The second term represents the accumulated intensity by the past posting events as a function of time elapsed, which may influence the occurrence of new posts [7].

Original Tweets. For the background term $\mu(t)$ we considered the dynamic of original PoS tweet posting actions, which are closely related to the mixture of various temporal factors occurring in the city [6]. In the proposed model, this background term $\mu(t)$ corresponded to $\mu(t) = exp(\beta T(t))$, where $T(t)$ is a time-dependent column vector containing factors that may influence PoS tweet's post intensity. $T(t)$ included the tweet posting weekday, the day timeslot of the tweet posting (0–12:00 pm to 12:00 am, 1–12:00 am to 12:00 pm). Besides, covariates related to events that may influence security in the city were also considered in $T(t)$. Particularly, two entries indicated a soccer game of one of the two major city teams (Millonarios or Santa Fe) occurring at time t^1. Two additional entries were also considered, the first one indicating if there were programmed citizen protests and another one if there was a festive or special date. The β corresponds to a weight vector, representing the amount each variable considered in $T(t)$ contributes to the intensity of the tweet posting occurrence [13].

Retweets. Unlike original tweets, retweets may result from a different dynamic-for instance, retweeting the content of highly ranked influencers, i.e., a tweet from

---

[1] The fans of these team shave previously caused disturbances in the city on the game dates.

a person or account that is influential, is more likely to be highly replicated in a short time. Also, the tweet's influence should be strong at the begging but may decrease over time as a function of the users who replicate the tweet and the importance it may have over time [12]. In the case of messages related to PoS, posts reflecting feelings of insecurity may have a more significant impact since they negatively affect people's perceptions. These messages may be more influential than positive news that does not significantly increase overall PoS [14]. Based on these observations, we model the function g that determines the intensity added by past events as:

$$g(t - t_i) = p_i(t)d_i\phi(t - t_i)$$

with

$$\phi(s) = \begin{cases} 0 & s < 0 \\ c_0 & 0 \le s \le s_0 \\ c_0(s/s_0)^{-(1+\theta)} & s > s_0 \end{cases}$$

and

$$p_i(t) = p_0^i \left[1 - (Sr_0)sin \left(\frac{2\pi}{T_m}(t + \phi_0) \right) \right] exp(-(t - t_0)/\tau_m)$$

where d_i represents the number of followers of the account that tweets or retweets at the instant t_i, the memory kernel $\phi(s)$ is a probability distribution of the time interval between a tweet and its retweet made by another user [9]. The parameters of $\phi(s)$ were set as $c_0 = 6.49x10^{-4}(/seconds)$, $s_0 = 300$ seconds and $\theta = 0.242$. $p_i(t)$ is a function that determines the influence of a tweet over time. This function considers S referring to sentiment given to an original tweet and provided by the sentiment analysis model. $p_i(t)$ corresponds to an exponentially decaying oscillating function modeled by the product of the exponential and sinusoidal functions where t_0 is the time when the original tweet was published, T_m is the period of oscillation set to one day, r_0 is the relative amplitude, ϕ_0 is the phase and τ_m is the characteristic decay time [9]. These last three variables (r_0, ϕ_0, τ_m) together with β correspond to parameters to be estimated in the training phase.

2.2 Estimating Model Parameters

The model parameters were estimated from historical data using the maximum likelihood method [13]. For this, the log-likelihood $\ell(\Theta)$ for a time interval $[t_0, T]$ was formulated as:

$$\ell(\Theta) = \sum_{i:Original} \beta T(t_i)$$

$$+ \sum_{i:Original} \sum_{j:RT \text{ from } i} log(p_i(t_j)d_j\phi(t_j - t_i)) - \int_{t_0}^{T} \lambda(t)dt$$

whose parameters to estimate are $\Theta = \{\beta, r_0, \phi_0, \tau_m\}$.

Note that β only affects the first line of the equation and the first part of the final integral when replacing $\lambda(t)$ and dividing the sum. In this way, calculating the derivative of $\ell(\Theta)$ with respect to β and setting it equal to zero we obtain:

$$\frac{\partial \ell}{\partial \beta} = \sum_{i:Original} T(t_i) - \frac{\partial \ell}{\partial \beta} \left(\int_{t_0}^{T} exp(\beta T(t))dt \right) = 0.$$

Due to the fact that the function $T(t)$ of temporal covariates can be considered as a piecewise function for a certain partition $\mathcal{P}$ of the interval $[t_0, T]$ its solution is calculable as an expression that depends on β as follow:

$$\sum_{i:Original} T(t_i) = \sum_{i \in \mathcal{P}} T_i T(i) exp(\beta T(i))$$

where T_i is the size of the partition element with value $T(i)$. From this expression β cannot be directly solved, therefore numerical methods were used to obtain at a good approximation of its value since the original likelihood expression is convex in β [13].

The assumption that the influence of tweets is constant for small-time windows was used to estimate the remaining parameters of the influence function [9]. For this, $p_i(t)$ was approximated by discrete version $\widehat{p}_i(t)$. Then, an error function was defined and that was minimized to obtain the remaining parameters.

Specifically, for a specific tweet i in a time window $[T_{st}, T_{end}]$ where the tweet has been retweeted R times, the value $\widehat{p}_i(t)$ was defined as:

$$\widehat{p}_i(t) = R \left(\sum_{j:RT \text{ from } i} d_j \int_{t_{st}-t_j}^{t_{end}-t_j} \phi(t)dt \right)^{-1}.$$

Starting from the estimated values for each tweet at the defined time instants, the parameters (r_0, ϕ_0, τ_m) were computed by minimizing the function:

$$E(r_0, \phi_0, \tau_m) = \sum_{i:Original} \sum_{t:t-estimated} ||\widehat{p}_i(t) - p_i(t)||.$$

With all the estimated parameters, it was possible to build a function that allows estimating the number of future tweets.

2.3 Predicting Future Tweets

The $\lambda(t)$ function can be evaluated at any $t \in [t_0, T]$. However, for a $t > T$ although it is possible to calculate the intensity of new tweets using the $\mu(t)$ function, its retweet times and quantity of accounts followers and the polarity of sentiment are not known. To estimate these values for new tweets and retweets, we calculate the expected value for $\lambda(t)$. First, samples of background events were generated using the thining algorithm [7]. Thus initial times where original tweets are produced were computed. These values were used to compute the expected intensity for their retweets by replacing S and d with samples extracted from their specific distributions obtained from the data with the following expression:

$$\mathbb{E}(\lambda(t)|t_1,\ldots) = \widehat{\lambda}(t) = \lambda(t) + \sum_{i:samples} \mathbb{E}(p_i(t)) \int_T^t \phi(t-s)ds.$$

Finally, knowing the value of the estimated intensity for a time t to be predicted, the thining algorithm was used again [7] to predict the number of tweets related to security in the interval $[T,t]$.

2.4   Experimental Settings

Two models provided the baseline approaches. The first model corresponded to a nonhomogeneous Poisson Process model. In this case, the same covariates were considered as independent variables that explain the number of final events, and both tweets and retweets equally counted [15]. The second model is an adaptation of the model mentioned above, in which the model predicts the number of retweets the original tweet can have over time by fitting a linear regression on the logarithm of retweets in the training data [9]. The original tweets were calculated using the background of the herein proposed model, and the retweets came from linear regression and an estimated value of retweets during the first hour from the data.

For quantitative comparison of the models, a cross-validation strategy with slicing time windows was used. The complete data set was divided into sliding blocks of time displaced each 15 days. Initial size of the training data was 30 days. Each block was chosen as training and the rest for testing. The mean absolute error (MAE) and the Pearson coefficient provided the quantitative quality errors.

Besides quantitative predictive performance, the covariates effect on the training data was also studied.

3   Results

Effect Covariates. Figure 1 shows an eight days prediction of tweets related to PoS resulted from using the proposed model. The training data included tweets between 24 September of 2019 to 23 November of 2019. This period was selected because there were many social protests in the city of Bogotá and few soccer matches.

The upper-left panel in Fig. 1 shows in blue the number of tweets, both original and retweets, related to security observed for the studied time-period together with the prediction in orange. The proposed model can adequately follow the oscillating pattern observed for the actual tweets. Nevertheless, the model does not capture the high amplitudes observed on the actual tweets. When comparing model predictions and the original tweets, both the oscillating pattern and the amplitude were correctly predicted by the model, as observed in panel upper-right. The bottom-left panel shows the time location for covariates for the period of prediction. Interestingly, the two covariates happening on 27 November of 2019 increased the number of tweets related to PoS predicted for this day.

Panel bottom-left in Fig. 1 shows the covariate values computed on the training used for the prediction. As observed, citizen protests increased the number of tweets related to PoS, while special days occurrence seems to help decrease this number. Remarkably, the influence of local teams' soccer matches appears to have minor importance on the number of tweets related to PoS, at least at this period.

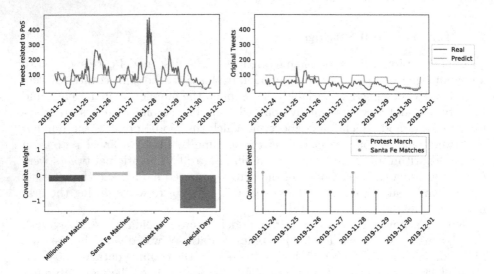

Fig. 1. Prediction obtained from the proposed model. The upper-left panel shows the actual (original and retweets) and predicted number of tweets related to PoS. The upper-right part shows the original and the predicted number of tweets related to PoS. In the lower-left panel, the covariates' weights were obtained using the model on the training data. The lower-right panel shows the covariates' events for the period of prediction.

Comparison with Other Models. Figure 2 shows the metrics calculated for the models evaluated under the proposed cross-validation scheme, on the different folds. Figure (a) shows the MAE and part (b) the Pearson coefficients. For Fig. 2a we obtained average values of 31.01 ± 23.68, 30.48 ± 21.44, 30.24 ± 23.02 for the proposed, Poisson and linear regression models, respectively. There were no significant differences between result performances provided by the different models. The average Pearson coefficients were 0.17 ± 0.12 for the proposed model, 0.16 ± 0.13 for the Poisson model, and 0.08 ± 0.05 for the linear regression model. However, the proposed model was the only one able to provide an interpretation about the covariates relevance.

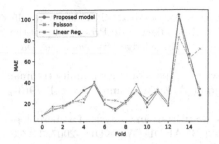 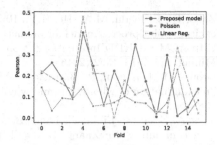

(a) MAE errors for different temporary partitions in cross validation configuration.

(b) Pearson coefficients for different temporary partitions in cross validation.

Fig. 2. Metrics evaluated in the different time partitions defined in the cross-validation configuration on the different folds.

4 Conclusions

In this work, a model to predict the number of tweets related to citizen security was proposed. The model considers factors that may contribute to the posting of texts related to PoS and the retweeting phenomena. Results suggest that the proposed model may provide competitive results when compared to baseline methods. In addition, the model may provide interpretability for the relevance of factors considered.

References

1. Skogan, W.G., Maxfield, M.G.: Coping with Crime: Individual and Neighborhood Reactions. Sage Publications, Beverly Hills (1981)
2. Rundmo, T., Moen, B.: Risk perception and demand for risk mitigation in transport: a comparison of lay people, politicians and experts. J. Risk Res. **9**, 623–640 (2006)
3. Skogan, W.: The Various Meanings of Fear, pp. 131–140. Enke (1993)
4. Boholm, A.: Comparative studies of risk perception: a review of twenty years of research. J. Risk Res. **1**(2), 135–163 (1998)
5. Pereira-Kohatsu, J.C., Quijano-Sánchez, L., Liberatore, F., Camacho-Collados, M.: Detecting and monitoring hate speech in twitter. Sensors **19**(21), 4654 (2019)
6. Curiel, R.P., Cresci, S., Muntean, C.I., Bishop, S.R.: Crime and its fear in social media. Palgrave Commun. **6**(1), 1–12 (2020)
7. Rizoiu, M.A., Lee, Y., Mishra, S., Xie, L.: A tutorial on hawkes processes for events in social media. arXiv preprint arXiv:1708.06401 (2017)
8. Chaparro, L., et al.: Sentiment analysis of social network content to characterize the perception of security. In: 2020 IEEE/ACM International Conference on Advances in Social Networks Analysis and Mining (ASONAM), pp. 685–691. IEEE Computer Society, Los Alamitos, December 2020
9. Kobayashi, R., Lambiotte, R.: Tideh: time-dependent hawkes process for predicting retweet dynamics. In: Proceedings of the International AAAI Conference on Web and Social Media, vol. 10 (2016)

10. Zhao, Q., Erdogdu, M.A., He, H.Y., Rajaraman, A., Leskovec, J.: Seismic: a self-exciting point process model for predicting tweet popularity. In: Proceedings of the 21th ACM SIGKDD International Conference on Knowledge Discovery and Data Mining, pp. 1513–1522 (2015)
11. Brown, M., Dustman, P., Barthelemy, J.: Twitter impact on a community trauma: an examination of who, what, and why it radiated. J. Community Psychol. 49(3), 838–853 (2020)
12. Java, A., Song, X., Finin, T., Tseng, B.: Why we twitter: an analysis of a microblogging community. In: Zhang, H., et al. (eds.) SNAKDD/WebKDD -2007. LNCS (LNAI), vol. 5439, pp. 118–138. Springer, Heidelberg (2009). https://doi.org/10.1007/978-3-642-00528-2_7
13. Reinhart, A.: Point process modeling with spatiotemporal covariates for predicting crime. Ph.D. thesis, Carnegie Mellon University (2016)
14. Prieto Curiel, R., Bishop, S.: Modelling the fear of crime. Proc. Roy. Soc. A: Mathe. Phys. Eng. Sci. 473(2203), 20170156 (2017)
15. Lawless, J.F.: Regression methods for poisson process data. J. Am. Stat. Assoc. 82(399), 808–815 (1987)

Metaheuristics

Fine-Tuning Dropout Regularization in Energy-Based Deep Learning

Gustavo H. de Rosa$^{(\boxtimes)}$ ⓘ, Mateus Roder ⓘ, and João P. Papa ⓘ

Department of Computing, São Paulo State University, Bauru, Brazil
{gustavo.rosa,mateus.roder,joao.papa}@unesp.br

Abstract. Deep Learning architectures have been extensively studied in the last years, mainly due to their discriminative power in Computer Vision. However, one problem related to such models concerns their number of parameters and hyperparameters, which can easily reach hundreds of thousands. Additional drawbacks consist of their need for extensive training datasets and their high probability of overfitting. Recently, a naïve idea of disconnecting neurons from a network, known as Dropout, has shown to be a promising solution though it requires an adequate hyperparameter setting. Therefore, this work addresses finding suitable Dropout ratios through meta-heuristic optimization in the task of image reconstruction. Several energy-based Deep Learning architectures, such as Restricted Boltzmann Machines, Deep Belief Networks, and several meta-heuristic techniques, such as Particle Swarm Optimization, Bat Algorithm, Firefly Algorithm, Cuckoo Search, were employed in such a context. The experimental results describe the feasibility of using meta-heuristic optimization to find suitable Dropout parameters in three literature datasets and reinforce bio-inspired optimization as an alternative to empirically choosing regularization-based hyperparameters.

Keywords: Machine learning · Deep learning · Regularization methods · Meta-heuristic optimization

1 Introduction

Throughout the last decade, there has been an increasing demand in Deep Learning (DL) algorithms [7,9,11,21], as they attempt to mimic the human visual system and learn hierarchical features. Amongst DL architectures, some energy-based models, such as Restricted Boltzmann Machines (RBM) [15], received recent spotlights due to their capacity in modeling the input data distribution. Even though RBMs are traditionally categorized as DL algorithms, they must be layer-stacked to be considered DL architectures, e.g., Deep Belief Networks (DBN) [4].

The authors would like to thank São Paulo Research Foundation (FAPESP) grants #2013/07375-0, #2014/12236-1, #2017/25908-6, #2019/02205-5, #2019/07825-1, and #2019/07665-4, and National Council for Scientific and Technological Development (CNPq) grants #307066/2017-7 and #427968/2018-6.

ⓒ Springer Nature Switzerland AG 2021
J. M. R. S. Tavares et al. (Eds.): CIARP 2021, LNCS 12702, pp. 99–108, 2021.
https://doi.org/10.1007/978-3-030-93420-0_10

One significant challenge regarding deep neural networks is their struggle to achieve low generalization errors (testing errors). Whenever there is a high discrepancy between training and testing errors, the model is prone to lose its generalization capacity, commonly known as overfitting [14], i.e., neurons tend to "memorize" the data instead of learning, thus producing a model that can not accurately predict unseen data. Additionally, they have a considerable number of parameters and hyperparameters, which might lead to an overfitted architecture if not accurately set.

Numerous attempts have been engaged to lessen the overfitting problem, such as early-stopping training or even introducing regularization methods such as soft-weight sharing [10], L_1 [18], and L_2 [5], among others. Alternatively, the best regularization would be to average the predictions of all possible parameter configurations, weighing the possibilities and checking out which would perform better. Nevertheless, such a methodology demands a cumbersome computational effort, only feasible for pitiful or non-complex models [23]. Some years ago, Srivastava et al. [16] proposed a regularization technique known as Dropout, where neurons are randomly dropped from neural networks during the training phase. Although some works addressed the problem of overfitting in DL architectures using Dropout regularization and obtained state-of-the-art results, their main disadvantage concerns fine-tuning the Dropout probability, which is usually accomplished by hand or set to a standard value (0.5).

This work aims to find an adequate threshold for dropping out units through meta-heuristic optimization tasks, where agents encode the Dropout ratios and find feasible solutions over a search space guided by the validation set loss function. As far as we are concerned, this is the first work that has attempted to select the Dropout ratio in energy-based models. Therefore, this work's main contributions are twofold: (i) introduce meta-heuristic optimization to Dropout-based regularization in energy-based models, and (ii) fill the lack of research regarding DL regularization and meta-heuristic optimization.

The remainder of this work is organized as follows. Section 2 presents a literature review regarding related works, while Sect. 3 introduces the theoretical background regarding Restricted Boltzmann Machines and Dropout-based Restricted Boltzmann Machines. Section 4 discusses the methodology and Sect. 5 presents the experimental results. Finally, Sect. 6 states conclusions and future works.

2   Related Works

Bergstra and Bengio [1] addressed that empirical- and grid-based searches were only feasible for low-dimensional spaces and minor precision due to their computational burden in evaluating thousands of possible solutions. Hence, they proposed a novel yet straightforward procedure that consists of simple random searches in the context of Deep Belief Networks hyperparameter optimization, concluding that their proposed approach was superior to grid searches. Furthermore, Papa et al. [13] extended such an approach by employing meta-heuristic optimization, such as Hyperopt and Harmony Search, in the context of Deep

Belief Networks hyperparameters optimization, achieving state-of-the-art results when compared to traditional random searches. Additionally, Papa et al. [12] employed the same methodology to fine-tune Discriminative Restricted Boltzmann Machines hyperparameters, obtaining better classification accuracies than naïve random searches.

Although several works have addressed fine-tuning hyperparameter optimization in energy-based models, only a few addressed Dropout regularization in such models. For instance, Wang et al. [20] introduced a fast version of Dropout, but not aiming RBMs as their primary focus, as their approach works by sampling from a Gaussian approximation instead of applying the Monte Carlo "optimization". Su et al. [17] tackled the problem of Dropout regularization in Restricted Boltzmann Machines in Field Programmable Gate Arrays. Even though they proposed a Dropout-based RBM, their work used the standard ratio (0.5) and did not employ hyperparameter optimization.

At the same time, Lee et al. [8] proposed a fixed Dropout regularization to improve the generalization capability of Deep Belief Networks and reduce their overfitting in self-training tasks. Additionally, Wang et al. [19] presented a review of regularization methods in RBMs, such as weight decay, network pruning, and Dropconnect. Finally, Rosa et al. [2] introduced meta-heuristic optimization to estimate reasonable Dropout probabilities in Convolutional Neural Networks applied to image classification. Although these works have obtained state-of-the-art results in some applications, they have not explored meta-heuristic optimization regarding Dropout regularization in energy-based models.

3   Theoretical Background

This section presents a brief theoretical background regarding Restricted Boltzmann Machines and Dropout-based Restricted Boltzmann Machines. Note that we opted not to include Deep Belief Networks as they are straightforward extensions of RBMs.

3.1   Restricted Boltzmann Machines

Restricted Boltzmann Machines are stochastic-based neural networks and inspired by physical concepts, such as energy and entropy. Usually, RBMs are trained through unsupervised algorithms and applied in image-based tasks, such as collaborative filtering, feature extraction, image reconstruction, and classification. An RBM comprises a visible layer $\mathbf{v}$ composed by m units, which deals directly with the input data, and a hidden layer $\mathbf{h}$ constituted of n units, which is in charge of learning features and the input data's probabilistic distribution. Also, let $\mathbf{W}_{m \times n}$ be the matrix that models the weights between visible and hidden units, where w_{ij} stands for the connection between visible unit v_i and hidden unit h_j. A deriving model from the RBM, often known as Bernoulli-Bernoulli RBM (BBRBM), assumes that all units in layers $\mathbf{v}$ and $\mathbf{h}$ are binary and sampled from a Bernoulli distribution [3], i.e., $\mathbf{v} \in \{0,1\}^m$ and $\mathbf{h} \in \{0,1\}^n$. Under such an assumption, Eq. 1 describes the energy function of a BBRBM:

$$E(\mathbf{v}, \mathbf{h}) = -\sum_{i=1}^{m} a_i v_i - \sum_{j=1}^{n} b_j h_j - \sum_{i=1}^{m}\sum_{j=1}^{n} v_i h_j w_{ij}, \tag{1}$$

where $\mathbf{a}$ and $\mathbf{b}$ stand for the visible and hidden units biases, respectively. Hence, the training algorithm of an RBM learns a set of parameters $\theta = (\mathbf{W}, \mathbf{a}, \mathbf{b})$ through an optimization problem, which aims at maximizing the product of probabilities derived from a training set $\mathcal{D}$.

3.2   Dropout-Based Restricted Boltzmann Machines

A Dropout-based Restricted Boltzmann Machine (DRBM) can be formulated as a simple RBM extended with one binary random vector $\mathbf{r} \in \{0, 1\}^n$. In such a formulation, $\mathbf{r}$ stands for the activation of hidden layer neurons, where each variable r_j defines whether neuron h_j is activated or not. Notice that $\mathbf{r}$ is re-sampled for every batch during learning. As units were dropped from the hidden layer, Eq. 1 can be rewritten as follows:

$$E(\mathbf{v}, \mathbf{h}, \mathbf{r}) = -\sum_{i=1}^{m} a_i v_i - \sum_{j=1}^{n} b_j h_j r_j - \sum_{i=1}^{m}\sum_{j=1}^{n} v_i h_j r_j w_{ij} \tag{2}$$

Therefore, a DRBM can be understood as a blend of several RBMs, each using different subsets of their hidden layers. As we are training the model with different subsets, the weight matrix $\mathbf{W}$ needs to be scaled at testing time, i.e., multiplied by p in order to adjust its weights.

4   Methodology

This section describes the proposed approach, the experimental setup, and the employed datasets[1].

4.1   Proposed Approach

We modeled the problem of selecting a feasible Dropout probability regarding energy-based architectures as a meta-heuristic optimization. As we are only interested in fine-tuning the Dropout ratio, we fixed every other hyperparameter according to Sect. 4.2 and optimized $p \in [0.0001, 1]$, guided by minimizing the loss function (minimum squared error) of the validation set. Additionally, we employed 15 agents over 25 convergence iterations. Algorithm 1 summarizes the pseudo-code of the proposed approach.

---

[1] Regarding the source-code, we have used the implementation available at GitHub: https://github.com/gugarosa/dropout_rbm.

Algorithm 1: Meta-heuristic optimization pseudo-code.

Input: Number of agents M, number of decision variables N, maximum
    number of iterations T, dropout probability p and validation
    reconstruction error $f(\cdot)$.
Output: Best solution g.
Objective function $f(p)$, $p = (p_1)$;
Initialize agents' positions $a_i \mid \forall i = (1, 2, \ldots, M)$;
$\hat{p} \leftarrow (\infty)$;
$t \leftarrow 1$;
while $(t <= T)$ **do**
    for i *from* 1 *to* M **do**
        Generate new solution $\mathbf{p}_i$ according to pre-determined heuristic;
        if $f(p_i) < f(a_i)$ **then**
            $a_i \leftarrow p_i$;
        if $f(a_i) < f(\hat{p})$ **then**
            $\hat{p} \leftarrow a_i$;
    $t \leftarrow t + 1$;
$g \leftarrow \hat{p}$;

To validate the proposed approach, we employed a subset of swarm-based algorithms, such as Particle Swarm Optimization (PSO) [6], Bat Algorithm (BA) [25], Cuckoo Search (CS) [26], and Firefly Algorithm (FA) [24]. Such algorithms have been selected due to their balance in exploring broad regions of the search space and their ability to refine promising solutions as swarm-based techniques commonly re-evaluate the whole swarm after updating their individuals' positions, allowing the re-positioning of sub-optimal agents to more likely optimal locations. Additionally, since we did not want to constrain the hyperparameters precision, we opted not to use the Grid Search algorithm, which requires a step interval to compose the grid and often produces large spaces due to its combinatorial approach[2].

Furthermore, Table 1 exhibits the parameter configuration for every metaheuristic technique[3]. Concerning BA, we need to set the minimum and maximum frequency ranges defined by f interval, the loudness parameter A, and the pulse rate r. Regarding CS, we have β as the Levy distribution's controller, p is the probability of replacing the worst nests, and α is the step size. FA requires the randomization parameter α, the attractiveness β, and the light absorption coefficient γ. Finally, PSO defines w as the inertia weight, and c_1 and c_2 as the social and cognitive constants.

---

[2] A ratio constrained between $[0.0001, 1]$ with a step size of 0.0001 generates $10,000$ possibilities.

[3] Note that these values were empirically chosen according to their author's definition to mitigate additional influences over the Dropout regularization.

Table 1. Meta-heuristic techniques hyperparameters configuration.

Meta-heuristic	Hyperparameters
BA	$f = [0,2], A = 0.5, r = 0.5$
CS	$\beta = 1.5, p = 0.25, \alpha = 0.8$
FA	$\alpha = 0.2, \beta = 1.0, \gamma = 1.0$
PSO	$c_1 = 1.7, c_2 = 1.7, w = 0.7$

4.2 Experimental Setup

We opted to employ two distinct DBN architectures varying in the number of hidden neurons as follows: $100 - 100 - 400$ and $500 - 500 - 2,000$[4]. Each RBM and DBN hyperparameter have been set as follows: $\eta = 0.1, \lambda = 0.0002$ e $\alpha = 0.5$. Furthermore, we have considered 25 epochs for their learning procedure with batches of size 20, while all models were trained using the Contrastive Divergence algorithm with $k = 1$ (CD-1). As we wanted to understand better how Dropout behaved in such a context, we decided not to vary the architectures' hyperparameters. On the other hand, the fixed hyperparameters have been chosen accordingly to the training guide proposed by Hinton et al. [3]. A rule of thumb is applied to choose the training rate, momentum, and weight decay in such a work, whereas the hidden neurons have been empirically decided to reflect a smaller and a large architecture.

To provide robust statistical analysis and acknowledging that the results of the experiments are independent and continuous over a particular dependent variable (e.g., number of observations), we identified the Wilcoxon signed-rank test [22] with 0.05 significance and 10 observations satisfied our obligations. It is a non-parametric hypothesis test that compares two or more related observations (in our case, repeated measurements of MSE) to assess whether there are statistically significant differences between them.

4.3 Datasets

Table 2 depicts the employed datasets along with their number of samples per set, size of samples, and the number of classes. Such datasets have been empirically selected as they reflect some well-known image reconstruction tasks, whereas MNIST depicts a more challenging problem than USPS, and USPS stands for a more complex problem than Semeion.

5 Experimental Results

This section describes the experimental results, where the most significant results according to the Wilcoxon-signed rank test are bolded.

---

[4] Note that a x-layer architecture uses x sequential values from $a-b-c$, being $x \in [1,3]$. Regarding RBMs, we opted to use 400 and 2,000 hidden neurons.

Table 2. Overview of employed datasets.

Dataset	# Training	# Validation	# Testing	Size	# Classes
MNIST	20.000	40.000	10.000	28×28	10
Semeion	200	400	993	16×16	10
USPS	2.406	4.855	2.007	16×16	10

5.1 Restricted Boltzmann Machines

Table 3 describes the mean reconstruction error, and standard deviation found over the testing set using the mean Dropout probability p found over the validation set. The first highlighted point concerns that more complex architectures (higher number of hidden neurons) seem to benefit from Dropout regularization, according to the #2 columns results. Even though such columns obtained statistically equal reconstruction errors than #1 columns, they achieved higher Dropout probabilities, which means that these tasks did not demand a vast amount of neurons to achieve feasible results. Regarding the meta-heuristics evaluation, we can point that CS attained a poor performance due to its inability to enhance promising solutions in the search space. Such a behavior lies in its nests' replacement procedure, which often abandons potential solutions and favors random ones.

Table 3. RBM mean reconstruction error and standard deviation [mean p] over MNIST, Semeion and USPS testing sets.

Model	MNIST #1	#2	Semeion #1	#2	USPS #1	#2
RBM	0.0145 ± 0.0002 [—]	0.0146 ± 0.0002 [—]	0.0548 ± 0.0015 [—]	0.0460 ± 0.0008 [—]	0.0233 ± 0.0004 [—]	0.0214 ± 0.0003 [—]
BA	0.0145 ± 0.0002 [0.0098]	0.0148 ± 0.0004 [0.0887]	0.0548 ± 0.0007 [0.0807]	0.0442 ± 0.0009 [0.2320]	0.0232 ± 0.0003 [0.0001]	0.0215 ± 0.0003 [0.0134]
CS	0.0161 ± 0.0003 [0.1204]	0.0159 ± 0.0016 [0.1559]	0.0562 ± 0.0019 [0.1498]	0.0443 ± 0.0009 [0.2174]	0.0256 ± 0.0032 [0.1096]	0.0227 ± 0.0020 [0.1563]
FA	0.0146 ± 0.0003 [0.0184]	0.0145 ± 0.0004 [0.0956]	0.0549 ± 0.0008 [0.0846]	0.0440 ± 0.0009 [0.2447]	0.0237 ± 0.0013 [0.0266]	0.0215 ± 0.0003 [0.0226]
PSO	0.0145 ± 0.0002 [0.0001]	0.0145 ± 0.0003 [0.0001]	0.0550 ± 0.0008 [0.0545]	0.0440 ± 0.0010 [0.2341]	0.0233 ± 0.0003 [0.0001]	0.0217 ± 0.0003 [0.0054]

As Semeion poses a more challenging task (higher reconstruction error), it is possible to perceive that Dropout played a meaningful role and that the meta-heuristics (BA, FA, and PSO) could achieve the best results, especially in the architecture with a more significant number of hidden units. Such a fact is explained due to the network's being more prone to overfit with a larger number of neurons. Additionally, the last point to highlight is that no meta-heuristic found Dropout probabilities higher than 0.2447, being far from the commonly employed Dropout ratio (0.5). Hence we opted not to train a network using such values as they would diminish the model's performance.

5.2   Deep Belief Networks

Table 4 describes the mean reconstruction error, and standard deviation found over the MNIST testing set using the mean Dropout probability p found over the validation set. It is possible to observe the maintenance of higher Dropout probabilities as we introduce more layers in the network architecture; however, too much Dropout inhibits the network from proper learning and often degrades its performance (CS and FA). On the other hand, it is possible to observe that the statistically best results by Wilcoxon (cells in bold) correspond to the meta-heuristics BA and PSO. Even though FA achieved comparable results in RBM experiments, its Dropout ratios are significantly higher than BA, and PSO, leading to a network that attained too much noise. Additionally, CS once again did not perform well due to its nest replacement approach.

Table 4. DBN mean reconstruction error and standard deviation [mean p] over MNIST testing set.

Model	# 1			# 2		
	1L	2L	3L	1L	2L	3L
DBN	0.0206 ± 0.0003 [-]	0.0374 ± 0.0005 [-, -]	0.0484 ± 0.0005 [-, -, -]	0.0147 ± 0.0004 [-]	0.0312 ± 0.0003 [-, -]	0.0420 ± 0.0004 [-, -, -]
BA	0.0214 ± 0.0020 [0.0188]	0.0376 ± 0.0004 [0.0001, 0.0013]	0.0488 ± 0.0012 [0.0219, 0.0335, 0.0244]	0.0144 ± 0.0002 [0.0022]	0.0321 ± 0.0023 [0.0604, 0.0979]	0.0440 ± 0.0046 [0.0883, 0.2546, 0.1147]
CS	0.0270 ± 0.0061 [0.1831]	0.0456 ± 0.0058 [0.1780, 0.3087]	0.0540 ± 0.0024 [0.1643, 0.3232, 0.3091]	0.0178 ± 0.0041 [0.1984]	0.0369 ± 0.0049 [0.2444, 0.3480]	0.0449 ± 0.0042 [0.1452, 0.3542, 0.1983]
FA	0.0205 ± 0.0003 [0.0001]	0.0409 ± 0.0033 [0.0744, 0.1402]	0.0503 ± 0.0017 [0.0749, 0.1218, 0.2124]	0.0147 ± 0.0005 [0.0334]	0.0322 ± 0.0012 [0.1036, 0.1538]	0.0413 ± 0.0006 [0.1258, 0.2903, 0.0641]
PSO	0.0209 ± 0.0003 [0.0001]	0.0375 ± 0.0005 [0.0001, 0.0004]	0.0483 ± 0.0004 [0.0001, 0.0001, 0.0044]	0.0144 ± 0.0002 [0.0198]	0.0313 ± 0.0003 [0.0057, 0.0035]	0.0417 ± 0.0007 [0.0357, 0.2173, 0.0267]

Table 5 describes the mean reconstruction error, and standard deviation found over the Semeion testing set using the mean Dropout probability p found over the validation set. Semeion is a more complicated task, thus having higher reconstruction errors. However, the introduction of more layers resulted in the opposite of what was expected, i.e., test errors increased considerably, and the regularization probabilities were not as efficient as desirable. Such behavior is explainable due to the datasets' small amount of samples and its size, thus allowing the network to extract the essential features in its first layer easily and not demand additional layers.

Table 5. DBN mean reconstruction error and standard deviation [mean p] over Semeion testing set.

Model	# 1			# 2		
	1L	2L	3L	1L	2L	3L
DBN	0.0769 ± 0.0007 [-]	0.1044 ± 0.0010 [-, -]	0.1200 ± 0.0017 [-, -, -]	0.0549 ± 0.0012 [-]	0.0828 ± 0.0013 [-, -]	0.0984 ± 0.0015 [-, -, -]
BA	0.0767 ± 0.0010 [0.0001]	0.1053 ± 0.0015 [0.0117, 0.0106]	0.1232 ± 0.0092 [0.0597, 0.0325, 0.0649]	0.0550 ± 0.0009 [0.0750]	0.0828 ± 0.0016 [0.0511, 0.0110]	0.0997 ± 0.0036 [0.0258, 0.0907, 0.0863]
CS	0.0788 ± 0.0014 [0.0471]	0.1151 ± 0.0062 [0.1438, 0.2402]	0.1346 ± 0.0060 [0.2503, 0.2311, 0.3239]	0.0568 ± 0.0055 [0.1352]	0.0893 ± 0.0051 [0.1738, 0.3174]	0.1059 ± 0.0080 [0.2026, 0.2574, 0.1979]
FA	0.0769 ± 0.0008 [0.0001]	0.1091 ± 0.0049 [0.0457, 0.1246]	0.1249 ± 0.0036 [0.0870, 0.1478, 0.1200]	0.0554 ± 0.0008 [0.0923]	0.0837 ± 0.0011 [0.0927, 0.1088]	0.1015 ± 0.0046 [0.1199, 0.2128, 0.2122]
PSO	0.0769 ± 0.0007 [0.0001]	0.1048 ± 0.0009 [0.0001, 0.0024]	0.1200 ± 0.0014 [0.0011, 0.0009, 0.0042]	0.0549 ± 0.0010 [0.0494]	0.0827 ± 0.0012 [0.0109, 0.0088]	0.0992 ± 0.0031 [0.0068, 0.0903, 0.0482]

Finally, Table 6 describes the mean reconstruction error, and standard deviation found over the MNIST testing set using the mean Dropout probability p found over the validation set. It is possible to observe that the introduction of layers to the architecture also attenuated the learning process, increasing the error rates in the testing set. Notably, the USPS also has a smaller sample resolution, withdrawing the necessity of employing deeper architectures to fulfill the task.

Table 6. DBN mean reconstruction error and standard deviation [mean p] over USPS testing set.

Model	#1			#2		
	1L	2L	3L	1L	2L	3L
DBN	0.0373 ± 0.0003 [-]	0.0549 ± 0.0004 [-,-]	0.0688 ± 0.0005 [-,-,-]	0.0233 ± 0.0003 [-]	0.0414 ± 0.0003 [-,-]	0.0535 ± 0.0004 [-,-,-]
BA	0.0373 ± 0.0003 [0.0001]	0.0568 ± 0.0067 [0.0054, 0.0517]	0.0751 ± 0.0096 [0.0995, 0.1580, 0.0940]	0.0232 ± 0.0002 [0.0004]	0.0412 ± 0.0004 [0.0001, 0.0007]	0.0578 ± 0.0056 [0.1776, 0.1571, 0.1252]
CS	0.0436 ± 0.0047 [0.1252]	0.0686 ± 0.0064 [0.2857, 0.1804]	0.0829 ± 0.0086 [0.2602, 0.2968, 0.1635]	0.0254 ± 0.0016 [0.1248]	0.0470 ± 0.0039 [0.1410, 0.3177]	0.0629 ± 0.0098 [0.2111, 0.2633, 0.2903]
FA	0.0374 ± 0.0007 [0.0035]	0.0580 ± 0.0029 [0.0339, 0.0776]	0.0733 ± 0.0036 [0.0844, 0.1003, 0.1550]	0.0234 ± 0.0002 [0.0030]	0.0419 ± 0.0018 [0.0439, 0.0444]	0.0556 ± 0.0016 [0.1374, 0.2254, 0.1508]
PSO	0.0372 ± 0.0003 [0.0001]	0.0545 ± 0.0004 [0.0001, 0.0001]	0.0691 ± 0.0005 [0.0105, 0.0001, 0.0001]	0.0232 ± 0.0003 [0.0001]	0.0412 ± 0.0005 [0.0001, 0.0001]	0.0534 ± 0.0004 [0.0243, 0.0371, 0.0001]

6 Conclusion

This work proposed to adequately select Dropout hyperparameters concerning RBMs and DBNs through meta-heuristic optimization algorithms, such as Particle Swarm Optimization, Bat Algorithm, Cuckoo Search, and Firefly Algorithm. The experiments were carried out over three literature datasets in the context of image reconstruction. The experimental section comprised different energy-based models, images with different resolutions, and datasets with distinct sizes. The results obtained by the meta-heuristics were compared against standard Dropout-less networks and showed to be promising alternatives to empirically setting Dropout ratios. On the other hand, the optimization task required a higher computational load than non-optimization models as each particle's fitness update needs to be evaluated under a DL architecture; thus, it took a longer time to find suitable Dropout hyperparameters. Regarding future works, we intend to investigate the proposed approach with other meta-heuristic algorithms, extend it to Deep Boltzmann Machines, and employ meta-heuristic optimization to select other regularization techniques hyperparameters.

References

1. Bergstra, J., Bengio, Y.: Random search for hyper-parameter optimization. J. Mach. Learn. Res. **13**(null), 281–305 (2012)
2. De Rosa, G.H., Papa, J.P., Yang, X.S.: Handling dropout probability estimation in convolution neural networks using meta-heuristics. Soft Comput. **22**(18), 6147–6156 (2018)
3. Hinton, G.E.: A practical guide to training restricted Boltzmann machines. In: Montavon, G., Orr, G.B., Müller, K.-R. (eds.) Neural Networks: Tricks of the Trade. LNCS, vol. 7700, pp. 599–619. Springer, Heidelberg (2012). https://doi.org/10.1007/978-3-642-35289-8_32
4. Hinton, G.E., Osindero, S., Teh, Y.W.: A fast learning algorithm for deep belief nets. Neural Comput. **18**(7), 1527–1554 (2006)
5. Hoerl, A.E., Kennard, R.W.: Ridge regression: biased estimation for nonorthogonal problems. Technometrics **12**(1), 55–67 (1970)
6. Kennedy, J., Eberhart, R.C., Russel, C., Kennedy, J.F., Shi, Y.: Swarm Intelligence. Morgan Kaufmann (2001)
7. LeCun, Y., Bottou, L., Bengio, Y., Haffner, P.: Gradient-based learning applied to document recognition. Proc. IEEE **86**(11), 2278–2324 (1998)
8. Lee, H.W., Kim, N.R., Lee, J.H.: Deep neural network self-training based on unsupervised learning and dropout. Int. J. Fuzzy Logic Intell. Syst. **17**(1), 1–9 (2017)

9. Mosavi, A., Ardabili, S., Várkonyi-Kóczy, A.R.: List of deep learning models. In: Várkonyi-Kóczy, A.R. (ed.) INTER-ACADEMIA 2019. LNNS, vol. 101, pp. 202–214. Springer, Cham (2020). https://doi.org/10.1007/978-3-030-36841-8_20
10. Nowlan, S.J., Hinton, G.E.: Simplifying neural networks by soft weight-sharing. Neural Comput. 4(4), 473–493 (1992). https://doi.org/10.1162/neco.1992.4.4.473
11. O'Mahony, N., et al.: Deep learning vs. traditional computer vision. In: Arai, K., Kapoor, S. (eds.) CVC 2019. AISC, vol. 943, pp. 128–144. Springer, Cham (2020). https://doi.org/10.1007/978-3-030-17795-9_10
12. Papa, J.P., Rosa, G.H., Marana, A.N., Scheirer, W., Cox, D.D.: Model selection for discriminative restricted Boltzmann machines through meta-heuristic techniques. J. Comput. Sci. 9, 14–18 (2015)
13. Papa, J.P., Scheirer, W., Cox, D.D.: Fine-tuning deep belief networks using harmony search. Appl. Soft Comput. 46(C), 875–885 (2016)
14. Roelofs, R., et al.: A meta-analysis of overfitting in machine learning. Adv. Neural Inf. Process. Syst. 32, 9179–9189 (2019)
15. Smolensky, P.: Parallel Distributed Processing: Explorations in the Microstructure of Cognition, vol. 1. MIT Press (1986)
16. Srivastava, N., Hinton, G.E., Krizhevsky, A., Sutskever, I., Salakhutdinov, R.: Dropout: a simple way to prevent neural networks from overfitting. J. Mach. Learn. Res. 15(1), 1929–1958 (2014)
17. Su, J., Thomas, D.B., Cheung, P.Y.K.: Increasing network size and training throughput of FPGA restricted Boltzmann machines using dropout. In: 2016 IEEE 24th Annual International Symposium on Field-Programmable Custom Computing Machines (FCCM), pp. 48–51 (2016)
18. Tibshirani, R.: Regression shrinkage and selection via the lasso. J. R. Stat. Soc. Ser. B (Methodol.) 58(1), 267–288 (1996)
19. Wang, B., Klabjan, D.: Regularization for unsupervised deep neural nets. In: Thirty-First AAAI Conference on Artificial Intelligence (2017)
20. Wang, S., Manning, C.: Fast dropout training. In: Proceedings of the 30th International Conference on Machine Learning, pp. 118–126 (2013)
21. Wang, X., Zhao, Y., Pourpanah, F.: Recent advances in deep learning (2020)
22. Wilcoxon, F.: Individual comparisons by ranking methods. Biometrics Bull. 1(6), 80–83 (1945)
23. Xiong, H.Y., Barash, Y., Frey, B.J.: Bayesian prediction of tissue-regulated splicing using RNA sequence and cellular context. Bioinformatics 27(18), 2554–2562 (2011)
24. Yang, X.S.: Firefly algorithm, stochastic test functions and design optimisation. Int. J. Bio-Inspired Comput. 2(2), 78–84 (2010)
25. Yang, X.S.: A new metaheuristic bat-inspired algorithm. In: González, J.R., Pelta, D.A., Cruz, C., Terrazas, G., Krasnogor, N. (eds.) Nature Inspired Cooperative Strategies for Optimization (NICSO 2010), pp. 65–74. Springer, Heidelberg (2010). https://doi.org/10.1007/978-3-642-12538-6_6
26. Yang, X.S., Deb, S.: Engineering optimisation by cuckoo search. Int. J. Math. Model. Numer. Optim. 1(4), 330–343 (2010)

Enhancing Hyper-to-Real Space Projections Through Euclidean Norm Meta-heuristic Optimization

Luiz Carlos Felix Ribeiro[ID], Mateus Roder[ID], Gustavo H. de Rosa[(⊠)][ID], Leandro A. Passos[ID], and João P. Papa[ID]

Department of Computing, São Paulo State University, Bauru, Brazil
{luiz.felix,mateus.roder,gustavo.rosa,leandro.passos,joao.papa}@unesp.br

Abstract. The continuous computational power growth in the last decades has made solving several optimization problems significant to humankind a tractable task; however, tackling some of them remains a challenge due to the overwhelming amount of candidate solutions to be evaluated, even by using sophisticated algorithms. In such a context, a set of nature-inspired stochastic methods, called meta-heuristic optimization, can provide robust approximate solutions to different kinds of problems with a small computational burden, such as derivative-free real function optimization. Nevertheless, these methods may converge to inadequate solutions if the function landscape is too harsh, e.g., enclosing too many local optima. Previous works addressed this issue by employing a hypercomplex representation of the search space, like quaternions, where the landscape becomes smoother and supposedly easier to optimize. Under this approach, meta-heuristic computations happen in the hypercomplex space, whereas variables are mapped back to the real domain before function evaluation. Despite this latter operation being performed by the Euclidean norm, we have found that after the optimization procedure has finished, it is usually possible to obtain even better solutions by employing the Minkowski p-norm instead and fine-tuning p through an auxiliary sub-problem with neglecting additional cost and no hyperparameters. Such behavior was observed in eight well-established benchmarking functions, thus fostering a new research direction for hypercomplex meta-heuristic optimization.

Keywords: Hypercomplex space · Real-valued projection · Euclidean norm · Meta-heuristic optimization · Benchmarking functions

The authors would like to thank São Paulo Research Foundation (FAPESP) grants #2013/07375-0, #2014/12236-1, #2017/25908-6, #2019/02205-5, #2019/07825-1, and #2019/07665-4, and National Council for Scientific and Technological Development (CNPq) grants #307066/2017-7 and #427968/2018-6.

© Springer Nature Switzerland AG 2021
J. M. R. S. Tavares et al. (Eds.): CIARP 2021, LNCS 12702, pp. 109–118, 2021.
https://doi.org/10.1007/978-3-030-93420-0_11

1   Introduction

Humanity sharpened their mathematical skills over several years of evolution by researching and studying formal and elegant tools to model world events' behavior. In such a context, when dealing with non-trivial problems, it is common to apply mathematical programming to overcome the before-mentioned tasks or even to streamline the process. Furthermore, once any prior knowledge might not be available, mathematical programming, commonly known as optimization [19], provides an attractive approach to tackle the burden of empirical setups.

In the past decades, a new optimization paradigm called meta-heuristic has been used to solve several optimization problems [21]. Essentially, a meta-heuristic is a high-level abstraction of a procedure that generates and selects a heuristic that aims to provide a feasible solution to the problem. It combines concepts of *exploration* and *exploitation*, i.e., globally searching throughout the space and enhancing a promising solution based on its neighbors, respectively, constituted of complex learning procedures and simple searches usually inspired by biological behaviors. Additionally, they do not require specific domain knowledge and provide mechanisms to avoid susceptibility to local optima convergence.

Although meta-heuristic techniques seem to be an exciting proposal, they still might perform poorly on challenging objective functions, being trapped in local optima and not achieving the most suitable solutions. Some attempts as hybrid variants [12], aging mechanisms [1], and fitness landscape analysis [3] try to deal with this issue. Relying on more robust search spaces, such as representing each decision variable as a hypercomplex number, is an alternative approach that is not fully explored in the literature.

One can perceive that handling hypercomplex spaces is based on the likelihood of having more natural fitness landscapes, although mathematically not proved yet. The most common representations are the quaternions [7] and octonions [6], which have compelling traits to describe the object's orientation in n-dimensional search spaces, being extremely useful in performing rotations in such spaces [8]. These representations have been successfully used in different areas, as in deep learning [14], feature selection [17], special relativity [11] and quantum mechanics [4]. Regarding meta-heuristic optimization, interesting results have been achieved in global optimization [5,13,16], although not yet mathematically guaranteed.

Notwithstanding, hypercomplex optimization also has its particular problems, i.e., before attempting to feed quaternions or octonions to a real-valued objective function, one needs to project their values onto a real-valued space, usually accomplished by the Euclidean norm function. However, to the best of our knowledge, there is no work in the literature regarding how using the standard Euclidean norm function might affect the loss of information when projecting one space onto another. Thus, we are incredibly interested in exploring the possibility of employing the p-norm function and finding the most suitable p value that minimizes the loss of information throughout the projection.

In this work, we investigate how employing the p-norm to refine the solution found by a standard hypercomplex meta-heuristic can affect the obtained result. In short, we optimize a real function using the standard quaternion-based variant of the Particle Swarm Algorithm (Q-PSO) [15], i.e., meta-heuristic operations are performed in the hypercomplex space. In contrast, decision variables are mapped to the real domain through the Euclidean Norm for function evaluation. Notwithstanding, the best solution is refined by finding a more suitable projection between domains using the p-norm. The rationale for this decision lies in the fact that this operation is a Euclidean norm generalization. Hence we resort to fine-tuning new, yet not explored, hyperparameter in the optimization procedure, thus allowing more robust solutions to be found. Regardless, such a procedure can be applied to any hypercomplex-based meta-heuristic. Therefore, this work's main contributions are twofold: (i) to introduce a generic and inexpensive procedure to refine solutions found by hypercomplex meta-heuristics; and (ii) to foster research regarding how to map better hypercomplex to real values in the context of meta-heuristic optimization.

The remainder of this paper is organized as follows. Sections 2 and 3 present the theoretical background related to hypercomplex-based spaces (quaternions and Minkowski p-norm) and meta-heuristic optimization, respectively. Section 4 discusses the methodology adopted in this work, while Sect. 5 presents the experimental results. Finally, Sect. 6 states conclusions and future works[1].

2 Hypercomplex Representation

A quaternion q is a hypercomplex number, composed of real and complex parts, being $q = a + bi + cj + dk$, where $a, b, c, d \in \mathbb{R}$ and i, j, k are fundamental quaternions units. The basis equation that defines what a quaternion looks like is described as follows:

$$i^2 = j^2 = k^2 = ijk = -1. \tag{1}$$

Essentially, a quaternion q is a four-dimensional space representation over the real numbers, i.e., $\mathbb{R}^4$. Given two arbitrary quaternions $q_1 = a + bi + cj + dk$ and $q_2 = \alpha + \beta i + \gamma j + \delta k$ and a scalar $\kappa \in \mathbb{R}$, we define the quaternion algebra [2] used throughout this work:

$$
\begin{aligned}
q_1 + q_2 &= (a + bi + cj + dk) + (\alpha + \beta i + \gamma j + \delta k) \\
&= (a + \alpha) + (b + \beta)i + (c + \gamma)j + (d + \delta)k,
\end{aligned} \tag{2}
$$

$$
\begin{aligned}
q_1 - q_2 &= (a + bi + cj + dk) - (\alpha + \beta i + \gamma j + \delta k) \\
&= (a - \alpha) + (b - \beta)i + (c - \gamma)j + (d - \delta)k,
\end{aligned} \tag{3}
$$

$$
\begin{aligned}
\kappa q_1 &= \kappa(a + bi + cj + dk) \\
&= \kappa a + (\kappa b)i + (\kappa c)j + (\kappa d)k.
\end{aligned} \tag{4}
$$

---

[1] The source code is available online at https://github.com/lzfelix/lio.

2.1 Minkowski p-norm

Another crucial operator that needs to be defined is the p-norm, which is responsible for mapping hypercomplex values to real numbers. Let q be a hypercomplex number with real coefficients $\{z_d\}_{d=0}^{D-1}$, one can compute the Minkowski p-norm as follows:

$$\|q\|_p = \left(\sum_{d=0}^{D-1} |z_d|^p \right)^{1/p} , \tag{5}$$

where D is the number of dimensions of the space (2 for complex numbers, and 4 for quaternions, for instance) and $p \geq 1$. Common values for the latter variable are 1 or 2 for the Taxicab and Euclidean norms, respectively. Hence, one can see the p-norm as a generalization of such norm operators.

3 Meta-heuristic Optimization

Optimization is the task of selecting a solution that best fits a function among a set of possible solutions. Several methods have been applied in this context, such as grid-search and gradient-based methods. Nevertheless, these methods carry a massive amount of computation, leading to burdened states in more complex problems, e.g., exponential and NP-complete problems.

An attempt to overcome such behaviors is to employ a meta-heuristic-based approach. Meta-heuristic techniques are nature-inspired stochastic algorithms that mimic an intelligence behavior, often observed in groups of animals, humans, or nature. Such approaches combine exploration and exploitation mechanisms in order to achieve sub-optimal solutions with low effort.

In this work, we employed the quaternion variant of the state-of-the-art Particle Swarm Optimization (PSO) [10] algorithm for function optimization. On the other hand, since fine-tuning the p hyperparameter is a single-variable optimization task with a small search interval, we resort to the hyperparameter-less Black Hole (BH) [9] algorithm.

4 Methodology

This section discusses how the presented meta-heuristics can be combined with quaternions to perform the so-called "hypercomplex-based meta-heuristic optimization". The proposed approach designated "Last Iteration Optimization" (LIO) is presented along with the considered benchmarking functions to evaluate it and the experimental setup.

4.1 Hypercomplex Optimization

In their original formulation, meta-heuristic algorithms were conceived to optimize real-valued target functions with multiple real parameters. However, one may decide to represent each decision variable as quaternions.

In this case, each decision variable is represented by a quaternion with its real coefficients randomly initialized from a uniform distribution in the interval $[0, 1]$. Furthermore, the mapping from quaternions to real numbers for function evaluation becomes a paramount operation, which is usually carried out through the Euclidean norm. Still, care must be taken to ensure that this transformation does not yield numbers outside the feasibility region. Hence, hypercomplex coefficients are clipped individually to the real interval $[0, 1]$ and the mapping for each decision variable is performed by the following mapping function:

$$\hat{q}_j = M(q_j, p)$$
$$= l_j + (u_j - l_j) \frac{\|q_j\|_p}{D^{1/p}}, \tag{6}$$

such that $j = \{1, 2, \ldots, n\}$, D is the number of hypercomplex dimensions (4 for quaternions), l_j and u_j are the lower and upper bounds for each decision variable, respectively, and $p = 2$ in this particular case.

4.2   Last Iteration Optimization

The main goal of this work consists of refining the solution found by a hypercomplex-based meta-heuristic using a low-cost procedure. To such an extent, given a fitness function $f : \mathbb{R}^n \to \mathbb{R}$, we first optimize it through the Q-PSO algorithm, which consists in representing each decision variable as a quaternion with the relations defined in Eqs. 2, 3, and 4. Once this step is finished, we have the best candidate solution $q^\star$ with a real representation $\hat{q}^\star \in \mathbb{R}^n$, which is obtained through Eq. 6 with $p = 2$. Shortly, one can compute the best solution fitness μ as follows:

$$\mu = f\Big(M(q_1^\star, 2), M(q_2^\star, 2), \ldots, M(q_n^\star, 2)\Big). \tag{7}$$

where $M(\cdot)$ is computed according to Eq. 6.

We propose a second phase to the optimization pipeline, where the best solution found is $q^\star$ is kept fixed, while the hyper-parameter p is unfrozen. Such an approach allows obtaining a better real representation of $q^\star$, which translates to an even smaller fitness value $\mu^\star$. Namely, we aim at solving the following auxiliary optimization problem:

$$p^\star = \arg \min_p f\Big(M(q_1^\star, p), M(q_2^\star, p), \ldots, M(q_n^\star, p)\Big),$$
$$\text{st. } 1 \leq p \leq p_{\max}, \tag{8}$$

where $p_{\max}$ denotes the maximum possible value for parameter p. If $p_{\max} = 2$, for instance, the problem consists in finding a suitable norm between the Taxicab and Euclidean ones.

Since the new search interval is usually small, as it is going to be discussed in Sect. 4.4, we resort to the traditional BH algorithm since it does not contain hyperparameters to be tuned, thus making the process even simpler. As this procedure is performed for a single decision variable in a small search space, the time spent in this phase is negligible compared to the Q-PSO step. Furthermore, since this new step is performed as the new last iteration of the optimization pipeline, we name it Last Iteration Optimization (LIO).

4.3   Benchmarking Functions

Table 1 introduces the eight benchmarking functions used to evaluate the proposed approach.

Table 1. Benchmarking functions.

Function	Equation	Bounds	f(x*)
Sphere	$f_1(x) = \sum\limits_{i=1}^{n} x_i^2$	$-10 \leq x_i \leq 10$	0
Csendes	$f_2(x) = \sum_{i=1}^{n} x_i^6 \left(2 + sin\frac{1}{x_i}\right)$	$-1 \leq x_i \leq 1$	0
Salomon	$f_3(x) = 1 - \cos(2\pi\sqrt{\sum_{i=1}^{n} x_i^2}) + 0.1\sqrt{\sum_{i=1}^{n} x_i^2}$	$-100 \leq x_i \leq 100$	0
Ackley #1	$f_4(x) = -20e^{-0.02\sqrt{n^{-1}\sum_{i=1}^{n} x_i^2}} - e^{n^{-1}\sum_{i=1}^{n} cos(2\pi x_i)} + 20 + e$	$-35 \leq x_i \leq 35$	0
Alpine #1	$f_5(x) = \sum_{i=1}^{n} |x_i sin(x_i) + 0.1x_i|$	$-10 \leq x_i \leq 10$	0
Rastrigin	$f_6(x) = 10n + \sum_{i=1}^{n} \left[x_i^2 - 10cos(2\pi x_i)\right]$	$-5.12 \leq x_i \leq 5.12$	0
Schwefel	$f_7(x) = \left(\sum\limits_{i=1}^{n} x_i^2\right)^{\sqrt{\pi}}$	$-100 \leq x_i \leq 100$	0
Brown	$f_8(x) = \sum_{i=1}^{n-1} \left[(x_i^2)^{(x_{i+1}^2+1)} + (x_{i+1}^2)^{(x_i^2+1)}\right]$	$-1 \leq x_i \leq 4$	0

4.4   Experimental Setup

The proposed approach divides function optimization into two parts: global and fine-tuning phases, which correspond to finding μ using Q-PSO and $\mu^\star$ by solving Eq. 8 through the BH algorithm.

Regarding the first phase, we use the same experimental setup from [16]. Namely, each benchmark function is optimized with $n \in \{10, 25, 50, 100\}$ decision variables, for $(2000 \cdot n)$ iterations using 100 agents. As the amount of iterations grows considerably fast, we adapt to the early stopping mechanism. Such a strategy allows detecting if the optimization is stuck for too long in a local optimum and unlikely to find a better solution, saving computational time. If the difference of fitness between two consecutive iterations is smaller than $\delta = 10^{-5}$ for 50 iterations or more, the optimization is halted, and the best fitness found so far is deemed the solution. Despite these values being determined empirically, they often present the same results as those obtained using all available iterations, despite using, at most 4% of all iterations for the extreme case when $n = 100$. For the Q-PSO hyperparameters we use $w = 0.7$, $c_1 = c_2 = 1.7$, as well established in the literature.

In the second phase, optimization is performed with $p_{max} = 5$, using 20 agents for 50 iterations, which were determined on preliminary experiments. Further, we do not rely on early stopping for this phase since it is performed much faster than the previous one. Finally, we compare the results obtained by Q-PSO and Q-PSO with LIO. Each experiment is executed 15 times, and the best results with significance smaller than 0.05, according to the Wilcoxon signed-rank [20], are highlighted in bold. Regarding the implementation, we used Opytimizer [18] library.

5   Experimental Results

Experimental results are presented in Table 2, where the average fitness values obtained by Q-PSO are compared against their refined versions, computed with LIO. More specifically, we ran Q-PSO, stored the results, and continued the LIO (denoted as Q-PSO + LIO).

5.1   Overall Discussion

Experimental results provided in Table 2 confirm the robustness of the proposed approach since the Q-PSO + LIO outperformed the standard Q-PSO in the massive majority of benchmarking functions and configurations. One can highlight, for instance, that LIO obtained the best results overall, considering all dimensional configurations, in half of the functions, i.e., Sphere, Csendes, Schwefel, and Brown. Besides, Alpine1 and Rastrigin can also be deliberated, although Q-PSO obtained similar statistical results. Further, LIO also obtained the best results considering all functions over three-out-of-four configurations, i.e., 25, 50, and 100 dimensions, being Q-PSO statistical similar in only two of them.

On the other hand, Q-PSO obtained the best results over two functions, i.e., Salomon and Rastrigin, considering a 10-dimensional configuration. Such behavior is very interesting since Q-PSO performed better over two functions who share similar characteristics: both are continuous, differentiable, non-separable, scalable, and multimodal, contemplating the same dimensionality, which may denote some specific constraint to the model.

Finally, as an overview, the proposed approach can significantly improve Q-PSO, with an almost insignificant computational burden, and whose growth is barely insignificant compared to the increase in the number of dimensions, as discussed in the next section.

Table 2. Best fitness found by varying the number of decision variables for each benchmark function.

Functions	Dimensions	Q-PSO	Q-PSO + LIO	p	Q-PSO time (s)	LIO time (s)
Sphere	10	$1.3447 \cdot 10^{-7} \pm 1.8964 \cdot 10^{-7}$	$\mathbf{1.2169 \cdot 10^{-7} \pm 1.6903 \cdot 10^{-7}}$	1.99 ± 5.78	5.66 ± 0.35	0.29 ± 0.01
	25	$3.1993 \cdot 10^{-1} \pm 1.9855 \cdot 10^{-1}$	$\mathbf{3.0657 \cdot 10^{-1} \pm 1.9018 \cdot 10^{-1}}$	1.98 ± 0.06	33.08 ± 3.93	0.30 ± 0.01
	50	$5.3962 \cdot 10^{0} \pm 2.0896 \cdot 10^{0}$	$\mathbf{5.3205 \cdot 10^{0} \pm 2.0266 \cdot 10^{0}}$	2.01 ± 0.19	77.55 ± 14.45	0.37 ± 0.01
	100	$2.3679 \cdot 10^{1} \pm 3.3931 \cdot 10^{0}$	$\mathbf{2.3433 \cdot 10^{1} \pm 3.4593 \cdot 10^{0}}$	1.84 ± 0.38	158.04 ± 24.79	0.40 ± 0.01
Csendes	10	$4.1345 \cdot 10^{-13} \pm 1.2771 \cdot 10^{-12}$	$\mathbf{2.3502 \cdot 10^{-13} \pm 6.5480 \cdot 10^{-13}}$	1.99 ± 0.00	2.43 ± 0.12	0.31 ± 0.01
	25	$6.2160 \cdot 10^{-7} \pm 6.2587 \cdot 10^{-7}$	$\mathbf{5.8670 \cdot 10^{-7} \pm 5.6680 \cdot 10^{-7}}$	1.97 ± 0.11	6.16 ± 0.35	0.34 ± 0.02
	50	$8.5587 \cdot 10^{-6} \pm 6.5273 \cdot 10^{-6}$	$\mathbf{8.3406 \cdot 10^{-6} \pm 6.2563 \cdot 10^{-6}}$	2.01 ± 0.05	13.31 ± 0.71	0.39 ± 0.01
	100	$4.9290 \cdot 10^{-5} \pm 2.3214 \cdot 10^{-5}$	$\mathbf{4.8460 \cdot 10^{-5} \pm 2.3491 \cdot 10^{-5}}$	1.94 ± 0.15	31.88 ± 2.95	0.48 ± 0.02
Salomon	10	$\mathbf{3.4654 \cdot 10^{-1} \pm 1.0242 \cdot 10^{-1}}$	$3.4655 \cdot 10^{-1} \pm 1.0243 \cdot 10^{-1}$	2.04 ± 0.08	4.97 ± 0.58	0.30 ± 0.02
	25	$2.0332 \cdot 10^{0} \pm 4.3919 \cdot 10^{-1}$	$\mathbf{2.0199 \cdot 10^{0} \pm 4.4000 \cdot 10^{-1}}$	2.07 ± 0.29	11.73 ± 1.78	0.31 ± 0.01
	50	$\mathbf{4.1332 \cdot 10^{0} \pm 5.6529 \cdot 10^{-1}}$	$4.1000 \cdot 10^{0} \pm 5.4903 \cdot 10^{-1}$	2.22 ± 0.52	24.11 ± 3.30	0.36 ± 0.01
	100	$6.3799 \cdot 10^{0} \pm 7.2682 \cdot 10^{-1}$	$\mathbf{6.3665 \cdot 10^{0} \pm 7.0016 \cdot 10^{-1}}$	2.07 ± 0.46	58.78 ± 9.67	0.44 ± 0.02
Ackley #1	10	$\mathbf{8.7839 \cdot 10^{-1} \pm 2.9911 \cdot 10^{-1}}$	$8.7843 \cdot 10^{-1} \pm 2.9914 \cdot 10^{-1}$	2.00 ± 0.00	7.61 ± 1.23	0.35 ± 0.02
	25	$1.2330 \cdot 10^{0} \pm 2.5165 \cdot 10^{-1}$	$\mathbf{1.2293 \cdot 10^{0} \pm 2.5283 \cdot 10^{-1}}$	1.99 ± 0.00	35.09 ± 5.11	0.38 ± 0.02
	50	$1.8135 \cdot 10^{0} \pm 1.8909 \cdot 10^{-1}$	$\mathbf{1.8062 \cdot 10^{0} \pm 1.9024 \cdot 10^{-1}}$	2.00 ± 0.01	61.51 ± 9.02	0.40 ± 0.02
	100	$2.2038 \cdot 10^{0} \pm 1.0338 \cdot 10^{-1}$	$\mathbf{2.2006 \cdot 10^{0} \pm 1.0280 \cdot 10^{-1}}$	2.00 ± 0.01	125.12 ± 14.76	0.49 ± 0.03
Alpine #1	10	$\mathbf{7.9560 \cdot 10^{-2} \pm 1.2551 \cdot 10^{-1}}$	$7.9515 \cdot 10^{-2} \pm 1.2565 \cdot 10^{-1}$	2.00 ± 0.00	9.95 ± 2.36	0.29 ± 0.02
	25	$2.2345 \cdot 10^{0} \pm 1.3898 \cdot 10^{0}$	$\mathbf{2.2240 \cdot 10^{0} \pm 1.3957 \cdot 10^{0}}$	2.01 ± 0.05	27.84 ± 3.47	0.31 ± 0.02
	50	$1.2124 \cdot 10^{1} \pm 4.1844 \cdot 10^{0}$	$\mathbf{1.2088 \cdot 10^{1} \pm 4.1675 \cdot 10^{0}}$	2.00 ± 0.05	63.12 ± 11.53	0.35 ± 0.01
	100	$\mathbf{2.7375 \cdot 10^{1} \pm 7.7322 \cdot 10^{0}}$	$2.7322 \cdot 10^{1} \pm 7.7351 \cdot 10^{0}$	1.95 ± 0.20	124.72 ± 20.72	0.42 ± 0.02
Rastrigin	10	$\mathbf{1.1608 \cdot 10^{1} \pm 5.1079 \cdot 10^{0}}$	$1.1608 \cdot 10^{1} \pm 5.1079 \cdot 10^{0}$	2.00 ± 0.00	7.21 ± 0.60	0.33 ± 0.02
	25	$3.7845 \cdot 10^{1} \pm 1.3993 \cdot 10^{1}$	$\mathbf{3.7673 \cdot 10^{1} \pm 1.4040 \cdot 10^{1}}$	1.99 ± 0.01	39.67 ± 5.66	0.33 ± 0.02
	50	$1.4677 \cdot 10^{2} \pm 2.3428 \cdot 10^{1}$	$\mathbf{1.4506 \cdot 10^{2} \pm 2.3524 \cdot 10^{1}}$	2.00 ± 0.04	94.89 ± 20.97	0.37 ± 0.02
	100	$4.8812 \cdot 10^{2} \pm 4.7400 \cdot 10^{1}$	$\mathbf{4.8437 \cdot 10^{2} \pm 4.7186 \cdot 10^{1}}$	1.99 ± 0.05	208.48 ± 39.45	0.44 ± 0.02
Schwefel	10	$4.9247 \cdot 10^{-9} \pm 9.7025 \cdot 10^{-9}$	$\mathbf{3.5510 \cdot 10^{-9} \pm 7.0349 \cdot 10^{-9}}$	1.99 ± 7.36	6.20 ± 0.43	0.31 ± 0.01
	25	$1.5213 \cdot 10^{3} \pm 2.5509 \cdot 10^{3}$	$\mathbf{1.1048 \cdot 10^{3} \pm 1.2858 \cdot 10^{3}}$	1.97 ± 0.07	64.56 ± 12.69	0.31 ± 0.01
	50	$9.4460 \cdot 10^{4} \pm 5.9571 \cdot 10^{4}$	$\mathbf{8.8915 \cdot 10^{4} \pm 5.3935 \cdot 10^{4}}$	1.88 ± 0.23	163.12 ± 52.00	0.35 ± 0.01
	100	$9.4876 \cdot 10^{5} \pm 3.8055 \cdot 10^{5}$	$\mathbf{9.0762 \cdot 10^{5} \pm 4.0064 \cdot 10^{5}}$	2.29 ± 0.56	349.40 ± 112.34	0.40 ± 0.02
Brown	10	$3.3230 \cdot 10^{0} \pm 4.3978 \cdot 10^{0}$	$\mathbf{1.0051 \cdot 10^{0} \pm 9.4007 \cdot 10^{-1}}$	1.58 ± 0.37	8.21 ± 3.22	0.33 ± 0.01
	25	$1.4622 \cdot 10^{1} \pm 7.9266 \cdot 10^{0}$	$\mathbf{5.6664 \cdot 10^{0} \pm 2.9834 \cdot 10^{0}}$	1.13 ± 0.11	56.16 ± 13.62	0.34 ± 0.02
	50	$2.8192 \cdot 10^{2} \pm 1.0655 \cdot 10^{2}$	$\mathbf{1.3212 \cdot 10^{2} \pm 5.4533 \cdot 10^{1}}$	1.01 ± 0.02	120.97 ± 36.77	0.41 ± 0.03
	100	$1.8173 \cdot 10^{3} \pm 3.1725 \cdot 10^{2}$	$\mathbf{1.2501 \cdot 10^{3} \pm 3.0196 \cdot 10^{2}}$	1.01 ± 0.01	245.45 ± 82.48	0.49 ± 0.02

5.2 Computational Burden

Germane to this aspect, the results in Table 2 show that LIO takes significantly less time than the main meta-heuristic to be evaluated. This phenomenon is expected since the latter involves solving an optimization problem with a single real variable in a small search interval. Nonetheless, despite this simplicity, our results show promising results by performing such a task. In the worst-case scenario, i.e., Csendes function with 10 variables, LIO takes only 12.6% of the time consumed by Q-PSO, which amounts to 0.31 seconds, while decreasing the fitness value by a factor of 1.7.

5.3 How Does p Influence Projections?

From the results in Table 2, one can highlight the variation in p-norm value. As expected, such a variable is highly correlated to the optimization performance, since small changes in its value resulted in better functions minima. On the other hand, one can notice that expressive changes in p may also support performance improvement, as in Brown function. Besides that, as p is changed, the mapping process, i.e., the projection, from the hypercomplex space to the real one becomes "less aggressive" to the latter, since the proposed approach gives margin to a smooth fit for the values obtained in the former space.

Therefore, examining the performance on the optimization functions, one can observe that employing LIO's projection, different optimization landscapes are achieved, and such a process provides better value's representation from the hypercomplex search space. It is worth observing that for Rastrigin, Alpine #1, and Ackley #1 functions, LIO found optimal p values with mean 2 and minimal standard deviations, thus showing this parameter's sensitiveness for some benchmarking functions. Moreover, only LIO optimization for the Schwefel function with 10 dimensions showed a large standard deviation for this hyperparameter. In contrast, in the remaining cases, there was no norm larger than 3, suggesting that in further experiments, and even smaller search intervals (with $p_{\max} = 3$, for instance) could be employed.

6 Conclusion

In this work, we introduced the Last Iteration Optimization (LIO) procedure, which consists of refining the solution found by a hypercomplex-based meta-heuristic optimization algorithm by solving a low-cost hyperparameter-less auxiliary problem after the primary heuristic has found the best candidate solution. Such a procedure provided robust results in various benchmarking functions, showing statistically significant gains in 24 out of 32 experiments, over functions with diverse characteristics. Since LIO has a low computational burden and is easy to implement, it can be readily incorporated into other works.

In future studies, we intend to investigate how changing the p parameter during the global optimization procedure can affect the obtained results. Furthermore, LIO can be extended to find a different p for each decision variable, making it more flexible, and even other functions can be employed (or learned) to perform the hypercomplex-to-real mapping process. Ultimately, fine-tuning the p hyper-parameter of the Minkowski norm opens new research directions for hypercomplex-based meta-heuristic function optimization methods.

References

1. Deng, L., Sun, H., Li, C.: JDF-DE: a differential evolution with Jrand number decreasing mechanism and feedback guide technique for global numerical optimization. Appl. Intell. **51**(1), 359–376 (2021)
2. Eberly, D.: Quaternion algebra and calculus. Tech. rep. Magic Software (2002)
3. Engelbrecht, A.P., Bosman, P., Malan, K.M.: The influence of fitness landscape characteristics on particle swarm optimisers. Natural Comput. 1–11 (2021). https://doi.org/10.1007/s11047-020-09835-x
4. Finkelstein, D., Jauch, J.M., Schiminovich, S., Speiser, D.: Foundations of quaternion quantum mechanics. J. Math. Phys. **3**(2), 207–220 (1962)
5. Fister, I., Yang, X.S., Brest, J., Fister Jr., I.: Modified firefly algorithm using quaternion representation. Expert Syst. Appl. **40**(18), 7220–7230 (2013). https://doi.org/10.1016/j.eswa.2013.06.070

6. Graves, J.T.: On a connection between the general theory of normal couples and the theory of complete quadratic functions of two variables. Philos. Mag. **26**(173), 315–320 (1845)
7. Hamilton, W.R.: On quaternions; or on a new system of imaginaries in algebra. London Edinburgh Dublin Philos. Mag. J. Sci. **25**(163), 10–13 (1844)
8. Hart, J.C., Francis, G.K., Kauffman, L.H.: Visualizing quaternion rotation. ACM Trans. Graph. (TOG) **13**(3), 256–276 (1994)
9. Hatamlou, A.: Black hole: a new heuristic optimization approach for data clustering. Inform. Sci. **222**, 175–184 (2013)
10. Kennedy, J.: Particle swarm optimization. In: Sammut, C., Webb, G.I. (eds.) Encyclopedia of Machine Learning, pp. 760–766. Springer, Boston (2011)
11. Leo, S.D.: Quaternions and special relativity. J. Math. Phys. **37**(6), 2955–2968 (1996)
12. Li, J., Lei, H., Alavi, A.H., Wang, G.G.: Elephant herding optimization: variants, hybrids, and applications. Mathematics **8**(9), 1415 (2020)
13. Papa, J., Pereira, D., Baldassin, A., Yang, X.-S.: On the harmony search using quaternions. In: Schwenker, F., Abbas, H.M., El Gayar, N., Trentin, E. (eds.) ANNPR 2016. LNCS (LNAI), vol. 9896, pp. 126–137. Springer, Cham (2016). https://doi.org/10.1007/978-3-319-46182-3_11
14. Papa, J.P., Rosa, G.H., Pereira, D.R., Yang, X.S.: Quaternion-based deep belief networks fine-tuning. Appl. Soft Comput. **60**, 328–335 (2017)
15. Papa, J.P., de Rosa, G.H., Yang, X.-S.: On the hypercomplex-based search spaces for optimization purposes. In: Yang, X.-S. (ed.) Nature-Inspired Algorithms and Applied Optimization. SCI, vol. 744, pp. 119–147. Springer, Cham (2018). https://doi.org/10.1007/978-3-319-67669-2_6
16. Passos, L.A., Rodrigues, D., Papa, J.P.: Quaternion-based backtracking search optimization algorithm. In: 2019 IEEE Congress on Evolutionary Computation (CEC), pp. 3014–3021. IEEE (2019)
17. de Rosa, G.H., Papa, J.P., Yang, X.S.: A nature-inspired feature selection approach based on hypercomplex information. Appl. Soft Comput. **94**, 106453 (2020)
18. de Rosa, G.H., Papa, J.P.: Opytimizer: a nature-inspired python optimizer (2019)
19. Törn, A., Žilinskas, A. (eds.): Global Optimization. LNCS, vol. 350. Springer, Heidelberg (1989). https://doi.org/10.1007/3-540-50871-6
20. Wilcoxon, F.: Individual comparisons by ranking methods. Biometrics Bull. **1**(6), 80–83 (1945)
21. Yang, X.S.: Review of meta-heuristics and generalised evolutionary walk algorithm. Int. J. Bio-Inspired Comput. **3**(2), 77–84 (2011)

Using Particle Swarm Optimization with Gradient Descent for Parameter Learning in Convolutional Neural Networks

Steven Wessels and Dustin van der Haar[✉]

Academy of Computer Science and Software Engineering,
University of Johannesburg, Gauteng, South Africa
swessels@jhb.dvt.co.za, dvanderhaar@uj.ac.za

Abstract. The use of gradient-based methods are ubiquitously used to update the internal parameters of neural networks. Problems commonly associated with gradient based methods are the tendency for the algorithms to get stuck in sub-optimal local minima, and their slow convergence rate. Efficacious solutions to these issues, such as the addition of "momentum" and adaptive learning rates, have been offered. In this paper, we investigate the efficacy of using particle swarm optimization (PSO) to help gradient-based methods search for the optimal internal parameters to minimize the loss function of a convolutional neural network (CNN). We compare the metric performance of traditional gradient-baseds method with and without the use of a PSO to either guide or refine the search for the optimal weights. The gradient-based methods we examine are stochastic gradient descent with and without a momentum term, as well as Adaptive Moment Estimation (Adam). We find that, with the exception of the Adam optimized networks, regular gradient-based methods achieve better metric scores than when used in conjunction with a PSO. We also observe that using a PSO to refine the solution found through a gradient-descent technique reduces loss better than when using a PSO to dictate that starting solution for gradient descent. Ultimately, the best solution on the MNIST dataset was achieved by the network optimized with stochastic gradient descent and momentum with an average loss score of 0.0092 when evaluated using k-fold cross validation.

Keywords: Particle swarm optimization · Meta-heuristics · Deep learning

1 Introduction

Convolutional neural networks use a cascade of trainable modules within the network to extract features from raw data input. The operations inside the modules are either linear or pointwise nonlinear in nature. By layering these modules appropriately and altering coefficients accordingly, you can closely approximate a function to achieve statistical generalization [1]. With the features created by

© Springer Nature Switzerland AG 2021
J. M. R. S. Tavares et al. (Eds.): CIARP 2021, LNCS 12702, pp. 119–128, 2021.
https://doi.org/10.1007/978-3-030-93420-0_12

the data passing through the layers of the model, a classification task is used to produce an output. The difference between the desired output and the actual output can be computed using an objective function which gets averaged over all the samples in the training data [2]. The model can then adjust its internal parameters, also known as weights, to minimize the objective function. The process of adjusting the weights of a neural network is considered to be a non-convex problem with many local optima and typically gradient-based methods are the preferred method when adjusting internal parameters [3]. Gradient-based methods compute a gradient vector for every internal parameter that specified the amount the misclassification error would change if the parameter was increased by that infinitesimal amount. The parameters are then adjusted in the opposite direction to the gradient vector. However, gradient-based methods tend to converge slowly towards the global optimum parameter set, and will often get stuck at sub-optimal local minima.

The use of meta-heuristic global optimization algorithms are frequently applied to multi-objective optimization problems prevalent in applied mathematics and engineering [3]. One such algorithm is particle swarm optimization, which has been shown to have fast convergence when locating the global minimum in optimization problems [4]. In this paper we investigate the effect of using a particle swarm to help guide the search for the optimum trainable parameters in a convolution network. We will study the affects of using a particle swarm within a CNN to 1) Find a more optimum starting point for the gradient descent algorithm, and 2) Refine the best gradient descent search solution.

The structure of this paper as follows: Sect. 2 introduces the traditional method of training a neural network with gradient descent and backpropagation, as well as the drawbacks of this approach. In Sect. 3 we discuss the PSO algorithm and how it may be used to in tandem with the established methods of training neural networks. Next, we review work related to the use of meta-heuristic optimization algorithms for training neural networks in Sect. 4. In Sect. 5 we outline the methods used to evaluate our model and make comparisons between different optimizer configurations. Section 6 presents the specific implementation details about the CNN used and about the PSO used. In Sect. 7 we report the results of the experiment and discuss the findings before concluding the paper with Sect. 8.

2   Gradient-Based Learning in Neural Networks

The types of functions used for activation and loss computations in neural networks are almost-everywhere differential w.r.t parameters θ. Gradient-based optimization methods are therefore the most commonly used optimization method used for updating θ [5]. Supposing there is a function $y = f(x)$ where $x, y \in \mathbb{R}$. This function's derivative $y' = f'(x)$ indicates how to scale incremental changes to the input in order to obtain an output with a corresponding factor of change such that $f(x + \epsilon) \approx f(x) + \epsilon f'(x)$. Knowing how to update x to minimize the value of y is an optimization technique known as gradient descent [6]. Partial differentiation is applied to functions with multiple inputs. We can measure how f

changes at point $\mathbf{x}$ with respect to variable x_i using $\frac{\partial}{\partial x_i} f(\mathbf{x})$. The gradient of f is represented by a vector $\nabla_x f(\mathbf{x})$ containing all the partial derivatives of $f(x_i)$ where $x_i \in \mathbf{x}$. The method of gradient descent proposes that we minimize f by moving in the direction where gradient is:

$$\mathbf{x}' = \mathbf{x} - \epsilon \nabla_x f(x) \tag{1}$$

where ϵ is a positive scalar value indicating the step size. When every value of the gradient is zero, the gradient descent algorithm converges.

Feedforward networks, such as the CNN, take an input $\mathbf{x}$ that will be propagated forward through the hidden layers and generate an output $\hat{\mathbf{y}}$. Then a scalar value for loss $J(\boldsymbol{\theta})$ is computed. The information of the loss calculation is then disseminated back down the network to make gradient computations using the backpropagation algorithm [7]. Backpropagation calculates the gradient of the loss function with respect to the trainable parameters $\nabla_\theta J(\boldsymbol{\theta})$ by utilizing the chain rule of calculus.

The point of $f(x)$ that gradient descent algorithms seek to obtain is known as the global minimum, where $f(x)$ produces the absolute minimum value. In deep learning problems, the functions that get optimized will contain many suboptimal local minimia and saddle points that are engulfed by flat regions. As the number of training examples and the number of connections in a network grow, so the complexity of the search space also grows [8]. A weakness of gradient-based methods is that they struggle to get unstuck from these flat regions in the search space. Methods to prevent gradient descent algorithms from getting stuck at local minima have been proposed. The addition of a hyperparameter, $\alpha \in [0, 1)$, that tracks the moving average of the gradients called momentum and is known to accelerate learning and quicken convergence [5]. If a gradient keeps pointing in the same direction, this will increase the size of the steps taken towards the minimum. Momentum also smooths out the variations in movement caused when the gradients continually change direction. Another method, called Adaptive Moment Estimation (Adam), is a variant on gradient descent where individual adaptive learning rates are computed for different parameters. Adam is effectively a combination of RMSprop and SGD with momentum, that uses squared gradients to scale the learning rate and using moving average of the gradient.

3   Particle Swarm Optimization

In 1995, Kennedy and Eberhart introduced a concept for the optimization of non-linear functions using a population-based search technique called particle swarm optimization (PSO) [9]. This technique was based off the social metaphor of a flock of birds that can fly in incredibly complex patterns while being able to change flight direction and regroup in a highly organised manner. Particles belonging to a swarm search around an n-dimensional search space where their position represents a solution. Changes in position are due to the particles velocity. This velocity is broken down into two components, a social component and a cognitive component. For particle swarms that disseminate information through a global topology, an individual particle p has it's velocity calculated using

$$v_{pj}(n+1) = v_{pj}(n) + c_1 \omega_{1_j}(n)[y_{pj}(n) - x_{pj}(n)] + c_2 \omega_{2_j}(n)[y_{j_{gbest}}(n) - x_{pj}(n)] \quad (2)$$

where $v_{pj}(n)$ is the velocity of particle p in at dimension $j = 1 \ldots N$ at iteration n. The particles current position is denoted as $x_{pj}(n)$. The personal best position achieved by p is found in the cognitive component term as $y_{pj}(n)$, while the current global best position $y_{j_{gbest}}(n)$ is found in the social component term. The acceleration constants c_1 and c_2 are selected before training and are used to balance the social and cognitive components. Stochastic behaviour is introduced into the algorithm using variables $\omega_{1_j}(n)$ and $\omega_{2_j}(n)$, which are sampled from a Normal distribution $\sim N(0, 1)$. After the new velocity for p is calculated, the position of p can be updated using the following equation

$$x_{pd}(t+1) = x_{pd}(t) + v_{pd}(t+1) \quad (3)$$

Once the positions of the particles have been updated, they are evaluated using an objective function to determine the quality of solution. The best positioned particle is the one that minimizes the loss score, and that position is designated as the new $y_{j_{gbest}}$. We also set the new personal best solution of each particle if applicable.

The gradient-based methods that are ubiquitous for training neural networks have issues overcoming local minima and often converge slowly. Meta-heuristic algorithms, such as the evolutionary algorithm and PSO, are inspired by naturally occurring phenomena and are often efficient solutions to finding global optima in complex search spaces. The PSO algorithm is known to converge faster than evolutionary algorithms, and is easy to comprehend and implement [4]. Furthermore, it is flexible by not requiring a good initial starting point. Therefore, we propose a hybrid algorithm that combines deterministic gradient descent techniques with high convergence rates to locate local minima while using stochastic meta-heuristic algorithms to escape from local minima and explore flat regions of interest.

4   Literature Review

Literature pertaining to the use of particle swarm optimization in neural network weight is scarce, particularly within the context of convolutional neural networks. A likely reason for this is that PSO's are not capable of optimizing the tens of thousands of variables found neural networks because of the particles tendency to deviate from good area's in the search space due to explosive velocities [10]. However, it is still feasible for PSO's to be used in high dimensional search spaces, such as neural network parameter optimization, to perform a highly exploitative and granular searches. As such the majority of existing literature regrading neural network parameter updates involve hybrid search strategies that combine meta-heuristic stochastic optimizers such as particle swarm optimization and genetic algorithms, with well established gradient-based techniques.

In 2009, Wang introduced a method of improving the performance of PSO's outside the realm of neural network optimization but still used gradient-based methods to initially guide the search [3]. The gradient search was used to find a local minimum in the search space of the objective function, From the local minimum point found, the PSO's initial population would be generated using a Gauss distribution. From there the PSO algorithm searches for a better local minima, before the algorithm repeats with the gradient-based search. To avoid early convergence, Wang used a repulsion technique to avoid getting stuck at previously found optima. The algorithm was benchmarked on various tests sets, of which the maximum dimensionality was 1000, and performed well, with a good computationally efficiency. Similarly, in 2010, Noel proposed a hybrid gradient-based local search algorithm combined with a modified PSO. This modified PSO, known as the GPSO, avoided the use of inertia weights and constriction coefficients [4]. Inertia weights and coefficient constants are used to bias between exploration of the search space, and the exploitation of seemingly good solutions. However, the GPSO presented sought to find a balanced exploration/exploitation trade off. First a PSO is used to find a close local minimum to the random starting point, then using that local minimum, a gradient based local search is done. The gradient search algorithm was the quasi Newton-Raphson (QNR) algorithm. The best solution found by the QNR search then becomes the best solution for the next PSO iteration, and the algorithm repeats until termination at a preset number of repetitions. The hybrid algorithm was benchmarked on the De Jong test suite of benchmark optimization problems and compared to a hybrid algorithm that used a standard PSO. The GPSO variant achieved a more accurate solution, and converged faster.

An early example of applying particle swarm optimization to neural network weight optimization was done by Mendes et al. [11]. The authors compared PSO's directly to backpropagation for training feed-forward neural networks on three classification tasks and two regression tasks. Additionally, the various topologies of PSO's, namely, global, local, and pyramid, were compared. The local topology performed the worst and converged slowly, while the pyramid and global topologies demonstrated efficacy on some of the benchmark tasks. Ultimately, backpropagation was shown to achieve better results for end to end training of a neural network. In 2011 Ding et al. authored another example of neural networking using meta-heuristics [12]. Here a genetic algorithm was used to address the slow rate of convergence associated with backpropagation. The genetic algorithm was used to start the initial search, with gradient descent used to refine the best solution found by the GA. The authors also compared models where the network used just gradient-based methods, and a genetic algorithm to update network weights respectively. The model was benchmarked on UCL's iris data set. It was found that the hybrid algorithm achieved a higher precision score than either of the individual methods.

Although the literature pertaining to the use of a meta-heuristic optimization algorithm for updating neural network internal weights is modest, there are still patterns that emerge. Firstly, in optimization tasks with high dimensionality,

PSO's are not performant as a standalone optimizer, particularly when compared to gradient-based methods on optimization problems where the function to optimize is everywhere differentiable. Secondly, the combination of gradient-based methods and PSO's was shown efficacy in highly dimensional problems, where gradient-based methods are used to find a good initial best solution for the swarm, or if a PSO guides the search to find a starting position for the gradient-based algorithm.

5   Experimental Setup

5.1   MNIST Database

The Modified National Institute of Standards and Technology (MNIST) database is a curated subset of the NIST Special database that has been historically used to benchmark the performance of image processing systems and computer vision algorithms [13]. It contains images of handwritten digits that have been centered and size normalized. As a data set, MNIST has low complexity with 70000 images sized at 28×28 pixels. The data set contains ten classes, namely the digits zero through nine, which can be seen in Fig. 1. In the figure we can see that the data set is comprised of various styles of handwritten digits that may have different defining characteristics for the same digit. MNIST is an effectively solved data set with accuracy scores of over 99% being regularly reported since the use of CNN's became common place in image processing models [14]. However, due to it's ubiquity and simplicity, MNIST is an excellent tool to evaluate experimental image processing algorithms. The MNIST data set splits into a training set of 60000 images and a test set of 10000 images.

Fig. 1. A sample of the example data elements from the MNIST Database

5.2   Evaluation

To evaluate the performance we used the k-fold cross validation resampling procedure with a k value of 5. The data set gets split into 5 separate "folds", and each fold will be used as the test set in 1 of the 5 training passes. The k-fold cross validation method of evaluating a model is considered to be fairer than the simpler train-test split method [15]. The metrics gathered for each of the five passes are averaged out to perform a comparison.

The results the we present in Sect. 7 are generated by training the network outlined in Sect. 6 with nine unique optimizers/optimizer combinations. We test stochastic gradient descent optimization both with and without momentum. We also evaluate a model trained using the Adam optimizer. We then evaluate combinations of these traditional optimizers with a particle swarm optimizer, with a variant where the PSO searches the solution space for a more optimal place to start the gradient descent process, and a variant where the PSO runs after gradient descent optimization for fine tuning.

5.3 Hyperparameters

A general heuristic for choosing batch size is that smaller batch sizes tend to perform better in terms of metric evaluation and train models to generalize better than larger batch sizes [16]. In heed of this accepted standard, we chose a batch size of 32. The learning rate of a gradient descent algorithm is a hyperparameter that determines the step size at each iteration. Given that each cross validation step runs for only 10 epochs, we decided to set the learning rate $l_r = 0.005$. A commonly used momentum values found in literature include 0.5, 0.9, and 0.99 [5]. We chose $\alpha = 0.9$. The PSO would be initialized with a swarm size of 25 to search the solution space for only 5 iterations to limit the computational requirement to perform classification while exploiting the PSO's fast convergence property.

6 Model

The network used is a small sequential CNN with a single convolutional layer consisting of eight 3×3 convolutional filters with ReLu activation. Next a max pooling layer is used to reduce the spatial dimensions and computation cost of classification, while also adding regularization. Finally, the output from the pooling layer is flattened and fed into a dense layer before predictions are made using a softmax layer. This CNN has 349,018 trainable weights which are initialized randomly drawing from a uniform distribution within $[-limit, limit]$ where $limit = \sqrt{(6/w)}$ and w is the number of input units in the weight tensor [17].

The PSO algorithm used to train our models is outlined in Algorithm 1. The particles are represented by a vector of length N, where N is the number of layers of trainable weights in the CNN. Within each element of this vector is another vector containing the weights of the layer. The use of the aforementioned particle encoding imposed some serious constraints on the training of the model. The swarm size used for optimization, and the number of iterations that the algorithm can run for will be minimal because of the extra memory requirements of having to keep "copies" of the weight vectors during training, and the exhaustive time required to run Eqs. 2 and 3, as well as calculating the loss for each solution. To initialize the swarm, we first create a particle using the current weights of the model, and the generate the remaining particles by multiplying each weight by a randomly sampled float in the range $(0, 1)$. We then find the best positioned particle in the swarm using the objective function before starting the first iteration. The objective function we will use will be the categorical cross-entropy loss function, commonly used to assess the quality of classification models used to perform multi-class classification tasks.

Algorithm 1. Particle swarm algorithm to update network parameters

Require: Hyperparameters $\omega, c_1, c_2, v_{max}$
Require: Loss function σ
Require: The trainable parameters θ
  Initialize swarm particles
  Find best position and set G_{best} for all particles
  while $n \leq n_{max}$ **do**
    for p in particles **do**
      set pbest
      if $\sigma(\theta_p) < \sigma(\theta_{pbest})$ **then**
        $\theta_p = \theta_{pbest}$
      end if
      set gbest
      if $\sigma(\theta_{pbest}) < \sigma(\theta_{gbest})$ **then**
        $\theta_{gbest} = \theta_{pbest}$
      end if
    end for
    for p in particles **do**
      for j in N **do**
        update velocity $v_{pj}(n+1)$
        update parameters $x_{pd}(t+1)$
      end for
    end for
  end while

---

7 Results

The results of the nine model variations that we outlined in Sect. 5.2 are summarized in Table 1. We can see that the most performant model across all metrics evaluated was the model that used stochastic gradient descent with the momentum term. The worst performing model was the stochastic gradient descent optimizer where the starting point for the gradient descent method had been determined by the particle swarm algorithm. The PSO variants failed to improve on the gradient-based methods in the SGD optimized models. However, the PSO refined solution of the Adam optimized network did outperform the network optimized only by Adam with regards to the loss score. Generally, networks where the solutions found by the gradient-based methods which then refined using a PSO algorithm performed slightly better than when a PSO was used to guide the starting point for gradient descent.

Table 1. Summary of results

Optimizer	Loss	Accuracy	Precision	Recall
SGD	0.0724	97.96%	98.35%	97.56%
SGD + PSO	0.0741	97.77%	98.10%	97.54%
PSO + SGD	0.0905	97.42%	97.93%	96.91%
Momentum	0.0092	99.78%	99.80%	99.74%
Momentum + PSO	0.0142	99.53%	99.56%	99.51%
PSO + Momentum	0.0132	99.56%	99.59%	99.55%
Adam	0.0161	99.67%	99.67%	99.66%
Adam + PSO	0.0157	99.55%	99.58%	99.53%
PSO + Adam	0.0216	99.49%	99.51%	99.47%

Ultimately, PSO's are not immune to becoming stuck at local minima. In the cases where the gradient descent algorithm started using the best solution of the PSO, the sub-optimal starting point for gradient descent may have inhibited the search for the best local minimum. This is seen in the metric scores of PSO + SGD and PSO + Adam versus their respective PSO last models. The PSO last models, while only being able to outperform the standard gradient-based method in the case of the Adam optimized models, did show promising results that warrant further investigation. Although, it is worth keeping in mind that any performance gains made in terms of loss reduction when using PSO optimizers, come at the expense of computation performance and added training time. As an example, the total training times of the SGD using momentum with and without a PSO were 1:29:52.677 and 1:15:55.145 respectively.

8   Conclusions and Future Work

This study was undertaken to determine if a PSO could be effective used to overcome the weaknesses of gradient-based methods traditionally used to train convolutional neural networks and reduce the validation loss score on a simple classification task. The best loss score was achieved by using stochastic gradient descent with a moment term to smooth out the gradient information and make the optimizer less susceptible to becoming stuck at local minima. Despite this, the using a PSO to refine the search for the best weights did show promising results, and is generally a better solution for weight optimization than using a PSO to initially find a starting position for the gradient descent algorithms.

For future consideration, an implementation of a PSO where the swarms degree of freedom is restricted either through particle coupling or velocity clamping could be considered. Running a similar experiment on larger data sets such as EMNIST or CIFAR10 may provide further clarity to the results presented here.

References

1. Lecun, Y.: The power and limits of deep learning. Res. Technol. Manage. **61**, 22–27 (2018)
2. Lecun, Y., Bengio, Y., Hinton, G.: Deep learning. Nature **521**(7553), 436–444 (2015)
3. Wang, Y.-J.: Improving particle swarm optimization performance with local search for high-dimensional function optimization. Optim. Methods Softw. **25**(5), 781–795 (2010)
4. Noel, M.M.: A new gradient based particle swarm optimization algorithm for accurate computation of global minimum. Appl. Soft Comput. **12**(1), 353–359 (2012)
5. Goodfellow, I., Bengio, Y., Courville, A.: Deep Learning (Adaptive Computation and Machine Learning series). MIT Press Ltd. (2017)
6. Cauchy, A.: Méthode générale pour la résolution des systèmes d'équations simultanées. Comptes Rendus **25**(2), 536–538 (1847)
7. Rumelhart, D.E., Hinton, G.E., Williams, R.J.: Learning representations by back-propagating errors. Nature **323**(6088), 533–536 (1986)
8. Montana, D.: Neural network weight selection using genetic algorithms. Intell. Hybrid Syst. **8**(6), 9–12 (1995)
9. Kennedy, J., Eberhart, R.: Particle swarm optimization. In: Proceedings of ICNN95 - International Conference on Neural Networks (1995)
10. Oldewage, E.T.: The perils of particle swarm optimization in high dimensional problem spaces. University of Pretoria (2017)
11. Mendes, R., Cortez, P., Rocha, M., Neves, J.: Particle swarms for feedforward neural network training. In: Proceedings of the 2002 International Joint Conference on Neural Networks. IJCNN 2002 (Cat. No. 02CH37290) (2002)
12. Ding, S., Su, C., Yu, J.: An optimizing BP neural network algorithm based on genetic algorithm. Artif. Intell. Rev. **36**(2), 153–162 (2011)
13. LeCun, Y., Cortes, C., Burges, C.: The mnist database, November 1998
14. "Papers with code - mnist benchmark (image classification)."
15. Pedregosa, F., et al.: Scikit-learn: machine learning in Python. J. Mach. Learn. Res. **12**, 2825–2830 (2011)
16. Keskar, N.S., Mudigere, D., Nocedal, J., Smelyanskiy, M., Tang, P.T.P.: On large-batch training for deep learning: Generalization gap and sharp minima (2017)
17. He, K., Zhang, X., Ren, S., Sun, J.: Delving deep into rectifiers: surpassing human-level performance on ImageNet classification. In: Proceedings of the IEEE International Conference on Computer Vision (ICCV), December 2015

Image Segmentation

Object Delineation by Iterative Dynamic Trees

David Aparco-Cardenas$^{(\boxtimes)}$, Pedro J. de Rezende, and Alexandre X. Falcão

University of Campinas, Campinas, Brazil
{rezende,afalcao}@ic.unicamp.br

Abstract. While semantic segmentation networks can approximate well the shape of objects (e.g., people, chairs, tables) in images, boundary adherence is still inaccurate. We present an unsupervised image segmentation approach, named *Iterative Dynamic Trees* (IDT), for improved object delineation. We intend to combine IDT and semantic segmentation networks in future work towards improving object segmentation. For a given number of objects in an image graph, the IDT algorithm (i) estimates one seed per object (and background), (ii) delineates each object as one optimum-path tree, (iii) improves seed estimation and repeats steps (ii)–(iii) for a preset number of iterations or until the seed set convergence is achieved. Then, (iv) the optimum-path forest found with the lowest total cost in the loop (ii)–(iii) is selected as final segmentation. The IDT algorithm is a new method based on the *Iterative Spanning Forest* (ISF) framework, in which the number of superpixels is drastically reduced to the number of objects. It adds step (iv) and exploits, for the first time, dynamic arc-weight estimation in ISF for unsupervised object segmentation. We show that IDT can outperform its counterparts in two image datasets – a result that motivates its combination with semantic segmentation networks in future work.

Keywords: Graph-based object delineation · Iterative spanning forest · Image foresting transform · Unsupervised object segmentation

1 Introduction

Object segmentation is a fundamental yet challenging research topic in Image Processing and Computer Vision, with a wide variety of applications [9]. Object segmentation requires two tightly coupled tasks: detection (the objects' whereabouts in the image) and delineation (the definition of their spatial extent). While semantic segmentation networks [5,13] can approximate quite well the shape of objects (e.g., people, chairs, tables) in images, solving object detection with inaccurate boundary adherence, they require user interaction to improve

This work was supported in part by grants from: *Brazilian National Council for Scientific and Technological Development* (CNPq), #313329/2020-6, #309627/2017-6, #303808/2018-7, *São Paulo Research Foundation* (Fapesp), #2020/09691-0, #2018/26434-0, #2014/12236-1.

© Springer Nature Switzerland AG 2021
J. M. R. S. Tavares et al. (Eds.): CIARP 2021, LNCS 12702, pp. 131–140, 2021.
https://doi.org/10.1007/978-3-030-93420-0_13

object delineation [11]. Hence, improvements in delineation are still crucial for automated object segmentation.

Superpixel segmentation methods can usually delineate object boundaries with high adherence, especially for a considerably higher number of superpixels than objects, but with no identification of which superpixels compose each object [3,8,14]. Unfortunately, the union of connected superpixels, with the same label assigned by semantic segmentation, does not solve the problem. On the other hand, their complementary properties motivate possible combinations between superpixel generation and semantic segmentation, our goal for a future work.

This paper focuses on object delineation as a superpixel segmentation task that defines each object by a single superpixel. The proposed object delineation method is named *Iterative Dynamic Trees* (IDT). The IDT algorithm consists of four steps: (i) initial seed estimation with one seed per object, (ii) object delineation as an optimum-path tree, (iii) seed set improvement and the loop of steps (ii)–(iii) for a preset number of iterations or up to seed set convergence. After that, step (iv) completes the process by selecting the optimum-path forest from loop (ii)–(iii) whose total path cost is minimum. The IDT algorithm is a new method based on the *Iterative Spanning Forest* (ISF) framework [14], which adds step (iv) and drastically reduces the number of superpixels to the number of objects. In ISF, more accurate delineation can be achieved by dynamic arc-weight estimation, as the optimum-path trees grow – a strategy that has been demonstrated for superpixel segmentation [3] and interactive object segmentation [4]. In IDT, we exploit this property in ISF for unsupervised object segmentation by the first time.

Another relevant work in this context is *Iterated Watersheds* [12] (IW). IW may also be implemented in the ISF framework. However, its presentation covers unsupervised object segmentation in the image domain and data clustering in the feature space with better performance than spectral clustering, isoperimetric partitioning and k-means. IW, IDT and all other ISF-based methods rely on the *Image Foresting Transform* (IFT) algorithm [7] for step (ii), wherein they use specific path-cost functions. These methods also differ in all other steps.

Our Contribution. We show that IDT can outperform the most recent ISF-based methods, Dynamic ISF [3] and IW [12], for object delineation. Our contribution extends the ISF framework with advancements summarized as follows: (1) an object delineation method based on multiple executions of the IFT algorithm from improved seed sets with dynamic arc-weight estimation; and (2) the identification of the forest with the lowest total cost among the iterations as the final segmentation.

Organization. The remainder of this work is organized as follows. In Sect. 2, we introduce the IDT algorithm. In Sect. 3, we evaluate variants of the IDT algorithm on two public datasets in comparison with their counterparts, IW and DISF. Finally, concluding remarks and potential future work comprise Sect. 4.

2   Iterative Dynamic Trees

The proposed Iterative Dynamic Trees (IDT) consists of four steps: (i) random selection of an initial seed set $\mathcal{S}$; (ii) object delineation by the Image Foresting Transform (IFT) algorithm [7] with dynamic arc-weight estimation [4]; (iii) recomputation of $\mathcal{S}$ as the geometric centers of the optimum-path trees in the image domain; and, after multiple iterations of the last two steps, (iv) selection of the forest with lowest total path cost among all executions.

Fig. 1. Object segmentation using the IDT algorithm.

Figure 1 depicts the application of the IDT algorithm from a randomly selected initial seed set and maximum number of iterations set to 20. The seed set is recomputed at the end of each single iteration and the optimum-path forest with the lowest total cost is returned as the final output.

2.1   Object Delineation by Image Foresting Transform

A two-dimensional image is a pair $(D_I, \mathbf{I})$, such that $\mathbf{I}(p)$ assigns local image features (e.g., color space components) for each pixel $p \in D_I \subset \mathbb{Z}^2$. An image can be rendered as a graph $(\mathcal{N}, \mathcal{A})$ under various configurations, depending upon how nodes $\mathcal{N} \subseteq D_I$ and *adjacency relation* $\mathcal{A} \subset \mathcal{N} \times \mathcal{N}$ are defined. In the present work, we define pixels as nodes ($\mathcal{N} = D_I$), such that $\mathbf{I}(p)$ represents the CIELab color components of pixel p, and the 8-neighborhood relation defines the arcs.

Given a seed set $\mathcal{S}$, we wish to partition the image into objects such that the pixels enclosed by an object are more closely connected to the seed within the object than to any other seed. A unique object identifier is given to each seed $p \in \mathcal{S}$ by a labeling function $\lambda(p) \in \{1, \ldots, c\}$, where c is the number of objects. Note, however, that multiple seeds per object could also receive a same label provided by a semantic segmentation network.

A simple path with terminus q is a sequence of distinct nodes $\pi_q = \langle p_1, p_2, \ldots, p_n = q \rangle$, $(p_i, p_{i+1}) \in \mathcal{A}$, $i = 1, 2, \ldots, n-1$, whereas $\pi_q = \langle q \rangle$ is called a *trivial path*. A *connectivity function* f stipulates a value to any path in the graph measuring how strongly connected the start and end nodes of the path are. Here, we employ $f_{\max}$ as connectivity function

$$f_{\max}(\langle q \rangle) = \begin{cases} 0 & \text{if } q \in \mathcal{S} \subset \mathcal{D_I} \\ +\infty & \text{otherwise} \end{cases}$$

$$f_{\max}(\pi_p \cdot \langle p, q \rangle) = \max\{f_{\max}(\pi_p), w(p, q)\}, \tag{1}$$

where $w(p, q)$ denotes the arc weight of $\langle p, q \rangle$ and $\pi_p \cdot \langle p, q \rangle$ is the concatenation of π_p and $\langle p, q \rangle$, with the two occurrences of p merged into one. The choice of $w(p, q)$ can considerably affect the results. Iterative Watersheds [12], for instance, adopts fixed arc weights (e.g., $w(p, q) = \|\mathbf{I}(q) - \mathbf{I}(p)\|$). For IDT and DISF [3], the arc weights are computed on-the-fly as proposed in [4]. This formulation uses dynamic sets $\mathcal{C}_i \subset \mathcal{D}_I$, $i = 1, 2, \ldots, c$, that contain the nodes of optimum-path trees as they grow to compose each object (or superpixel) $\mathcal{C}_i$. The arc weights $w(p, q)$ are estimated when q is reached by p and evaluated to whether be part of $\mathcal{C}_i$:

$$w(p, q) = \|\mathbf{I}(q) - \mu_{L(p)}\|, \tag{2}$$

$$\mu_{L(p)} = \frac{1}{|\mathcal{C}_{L(p)}|} \sum_{\forall r \in \mathcal{C}_{L(p)}} \mathbf{I}(r) \tag{3}$$

where $\mathcal{C}_{L(p)}$ is the growing object/superpixel (optimum-path tree) that contains p by the time a path $\pi_p \cdot \langle p, q \rangle$ reaches a node $q \in \mathcal{D}_I \setminus \cup_{i=1}^{c} \mathcal{C}_i$ under evaluation.

A path π_q is called *optimum* if $f(\pi_q) \leq f(\tau_q)$ for any other path τ_q, regardless of its starting node. The IFT algorithm minimizes a path-cost map C,

$$C(q) = \min_{\forall \pi_q \in \Pi_q} \{f(\pi_q)\}, \tag{4}$$

so that Π_q is the set of all possible paths in the graph with terminus q, while it outputs an *optimum-path forest* P, i.e., an acyclic map that attributes to each node q either a predecessor $P(q) \in \mathcal{D}_I$ or a distinct marker $P(q) = nil \notin \mathcal{D}_I$ if $q \in \mathcal{S}$. The current path π_q is replaced by $\pi_p \cdot \langle p, q \rangle$ whenever $f(\pi_q) > f(\pi_p \cdot \langle p, q \rangle)$.

Each seed $r \in \mathcal{S}$ grows to an optimum-path tree by computing optimum paths to the remaining nodes. Moreover, each of these optimum-path trees is defined as a unique object in the image. Let $\mathcal{T}_i \subset \mathcal{D}_I$, $i = 1, 2, \ldots, c$, be the c objects defined by the c optimum-path trees in P (the nodes of $\mathcal{T}_i$ form the final set $\mathcal{C}_i$) . Note that the optimum-path forest P is a partition of $\mathcal{D}_I$, i.e., $\cup_{i=1}^{c} \mathcal{T}_i = \mathcal{D}_I$ and $\cap_{i=1}^{c} \mathcal{T}_i = \emptyset$. By assigning a distinct object label $i \in \{1, 2, \ldots, c\}$ to each seed $p \in \mathcal{S}$, the IFT algorithm can also propagate the corresponding label $L(p) \in \{1, 2, \ldots, c\}$ to its most closely connected pixels in $\mathcal{D}_I$, creating a label map L.

2.2   The IDT Algorithm

Algorithm 1 presents the Iterative Dynamic Trees for $f_{\max}$. The algorithm starts off by initializing the cost map C^*, label map L^* and predecessor map P^* in

Lines 1–2. These maps are updated throughout the execution of the algorithm aiming to minimize the total path-cost value derived from C^*. Later, in Line 3, c seeds $r_i \in D_I, i = 1, 2, \ldots, c$ are picked, such that each is uniquely identified as belonging to one among c objects.

Algorithm 1: Iterative Dynamic Trees for f_{max}

Input : Image $(D_I, \mathbf{I})$, adjacency relation $\mathcal{A}$, seed set $\mathcal{S}$ with labeling function λ, number of seeds $c \geq 1$ and maximum number of iterations $T \geq 1$

Output : Cost C^*, label L^* and predecessor P^* maps

Auxiliar: Priority queue Q, dynamic sets $\mathcal{C}_i, \forall r_i \in \mathcal{S}, i = 1, 2, \ldots, c$, maps C, L and P, and variables tmp and $converged$

1   **foreach** $p \in D_I$ **do**
2     $C^*(p) \leftarrow +\infty, L^*(p) \leftarrow 0, P^*(p) \leftarrow nil$
3   Pick c seeds from $D_I : \mathcal{S} = \{r_1, r_2, \ldots, r_c\}$, iter $\leftarrow 1$, converged $\leftarrow$ **false**
4   $Q = \emptyset$
5   **while** iter $\leq T$ *and* converged $=$ **false do**
6     $\mathcal{C}_i \leftarrow \emptyset, \forall i \in \{1, 2, \ldots, c\}$
7   **foreach** $p \in D_I$ **do**
8       $C(p) \leftarrow +\infty, L(p) \leftarrow 0, P(p) \leftarrow nil$
9   **if** $p = r_i \in \mathcal{S}, i \in \{1, 2, \ldots, c\}$ **then**
10        $C(p) \leftarrow 0, L(p) \leftarrow i$
11   Insert p in Q
12   **while** $Q \neq \emptyset$ **do**
13   Remove p from Q, so that $p = \operatorname{argmin}_{q \in Q}\{C(q)\}$ and $\mathcal{C}_{L(p)} \leftarrow \mathcal{C}_{L(p)} \cup \{p\}$
14   **foreach** $(p, q) \in A \mid q \in Q$ **do**
15        $tmp \leftarrow \max\{C(p), \|\mathbf{I}(q) - \mu_{L(p)}\|\}$
16   **if** $tmp < C(q)$ **then**
17         $C(q) \leftarrow tmp, L(q) \leftarrow L(p), P(q) \leftarrow p$
18     $\mathcal{S}_{prev} \leftarrow \mathcal{S}, \mathcal{S} \leftarrow \emptyset$
19   **foreach** $i \in \{1, 2, \ldots, c\}$ **do**
20       $r_i \leftarrow \operatorname{argmin}_{p \in \mathcal{C}_i}\{\|p - \frac{1}{|\mathcal{C}_i|} \times \sum_{\forall q \in \mathcal{C}_i} q\|\}$ and $\mathcal{S} \leftarrow \mathcal{S} \cup \{r_i\}$
21   converged $\leftarrow (\mathcal{S} = \mathcal{S}_{prev})$
22   **if** $\sum_{\forall p \in D_I} C(p) < \sum_{\forall p \in D_I} C^*(p)$ **then**
23       $(C^*, L^*, P^*) \leftarrow (C, L, P)$
24   iter $\leftarrow$ iter $+ 1$
25 **return** (C^*, L^*, P^*)

In Lines 5–24, Algorithm 1 computes c optimum-path trees (objects) from a seed set $\mathcal{S}$, recomputes the seed set $\mathcal{S}$ in Lines 18–20 and repeats both operations until either the convergence criterion is met or the maximum number of iterations $T \geq 1$ is reached. In Line 6, it sets the dynamic sets to empty. In Lines 8–11, it initializes cost map C, label map L and predecessor map P, and inserts all nodes in a priority queue Q. In Lines 12–17, the algorithm maintains the dynamic sets $\mathcal{C}_1, \mathcal{C}_2, \ldots, \mathcal{C}_c$, cost map C, label map L and predecessor map P. At each iteration of this loop, a node p of minimum cost $C(p)$ is removed from Q and inserted into the corresponding dynamic set $C_{L(p)}$ in Line 13. At this moment, the current path π_p is optimum (*i.e.*, its cost is minimum among all possible paths from $\mathcal{S}$). In Lines 14–17, node p offers an extended path $\pi_p \cdot \langle p, q \rangle$ to a node $q \in Q$ (*i.e.*, $q \in D_I \setminus \cup_{i=1}^{c} C_i$). The path value $f_{\max}(\pi_p \cdot \langle p, q \rangle)$ is computed and stored in tmp in Line 15. If tmp is less than the cost $C(q)$ of the current path π_q in P, then π_q is replaced by $\pi_p \cdot \langle p, q \rangle$ in Line 17 by updating the values of cost $C(q)$, label $L(q)$ and predecessor $P(q)$ of q to tmp, $L(p)$ and p, respectively.

In Lines 18–20, the algorithm saves the current seed set $\mathcal{S}$ into $\mathcal{S}'$, resets the seed set $\mathcal{S}$ to empty and then recomputes it by inserting the nodes $r_i \in C_i$ that are closest to the mean pixel of their resulting optimum-path tree $\mathcal{T}_i, i = 1, 2, \ldots, c$. The mean pixel is defined as the arithmetic mean of pixel coordinates of the elements of C_i. Next, in Line 21, $\mathcal{S}'$ is compared to $\mathcal{S}$ to test for convergence and the result of this comparison is saved in the variable *converged*. In Lines 22–23, we test whether the map C provides a smaller total path-cost value than the map C^*; if that is so, then, C^*, L^* and P^* are updated with the maps C, L and P, respectively. Lastly, the tuple (C^*, L^*, P^*) with minimum total path-cost value among all iterations is returned in Line 25.

In order to reduce the complexity of the algorithm, we can store the mean feature vector of each dynamic set and its size, so that these measures can efficiently be updated during label propagation. Therefore, each time a new element p is added to the dynamic set, the mean feature vector and the dynamic set size are updated as in Eq. 5, where $\mathbf{I}(p)$ represents the color components of p, μ_{prev} and μ_{next} are the previous and next mean feature vectors, while n_{prev} and n_{next} represent the size of the dynamic set before and after the update.

$$\mu_{next} = \mu_{prev} + \frac{\mathbf{I}(p) - \mu_{prev}}{n_{prev} + 1}$$

$$n_{next} = n_{prev} + 1$$

(5)

3   Experimental Results

To demonstrate the advantages of step (iv) and random seed sampling in step (i), we compare three versions of IDT. IDT_1 is the proposed version, as described in the previous section. IDT_2 is IDT_1 without step (iv), it selects the last optimum-path forest after T iterations, as proposed in the original ISF framework and adopted by all previous ISF-based methods, such as DISF [3]. IDT_3 is IDT_1 with grid sampling (seed sampling with uniform distance among seeds) in step (i), as used in some ISF-based approaches, such as DISF [3] and most superpixel segmentation methods. To demonstrate the improvement of IDT for object delineation, we compare it against DISF and IW [12] with two path-cost functions in the IFT algorithm: IW-max computes the cost of a path as the maximum arc weight along it, for fixed arc weights $\|\mathbf{I}(q) - \mathbf{I}(p)\|$, and IW-sum computes the cost of a path as the sum of its arc weights. Like IDT_1 and IDT_2, both IW-based methods start from a random seed set of size equal to the number of

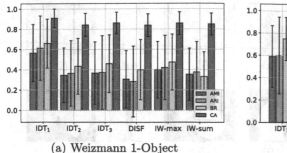
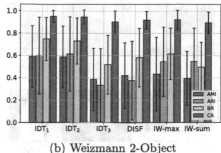

(a) Weizmann 1-Object   (b) Weizmann 2-Object

Fig. 2. Results obtained in each dataset for AMI, ARI, BR and CA. (a) Weizmann 1-Object dataset, (b) Weizmann 2-Object dataset.

desired objects (and background) [12]. DISF starts from a set with 150 seeds selected by grid sampling for all images and reduces the seed set size at every iteration until it reaches the number of desired objects [3] (Fig. 2).

For evaluation of object segmentation, we use the Weizmann 1-Object and 2-Object datasets [2], containing 100 images each, along with ground-truth segmentations. Images in these datasets (available at http://www.wisdom.weizmann. ac.il/~vision/Seg_Evaluation_DB/) depict one or two objects in the foreground.

Table 1. AMI, ARI, Boundary Recall and Cluster Accuracy (Mean +/- Std. Deviation) for Weizmann 1-Object and 2-Object datasets for IDT variants, DISF, IW-max and IW-sum.

	Method	AMI	ARI	BR	CA
1-Object	IDT$_1$	**0.564673 ± 0.283**	**0.613058 ± 0.317**	**0.657833 ± 0.241**	**0.908387 ± 0.091**
	IDT$_2$	0.344623 ± 0.270	0.363208 ± 0.323	0.433819 ± 0.276	0.841895 ± 0.114
	IDT$_3$	0.366932 ± 0.307	0.372370 ± 0.363	0.458131 ± 0.285	0.860064 ± 0.107
	DISF	0.304520 ± 0.282	0.282088 ± 0.347	0.398606 ± 0.296	0.836631 ± 0.112
	IW-max	0.397320 ± 0.278	0.419055 ± 0.318	0.473212 ± 0.276	0.856288 ± 0.112
	IW-sum	0.352781 ± 0.257	0.373990 ± 0.300	0.330048 ± 0.243	0.847699 ± 0.108
2-Object	IDT$_1$	**0.589247 ± 0.278**	0.600024 ± 0.345	**0.748527 ± 0.194**	**0.953605 ± 0.054**
	IDT$_2$	0.587252 ± 0.278	**0.614408 ± 0.333**	0.730065 ± 0.207	0.946522 ± 0.064
	IDT$_3$	0.386087 ± 0.279	0.334149 ± 0.328	0.518125 ± 0.263	0.902305 ± 0.100
	DISF	0.420036 ± 0.295	0.376453 ± 0.352	0.582483 ± 0.263	0.919615 ± 0.078
	IW-max	0.435559 ± 0.330	0.544933 ± 0.311	0.615948 ± 0.231	0.921671 ± 0.086
	IW-sum	0.395757 ± 0.242	0.347743 ± 0.299	0.496769 ± 0.224	0.895421 ± 0.097

For assessment of the methods, we use four popular effectiveness measures: (*i*) *Adjusted Mutual Information* (AMI) [15], which is an adjustment of the Mutual Information (MI) score to account for chance, (*ii*) *Adjusted Rand Index* (ARI) [10], which determines the Rand index (RI) score adjusted for chance, (*iii*) *Boundary Recall* (BR), which measures boundary adherence [1], and (*iv*) *Cluster*

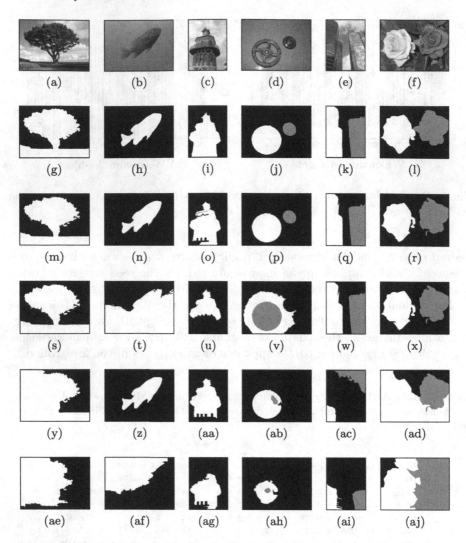

Fig. 3. Segmentation results for Weizmann 1-Object and 2-Object datasets. (a)–(f) Original images, (g)–(l) Ground-truth, (m)–(r) IDT_1, (s)–(x) DISF ($N_o = 150$), (y)–(ad) IW-max, and (ae)–(aj) IW-sum.

Accuracy [6], which measures the degree of intersection between predicted and ground-truth segmentation.

The experiments were conducted using the same sets of initial seeds for IDT_1, IDT_2, IW-max and IW-sum. To guarantee the best result from each algorithm, they are executed 20 times for each image, from which the best object segmentation is selected according to the evaluation metrics. Next, mean and standard deviation are computed from these values across all images for each dataset.

Table 1 shows the effectiveness of object segmentation for all methods according to four different metrics (AMI, ARI, BR and CA). IDT_1 is the best approach, being worse than IDT_2 in only a single case, according to ARI for the 2-object segmentation task. An important finding from the experiments showed that DISF relies heavily on the size of the initial seed set, imparting outstanding results for some images while failing for others. The results also show that random sampling suffices for step (i), and step (iv), added by the proposed approach to the ISF framework, is vital for improved object segmentation. The results raise the question of how good it would be IDT for superpixel segmentation (when the number of seeds is higher than the number of desired objects), which we will leave for future work.

Figure 3 shows the segmentation results for IDT_1, DISF, IW-max and IW-sum on some images of Weizmann 1-Object and 2-Object datasets.

4 Conclusion

In this work, we have introduced a novel iterative procedure, called Iterative Dynamic Trees (IDT) that, through a sequence of executions of IFT with dynamic arc-weight estimation, each of which followed by a seed re-computation stage, achieves a better object segmentation in comparison to other similar graph-based algorithms. The carried out experiments show that IDT attains considerably better performance than its counterparts assessed by four popular metrics. We intend to investigate new techniques for seed set sampling and re-computation to further improve its performance on object segmentation. Another future work direction is to explore IDT for superpixel segmentation. Furthermore, IDT as a complementary tool for semantic segmentation networks to improve object delineation seems a promising research direction as well.

References

1. Achanta, R., Shaji, A., Smith, K., Lucchi, A., Fua, P., Süsstrunk, S.: SLIC superpixels compared to state-of-the-art superpixel methods. IEEE Trans. Pattern Anal. Mach. Intell. **34**(11), 2274–2282 (2012)
2. Alpert, S., Galun, M., Basri, R., Brandt, A.: Image segmentation by probabilistic bottom-up aggregation and cue integration. In: Proceedings of the IEEE Conference on Computer Vision and Pattern Recognition, June 2007
3. Belém, F.C., Guimarães, S.J.F., Falcão, A.X.: Superpixel segmentation using dynamic and iterative spanning forest. IEEE Signal Process. Lett. **27**, 1440–1444 (2020)
4. Bragantini, J., Martins, S.B., Castelo-Fernandez, C., Falcão, A.X.: Graph-based image segmentation using dynamic trees. In: Vera-Rodriguez, R., Fierrez, J., Morales, A. (eds.) CIARP 2018. LNCS, vol. 11401, pp. 470–478. Springer, Cham (2019). https://doi.org/10.1007/978-3-030-13469-3_55
5. Chen, L., Papandreou, G., Kokkinos, I., Murphy, K., Yuille, A.L.: DeepLab: semantic image segmentation with deep convolutional nets, atrous convolution, and fully connected CRFs. IEEE Trans. Pattern Anal. Mach. Intell. **40**(4), 834–848 (2018)

6. Fahad, A., et al.: A survey of clustering algorithms for big data: taxonomy and empirical analysis. IEEE Trans Emerg. Top. Comput. **2**(3), 267–279 (2014)

7. Falcão, A.X., Stolfi, J., de Alencar Lotufo, R.: The image foresting transform: theory, algorithms, and applications. IEEE Trans. Pattern Anal. Mach. Intell. **26**(1), 19–29 (2004)

8. Galvão, F.L., Guimarães, S.J.F., Falcão, A.X.: Image segmentation using dense and sparse hierarchies of superpixels. Pattern Recognit. **108**, 107532 (2020)

9. Hafiz, A.M., Bhat, G.M.: A survey on instance segmentation: state of the art. Int. J. Multimedia Inf. Retr. 1–19 (2020)

10. Hubert, L., Arabie, P.: Comparing partitions. J. Classif. **2**(1), 193–218 (1985). https://doi.org/10.1007/BF01908075

11. Sofiiuk, K., Petrov, I., Barinova, O., Konushin, A.: F-BRS: rethinking backpropagating refinement for interactive segmentation. In: 2020 IEEE/CVF Conference on Computer Vision and Pattern Recognition (CVPR), pp. 8620–8629 (2020)

12. Soor, S., Challa, A., Danda, S., Daya Sagar, B., Najman, L.: Iterated watersheds, a connected variation of k-means for clustering GIS data. IEEE Trans. Emerg. Top. Comput. (2019)

13. Sultana, F., Sufian, A., Dutta, P.: Evolution of image segmentation using deep convolutional neural network: a survey. Knowl.-Based Syst. **201–202**, 106062 (2020)

14. Vargas-Muñoz, J.E., Chowdhury, A.S., Alexandre, E.B., Galvão, F.L., Vechiatto Miranda, P.A., Falcão, A.X.: An iterative spanning forest framework for superpixel segmentation. IEEE Trans. Image Process. **28**(7), 3477–3489 (2019)

15. Vinh, N.X., Epps, J., Bailey, J.: Information theoretic measures for clusterings comparison: Variants, properties, normalization and correction for chance. J. Mach. Learn. Res. **11**, 2837–2854 (2010)

Low-Cost Domain Adaptation for Crop and Weed Segmentation

Gustavo J. Q. Vasconcelos[1], Thiago V. Spina[2], and Helio Pedrini[1(✉)]

[1] Institute of Computing, University of Campinas, Campinas, SP, Brazil
helio@ic.unicamp.br
[2] Brazilian Synchrotron Light Laboratory (LNLS), Brazilian Center for Research in Energy and Materials (CNPEM), Campinas, SP, Brazil

Abstract. The use of new and more sustainable technologies in agriculture is important to reduce the need for agrochemicals and improve energy efficiency. Many of the recent approaches in this area are based on computer vision algorithms. However, due to the great variability of the scenes that can occur in an agriculture field, the domain shift phenomenon is a relevant problem in this area. Domain adaptation attempts to mitigate this data variability problem. In this work, we propose a low-cost domain adaptation method between agriculture domains for image segmentation. Our approach performs domain adaptation by changing the amplitude of the low-frequency spectrum of images along with contrast limited adaptive histogram equalization (CLAHE), which is an efficient replacement for a commonly used image-to-image translation methods, such as Cyclegan, drastically reducing the number of parameters and improving quality in cases where these models do not maintain semantic consistence between translations.

Keywords: Domain adaptation · Image segmentation · Deep learning

1 Introduction

Intensive crop production is fundamental for today's society, however, the resulting massive application of agrochemicals is a major productive and environmental problem [5]. New technologies are emerging to reduce this issue, such as the application of robots and intelligent systems in the field, many of them based on computer vision algorithms.

The use of semantic segmentation algorithms in agriculture can introduce numerous benefits, such as the reduction of the amount of herbicides used and the detailed analysis of the phenotypic characteristics of the plants [15]. Currently, the state of the art for semantic segmentation problems is based on deep neural networks [6,7]. However, due to the large number of parameters found in this type of model, a substantial amount of labeled data is required for training [22].

This work was partially supported by CAPES, FAPESP (grant #2017/12646-3) and CNPq (grant #309330/2018-1).

© Springer Nature Switzerland AG 2021
J. M. R. S. Tavares et al. (Eds.): CIARP 2021, LNCS 12702, pp. 141–150, 2021.
https://doi.org/10.1007/978-3-030-93420-0_14

Therefore, the requirement to have labels for different plant species at different stages of growth is currently one of the biggest challenges in this area.

The inherent complexity of the image acquisition process in the field and the manual annotation of images make the creation of this type of data laborious and expensive. Furthermore, the domain shift phenomenon is a critical problem, occurring when there is a change in the probability distribution from the training data set to the test set [18]. An apparent example of this problem is observed when the model is trained using images of plants extracted with a different sensor or lighting condition than the one used during the test phase.

Recently, domain adaptation for semantic segmentation has made great progress by separating it into two steps. First, it performs the mapping of images from the source domain to the target domain, and then the semantic segmentation model in the target domain is trained with the images transferred from the domain of origin and their respective annotations [11]. However, the training of domain mapping models, such as Cyclegan [21], uses a large set of images to have a satisfactory result. In addition, the translation does not guarantee the semantic fidelity of the images in both domains, which can cause confusion during the training of the semantic segmentation model, especially when there are small objects, such as weeds in images of agricultural robots. Figure 1 shows a mapping adaptation with a full image using Cyclegan. It may be observed that the mapping presents semantic errors, translating plants and weeds to soil. Some approaches in the literature mitigate this problem by dividing the image into many patches [3]; however, this strategy may severely limit the application for real-time inference, which is critical in targeted herbicide spraying using robots.

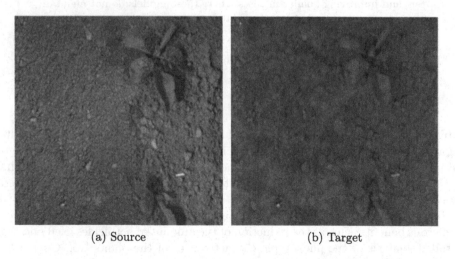

(a) Source   (b) Target

Fig. 1. Domain translation using a Cyclegan architecture. Some errors in this translation can be seen in weed and crop pixels being transformed into soil.

In this work, we propose a new approach to perform the mapping in agricultural images by swapping the low-frequency spectrum of images along with contrast limited adaptive histogram equalization (CLAHE) [13]. The main contribution of this work is showing that the application of CLAHE in conjunction with adaptation of the amplitudes in the Fourier space is an efficient replacement for adversarial generative network models, such as Cyclegan, drastically reducing the number of parameters and obtaining better results for cases where the weed occupies a small area in the image. In addition, the possibility of inference in the complete image improves the suitability of our technique for inference in real time. Our source code and dataset splits will be made available upon acceptance for fair comparison with future works.

This text is organized as follows. Section 2 reviews some concepts and works related to the topic under investigation. Section 3 describes the proposed method. Section 4 presents and discusses the obtained results. Finally, Sect. 5 encompasses some concluding remarks and directions for future work.

2   Background

In the computer vision area, different probability distributions between training and test sets are a common scenario [17], also known as a domain shift problem. This type of problem often leads to a performance degradation of the system under test [16]. Domain adaptation aims to mitigate this problem [9,12], using only images x^s from source domain and their respective y^s labels together with images x^t from target but without any labels in this domain. The goal is to develop a system that can correctly predict the label for the x^t target data.

Following the notation used by Patel et al. [12], let X be the random input variable and Y its respective label, $P(x)$ is the joint probability distribution of X and Y. The source and target distributions are called $P_s(X)$ and $P_t(X)$, such that, in cases involving domain shift, the target and source distribution are different. $\mathcal{S} = \{(x_i^s, y_i^s)\}_{i=1}^{N_s}$ denotes the samples from the source set with $x^s \in \mathbb{R}^N$ being a sample extracted from the domain and y^s its respective label. The unlabeled target is given by $\mathcal{T}_u = \{x_i^{tu}\}_{i=1}^{N_{tu}}$. The sets $\mathcal{S}$ and $\mathcal{T}$ are not known in their entirety, so the subsets $S = [x_1^S, \cdots, x_{N_s}^S]$ and $T = [x_1^{tu}, \cdots, x_{N_{tu}}^{tu}]$ are used to minimize the parameters of function $f(X)$, in the case of this work, a convolutional semantic segmentation model.

Image-to-image translation is one of the techniques used in several state-of-the-art works in domain adaptation in semantic segmentation [8,11,14]. The goal is to copy the visual style of the images from one domain to another. This technique individually is not specifically a domain adaptation technique, however, it comes to be considered when it is used to perform a specific task in a new domain [18].

The deep learning architecture named Cyclegan presented a solution based on generative adversarial networks [4], which translates visual styles without the need for paired images between the source and target domains, something that previously limited the possibilities of applying this type of technique, since the existence of images paired between domains is scarce or even impossible [10].

In this sense, several works have used the Cyclegan architecture and its variations for domain adaptation. Hoffman et al. [8] presented an architecture called Cycada that adapts representations at both pixel and coded levels, in addition to adding cost functions that reinforce the semantic consistency of adaptations between domains. Ramirez et al. [14] proposed an approach to reconstruct the appearance in the domain mapping process in order to help preserve small details in adapting the domain.

Li et al. [11] developed an architecture, called Bidirectional Learning for Domain Adaptation of Semantic Segmentation. They presented a technique for training the domain mapping model in conjunction with the semantic segmentation model, in contrast to other approaches that performed the training separately. Gogoll et al. [3] used a similar architecture for adapting the domain of agricultural images.

The computational cost and the need to have enough data to train Cyclegan-based domain adaptation architectures have been minimized by the Fourier Domain Adaptation (FDA) for semantic segmentation [19]. Its authors used a simple method, switching the low-frequency spectrum of images from the source to target domain. Thus, they were able to replace the four models normally present in a Cyclegan-based network with a simple (Fast) Fourier Transform (FFT) and its inverse FFT (iFFT).

3   Methodology

The approach proposed in this work uses only a convolutional model trained with data from the source domain. In order for the target images to be inferred in this model, the target images are translated into source using an unsupervised technique without the need for the images to be paired. The image-to-image translation technique is based on the replacement of low frequencies in the Fourier space [19] and no training is required for its use, which opens the possibility of using this method for domains with a reduced number of images. The method performs FFT for each input image and replaces the low-level frequencies of the target images into the source images before reconstituting them via the inverse FFT. Figure 2 illustrates this process.

The main motivation for this technique comes from the observation that low frequencies can carry more general information about the environment, such as lighting, image quality and sensor characteristics [19]. On the other hand, high frequencies are responsible for details of plant and weed phenotype, thus preserving semantic information during adaptation.

The technique of replacing low frequencies, although reduces the visual difference between the images of the source and target domains, it is not sufficient to guarantee that the model trained with the source data will have a good performance when inferred with the data mapped from the target domain to the source domain. In this sense, our approach demonstrates that the use of the CLAHE technique with different clipping limits (CL), when carefully chosen both for the source images during model training and for the target images during inference,

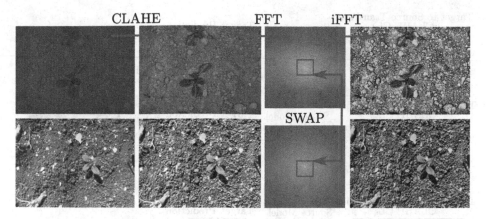

Fig. 2. Image-to-image translation method. Initially, the image is pre-processed with CLAHE. Once FFT is applied, the low-frequency amplitudes are replaced between source and target domains. The inverse FFT is then performed.

can further reduce the distance between distributions, causing a better result in adapting the domain.

Thus, the proposed technique performs a search for the best CLAHE's CL values of the pre-processed images from the source domain for training the model and for the target images at the inference time. These limits are chosen by statistically comparing the masks of the target image predictions with the masks from the source domain. The best CL parameters are chosen based on similar values of the pixel rate of the soil class in relation to the crop and weed classes. Figure 3 illustrates the inference process and the comparison between labels.

4   Results

We perform the evaluation of four methods on two datasets. The Sugar Beets [1] contains images of beet plantation from a farm near the city of Bonn, Germany, which were obtained from photographs captured by a terrestrial robot 3 times a day for 6 weeks. The Sunflower dataset [2] was recorded using a robot near Ancona, Italy. The growth stage in both datasets contain 4–6 leaf plants. Figure 2 illustrates examples of both datasets, where the first row presents some samples of the Sugar Beets dataset and the second row presents some samples of the Sunflower dataset. The datasets were split into training and testing in the proportion of 80% and 20%, respectively.

We fixed the search for CLAHE CL parameters at the ends of the possible limits, more precisely, at 0.01 and 0.9. We also compared the results with models without adaptation. The model chosen to evaluate the method was the Bisenet convolutional network [20], due to its computational efficiency.

The dataset images were scaled to 800 × 800 pixels, requiring only one inference for the entire image, in contrast to the state-of-the-art approaches on adapting these two domains, which divided the inference into 240 × 240 pixel patches.

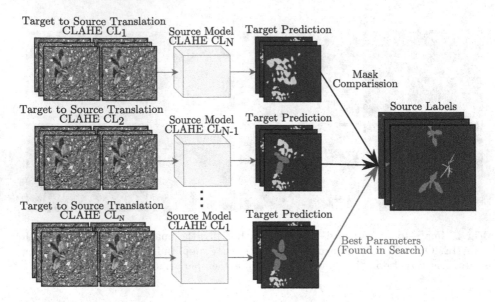

Fig. 3. Inference process of a target image. First, the translation described in Fig. 2 is performed with different CLAHE CL parameters in the target image, then the inference is performed using different source models trained with images and labels of the source domain, also with different CLAHE parameters. The best CLAHE parameters for both source and target are chosen comparing the statistics of the predicted masks with the masks of the source domain.

The original resolution of the Sugar Beets dataset is 1296×966, whereas the resolution of the Sunflower dataset is 1296×964. The optimizer used for training the models was Adam, with a $1e - 5$ learning rate, 0.5 momentum and $1e - 4$ weight decay. Data augmentation was performed only by mirroring the images vertically and horizontally. In all replacement cases of the low frequency amplitudes of the Fourier amplitudes, we used $\beta = 0.001$. Other values were tested, however, this demonstrated to provide the best results.

For the analysis of the statistical variable used to choose the best model, some parameters were evaluated based on the ground-truth masks of the training set of the two datasets, and the index with the least variation was chosen. The mean of the ratio between soil and plant is a factor that remains without great oscillation between the datasets and was the index selected for evaluating the masks after adaptation. If the adaptation is not carried out correctly, the values vary greatly from this index found in the ground truth both in the source and target domains. In this sense, the parameters that generate indices closer to the mean of the ratio between soil and plants are the index chosen for CL in the CLAHE algorithm. These proportions in the training set for the Sugar Beets and Sunflower datasets were 27.08 and 19.98, respectively.

Table 1 illustrates the F1 metric for the three classes obtained in the source models after training the models with different CLAHE CL values. It can be observed that there is no deterioration in the application of the CLAHE in the models. In fact, there was a small improvement in all cases. It is also observed that the Sunflower dataset is slightly easier than the Sugar Beets dataset, especially when we evaluate the metric in the weed class.

Table 1. F1 metric of the models trained with labels from their own domain with different parameters of CLAHE clipping limit.

Method	Resolution	Sugar Beets				Sunflower			
		Soil	Crop	Weed	Mean	Soil	Crop	Weed	Mean
NO CLAHE	Resize 800	0.993	0.868	0.430	0.764	0.984	0.840	0.565	0.796
CL = 0.01	Resize 800	0.994	0.891	0.475	**0.787**	0.985	0.842	0.565	0.797
CL = 0.90	Resize 800	0.992	0.863	0.442	0.766	0.986	0.847	0.576	**0.803**

Table 2 shows the results used as reference for the search for the best method parameters. The baseline result is obtained by training the domains with their respective training labels. The other results are obtained only with training data from the other domain, that is, the results illustrated for the Sugar Beets dataset come from a model trained only with the data from the Sunflower dataset and vice versa. This table also shows the ablation gain of the two main components of the method, the pre-processing using the CLAHE algorithm and the swapping between the amplitudes of the low Fourier frequencies.

It is observed that, in the Sunflower domain as target, the use of the Fourier transfer amplitudes already provides a relative gain of more than 0.5 in the F1 metric. However, the gain does not hold for the Sugar Beets domain as target. On the other hand, the gain using the CLAHE alone is also very small in both adaptations, where the best combination in this case obtained an increase of 0.13 in F1 for Sugar Beets as target and 0.14 for Sunflower as target. When analyzing the combined use of both techniques, there are gains of more than 0.62 in F1 for the Sunflower dataset for the worst combination of parameters, whereas the smallest gain is 0.24 in F1 metric for the Sugar Beets domain.

The selection of the best parameters in the method is performed by comparing the averages of the ratio between the total number of pixels in the soil class in relation to the number of pixels in the plant and weed classes. In the case of the Sunflower domain as a target, the best result was obtained with a CL of 0.9 for source images and also 0.9 for target images. The predictions of this adaptation obtained an average soil/plant ratio of 26.81, the closest to the ratio found in the source Sugar Beets domain, which is 27.08. In the Sugar Beets domain as a target, the best result was a CL of 0.9 for source images and 0.01 for target images. The predictions of this adaptation obtained an average soil/plant ratio of 21.62, the closest to the ratio found in the source Sunflower domain, which is 19.98. The left mask in Fig. 4 is an example of prediction with the best

Table 2. F1 metrics in the domain adaptation between Sugar Beets and Sunflower datasets using different combinations of CLAHE clipping limit parameters.

	Source CL	Target CL	Sugar Beets				Sunflower			
			Soil	Crop	Weed	Mean	Soil	Crop	Weed	Mean
Baseline	–	–	0.993	0.868	0.430	0.764	0.984	0.840	0.565	0.796
No adaptation	–	–	0.991	0.021	0.061	0.358	0.081	0.022	0.004	0.036
Only CLAHE	0.01	0.01	0.985	0.360	0.076	0.474	0.146	0.170	0.011	0.109
Only CLAHE	0.90	0.90	0.974	0.100	0.039	0.371	0.041	0.469	0.020	0.177
Only CLAHE	0.01	0.90	0.977	0.207	0.046	0.410	0.157	0.187	0.011	0.118
Only CLAHE	0.90	0.01	0.982	0.408	0.080	0.490	0.038	0.428	0.020	0.162
Only FDA	–	–	0.986	0.076	0.067	0.376	0.976	0.667	0.057	0.567
CLAHE + FDA	0.01	0.01	0.977	0.478	0.079	0.511	0.974	0.689	0.300	0.654
CLAHE + FDA	0.90	0.90	0.931	0.246	0.027	0.401	**0.981**	**0.805**	**0.397**	**0.728**
CLAHE + FDA	0.01	0.90	0.932	0.324	0.029	0.428	0.978	0.724	0.300	0.667
CLAHE + FDA	0.90	0.01	**0.987**	**0.726**	**0.190**	**0.634**	0.973	0.745	0.367	0.695

parameters of the Sugar Beets domain as a target, while the mask in the center is an unsuccessful adaptation. The image also shows its impact on the soil/plant ratio.

Source CL: 0.9
Target CL: 0.01
S/P RATIO: 21.62

Source CL: 0.9
Target CL: 0.9
S/P RATIO: 8.79

Ground Truth
S/P RATIO: 27.08

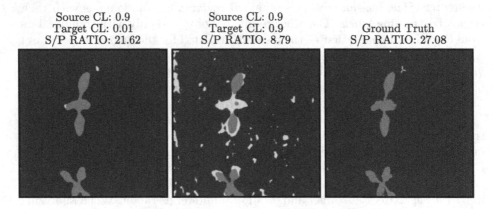

Fig. 4. Example of predictions of two different parameter combinations in the Sugar Beets domain as a target. The left is the best combination and, in the center, an unsuccessful combination. It is observed in the last case that the soil/plant ratio (S/P ratio) is significantly reduced due to artifacts present in the mask.

It is also worth mentioning that the method presented in this work could also be used in combination with very few images annotated in the target domain. In this case, it would not be necessary to choose the parameters based on the soil/plant ratio, as the mask itself could be used to select the adaptation to generate the best results in the annotated target images. This method could also be used as a pseudo-mask to retrain a new model in the target domain.

Although the approach developed by Gogoll et al. [3] reported results on the adaptation of the two domains presented in our work, where the adaptation of the Sunflower domain to Sugar Beets is reported as 0.67 mean F1 metric and Sugar Beets for Sunflower as 0.70, a fair comparison was not possible because the authors did not provide their used dataset splits. However, despite the competitive results, their work uses small patches of 240×240 pixels for adaptation using Cyclegan, thus requiring numerous inferences in the same image, which restricts its use in real-time cases.

5   Conclusions and Future Work

In this work, we presented an unsupervised domain adaptation approach to the crop and weed semantic segmentation problem. Our method is based on the source domain data to perform domain adaptation using simple FFT and CLAHE techniques.

According to the experiments conducted in our work, we achieved competitive results with a much lighter approach than other methods available in the literature based on Cyclegan. In addition, we also improved the results for full image-to-image translation while maintaining semantic consistency in regions that are difficult for Cyclegan methods, such as domain translation of small weeds. Therefore, the proposed method allows significant advances in the application of real-time domain adaptation.

As directions for future work, the use of the proposed method could be used as a pseudo-label for training a new model in the target domain. New parameter selection techniques could also be be evaluated and compared with the use of a few images in the target domain to select parameters directly through the target masks. Other data sets could be evaluated, including domains of images captured by drones due to their greater availability in the literature compared to terrestrial images.

References

1. Chebrolu, N., Lottes, P., Schaefer, A., Winterhalter, W., Burgard, W., Stachniss, C.: Agricultural robot dataset for plant classification, localization and mapping on sugar beet fields. Int. J. Robot. Res. **36**(10), 1045–1052 (2017)
2. Fawakherji, M., Potena, C., Pretto, A., Bloisi, D.D., Nardi, D.: Multi-Spectral Image Synthesis for Crop/Weed Segmentation in Precision Farming. arXiv preprint arXiv:2009.05750 (2020)
3. Gogoll, D., Lottes, P., Weyler, J., Petrinic, N., Stachniss, C.: Unsupervised domain adaptation for transferring plant classification systems to new field environments, crops, and robots. In: IEEE/RSJ International Conference on Intelligent Robots and Systems (IROS), October 2020
4. Goodfellow, I., et al.: Generative adversarial nets. In: Advances in Neural Information Processing Systems, pp. 2672–2680 (2014)
5. Hamill, A.S., Holt, J.S., Mallory-Smith, C.A.: Contributions of weed science to weed control and management. Weed Technol. **18**(1), 1563–1565 (2004)
6. Hao, S., Zhou, Y., Guo, Y.: A brief survey on semantic segmentation with deep learning. Neurocomputing **406**, 302–321 (2020)

7. Herrera, A., Cuadros-Vargas, A., Pedrini, H.: Semantic segmentation of volumetric medical images with 3D convolutional neural networks. CLEI Electron. J. **23**(1), 1–15 (2020)
8. Hoffman, J., et al.: CyCADA: cycle-consistent adversarial domain adaptation. In: 35th International Conference on Machine Learning, pp. 1–10, Stockholm, Sweden (2017)
9. Jiang, J.: A Literature Survey on Domain Adaptation of Statistical Classifiers, pp. 3:1–12 (2008). http://sifaka.cs.uiuc.edu/jiang4/domainadaptation/survey
10. Kim, H., Kim, J., Won, S., Lee, C.: Unsupervised Deep Learning for Super-resolution Reconstruction of Turbulence. arXiv preprint arXiv:2007.15324 (2020)
11. Li, Y., Yuan, L., Vasconcelos, N.: Bidirectional Learning for Domain Adaptation of Semantic Segmentation. CoRR, abs/1904.10620:1–10 (2019)
12. Patel, V.M., Gopalan, R., Li, R., Chellappa, R.: Visual domain adaptation: a survey of recent advances. IEEE Signal Process. Mag. **32**(3), 53–69 (2015)
13. Pizer, S.M., et al.: Adaptive histogram equalization and its variations. Comput. Vis. Graph. Image Process. **39**(3), 355–368 (1987)
14. Ramirez, P.Z., Tonioni, A., Di Stefano, L.: Exploiting semantics in adversarial training for image-level domain adaptation. In IEEE International Conference on Image Processing, Applications and Systems, pp. 49–54. IEEE (2018)
15. Rask, A.M., Kristoffersen, P.: A review of non-chemical weed control on hard surfaces. Weed Res. **47**(5), 370–380 (2007)
16. Tzeng, E., Hoffman, J., Saenko, K., Darrell, T.: Adversarial discriminative domain adaptation. In: IEEE Conference on Computer Vision and Pattern Recognition, pp. 7167–7176 (2017)
17. Wang, M., Deng, W.: Deep visual domain adaptation: a survey. Neurocomputing **312**, 135–153 (2018)
18. Wilson, G., Cook, D.J.: A survey of unsupervised deep domain adaptation. ACM Trans. Intell. Syst. Technol. **11**(5), 1–46 (2020)
19. Yang, Y., Soatto, S.: FDA: fourier domain adaptation for semantic segmentation. In: IEEE/CVF Conference on Computer Vision and Pattern Recognition, pp. 4085–4095 (2020)
20. Yu, C., Wang, J., Peng, C., Gao, C., Yu, G., Sang, N.: BiSeNet: bilateral segmentation network for real-time semantic segmentation. In: Ferrari, V., Hebert, M., Sminchisescu, C., Weiss, Y. (eds.) ECCV 2018, Part XIII. LNCS, vol. 11217, pp. 334–349. Springer, Cham (2018). https://doi.org/10.1007/978-3-030-01261-8_20
21. Zhu, J.-Y., Park, T., Isola, P., Efros, A.A.: Unpaired image-to-image translation using cycle-consistent adversarial networks. In: IEEE International Conference on Computer Vision, pp. 1–18 (2017)
22. Zhuang, F., et al.: A comprehensive survey on transfer learning. Proc. IEEE **109**(1), 43–76 (2021)

Databases

MIGMA: The Facial Emotion Image Dataset for Human Expression Recognition

Jhennifer Cristine Matias[1]([✉]), Tobias Rossi Müller[1], Felipe Zago Canal[1], Gustavo Gino Scotton[1], Antonio Reis de Sa Junior[3], Eliane Pozzebon[1,2], and Antonio Carlos Sobieranski[1,2]

[1] Department of Computing (DEC), Federal University of Santa Catarina, Florianópolis, Brazil
j.matias@grad.ufsc.br
[2] Post Graduate Program in Information Technology and Communication (PPGTIC), Federal University of Santa Catarina, Florianópolis, Brazil
[3] Department of Medical Clinic (DCM), Federal University of Santa Catarina, Florianópolis, Brazil

Abstract. Recognition of emotions from facial information is a simple task well-performed by humans, but very complex to be executed computationally. Since many of the computational trials to solve this problem lead to studies for a generic approach, it needs to be comprehensive to provide a solution analytically possible. Several approaches were proposed over the past few years, and apart of the chosen model, a considerable amount of input samples must be used to train the computational approach properly. Over the literature image datasets for facial recognition can be found, and despite the fact that several of them are in public domain available, the presented images are usually very restricted, with slight variations and limited number among participants, low miscegenation mixing and short age ranges, which ends up making the studies and new algorithms very specific. For this purpose, this paper has as main goals (i) to present a newly designed dataset entitled MIGMA for human expression recognition from facial images and (ii) to address essentials features of this dataset such as high-quality spatial resolution images, varied ethnicity, ages and genders, and including non-induced and induced expressions via emoji. The dataset has 323 participants and 15k images on its current first version, taking into account 8 distinct emotions, photographed in an academical environment in a south American country. In contrast, all the participants completed questionnaires for anxiety and depression, allowing to address further studies in this area with facial emotions. The obtained dataset was tested in an experimental environment using a Convolutional Neural Network recognizer for general recognition overall and class separability estimation.

Keywords: MIGMA · Facial emotion dataset · Feature recognition · Computer vision dataset

© Springer Nature Switzerland AG 2021
J. M. R. S. Tavares et al. (Eds.): CIARP 2021, LNCS 12702, pp. 153–162, 2021.
https://doi.org/10.1007/978-3-030-93420-0_15

1   Introduction

Emotion recognition from facial expressions, besides instinctive and natural for humans, is a very complex task to be performed computationally. This fact is due its natural sense, with hard description and explicitness, and as a consequence, it is difficult to be reproduced in analytical terms. The achievement of a good general solution for this problem has several practical implications [1], being applied by medicine education, robotics, driver safety, games and the educational area. However, emotion recognition has been still an open problem in the computational area. Lately, thanks to the recently developed hardware architectures and computational models, approaches were designed to be dynamic in terms of recognition and generalization. This generalization, on the other hand, can be achieved at the cost of requirements for an large input dataset, used to train the recognition model properly.

In fact, the emerging convolutional neural network models (CNN's) are paving the way for very interesting generic solutions, the datasets available are not evolving at the same compass. Larger datasets are required nowadays to properly train the CNN's in order to be effective and provide the expected results. Over the past few years, several image sets for recognizing human emotions in facial expressions have been proposed, as demonstrated in our related works section. However, the found databases are not always able to meet minimal requirements to develop a robust facial emotion recognition system. As a consequence, it is common to see a large amount of projects including their own dataset (but not for public domain) to develop their computational approaches. For instance, JAFFE and Cohn-Kanade both have low miscegenation and a higher number of photos of people of the same gender; JAFFE and MMI have a low number of participants in their bases, being corresponding to 10 and 75, respectively. There are datasets in addition with images in an uncontrolled environment, such as FER-2013, where the images were collected from the WEB. All of these drawbacks result in a problem for the training of classifiers and the new convolutional neural networks, making the computational model restricted to very specific domains.

In the present study, a new dataset for emotion recognition from facial images is proposed. The dataset was developed taking into account the aforementioned limitations and providing higher resolution images, in a well-controlled environment, good number of participant and images. Unlike others, our dataset presents a very high miscegenation of participants, since it was built in an academic environment, including people from the most varied ethnic groups. Additionally, as far as we know, this is the very first time a emotion facial image dataset is provided taking a higher miscegenation from a South American nation for open domain, high-resolution images, and associated with a medical scale of psychiatric symptoms.

2   Related Works

Given the importance and need for automated emotion recognition systems for facial images, it became necessary to create datasets that allow us to both training and testing the effectiveness for such in developing algorithms. Over the past few years, several datasets have been appeared in different countries with varied purposes and characteristics, such as those presented in the Tables 1 and 2.

The Japanese Female Facial Expression (JAFFE), despite being developed in 1998, it is still widely used since its creation by a group of Japanese researchers [2]. The images in this dataset have a spatial resolution of 256 × 256 pixels and were taken in gray scale [3]. The base has a total of 213 images of 10 Japanese women, in which they express 7 emotions: happiness, sadness, surprise, disgust, fear, anger and neutrality. However, emotions are not considered pure, and were classified according to their predominant emotion.

Another well-known dataset is Cohn-Kanade (CK), and it had its first version developed in 2000, whose project was carried out by [4] and is currently active and managed by the University of Pittsburgh. According to [3] in its first version, it already had 486 image sequences, which, unlike static images, had information about the process of creating emotion. Initially, this dataset had 97 participants of both gender and ages ranging from 18 to 30 years, the majority of participants were Euro-American or African-American, only 3% were Latin or Asian. In 2010, this dataset was updated, and the newest version became known as CK+, having 107 new sequences added and 26 new participants representing the following 6 emotions: anger, contempt, heartbreak, fear, happiness, fear and surprise [5].

MMI is a dataset that started to be developed in 2002, being composed by images and videos of 75 people until the present moment [6]. Participants are men and women, aged from 19 to 62 years old and from different ethnicities. This dataset has static images and sequences of images of the participants in which they express the six basic expressions: happiness, anger, sadness, disgust, fear and surprise. Furthermore, according to [3] this is the first web-based facial expression database, and has as main drawback a controlled environment.

BU-3DFE was created in 2006 by [7], being different from the other ones since its has 3D image sequences. This database is divided into 7 emotions and composed by a total of 2500 sequences with a spatial resolution of 1040 × 1329. As presented by [3], the base has 100 participants, men and women, aged from 18 to 70 years old, of different ethnicities including whites, blacks, Indians, Asians, among others, they performed 6 emotions: sadness, happiness, fear, angry, surprise and disgust.

FER-2013 is a dataset developed for the International Conference on Machine Learning (ICML - 2013), being shared publicly afterwards. The dataset has a total of 35887 images divided into 7 emotions. Unlike other datasets, it is not known for sure about the source participants because the images were obtained through google images API. The resolution of the photos are 48 × 48 and they are in gray scale [8], being already normalized to be used as input layer in training procedures.

In addition to the datasets already mentioned, there are countless others that can be found over the internet. It has been observed that many computational approaches for emotion recognition are closed, since they were developed to provide artifacts for their own computational approaches [9–11], indicating a real research demand and that a new one is still required.

Table 1. Comparison of datasets proposed over time, and the positioning of the presented dataset MIGMA in quantitative terms.

Name	No. of images	Resolution	Color	No. of emotions
JAFFE	213	256 × 256	Gray	7
CK	486	640 × 490	Gray	6
MMI	545	720 × 576	Color	6
BU-3DFE	2500	1040 × 1329	Color	7
CK+	593	640 × 490	Color and Gray	7
FER-2013	35.887	48 × 48	Gray	7
MIGMA	15.071	1920 × 1080	Color	8

Table 2. (Cont.) Comparison of datasets proposed over time, and the positioning of the presented dataset MIGMA in quantitative terms.

Name	Images/Images sequence	Year	Environment	No. of participants
JAFFE	Image	1998	Controlled	10
CK	Image sequence	2000	Controlled	97
MMI	Image, Image sequence	2002	Controlled	75
BU-3DFE	Image, 3D models	2006	Controlled	100
CK+	Image sequence	2010	Controlled	123
FER-2013	Image	2013	In-the-wild	–
MIGMA	Image	2020	Controlled	323

3   Methodology: Proposed Dataset Environmental Protocol

Figure 1 illustrates the computational flow adopted in our approach and an example of the computational environment prepared to acquire the images, respectively in (a) and (b). In (a), the protocol flow is demonstrated in five stages: (1) participant information collection, (2) psychological profile gathering, (3) image acquisition in unsupervised manner and (4) image acquisition in supervised manner, and (5) images verification and eventual corrections.

A web-based system was developed in order to conduct the participant information and images acquisition, following the five steps aforementioned in sequence. Concurrent participants of the dataset are allowed in this web system. At the first step (Fig. 1-(a)) the presented system required that the user provides his/her identification such as register, and afterwards, answering a questionnaire elaborated to assess demographic information (age, gender, marital status, ethnicity and religion), as well psychiatry symptoms of the participants.

At the second stage, participants were submitted to a test to track their psychiatric symptoms, where questions about depression, anxiety and stress were

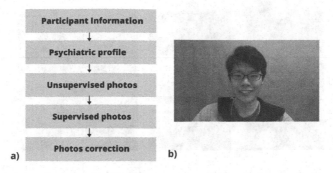

Participant Information

↓

Psychiatric profile

↓

Unsupervised photos

↓

Supervised photos

↓

Photos correction

a)   b)

Fig. 1. General overview of the proposed dataset, having its computational flow in (a), and an example of the acquisition process illustrated in (b).

answered. The following instrument for screening were used with the supervision of a psychiatrist: Beck Depression Inventory (BDI) [12]; Beck Anxiety Inventory (BAI) [13]; Adult ADHD Self-Report Scale (ASRS 18) [14]; Obsessive-Compulsive Inventory-Revised (OCI-R) [15] and Self Reporting Questionnaire (SRQ-20) [16,17].

The final step of our protocol is the acquisition itself, in which before the images capture themselves, a script is executed, so that the characteristics of the images such as color and quality are kept. Thus, the 6 capture sessions of 8 images each, are taken by the participant himself through the system. The third stage is removed from the first 3 sessions, which participants alone must imagine each emotion and save their image.

During the fourth stage, 3 more sessions are held with 8 photos each, but different from the previous one, the participant now has an emoticon as support to reproduce the emotion presented. Due to the fact that the participants take their own photos, in the fifth stage a supervisor was chosen to review possible flaws, such as reddish or blurred images, so if an inconsistencies is found, the images can be retaken.

Regarding the capture environment shown in the Fig. 1-b, a controlled environment was chosen for the realization of this data set, so the images contained in it have a standard background: all photographs were taken with the same camera as model being HP hd-4110, with 13 mega-pixels and connected to a conventional computer via USB. During the acquisition procedure, only one face appears at a time for the participant, avoiding induction for the next emotion expressions. Distinct eight emotion expressions from facial images are shown in the Fig. 2, where the expressions sadness, fear, happiness, disgusting, angry, surprised, contempt and neutral are demonstrated from (a) to (h), respectively.

The full dataset can be downloaded in our institutional web site migma.ufsc.br. To download the dataset, a research term must be signed and submitted electronically. No participant information is released, except photos categorized into expressions. Psychiatric symptoms, on the other hand, is protected under ethical terms and can be released only in the form of meta-data,

Fig. 2. Distinct expressions considered for the proposed dataset.

excluding names and mentions to images. It can be used for further assays involving computer vision associated with the corresponding medical area, opening a wide range of possible studies whose median faces can be correlated to its medical scales to track possible depressive and anxiety symptoms.

4   Results

4.1   Dataset Properties

The target population chosen was college students from different courses and university staffs, reaching a total of 323 participants. All participants completed a structured and self-administered form composed of closed questions about depression, anxiety and attention deficit and hyperactivity symptoms (ADHD). The demographic characteristics of the sample were 200 men (61.91%), 119 women (36.84%) and 3 no reply (0.92%), with an average age of 26.54 years (STD = 8.12; range 18 to 55 years). In additionally, the self-reported skin color showed that most of the sample consisted of self-declared white students (86.06%), followed by mulatto (2.47%), black (5.88%), Asian (1.23%), other mixed colors (0.30%), and 2.78% no declared.

4.2   Dataset Statistical Analysis

Over the obtained dataset, some statistics were obtained in order to validate some premises such as regularity of the expressions and standard deviation. The first analysis performed was the acquisition of the mean faces for each expression, corresponding to a simply summation of the dataset images for each category of emotion, as demonstrated in the Fig. 3. The mean faces need to be registered to a reference planar domain to be representative and avoid deformations and blurring effects. The eyes for each participant were chosen to perform this alignment.

To provide the mentioned statistics, a pre-processing step was applied to extract relevant data features from facial images and correlate them by spatial transformations in an accurate manner. Statistics were extracted from the image

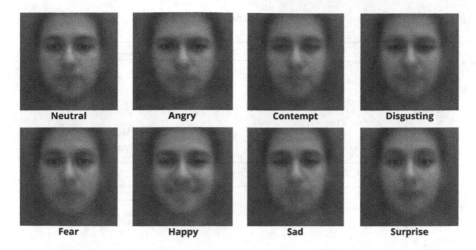

Fig. 3. Images generated summing the faces aligned by eyes for each emotion. The aligner algorithm used is from Dlib package for python 3.7.

domain and converted to a set of 2-dimensional sparse space points, allowing the use of descriptive and inferential statistics metrics, such as central tendency, measures of variance, and permutation tests. For this purpose, the Dlib library was used, being a popular set of algorithms in python package used as feature extractor of facial points. The algorithm was trained from the iBUG 300-W dataset which is used in some machine learning competitions [18]. The data returned by the function provides a cardinality of 68 two-dimensional points of relevant parts of the human face, such as mouth, nose, eyebrows, eyes and the contour of the face.

4.3 Case-Study: Dataset Performance in a Convolutional Neural Network Framework

The second way used to validate this dataset was a case-study, which aims to assess the separability of the classes belonging to the dataset with a trained convolutional neural network.

The processing pipeline performed was initiated with the transformation of the images to gray scale, then the standard facial detection algorithm from the Dlib library was used to perform the alignment. After that, a resizing using cubic interpolation was used to bring the images to 70 × 70 resolution. After the first resizing, a 12-pixel cut for all sides was performed, transforming the images to a 46 × 46 resolution. At that moment, normalization and equalization of the image were carried out, with the intention of reducing the input domain.

The network architecture was implemented using *Tensorflow 2*, with a pattern of *Conv2D* folowed by Batch Normalization and *Max Pooling*. This pattern was used 3 times followed by flatten, a MLP with 2 layers and a softmax for 8 classes. The training parameters were the Adam optimizer, batch size of 128,

Accuracy and Error History

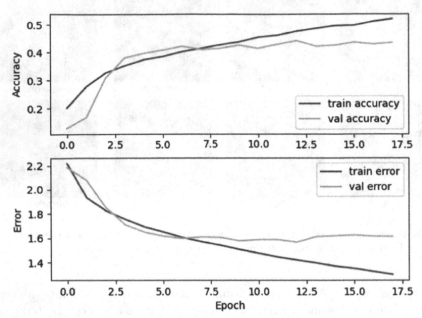

Fig. 4. History of the model training plotted with Matplotlib, a very popular python package.

learning rate of 0.0001. Two callback functions were used in the training of the network, *ModelCheckpoint* for save the model with best accuracy on validation set, and *EarlyStopping* with patience 5.

Sets of training, testing and validation were separated with size respectively 70%, 15% and 15% of the original. The samples of each participant are present in only one of the three sets, making the analysis more robust. The training history can be seen in the Fig. 4. The best model saved has 44% of accuracy on validation set and 41% in the test set.

5   Conclusion and Discussions

In present study a new dataset for emotion recognition from facial images containing more than 15k images and 323 participants was presented. The dataset entitled ******* was designed taking into account higher spatial resolution in a frontal well-controlled environment aiming to cover the most common and widely-used expressions from facial images. Differently from the existing datasets, here we focus in a high-miscegenated source, presenting facial images from the most distinct and rich types, ages and genders, since it was developed in a south American country in an academical environment. Although, this dataset includes induced and non-induced expressions associated with a psychiatric profile for each participant, performed by a physician in the area.

The preliminary statistics we obtained reveals interesting aspects in terms of regularity of the dataset when distinct emotions are compared, indicating those expressions that are consensus and ambiguity among participants. Also, we presented the mean faces obtained in our dataset, which can be used as a general pattern distribution for those expressions in classifiers or decision-making systems.

There are some important points that can limit the aforementioned classifier to achieve a higher performance when compared to other models available over the literature. A first aspect is the larger number of classes presented in our dataset, increasing the number of cluster combinations in the feature-space and the complexity. Moreover, one can observe the inclusion of some properties that may increase the dataset variance, such as the high miscegenation and larger number of participants. Nonetheless, the proposed dataset was designed taking into account a significant increment of the spatial resolution used during the acquisition, which infers directly in terms of execution time for classifiers. An example of a classifier presented a high accuracy is presented in [19], in this article the test is made with dataset JAFFE and the accuracy is 100% however, it has 213 images and 10 participants, which ends up affecting the generalization of the solution.

Acknowledgment. To the CNPq/PIBIC program and UFSC for the grant of the Scientific Initiation scholarship for the research project. This study was approved by the Ethics Committee of Plataforma Brasil, under number 13300919.6.0000.0121. All participants signed an informed consent form before participating in data collection.

References

1. Happy, S., Routray, A.: Automatic facial expression recognition using features of salient facial patches. IEEE Trans. Affect. Comput. **6**(1), 1–12 (2014)
2. Lyons, M.J., Akamatsu, S., Kamachi, M., Gyoba, J., Budynek, J.: The Japanese female facial expression (JAFFE) database. In: Proceedings of Third International Conference on Automatic Face and Gesture Recognition, pp. 14–16 (1998)
3. Anitha, C., Venkatesha, M., Adiga, B.S.: A survey on facial expression databases. Int. J. Eng. Sci. Technol. **2**(10), 5158–5174 (2010)
4. Kanade, T., Cohn, J.F., Tian, Y.: Comprehensive database for facial expression analysis. In: Proceedings Fourth IEEE International Conference on Automatic Face and Gesture Recognition (Cat. No. PR00580), pp. 46–53. IEEE (2000)
5. Lucey, P., Cohn, J.F., Kanade, T., Saragih, J., Ambadar, Z., Matthews, I.: The extended cohn-kanade dataset (ck+): a complete dataset for action unit and emotion-specified expression. In: 2010 IEEE Computer Society Conference on Computer Vision and Pattern Recognition-Workshops, pp. 94–101. IEEE (2010)
6. Pantic, M., Valstar, M., Rademaker, R., Maat, L.: Web-based database for facial expression analysis. In: 2005 IEEE International Conference on Multimedia and Expo, pp. 5-pp. IEEE (2005)

7. Yin, L., Wei, X., Sun, Y., Wang, J., Rosato, M.J.: A 3d facial expression database for facial behavior research. In: 7th International Conference on Automatic Face and Gesture Recognition (FGR06), pp. 211–216. IEEE (2006)
8. Goodfellow, I., et al.: Challenges in representation learning: a report on three machine learning contests (2013). http://arxiv.org/abs/1307.0414
9. Jazouli, M., Majda, A., Zarghili, A.: A $ p recognizer for automatic facial emotion recognition using kinect sensor. in: Intelligent Systems and Computer Vision (ISCV), pp. 1–5. IEEE (2017)
10. Zhang, Y., Ji, Q.: Active and dynamic information fusion for facial expression understanding from image sequences. IEEE Trans. Pattern Anal. Mach. Intell. **27**(5), 699–714 (2005)
11. Tarnowski, P., Kolodziej, M., Majkowski, A., Rak, R.J.: Emotion recognition using facial expressions. In: ICCS, pp. 1175–1184 (2017)
12. Beck, A.T., Steer, R.A., Carbin, M.G.: Psychometric properties of the beck depression inventory: twenty-five years of evaluation. Clin. Psychol. Rev. **8**(1), 77–100 (1988)
13. Beck, A.T., Epstein, N., Brown, G., Steer, R.A.: An inventory for measuring clinical anxiety: psychometric properties. J. Consult. Clin. Psychol. **56**(6), 893 (1988)
14. Kessler, R.C., et al.: The world health organization adult ADHD self-report scale (ASRS): a short screening scale for use in the general population. Psychol. Med. **35**(2), 245 (2005)
15. Foa, E.B., et al.: The obsessive-compulsive inventory: development and validation of a short version. Psychol. Assess. **14**(4), 485 (2002)
16. Harding, T.W., et al.: Mental disorders in primary health care: a study of their frequency and diagnosis in four developing countries. Psychol. Med. **10**(2), 231–241 (1980)
17. de Jesus Mari, J., Williams, P.: A validity study of a psychiatric screening questionnaire (SRQ-20) in primary care in the city of Sao Paulo. Br. J. Psychiatry **148**(1), 23–26 (1986)
18. Sagonas, C., Tzimiropoulos, G., Zafeiriou, S., Pantic, M.: 300 faces in-the-wild challenge: the first facial landmark localization challenge. In: Proceedings of the IEEE International Conference on Computer Vision Workshops, pp. 397–403 (2013)
19. Chen, T., et al.: Emotion recognition using empirical mode decomposition and approximation entropy. Comput. Electr. Eng. **72**, 383–392 (2018)

Construction of Brazilian Regulatory Traffic Sign Recognition Dataset

Rafael Silva[1]($\boxtimes$), Bruno Prado[1], Leonardo N. Matos[1], Flávio Santo[2], Cleber Zanchettin[2], and Paulo Novais[3]

[1] Department of Computing, Federal University of Sergipe, São Cristóvão, Brazil
{rafael.silva,bruno,leonardo}@dcomp.ufs.br
[2] Center of Computing, Federal University of Pernambuco, Recife, Brazil
{faos,cz}@cin.ufpe.br
[3] Department of Computing, Minho University, Braga, Portugal
pjon@di.uminho.pt

Abstract. In this article, we present the Brazilian Regulatory Traffic Sign Recognition Dataset, following the style of the CIFAR10 dataset. A convolutional neural network is also proposed to recognize and identify these traffic signs as a possible aid for ADAS (Advanced Driver Assistance Systems). The developed architecture has thirteen layers, selected after attempts to search for a sufficiently efficient organization. CNN used the RMSProp optimizer, a variant of the stochastic gradient descent technique (SGD), reaching 99.31% accuracy in training and 93.73% in the validation set. This document covers the dataset development process, convolutional neural network architecture, discussions about operation and results.

Keywords: Self-driving car · Traffic sign recognition · Convolutional neural network

1 Introduction

Regulatory signs and guide signs are a road safety item in everyone's daily life of everyone who drives through the streets of large and small cities. With the promise of making traffic safer and less chaotic, autonomous cars emerged, providing support to navigation and identification of danger or attention zones. One of the systems present in autonomous navigation and extensively researched is the Advanced Driver Assistance System (ADAS). Along with ADAS, we have Traffic Sign Recognition and Recognition (TSDR), responsible for identifying and recognizing traffic signs in the vehicles sensors images. In Brazil, the National Transport Council (CONTRAN) regulates the road network's signs through the Brazilian Road Signs Guide.

The present work will focus on vertical regulatory signaling, standardized through the Brazilian Manual of Road Signs Volume I [6]. This manual identifies the colors, shapes, and symbols used to define each road sign. A total of 51 regulatory traffic signs are defined (see Fig. 1). The manual definitions are

© Springer Nature Switzerland AG 2021
J. M. R. S. Tavares et al. (Eds.): CIARP 2021, LNCS 12702, pp. 163–172, 2021.
https://doi.org/10.1007/978-3-030-93420-0_16

Fig. 1. Brazilian traffic sign classes, from class 0 to class 50.

sufficient for human classification, but for an algorithm in a TSDR device, this convention may not be enough. TSDR systems are intended to warn drivers of road signs proximity, even in partial occlusions or bad weather. However, there can still be challenging conditions where the system is not functioning correctly, e.g., the blur caused by the vehicle's movement during capturing an image, lighting, or weather conditions [7]. According to [4], the methods most commonly used by a TSDR system to detect a road sign are defined in three steps: color segmentation, detection, and recognition of the sign. Color segmentation generates tone maps from given color characteristics of the object to be studied. The detection works by selecting the color map areas where the pixels correspond to the searched attribute. The sign recognition stage classifies traffic sign's content, and in this article, a Convolutional Neural Network (CNN) will be used for this task. Road sign recognition has a wide variety of applications, such as driver assistance systems, autonomous driving, navigation systems, and conservation mapping of traffic signs for federal agencies.

We present the Brazilian Regulatory Traffic Sign Recognition Dataset (BRT-SRD), a dataset of more than 24,000 regulation traffic sign images in 51 classes. The choice for creating a new dataset instead of using the existing ones, e.g., GTSRB or BTSD, is due to the peculiarities of Brazilian traffic signs. The German and Belgian plates have colors that are not part of the Brazilian standard and some of their meanings. The article is structured as follows. Section 2 provides an overview of published work on TSDR systems built using CNN. Section 3 describes the proposed architecture, information about the dataset used, and other information about the proposed system. Section 4 lists the experiments and their results. Finally, conclusions and future works are given in Sect. 5.

2   Related Works

Convolutional neural networks are widely used for image classification. It is no different in the field of road sign recognition. In [8], a system is proposed for recognizing traffic signals using convolutional neural networks. In this model, CNN is responsible for learning the characteristics of road signs. The proposed model has three convolutional layers, three max pooling layers, and two fully connected layers. Also, is presented a discriminatory method using Max Pooling Positions (MPPs) as an effective alternative in predicting categorical labels. Y. Sun, P. Ge, and D. Liu [10] used color segmentation to preprocess areas that may contain a road sign. Then, they applied the Hough transformation to detect forms similar to traffic signage. They also used TensorFlow to implement a CNN that could locate road signs with 98.2% accuracy. The authors used the GTSRB (German Traffic Sign Recognition Benchmark) dataset for training and testing. In [8], the substantial difference from the previous one is using a panoramic camera with a 360-degree aperture to capture the surroundings. They also used a CNN for image detection, and they added some image processing techniques to improve image detection accuracy. The overall accuracy of the method was 88.61%. Recently, S. Mehta, C. Paunwala, and B. Vaidya [7] proposed a deep convolutional network to classify road signs. The dataset used to perform the training and tests was the Belgian Traffic Sign Dataset (BTSD). The architecture consisted of three convolutional layers, each accompanied by a max pooling layer, and two fully connected layers. We use a dropout layer between the two fully connected layers. The study showed that with only ten training epochs and the Adam optimizer and Softmax activation, it was possible to achieve 97.06% accuracy in tests with the dataset. In [2], Wael Farag modeled a CNN, which he called WAF-LeNet, to recognize and identify traffic signs. The CNN is a deep fifteen-layer network trained with the Adam optimization algorithm. They added two dropout layers after the fully connected layer. The data set used for training the learning algorithm was the German Road Sign Benchmarks (GTSRB), with some additional signs images collected from the web. The proposed approach proved successful in identifying correctly 96.5% of the testing dataset.

3   Proposed Architecture

3.1   Image Pre-processing

The images used to compile the dataset were downloaded from the Internet and captured by other means. Many of them contained unnecessary artifacts and information, e.g., vehicles, pedestrians, objects of no interest, and had different resolutions. Therefore, before adding the images to the final dataset, we need to preprocess that. The steps used in the preprocessing of the database, depicted in Fig. 2, are as follows:

- **Noise reduction**: A Gaussian Blur with kernel size three is used for this purpose. The Gaussian filter reduces the image noise and eliminates insignificant details.
- **Map to HSV color space**: The HSV color space is a suitable space to distinguish objects with very distinctive colors, such as regulatory road signs.
- **Color detection**: Brazilian traffic signs have predominant colors that allow them to stand out against the background landscape, as shown in Table 1. This enables the segmentation of targets by thresholding. The operation result

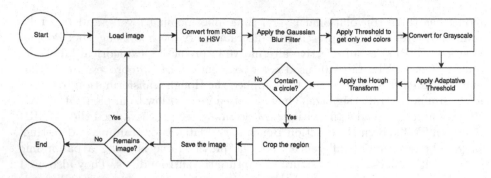

Fig. 2. Diagram showing the operation to obtain traffic sign images from any raw image.

is a binary image in which the values hue segments the targets in the HSV space.

- **Gray scale**: Images are converted from RGB to grayscale.
- **Image enhancement**: As described by [11], captured images under different perspectives and environmental conditions, so they can show significant differences even though they belong to the same class. In order to improve the distinguishing features, histogram equalization is applied to the images.
- **Adaptive threshold**: The threshold acts like an image segmentation, activating only the regions of interest.
- **Hough transform**: Use this transform to capture the position of elements with specific shapes in an image [11], in this article, circles. The segmented circles are scaled to 32 × 32 pixels, creating the database's image.

Table 1. The use of colors in Brazilian regulatory signs must be done according to the criteria below and the Munsell standard indicated.

Color	Munsell pattern	Regulatory signs use
Red	7, 5 Red 4/14	R-1 background; border and stripe
Black	Neutral (absolute) 0, 5	Regulatory symbols and legends
White	Neutral (absolute) 9, 5	Regulatory signs fund; letters of the R-1 sign

3.2   The Dataset

The dataset used in this work is composed of images obtained manually using the Google Street View tool[3], through Google Images[4] and frames extracted from videos recorded on smartphone cameras. The raw images obtained are taken from different Brazilian scenarios (rural and urban). It describes some climatic

---

[3] https://www.google.com.br/intl/pt/streetview/.
[4] https://www.google.com/imghp?hl=en.

situations, e.g., sun and rain, in addition to the movies of the mobile devices taken at different times, e.g., dawn, daytime, and twilight, which gives the base good homogeneity.

Dataset Structure. The structure being considered for the proposed dataset follows the standard used by the CIFAR10 dataset. Each image has a resolution of 32 × 32 pixels with three color channels. Each entry in the dataset consists of a 3073 integer vector where the first index of each corresponds to the label, and the other to the image, where every 1024 elements have a channel color, in order (i) red channel; (ii) green channel; and (iii) blue channel.

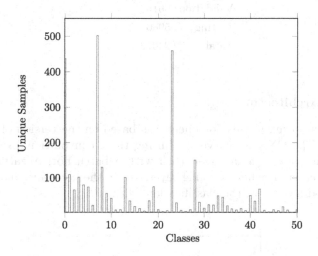

Fig. 3. Distribution of number of instances over classes in the dataset.

The data obtained after the preprocessing step consists of 3076 traffic regulation signs, and the distribution of the images per class can be seen in Fig. 3. The traffic signs are not centering with the image border. Labeling of classes in the dataset was done manually. For each image, the corresponding label has been assigned according to as follows: for the first traffic sign defined in [6] (traffic sign R-1), the initial class in the dataset is defined as zero, from this point, the initial value of the class is increased by one with each new traffic sign class is added, until all the traffic sigs classes have been identified.

However, in the initial dataset, many classes did not have sufficient samples for representing a traffic sign during the network training, in addition to class imbalance. Hence, it was necessary to use Data Augmentation (DA) to generate new samples. DA is a process in which new images are generated through various transformations and filters, e.g., rotations, blurs, inversions, and rotations [3]. The technique simulated environmental situations, e.g., ambient light, rain, snow, morphological operations such as image blur, salt and pepper filter, and contrast adjustment. At the end of this phase, the dataset is formed by a total of 24072 images of traffic signs. In the Fig. 4 we can observe the new distribution of images by classes after the application of the DA step in the initial dataset. We randomly divide all images into three blocks for the final assembly of the dataset: training, validation, and testing images. The distribution of the instances in the dataset can be seen in Table 2.

Table 2. Dataset images distribution.

Dataset	No. of images
Training	12071
Validation	6015
Testing	5986
Total	24072

3.3 CNN Architecture

The architecture used in this document was based on the version of LeNet presented in [5]. The CNN was developed using three convolutional sets arranged as follows: one convolutional layer, each with a batch normalization layer, an activation layer, and a max pooling layer. After the convolutional block, one fully connected layer does the classification work.

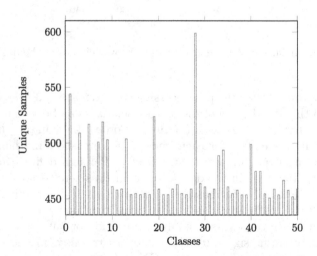

Fig. 4. Distribution of number of instances over classes in the dataset after the DA step.

This article uses the softmax activation function in the last layer and the ReLU function in the convolutional layers. ReLU, short for Rectified Linear Unit, produces results in the range $[0, \infty[$. The Softmax function is usually used in the output layer because it produces probabilistic values for each attribute. The optimization algorithm chosen for the work was Root Mean Square Propagation. RMSProp is a method used for optimization with an adaptive learning pace. This algorithm divides the learning rate by an exponentially decreasing mean of the square of the gradients [9], addressing the disappearing gradient problem. In Table 3, rho and momentum are parameters of optimizer RMSProp. The regularization process introduces a new term to prevent *overfitting*. It helps to avoid the linear models overfitting with the training dataset penalizing the extreme weight values. At this work, we utilize the L2 Regularization at the convolutional and dense layers.

Table 3. This table shows the hyperparameters (h-params) values choosed by the Bayesian Optmization.

H-Params	Min. value	Max. value	Best value
Conv0 Filters	32	64	32
Conv0 L2 Regularization	1e-5	1e-3	1e-3
Conv1 Filters	32	64	64
Conv1 L2 Regularization	1e-5	1e-3	1e-05
Conv2 Filters	32	64	32
Conv2 L2 Regularization	1e-5	1e-3	1e-05
L2 Regularizarion Dense	1e-5	1e-3	1e-3
Rho	0.75	0.9	0.90
Momentum	0	0.2	0.0
Learning rate	1e-4	1e-3	1e-3

Hyperparameters are training control variables, basically neural network configuration variables. The hyperparameters, such as learning rate or regularization, are selected by Bayesian Optimization (BO). BO is an approach that uses Bayes Theorem to direct the search to find the minimum or maximum of an objective function. Bayesian Optimization is a strategy to find the extreme of a function that may not have a closed expression but can obtain observations in samples. This method is advantageous when the cost of function evaluation is high when there is no access to its derivatives and/or when the problem is not convex [1]. As the library chose for this work was the TensorFlow[1], we use the Keras Tuner to apply the BO. The choice of the search space is a subjective task. How the BO operates and probability distributions for each parameter have to be set by a user. We are chosen to define a range of hyperparameters and let

---

[1] https://www.tensorflow.org/.

the optimization algorithm choose the best set. For the initial model search, we defined the baseline structure as in Table 3.

The BO objective is to locate the set of hyperparameters that would maximize the validation accuracy. Therefore, we are configuring the bayesian optimizer to perform 20 searches for the best set of configurations, executing each set for up to 50 epochs. For training, we used the available hardware in Google Colab[2] free account. We use the step decay technique for the learning rate parameter. During the first 50 epochs, the value remained unchanged, following the best value indicated by the BO. However, the learning rate value was reduced from this point on, using the values 0.1, 1-e2, and 1-e3, until the total of 200 epochs was completed. The batch size selected for training was 32.

4   Results and Discussion

In this section, we will present the results obtained after training the proposed CNN model. The dataset developed as described in Sect. 3.2 was used to evaluate the architecture proposed in this paper. This dataset has different images with all 51 traffic regulation signs defined by Brazilian legislation. Table 4 displays the average of the values obtained by evaluating the metrics accuracy, precision, recall, and F1-Score. The model presented reached a maximum accuracy of 99,31% in the training set. The accuracy achieved was 93,73% in the validation set, indicating that the model was properly trained. The best loss values obtained were 0.3737 during training and 0.5574 for the validation set.

Table 4. Training, validation and testing set results (in percentage).

Metric	Training	Validation	Testing
Accuracy	99,31	93,73	91,41
Precision	99,52	96,37	91,55
Recall	99,02	92,45	89,28
F1-Score	99,27	94,34	99,27

The model predicted which test traffic signs were accurate to 91.41%. The confusion matrix displays the classification frequencies for each class of the model. It will show us the frequencies of true positives, false positives, false true, and false negatives. In the end, we have the confusion matrix generated by the model in Fig. 5. Looking at the matrix, it appears that classes 20, 39, 44, and 46 were the ones that obtained the lowest accuracy during classification, with a percentage below 75%. Still observing the confusion matrix, it is noted that the neural model encountered difficulties in recognizing some other classes. Classes 39–44, 46–45, and 11–37 are very similar. In this way, the neural network found it challenging to distinguish them correctly.

---

[2] https://colab.research.google.com/.

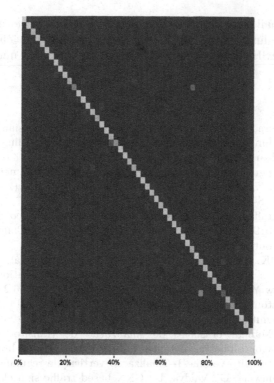

Fig. 5. Confusion matrix.

5   Conclusion and Future Work

In this paper, we describe a pipeline capable of accurately recognizing traffic signs. The proposed approach applies different robust methods used by the scientific community, obtaining good results in traffic signs classification. CNN's architecture allowed the detection of traffic signs in various conditions with a test accuracy of 91.42%. In training the neural network, we find it difficult to optimize hyperparameters correctly, so we resort to Bayesian Optimization. Although the parameters described in Table 3 were the best during the optimization attempts, it is understood that there is still room for improvement. Besides, a new set of data was presented regarding the Brazilian traffic regulation signs. Another point to be developed is the dimensioning of the CNN model to be ported to embedded devices without degradation of accuracy.

We have identified some lines that we are currently exploring as future work. In neural network training, hyperparameter optimization is developed using systematic methods. We are currently using the Bayesian Optimization method. We are also developing model compression so that it can be transferred to embedded devices. In particular, we intend to generate a compact enough model to be delivered in an ESP32 MCU device that, due to its small size, can be attached to the chassis of a vehicle and used to assist drivers in real-time. On the other hand, the great challenge is to make decisions with a good performance given the existing hardware limitations. Specifically, this device only has 520 KB of RAM and 4 MB of FLASH, in which both the model and the data must be stored.

References

1. Brochu, E., Cora, V.M., de Freitas, N.: A tutorial on bayesian optimization of expensive cost functions, with application to active user modeling and hierarchical reinforcement learning (2010)
2. Farag, W.: Recognition of traffic signs by convolutional neural nets for self-driving vehicles. Int. J. Knowl. Based Intell. Eng. Syst. **22**, 205–214 (2018). https://doi. org/10.3233/KES-180385
3. Islam, M.T.: Traffic sign detection and recognition based on convolutional neural networks. In: 2019 International Conference on Advances in Computing, Communication and Control (ICAC3), pp. 1–6 (2019)
4. Kim, J., Lim, K., Uh, Y., Kim, S., Choi, Y., Byun, H.: Real-time korean traffic sign detection and recognition. In: ICUIMC 2013, Association for Computing Machinery, New York (2013). https://doi.org/10.1145/2448556.2448575
5. Lecun, Y., Bottou, L., Bengio, Y., Haffner, P.: Gradient-based learning applied to document recognition. In: Proceedings of the IEEE, pp. 2278–2324 (1998)
6. LTDA, V.I.: Manual brasileiro de sinalização de trânsito volume i. www. capacidades.gov.br/biblioteca/detalhar/id/118/titulo/manual-brasileiro-de-sinalizacao-de-transito-volume-1--sinalizacao-vertical-de-regulamentacao
7. Mehta, S., Paunwala, C., Vaidya, B.: CNN based traffic sign classification using adam optimizer. In: 2019 International Conference on Intelligent Computing and Control Systems (ICCS), pp. 1293–1298 (2019)
8. Qian, R., Yue, Y., Coenen, F., Zhang, B.: Traffic sign recognition with convolutional neural network based on max pooling positions. In: 2016 12th International Conference on Natural Computation, Fuzzy Systems and Knowledge Discovery (ICNC-FSKD), pp. 578–582 (2016)
9. Ruder, S.: An overview of gradient descent optimization, pp. 1–14 (2016)
10. Sun, Y., Ge, P., Liu, D.: Traffic sign detection and recognition based on convolutional neural network. In: 2019 Chinese Automation Congress (CAC), pp. 2851–2854 (2019)
11. Yang, Y., Luo, H., Xu, H., Wu, F.: Towards real-time traffic sign detection and classification. IEEE Trans. Intell. Transp. Syst. **17**(7), 2022–2031 (2016)

Japanese Kana and Brazilian Portuguese Manuscript Database

Luiz Fellipe Machi Pereira[1]([✉]), Fabio Pinhelli[1], Edson M. A. Cizeski[1],
Flávio R. Uber[1], Diego Bertolini[1,2], and Yandre M. G. Costa[1]

[1] State University of Maringá (UEM), Maringá, PR, Brazil
{fuber,yandre}@din.uem.br
[2] Universidade Tecnologica Federal do Paraná (UTFPR), Campo Mourão, PR, Brazil
diegobertolini@utfpr.edu.br

Abstract. This work introduces a Japanese Kana and Brazilian Portuguese database (JKBP), a novel database made freely available aimed at supporting the development of manuscript-based recognition tasks. As far as we know, this is the first database created and made available to the research community composed of Portuguese and Japanese manuscripts. The samples were collected from 57 volunteers. Each volunteer provided five manuscript samples written in Brazilian Portuguese, and five more samples written in Japanese Kana. We also describe baseline experiments accomplished on JKBP exploring some of its different possibilities of classification tasks, such as syllabary and writer classification. We conducted the experiments using Speed Up Robust Features (SURF), a texture descriptor which has been successfully used to capture the textural content in this application domain. Support Vector Machine (SVM) was used for classification, given its good performance in other works available in the literature involving similar tasks. The obtained results showed that the zoning process and use of late fusion provide a meaningful increase of the performance in some scenarios. The best accuracy rates achieved were 97.98% and 83.77% on writer identification task using Portuguese and Japanese manuscripts respectively, and 100% in syllabary classification.

Keywords: Manuscripts database · Multi-script handwritten database · Writer identification · Pattern recognition

1 Introduction

Identifying the writer is a recurring task, commonly needed in some specific application domains, such as forensic science, for example. In this type of application domain, the need for an accurate identification of the person who wrote a particular manuscript is a common problem. Automatic identification of the writer is a topic widely discussed by the pattern recognition research community. With the increase in research efforts on this problem, the need for databases of manuscripts arises, especially if we take into account databases made publicly available, considering diverse languages and scripts, and created under conditions that emulate as much as possible the real-world scenario.

© Springer Nature Switzerland AG 2021
J. M. R. S. Tavares et al. (Eds.): CIARP 2021, LNCS 12702, pp. 173–183, 2021.
https://doi.org/10.1007/978-3-030-93420-0_17

At this point, it is important to remind some basic concepts regarding classification tasks based on manuscript images. One of this concepts regards to online *versus* offline modes [19]. The online mode is characterized when the manuscript is acquired with additional information, beyond the manuscript image itself. In this case, information regarding the writing movement, like direction and/or pen pressure is also collected. Online manuscripts can be acquired by means of specific electronic devices, like a graphic tablet for example. In another vein, the offline mode refers to those situations in which only the manuscript image itself is collected and made available. Obviously, the use of offline manuscripts is much more challenging, since we have less information, but we must not to lose sight that this scenario is much more realistic, since it is much more common to have only the manuscript image itself available. Moreover, there are still other external factors that can introduce a bias affecting the writing on the offline manuscript classification mode, such as the paper texture, and the pen flow. In addition, the way how the manuscripts originally described in the paper are transformed to a digital format can introduce some noise as well.

Text-dependence is another important concept in this field of investigation [2]. A manuscript database is considered text-dependent when all the volunteers who contributed to write the manuscripts are asked to make copies of previously defined texts. On the opposite way, a database is classified as text-independent when the writes are supposed to freely describe the manuscript, without a previous definition of the textual content. Text-dependent databases use to be much more suitable to the accomplishment of investigations on this subject, once in this mode the volunteers use to contribute with a more homogeneous amount of manuscript content, making more feasible the creation of more reliable classifiers.

There are still some important concepts about the modern Japanese writing system to take into account here. In this type of writing, a combination of logographic kanji (adopted Chinese characters) and Kana is used. Kana itself is formed by two syllabaries: Hiragana, that contains 46 base characters used mainly for native or naturalized Japanese words and grammatical elements, and Katakana, also with 46 base characters and used mainly for foreign words and names, scientific names, transliterations, onomatopoeia and, sometimes, for emphasis. Most phrases written in Japanese use a mixture of Kanji and Kana [8].

After presenting some important concepts about this field of research, we believe the reader will be more comfortable to follow the remaining of this paper. In this work, we introduce a new offline, multi-script and text-dependent manuscripts database. The database is composed of more than 500 images of manuscripts written in Brazilian Portuguese and Japanese Kana, provided by 57 volunteer writers. Each volunteer contributed with five samples in Brazilian Portuguese and five more samples in Japanese Kana. The creation of this database is justified by the scarcity of handwriting databases available for Japanese Kana, and also the scarcity of databases designed with different scripts available for all the volunteers, such a way that the database can also support investigations on writer identification on the multi-script scenario. The database introduced here was especially curated for investigations on manuscript classification tasks on both aforementioned languages. Beyond the writer identification task, other tasks can be explored, since there are also labels for age group and gender.

2   The Dataset

The Japanese Kana and Brazilian Portuguese (JKBP) is a public database, made available under request[1], that supports the development of research on manuscript classification tasks, like writer identification. The database is composed of 570 samples taken from 57 different contributors, all of whom are proficient in writing in Portuguese and Japanese, with the vast majority having Portuguese as mother tongue, and Japanese as second language. They are also from different genders, age groups and levels of education. The distribution of the volunteers among these categories was made balanced as much as possible, and it can be seen in Table 1. A complete balancing was not possible due to the lack of volunteers for some specific categories (e.g.: men aged over 55 years old).

Table 1. JKBP contributors and samples distribution by gender and age group.

Contributor ID	Gender	Age group
1 to 8	Female	0 to 14
9 to 15	Male	
16 to 26	Female	15 to 24
27 to 36	Male	
37 to 46	Female	25 to 54
47 to 51	Male	
52 to 56	Female	55 over
57	Male	

The manuscript samples were obtained such a way that the volunteers were asked to copy ten different reference texts, five written in Portuguese and five in Japanese. We concentrate on using only the Kana syllabary for Japanese texts given its reduced number of characters in relation to Kanji and, in general, being the first syllabary taught in language schools. The ones written in Japanese were divided as follows: two written only in Hiragana, one written only in Katakana and two texts formed by the combination of Hiragana and Katakana. It is important to emphasize that the texts written in Portuguese are not translations of the texts in Japanese, that is, all chosen texts are unique.

All the manuscript samples collected from the collaborators were digitized, using the same scanner with 300 dpi, RGB color space, and they were saved in PNG format with identification of age group, gender, age and a unique (anonymous) identifier of the author. The available database includes the original samples and cut versions of the handwritten regions, highlighting the content of the writing and excluding blank areas that correspond to the margins.

As we can see in Fig. 1a and 1b, samples of the same author, on Portuguese or Japanese scripts are very different, whether in syllabary or even in spacing between words. This is something that naturally happens on the writing process, and makes the multi-script scenario even more complex.

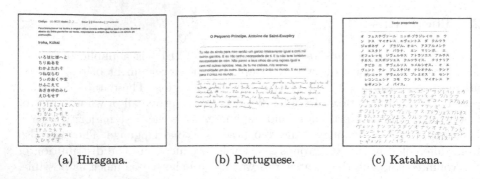

(a) Hiragana.   (b) Portuguese.   (c) Katakana.

Fig. 1. Examples of parts of manuscripts written by the same author.

Table 2 shows the transcript of excerpts taken from the database. In texts 1, 2 and 10, originated from poems, the presence of the character "/" indicates a mandatory line break. Due to space restrictions, aiming to suitably fulfil the table, we decided to keep this notation here. It is worth mentioning that texts 1, 2 and 6 are pangrams, that is, a sentence in which all the letters of the alphabet of a given language are used and no repeated. In case of texts 1 and 2, all modern ideograms of Hiragana syllabary and in case of text 6, all letters, including accented ones, of Brazilian Portuguese alphabet. Text 3, written entirely in Katakana, describes the Japanese-Brazilian Festival that takes place annually in

Table 2. Transcription of selected excerpts used to conduct the writing.

# id.	Syllabary	Transcription	Extracted from
1	Hiragana	いろはにほへと／ちりぬるを／わかよたれそ／つねならむ／ういのおくやま／けふこえて／あさきゆめみし／えひもせす	Iroha by Kūkai
2	Hiragana	あめ つち ほし そら／やま かは みね たに／くも きり むろ こけ／ひと いぬ うへ すえ／ゆわ さる おふせよ／えのいえを なれいて	Ametsuchi no Uta by 9th century AD
3	Katakana	オ フェスチヴァール ニッポ-ブラジレイロ エ ウン ドス マイオレス エヴェントス ダ クルツラ ジャポネザ ノ ブラジル. オコヘ アヌアルメンテ ノ エスタド ド パラナ, エン マリンガ, エ オフェレッセ ジヴェルサス アトラショエス アルチスチカス, エスポジショエス クルツライス, クリナリア チピカ エ ヂヴェルソス コメルシオス. オ エヴェント テン プレスチジオ ナシオナル, テンド ガンニャド ヂヴェルソス プレミオス エ セント レコンニェシド コモ ウン ドス マイオレス ド セギメント ノ パイス.	Created by the authors
4	Kana	ミラーさんは まいあさ 7じに おきます. あさごはんは いつも パンと コーヒーです. でんしゃ で かいしゃへ いきます. かいしゃは 9じから 5じまでです. 7じに うちへ かえります. 7じはん に ばんごはんを たべます. それから テレビを みます. えいごの しんぶんを よみます. よる 12じに ねます.	Minna no Nihongo by 3a Corporation book
5	Kana	この どうぶつの なまえをしって いますか. カンガルーです. オーストラリアに すんで います. 1778ねんに イギリスの キャプテン・クックはふねで オーストラリアへ いきました. そして, はじめて この どうぶつをみました. クック は オーストラリアの ひとに この どうぶつの なまえ を しりたい と いいました. その ひとは オーストラリアの ことば 「カンガルー（わたしはしらない）」といいました. それを きいて, イギリスじんは みんな この どうぶつの なまえは 「カンガルー」だと おもいました. それから, この どうぶつの なまえは 「カンガルー」に なりました.	Minna no Nihongo by 3a Corporation book
6	Roman	À noite, vovô Kowalsky vê o fmã cair no pé do pinguim queixoso e vovó põe açúcar no chá de tâmaras do jabuti feliz.	Unknown author
7	Roman	O amor não mira cumprimentos nem guarda termos de razão em seus discursos, e tem a mesma condição da morte: que assim acomete os grandes palácios dos reis como as humildes cabanas dos pastores, e quanto toma posse de uma alma, o primeiro que faz é tirar o medo e a vergonha.	Dom Quixote by Miguel de Cervantes
8	Roman	Cada estação da vida é uma edição, que corrige a anterior, e que será corrigida também, até a edição definitiva, que o editor dá de graça aos vermes.	Memórias Póstumas de Brás Cubas by Machado de Assis
9	Roman	Tu não és ainda para mim senão um garoto inteiramente igual a cem mil outros garotos. E eu não tenho necessidade de ti. E tu não tens também necessidade de mim. Não passo a teus olhos de uma raposa igual a cem mil outras raposas. Mas, se tu me cativas, nós teremos necessidade um do outro. Serás para mim o único no mundo...	O Pequeno Príncipe* by Antoine de Saint-Exupèry
10	Roman	E eu vos direi: "Amai para entendê-las!/ Pois só quem ama pode ter ouvido/ Capaz de ouvir e de entender estrelas."	Via Láctea by Olavo Bilac

* Brazilian version of Le Petit Prince.

1 https://lfmp.github.io/JKBP/.

the city of Maringá, in the south of Brazil. Texts 4 and 5 were extracted from the book "Minna No Nihongo", because it is commonly used in the teaching of Japanese in language schools that contributed providing volunteers to the development of this research. Texts 7 to 10 were taken from the classical Brazilian literature, they are often considered mandatory reading for entrance exams.

3 Related Databases

In recent years, several databases focusing on the writer identification task have been published in the literature [11–16]. However, few of this databases can be used on the multi-script writer identification scenario. In this section, we present an overview of some databases composed of handwritten texts in Japanese or Portuguese and works that presented multi-script databases.

The JEITA-HP, presented in [9], is a database made up of handwritten kanji samples collected from 570 writers. In this database, each of the writers contributed 3,306 images from 3,214 categories, in which each Kanji character was written once, while each of the kana/alphanumeric characters was written twice.

A handwritten Kanji characters database was introduced by [17] in literature. The database was composed by 100 kanji classes frequently used for Japanese place names, with 50 samples for each class, written by 100 volunteers, totaling 5,000 samples. It was properly organized for author identification investigations.

The "ETL Character Database" is a database of images composed of about 1.2 million handwritten and machineprinted numerals, symbols, Latin alphabets and Japanese characters, compiled in nine datasets. This database has been prepared by the Electrotechnical Laboratory (currently reorganized as the National Institute of Advanced Industrial Science and Technology (AIST)), under cooperation with the Japan Electronic Industry Development Association (currently reorganized as Japan Electronics and Information Technology Industries Association), universities and other research organizations for character recognition researches from 1973 to 1984. The ETL-1 dataset can be used for writer identification, this set contains 99 character classes written by 1,445 writers.

The Brazilian Forensic Letter Database (BFL) [7] is composed of manuscripts in Portuguese, great part of them following the text-dependent mode. A total of 315 writers contributed with three letters each, written on plain A4 sheets.

An important contribution on the multi-script scenario was the creation of the CVL database [11], this database consists of RGB images with handwritten texts described in German and English. In total, 311 different writers copied six different texts in English and one text in German.

The QUWI database [13] contains manuscript documents taken from 1017 writers who wrote a total of 4068 documents in Arabic and English languages. Part of this public database was used in the ICDAR 2015 Competition on Multi-script Writer Identification and Gender Classification [6].

LAMIS-MSHD database [5] is a multi-script offline handwriting database comprising 600 text samples in Arabic and 600 in French, 1300 signatures and 21,000 digits taken from 100 volunteers selected at random in Tebessa (Algeria).

The ALTID database [4] includes texts printed in Arabic and Latin with various fonts and sizes, as well as handwritten texts in Arabic and Latin. The handwritten part consists of 460 blocks of text in Arabic and 582 in Latin, written by seventeen individuals of different ages and educational levels.

4   Experimental Settings

In this section we describe the protocol used for feature extraction in Subsect. 4.1, and details of the classifiers used for identification in Subsect. 4.2.

4.1   Feature Extraction

Digital images in gray scale can be defined as an array of pixels, each pixel having a brightness value between 0 and 255. The images that present a low value of brightness, mainly in the proximity of words or ideograms, may indicate aspects such as the pen pressure applied at the time of writing. In this context, there are inherent characteristics that may contain a rich textural content, allowing to investigate a wide variety of approaches.

In this work we use Speed Up Robust Features (SURF) [1] as the texture descriptor to be applied on the manuscript images. This texture descriptor was chosen to carry out the experiments due to the good results previously obtained on the writer identification task, as shown in [18] using Document Filter[2] proto-col and in [20] presenting results from the use of texture descriptors with entire documents, texture compression and dividing the image into blocks. The idea behind it is to extract points of interest from different regions of a given image. Points of interest are detected based on the Hessian matrix and the construction of a square region around each identified points of interest. Next, SURF gen-erates a feature vector with 257 features, formed by concatenating information from the sub-regions around each point of interest. More details of SURF in [1].

Experiments dividing the images into blocks without overlapping were made inspired by [20]. In our tests, we limited the number of vertical and horizontal divisions to one, that is, combinations of divisions were experimented with the number of divisions varying from zero to one, so the document was divided into a maximum of four blocks, as shown in Fig. 2. After dividing the image into blocks, the SURF descriptor was applied to each block in order to extract features. When performing this procedure we can find good local points of interest that would not be considered in the entire image.

---

[2] Document filter is a protocol which determines that all the blocks taken from the same manuscript sample must be placed on the same fold.

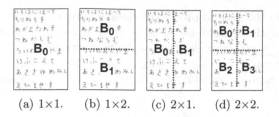

(a) 1×1.   (b) 1×2.   (c) 2×1.   (d) 2×2.

Fig. 2. Examples of block division.

4.2 Classifiers

The classification was performed using Support Vector Machine (SVM) because it has already been successfully used for classification tasks in this scenario, as shown in the works [2,18,20]. For experiments, the classification was performed using C-SVM present in the LibSVM framework [3] with radial basis function (RBF) kernel and the optimization of hyper-parameters made using grid search.

To increase accuracy results and to make it more reliable, K-fold Cross-Validation (CV) was applied on all tests, but with different configurations. It is worth mentioning that in the writer identification task performed here, all the samples of the same author were distributed to different folds, in other words, the M_1 manuscript of individual I_k was allocated to fold F_1, the M_2 manuscript of the same author was allocated to fold F_2 and so on until the fold F_5 with manuscript M_5[3]. On the other hand, in the script identification task, the separation of samples used a different criterion, all samples from the same author remained in the same fold. So, samples from the same author never participate at the same time on the training stage and on the testing stage. It was made in script identification aiming to avoid a bias by author.

In this work, we have adopted the Document Filter (DF) protocol. DF introduces a mandatory restriction that prohibits the use of data taken from different fragments of the same manuscript at the same time in the training and test sets. Investigations carried out by [18] showed that the use of the DF protocol implied, in the most extreme case, a decrease in the rate obtained by 13.95% percentage points in relation to the same test without the use of the protocol. In order to avoid biased results, the DF protocol was applied to the experiments that used block divisions performed here. To ensure that the protocol was properly applied, all P_t parts taken from a document P were placed into the same fold. In this case t refers to the total number of blocks, given by $1+m$ vertical divisions multiplied by $1+n$ horizontal divisions $(1+m \times 1+n)$. The SVM predictions were obtained for each block individually, and we merged the predictions obtained from them to get a single final decision, for the document as a whole. We performed tests using Sum, Product and Max fusion rules [10].

---

[3] It is important to remark that the document filter has not been broken. Do not to confuse the placement of samples from the same author in different folds, with the placement of data taken from blocks of the same document in different folds.

Figure 3 shows the general process of classification used in this work. As seen in this figure, the JKBP database step by pre-processing, in which the samples are transformed into gray scale, then go through the process of dividing into blocks and later have their features extracted. The classification process and fusion of classifiers to obtain the final decision occurs as previously presented.

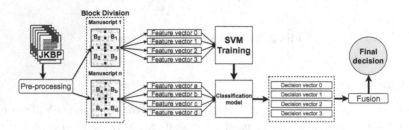

Fig. 3. Classification scheme for process with four blocks division and late fusion.

5   Experimental Results and Discussion

Aiming at exploring the different possibilities of classification using the JKBP database, we have conducted experiments on syllabary and writer recognition tasks. Firstly, the results obtained on experimental variations for the writer identification task are presented in Subsect. 5.1. After that, the protocols used to obtain the higher accuracy rates on syllabary identification are presented in details. The rates presented following range from zero to one.

5.1   Writer Identification

For writer identification experiments, two different sets of experiments were accomplished. The first set refers to tests with manuscripts in Portuguese only and the second set refers to tests performed using only Japanese manuscripts. In these sets, the classification setup was evaluated considering the following variations: i) feature extraction using the entire document; ii) feature extraction using image zoning followed by late fusion; iii) different amount of samples;

We will start by exposing the results obtained on experiments in which the images have not been zoned. In this scenario the SURF descriptor was applied in each sample of each author and a 5-fold CV was made, such a way that each fold has no more than one manuscript of the same author and taking care that each fold has the same amount of samples. This configuration produced adequate results in author recognition task using samples in Portuguese, with an accuracy of 0.9509 ($\pm$ 0.0288), while the results for samples in Japanese were substantially lower, 0.5895 ($\pm$ 0.2179) of accuracy.

In order to improve the results, the process of classification described in Sect. 4.2, using block division, was applied using 5-fold CV again. In tests with both languages, the accuracy rates produced using the Product rule to support late fusion surpassed those obtained using Max and Sum rules. On one hand, the accuracy for writer recognition using manuscripts in Portuguese slightly increased to 0.978947 ($\pm$ 0.0288). On the other hand, this new protocol has increased the accuracy for Japanese tests, becoming equal to 0.7754 ($\pm$ 0.2516).

The previous results, the confusion matrix and the results per fold on writer identification with Japanese manuscripts, led us to notice that when the folder selected for test was the folder with only katakana samples, the obtained rates were the worst by far. It can be explained by the imbalance in the number of samples of each syllabary, while there are two texts completely written in Hiragana, and another two texts with much more characters in Hiragana than in Katakana, there is only one text completely written in Katakana. To validate this suspicion, an experimental variation was carried out. In this new experiment, all Katakana samples are removed from test and the number of folds was decreased to four. The best accuracy rate obtained in this variation was 0.8837 ($\pm$ 0.043860), using four blocks division and the Product as fusion rule. These results showed that the suspicion raised earlier seems to make sense.

5.2 Syllabary Identification

In this Subsection we will describe the procedures adopted to classify the manuscripts among three classes according to their syllabary, that is, Hiragana, Katakana or Roman, the latter corresponding to the manuscripts in Portuguese. For carrying out the current tests it is important to note that the class referring to Hiragana was composed of the manuscripts in Hiragana, and the manuscripts that used Hiragana with Katakana.

For the experiments with four and five folds, the presented results were not satisfactory since the classifiers always predicted the same class (Roman) for all samples. This result probably happened due to the imbalance between the number of samples per class. Based on this hypothesis, we decided to perform new tests making undersampling on the Roman class. For that, we randomly excluded three of five samples for that class, taking care to select always the same samples for all the writers. When applying this strategy, the 570 manuscripts were reduced to 399. These in turn were divided into three-folds, that is, each fold composed of 133 samples, this set being composed of all seven manuscripts of each of the nineteen participants per fold.

Using three folds and dividing each image in four blocks, the setup with best values of accuracy and F-Measure was the one in which the manuscripts 6, 7 and 8 of each writer were not used. The best value of accuracy, equal to 1 and with standard deviation equal to 0, was obtained using the Sum rule.

6   Concluding Remarks

In this work, we presented the JKBP, a novel database made freely available to the research community, aimed at supporting the development of manuscript-based recognition tasks. The database is composed of ten handwriting texts, five in Brazilian Portuguese and five in Japanese, taken from 57 volunteers, totalling 570 images of manuscripts.

Experiments with author and syllabary recognition and their results were also presented. Features were extracted using the SURF descriptor, and SVM was used for classification. Aiming to get good performances and unbiased results, we have tried block division of images followed by late fusion, and we have also taken into account the restrictions imposed by the Document Filter (DF) protocol.

In future, we intend to expand the JKBP, increasing the number of contributors, introducing samples with kanji, and more samples with Katakana. We also plan to carry out new experiments, this time making use of dissimilarity to identify the author in the multi-script scenario and making tests with others classifiers, in order to compare the results and increase the classification rates.

Acknowledgement. The authors thank Professor Kiyomi Kimura Fugie, from the Institute of Japanese Studies (IEJ) of the State University of Maringá (UEM) for her assistance in the preparation and revision of the texts, to the IEJ and the Japanese Language School of Maringá, part of the Cultural and Sports Association of Maringá (ACEMA), for providing space and students for the collection of manuscripts, to the Tutorial Education Program, developed by Ministry of Education, in particular the PET-Informática/UEM, and to the National Council for Scientific and Technological Development (CNPq), for having provided financial support.

References

1. Bay, H., Tuytelaars, T., Van Gool, L.: SURF: speeded up robust features. In: Leonardis, A., Bischof, H., Pinz, A. (eds.) ECCV 2006. LNCS, vol. 3951, pp. 404–417. Springer, Heidelberg (2006). https://doi.org/10.1007/11744023_32
2. Bertolini, D., Oliveira, L., Justino, E., Sabourin, R.: Texture-based descriptors for writer identification and verification. Expert Syst. Appl. **40**, 2069–2080 (2013)
3. Chang, C.C., Lin, C.J.: Libsvm: A library for support vector machines. ACM Trans. Intell. Syst. Technol. **2**(3), 1–27 (2011)
4. Chtourou, I., Rouhou, A.C., Jaiem, F.K., Kanoun, S.: Altid : arabic/latin text images database for recognition research. In: 2015 13th International Conference on Document Analysis and Recognition (ICDAR), pp. 836–840 (2015)
5. Djeddi, C., Gattal, A., Souici-Meslati, L., Siddiqi, I., Chibani, Y., El Abed, H.: LAMIS-MSHD: a multi-script offline handwriting database. In: 2014 14th International Conference on Frontiers in Handwriting Recognition, pp. 93–97 (2014)
6. Djeddi, C., Al-Maadeed, S., Gattal, A., Siddiqi, I., Souici-Meslati, L., El Abed, H.: ICDAR2015 competition on multi-script writer identification and gender classification using 'quwi' database. In: Proceedings of the 2015 13th International Conference on Document Analysis and Recognition (ICDAR), ICDAR 2015, pp. 1191–1195. IEEE Computer Society, USA (2015)
7. Freitas, C.O.A., Oliveira, L.S., Sabourin, R., Bortolozzi, F.: Brazilian forensic letter database. In: 11th International Workshop on Frontiers on Handwriting Recognition (2008)
8. Iwasaki, S.: Japanese: Revised Edition, vol. 17. John Benjamins Publishing, Amsterdam (2013)

9. Kawatani, T., Shimizu, H.: Handwritten kanji recognition with the LDA method. In: Proceedings. Fourteenth International Conference on Pattern Recognition (Cat. No.98EX170), vol. 2, pp. 1301–1305 (1998)

10. Kittler, J., Hater, M., Duin, R.P.W.: Combining classifiers. In: Proceedings of 13th International Conference on Pattern Recognition, vol. 2, pp. 897–901 (1996)

11. Kleber, F., Fiel, S., Diem, M., Sablatnig, R.: Cvl-database: an off-line database for writer retrieval, writer identification and word spotting. In: Proceedings of the 2013 12th International Conference on Document Analysis and Recognition, ICDAR 2013, pp. 560–564. IEEE Computer Society, USA (2013)

12. Liu, C., Yin, F., Wang, D., Wang, Q.: Casia online and offline chinese handwriting databases. In: International Conference on Document Analysis and Recognition (2011)

13. Maadeed, S.A., Ayouby, W., Hassaïne, A., Aljaam, J.M.: Quwi: an arabic and english handwriting dataset for offline writer identification. In: 2012 International Conference on Frontiers in Handwriting Recognition, pp. 746–751 (2012)

14. Mahmoud, S.A., et al.: Khatt: an open arabic offline handwritten text database. Pattern Recogn. **47**(3), 1096–1112 (2014)

15. Marti, U.V., Bunke, H.: The iam-database: an english sentence database for offline handwriting recognition. Int. J. Doc. Anal. Recogn. **5**(1), 39–46 (2002)

16. Mezghani, A., Kanoun, S., Khemakhem, M., Abed, H.E.: A database for arabic handwritten text image recognition and writer identification. In: 2012 International Conference on Frontiers in Handwriting Recognition, pp. 399–402 (2012)

17. Misaki, K., Honjou, D., Umeda, M.: A database of handwritten characters for writer recognition study and software for display and analysis. Jpn. J. Sci. Technol. Ident. **7**(1), 71–81 (2002)

18. Pinhelli, F., A.S.B., Oliveira, L.S., Costa, Y.M.G., Bertolini, D.: Single-sample writers - "document filter" and their impacts on writer identification. arXiv preprint arXiv:2005.08424 (2020)

19. Plamondon, R., Lorette, G.: Automatic signature verification and writer identification - the state of the art. Pattern Recogn. **22**(2), 107–131 (1989)

20. Roberto e Souza, M., Bertolini, D., Pedrini, H., Costa, Y.M.G.: Offline handwritten script recognition based on texture descriptors. In: 2019 International Conference on Systems, Signals and Image Processing (IWSSIP), pp. 57–62 (2019)

Skelibras: A Large 2D Skeleton Dataset of Dynamic Brazilian Signs

Lucas Amaral, Victor Ferraz, Tiago Vieira, and Thales Vieira(✉)

Institute of Computing, Federal University of Alagoas, Maceió, Brazil
{lafa,vmhf,tvieira,thales}@ic.ufal.br

Abstract. Dynamic sign language recognition is a challenging and active research topic over the last years. Due to the lack of high quality annotated datasets, it is currently unfeasible to train neural networks to perform neural machine translation of large dictionaries of dynamic signs from RGB videos. In this work, we present Skelibras: a large annotated skeleton dataset of dynamic Libras signs comprised of unsegmented and segmented conversations. Skelibras was built by exploiting an existing large and public RGB video dataset. We adopted an automatic approach to properly extract poses from each recorded individual, and then manually filter inconsistent representations and/or subtitles. We also present results of experiments performed on variants of recurrent neural networks to provide baseline benchmarks for future research on dynamic Libras signs recognition.

Keywords: Brazilian sign language · Dynamic sign language recognition · Deep learning

1 Introduction

Sign languages are visual languages known to promote social inclusion mainly among deaf individuals and those with hearing disabilities [7]. Nevertheless, hearing individuals are generally not educated to learn sign languages, thus limiting the social interactions of that impaired community [3].

Fortunately, during the last decade, remarkable improvements in hardware and Computer Vision algorithms have been achieved. Besides, Deep Learning approaches have been shown to outperform human professionals in tasks such as language translation [14] and image recognition [19]. Consequently, real-time Sign Language Recognition (SLR) has become a prominent research topic in the recent years [4].

However, training deep neural networks to recognize and translate a specific sign language requires a huge amount of supervised visual data, such as raw videos or skeletons. In practice, collecting and organizing such datasets may be as relevant as, for instance, investigating novel neural network architectures. On the one hand, collecting or exploiting previously recorded raw RGB videos is easier than employing Motion Capture (*MoCap*) systems that directly provide human skeletons. On the other hand, when compared to high dimensional RGB videos,

© Springer Nature Switzerland AG 2021
J. M. R. S. Tavares et al. (Eds.): CIARP 2021, LNCS 12702, pp. 184–193, 2021.
https://doi.org/10.1007/978-3-030-93420-0_18

skeletons provide a higher level of human pose representation that substantially accelerates the machine learning task due to a much lower dimension.

In this paper, we fill this gap by adopting an approach based on the Open-Pose [5] algorithm to automatically extract skeletons from raw RGB video. More importantly, we introduce *Skelibras*: a dataset comprised of thousands of annotated 2d skeleton sequences representing executions of 9,080 distinct dynamic signs of the Brazilian Sign Language (Libras).

The compilation of Skelibras was motivated by the lack of publicly available large datasets of dynamic signs of Libras that could be exploited for SLR. We take advantage of *Corpus de Libras* [16]: a huge dataset comprised of RGB videos exhibiting conversations between pairs of individuals. However, as the dataset was built for cataloging and educational purposes, it is not suitable for Computer Vision tasks in its original form.

Skelibras was built through a sequence of steps that includes: 1. the extraction of 2d skeletons from raw RGB videos of *Corpus de Libras*; 2. a manual temporal segmentation step for each sign execution; 3. a manual annotation enhancement step, and; 4. a quality filtering step, to remove inconsistencies. We also performed experiments with baseline neural network architectures to validate the Skelibras dataset, and to provide baseline performance for future research.

2   Related Work

Human Action Recognition (HAR) assigns a specific label corresponding to the specific action being executed. It is commonly done using RGB or/and depth videos directly or skeleton joints [10,21]. Sign Language Recognition (SLR) is comprised in HAR and has been addressed in the scientific literature for years as an attempt to provide better communication capabilities to deaf individuals.

Initially, researchers were focused on identifying static hand poses (finger spelling) using feature engineering and traditional image processing techniques, such as hand segmentation and classification [13]. A similar approach was applied to depth images with hand alignment, feature extraction and classification [6].

Next, the scientific community started focusing on *dynamic* Sign Language Recognition (SLR) due to its potential applicability towards deaf communities. Indeed, many efforts has been dedicated to continuously reading the subject's hand pose and position using dedicated hardware such as Glove-Talk-II [8,11,20]. Wearing gloves, however, can be disruptive, thus leading to non-standard gestures, and consequently compromise the accuracy of Sign Language Recognition (SLR) systems even through wireless devices.

A second methodology is based on Computer Vision and image processing techniques. Due to its ability to recognize sequences, Hidden Markov Models (HMM) or derivations have been widely used in SLR in the past. More recently, with the popularization of specific Recurrent Neural Networks (RNN) architectures such as the Long Short-Term Memory (LSTM) and high efficacy dealing with sequences, one can observe many applications of Neural Networks in SLR systems [17].

Many works use devices such as the Kinect sensor [9] to obtain a numeric representation of human's joints [12] and estimate the 3d skeleton. However, even though sensors can work well for many real applications and improve the system's accuracy by fusing information with the 2d image acquired, in general, their performance depend on environmental conditions such as lighting. Therefore, devices must be tailored to the environment in which the application must operate.

Many databases for SLR with different characteristics have been presented in the last decade with a higher variety of languages [17]. This is balancing the traditional predominance of databases containing signs of the American Sign Language (ASL) and promote more diverse research [17]. Amaral *et al.* for instance, evaluated the applications of Deep Neural Networks in labeling hand poses using depth sensors [1] for the Brazilian Sign Language (BSL).

Dynamic inputs consider sequential information which has the potential to improve the classification accuracy as opposed to static signs. In addition, dynamic strategy is further divided into isolated (where tokenization happens in word level) and continuous (with sentence-level tokenization). In this work we use the former.

3 *Corpus de Libras* Dataset

The database includes a high diversity of actors from different Brazilian states, leading to a rich variety of distinct accents. The recorded performances serve distinct purposes, such as dialogs, spoken poetry, interviews, and basic vocabulary. It is organized as follows: for each Brazilian state, there are several distinct projects which are comprised of many recordings. Performances that include two speakers were acquired from four different points of view, resulting in four videos per performance: one recorded from a lateral view and one recorded from a top view showing both speakers; and two from frontal views of each speaker, as illustrated in Fig. 1. Performances with a single speaker were recorded in a single video from a single frontal viewpoint. Since videos were recorded at different Brazilian regions and distinct projects, video resolution and quality is not uniform. It is worth mentioning that a minority of the videos were annotated with synchronized Libras subtitles. Such subtitles were annotated using the open-source ELAN annotation tool [15] and are publicly available in the ELAN Annotation Format (EAF) at the official website[1].

Each annotated video holds separate subtitle tracks for right-hand signs and left-hand signs. When the sign is performed by both hands, the subtitle is equally annotated in both tracks, although we found cases where they were not the same. Additionally, since there may be two speakers per conversation, there is a total of four tracks (two for each speaker).

---

[1] http://corpuslibras.ufsc.br/.

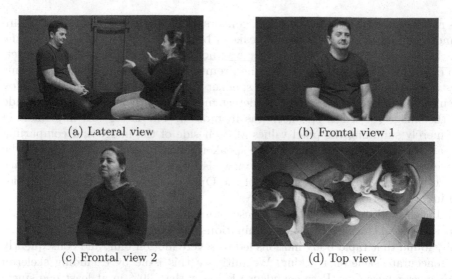

(a) Lateral view (b) Frontal view 1

(c) Frontal view 2 (d) Top view

Fig. 1. Frame taken from the original *Corpus de Libras* dataset: performances were recorded from four distinct points of views.

4 *Skelibras* Dataset

Since we aimed at compiling a supervised dataset, we only made use of the annotated videos from the original repository. Unfortunately, we found out that many files were corrupted, thus requiring an exhaustive manual check of every video file to exclude unusable files. Subsequently, the remaining videos were also manually checked and filtered to discard incorrect, inconsistent, or out of sync subtitles.

In the following step, we extracted pose representations of the speakers, in the form of body skeletons, for each video of the filtered video dataset. With this aim, we adopted the well-known OpenPose [5] algorithm to extract poses from the RGB videos, and then sync the skeleton sequences to the corresponding videos' subtitles. The OpenPose skeleton representation is a graph comprised of 18 joints as nodes, including 21 hand joints, as illustrated in Fig. 2.

It is worth mentioning that, for conversations with two speakers, we extract and exploit the skeletons of both speakers. Since the OpenPose algorithm does not provide any ordering among the recognized skeletons, we use the horizontal coordinate of the center of mass of each skeleton to identify each speaker as *left* or *right* speaker. Note that this is a required step since each skeleton is individually extracted from each frame of the video. Then, we performed a temporal segmentation of the signs according to the corresponding subtitles, in such a way that each skeleton sequence represents a single sign execution.

The Skelibras dataset was also intended to be comprised of skeletons from both frontal and lateral viewpoints, in a synchronized fashion. With this aim, it was first necessary to associate each speaker skeleton to a specific subtitle track.

As mentioned in the previous section, conversations with a couple of speakers are annotated in four tracks (two per speaker), but the correspondence between the tracks and the (left or right) speaker was not provided in the original dataset. To overcome this issue, we exploit the premise that only one speaker performs gestures at a time. To detect which speaker was performing a sign, we employ a simple frame differencing filter to detect motion and then evaluate which side of the image (or which speaker) was in motion. The latter was implemented by merely summing the pixel values of each side of the image and comparing. Finally, it was required to match the speakers of each frontal view video to the corresponding speaker of the lateral view. For this purpose, we adopt the face detection and recognition approach of the OpenFace library [2], which provides an implementation of FaceNet [18].

The last filtering step of the dataset compilation is meant to exclude mistracked poses. During a manual examination of the resulting dataset, we empirically found that rapid body motions led to severe motion blur, and consequently to inaccurate skeleton tracking. We mitigate this issue by excluding skeleton sequences whose OpenPose confidence is lower than 30% in at least one single joint of at least one frame of the sequence.

The average length of unsegmented conversations of the Skelibras dataset is 3 min, with the lengthier ones achieving up to 29 min, at 30 fps. On average, segmented sequences duration is 707 ms. The lengthier sequences, however, may achieve up to 19000 ms.

Skelibras currently has 9,080 distinct classes of signs, and 85,229 segmented signs (each sign has 3 samples taken from different points of view – POV, thus resulting in 255,687 sequences). The dataset includes many daily words; pointing gestures such as: "my mother", "we", "three", "you", "I"; demonstrative and possessive pronouns; verbs and classifying verbs, which describe actions such as "the man touched his feet", "the man grabbed the apple"; negations like "didn't

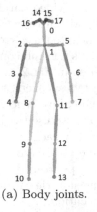

(a) Body joints.

(b) Hand skeleton joints.

Fig. 2. OpenPose skeleton output format. Source: edited from the original OpenPose repository [5].

like", "don't want"; and nominal signs representing proper nouns, such as people names. Skelibras is publicly available on the official website, which also includes more specific details of the dataset[2].

5 Baseline Classifiers

In this section, we briefly describe three classes of recurrent neural network (RNN) architectures commonly employed for human action recognition. We experimented on variations of these classifiers to validate our dataset, investigate whether dynamic Libras recognition is practical, and provide a baseline for future research.

Input. Four distinct skeleton representations were examined as input for the baseline classifiers: normalized cartesian coordinates (NCC); oriented joint angles (OJA); and OJA combined with their time derivatives (OJATD).

Cartesian pose normalization was performed to achieve both translation and cross-subject invariance, for each individual frame t. Let $S^t = \{x_1^t, x_2^t, \ldots, x_{61}^t\}$ be a vector of 2d coordinates of the 61 joints of the skeleton S^t. We compute the center of mass (μ^t) and standard deviation (σ^t) of S^t, and normalize each joint x_i^t as

$$z_i^t = \frac{(x_i^t - \mu^t)}{\sigma^t}, \tag{1}$$

where z_i^t is the vector of normalized 2d coordinates of joint i at time t.

Oriented angles were also considered due to their appropriate invariance properties. For each adjacent pair of segments of joints, here denoted by v_i^t and v_j^t, we compute their oriented angle of incidence:

$$\theta_{ij}^t = \text{sign}(v_i^t \times v_j^t) \cdot \arccos{(v_i^t \cdot v_j^t)}, \tag{2}$$

where $\times$ is the 2D cross product.

Time derivatives were computed at each frame as the difference between the following and current features (Cartesian normalized poses or oriented angles). For each of the four input formats, the resulting feature vectors are concatenated to compose a time series.

Architectures. Due to the sequential nature of the data, we chose as baselines three classes of architectures based on the well-known Long Short-Term Memory (LSTM) recurrent networks, as revealed in Fig. 3. Such architectures are known to be effective in learning temporal features from time series. The baseline architecture b_1 is a naive LSTM multi-class neural network with softmax activation in the output layer, possibly with more than one consecutive LSTM layer. We also experimented with learning, at each frame, local and global features from the input feature vectors. To this purpose, we inserted a dense layer (b_2) or an unidimensional convolutional layer (b_3) before the LSTM layer. It is worth

---

[2] https://github.com/luqsthunder/Skelibras.

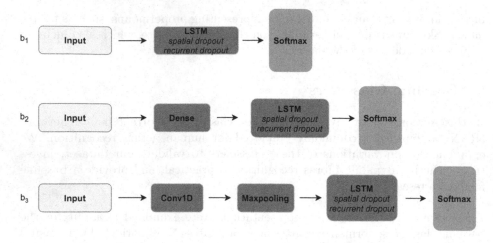

Fig. 3. Baseline architectures experimented on Skelibras: the inputs are given directly to LSTM layers (b_1); to an intermediate dense layer (to learn global skeleton features) followed by a LSTM layer (b_2); or to an intermediate convolutional layer (to learn local skeleton features) before following to the LSTM layer (b_3). Note that the LSTM box may represent multiple concatenated LSTM layers, and dropouts are optional, at 35%.

mentioning that, for all architectures, the input is sequentially given, one frame per time, making use of their cells' states. Thus, temporal features of each sign are learned and considered for classification. The use of spatial dropout and recurrent dropout was also investigated to avoid overfitting. With this aim, both dropouts were experimented in the LSTM layer.

6   Experiments

In this section, we present results of the experiments carried out to validate the Skelibras dataset and to provide baseline performance for future evaluation. We also aimed to answer the following research questions: 1) which of the four input formats (see Sect. 5) is more appropriate to perform dynamic Libras recognition through LSTMs? 2) is spatial or recurrent dropout effective to improve the accuracy of LSTMs? 3) are dense or convolutional layers effective to learn and provide higher-level global or local features to the recurrent layers, outperforming the accuracy of b_1 architectures?

For practical reasons, due to the huge size of the dataset, we selected a subset comprised of 1,000 samples of 10 distinct segmented signs: "BOM", "COMO", "E", "ESTUDAR", "HOMEM", "NÃO", "PORQUE", "TER", "TRABAL-HAR" and "VER". For each sign, 100 samples were randomly chosen.

We performed hyperparameter tuning in a comprehensive manner to search for the top architectures and perform fair comparisons. An uniform grid search was conducted by experimenting combinations of the following hyperparameters:

- Cass of architecture: b_1, b_2 or b_3;
- Input format: NCC, OJA, or OJATD;
- # of LSTM layers: 1, 2 or 3;
- # of LSTM units per layer: 45 or 90 units for the first, 20 or 30 for the second, and 10 or 15 for the third;
- Spatial dropout: 0% or 35%;
- Recurrent dropout: 0% or 35%.

For each configuration, we carried out cross-validation experiments by randomly assigning 70% of the samples for training and 30% for testing. The results are revealed in 6 separate histograms describing distinct combinations of inputs and architectures, as shown in Fig. 4. It was possible to find configurations that achieved more than 75% of accuracy, for all histograms.

In answer to research question 1, the NCC input outperformed the angle-based representations. Nevertheless, the latter also achieved accuracies above 80% for the top trained networks. Research question 2 is answered in Table 1, where one can see that the top networks applied spatial dropout, but not recurrent dropout. This may be explained by the velocity of signs executions, in which case every frame is relevant. Last, research question 3 is answered in Table 2: NCC+b_3 achieved the best result, followed by b_2 networks. Thus, we conclude that it is appropriate to apply convolutions in the input NCC representation. Even dense layers achieved better results than just feeding LSTM layers with the input data.

Finally, it is also worth mentioning that, by achieving up to 88.4% of accuracy, the experimented neural networks reveal that Skelibras is an appropriate dynamic Libras signs dataset to train classifiers for sign recognition.

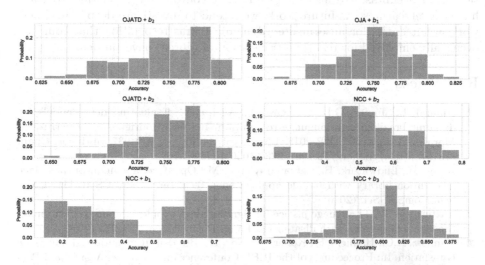

Fig. 4. Histograms of accuracies of distinct configurations of neural networks, according to the input and the class of architecture.

Table 1. Top 5 configurations achieved in our experiments: architectures with a convolutional layer, spatial droupout and no recurrent dropout prevailed.

Arch	Acc	# LSTM layers	# LSTM units	Spatial dropout	Recurrent dropout
b_3	**88.40%**	1	45	35%	0%
b_3	87.99%	2	45 e 20	35%	0%
b_3	87.59%	2	90 e 30	35%	0%
b_3	87.19%	1	90	35%	0%
b_3	87.19%	3	45, 20 e 10	35%	0%

Table 2. Top performing networks per input format and class of architecture.

Architecture	OJATD + b_2	OJATD + b_1	OJA + b_2	NCC + b_2	NCC + b_1	NCC + b_3
Accuracy	82.19%	80.80%	82.80%	79.19%	75.59%	**88.40%**

7   Conclusion

In this work, we presented Skelibras: a large dataset of dynamic Libras signs aimed at training classifiers for Libras recognition from 2d skeletons. We also experimented with baseline neural networks on a small sample of the dataset to provide benchmarks for future research on the recognition of dynamic Libras signs. By examining the accuracy of the top neural networks, which reached more than 88%, we mainly conclude that it is practical to perform gesture recognition from Skelibras. We also found out that convolution is relevant to learn higher-level features. As future work, we intend to investigate deep neural network architectures for unsegmented dynamic Libras recognition that could be eventually employed for real-time translation of Libras conversations.

References

1. Amaral, L., Júnior, G.L.N., Vieira, T., Vieira, T.: Evaluating deep models for dynamic brazilian sign language recognition. In: Vera-Rodriguez, R., Fierrez, J., Morales, A. (eds.) CIARP 2018. LNCS, vol. 11401, pp. 930–937. Springer, Cham (2019). https://doi.org/10.1007/978-3-030-13469-3_107
2. Amos, B., Ludwiczuk, B., Satyanarayanan, M.: Openface: a general-purpose face recognition library with mobile applications. Tech. rep., CMU-CS-16-118, CMU Sch. Comput. Sci. (2016)
3. Batterbury, S.C.: Language justice for sign language peoples: the un convention on the rights of persons with disabilities. Lang. Policy **11**(3), 253–272 (2012)
4. Camgoz, N.C., Hadfield, S., Koller, O., Ney, H., Bowden, R.: Neural sign language translation. In: Proceedings of the IEEE Conference on Computer Vision and Pattern Recognition, pp. 7784–7793 (2018)
5. Cao, Z., Hidalgo Martinez, G., Simon, T., Wei, S., Sheikh, Y.A.: Openpose: real-time multi-person 2d pose estimation using part affinity fields. IEEE Trans. Pattern Anal. Mach. Intell. **43**, 172–186 (2019)

6. Cardenas, E.J.E., Chávez, G.C.: Finger spelling recognition from depth data using direction cosines and histogram of cumulative magnitudes. In: SIBGRAPI, pp. 173–179. IEEE (2015)

7. Das, M., Kyte, R., Fisiy, C.F.: Inclusion Matters: The Foundation for Shared Prosperity. World Bank, Bretton Woods (2013)

8. Fels, S.S., Hinton, G.E.: Glove-talk ii - a neural-network interface which maps gestures to parallel formant speech synthesizer controls. IEEE Trans. Neural Netw. **8**(5), 977–984 (1997). https://doi.org/10.1109/72.623199

9. Han, J., Shao, L., Xu, D., Shotton, J.: Enhanced computer vision with microsoft kinect sensor: a review. IEEE Trans. Cybern. **43**(5), 1318–1334 (2013)

10. Herath, S., Harandi, M., Porikli, F.: Going deeper into action recognition: a survey (2017)

11. Kudrinko, K., Flavin, E., Zhu, X., Li, Q.: Wearable sensor-based sign language recognition: a comprehensive review. IEEE Rev. Biomed. Eng. **14**, 82–97 (2021). https://doi.org/10.1109/RBME.2020.3019769

12. LaViola, J.J., Kruijff, E., McMahan, R.P., Bowman, D., Poupyrev, I.P.: 3D User Interfaces: Theory and Practice. Addison-Wesley Professional, Boston (2017)

13. Pizzolato, E.B., dos Santos Anjo, M., Pedroso, G.C.: Automatic recognition of finger spelling for libras based on a two-layer architecture. In: Proceedings of the 2010 ACM Symposium on Applied Computing, pp. 969–973. ACM (2010)

14. Popel, M., et al.: Transforming machine translation: a deep learning system reaches news translation quality comparable to human professionals. Nature Commun. **11**(1), 1–15 (2020)

15. Nijmegen: Max Planck Institute for Psycholinguistics, T.L.A.: Elan (version 6.0) [computer software] (2020). https://archive.mpi.nl/tla/elan

16. Quadros, R.M.D., Schmitt, D., Lohn, J.T., Leite, T.d.A.: Corpus de libras. http://corpuslibras.ufsc.br/

17. Rastgoo, R., Kiani, K., Escalera, S.: Sign language recognition: a deep survey. Expert Syst. Appl. **164**, 113794 (2021). https://doi.org/10.1016/j.eswa.2020. 113794, https://www.sciencedirect.com/science/article/pii/S095741742030614X Survey. Expert Systems with Applications **164**, 113794 (2021). https://doi.org/ 10.1016/j.eswa.2020.113794, https://www.sciencedirect.com/science/article/pii/ S095741742030614X

18. Schroff, F., Kalenichenko, D., Philbin, J.: Facenet: a unified embedding for face recognition and clustering. In: Proceedings of the IEEE Conference on Computer Vision and Pattern Recognition, pp. 815–823 (2015)

19. Simonyan, K., Zisserman, A.: Very deep convolutional networks for large-scale image recognition. arXiv preprint arXiv:1409.1556 (2014)

20. Sun, Z.: A survey on dynamic sign language recognition. In: Bhatia, S.K., Tiwari, S., Ruidan, S., Trivedi, M.C., Mishra, K.K. (eds.) Advances in Computer, Communication and Computational Sciences. AISC, vol. 1158, pp. 1015–1022. Springer, Singapore (2021). https://doi.org/10.1007/978-981-15-4409-5_89 Computer, Communication and Computational Sciences. pp. 1015–1022. Springer Singapore, Singapore (2021)

21. Xia, Z., et al.: Vision-based hand gesture recognition for human-robot collaboration: a survey. In: 2019 5th International Conference on Control, Automation and Robotics (ICCAR), pp. 198–205 (2019). https://doi.org/10.1109/ICCAR. 2019.8813509

Deep Learning

Cricket Scene Analysis Using the RetinaNet Architecture

Tevin Moodley[ID] and Dustin van der Haar[✉][ID]

University of Johannesburg, Kingsway Avenue and University Rd,
Auckland Park, Johannesburg 2092, South Africa
tevin@uj.ac.za, dvanderhaar@uj.ac.za

Abstract. The increased challenges surrounding object detection within the sporting environment have been studied with multiple proposed solutions for various domains. The increasing environmental changes with moving objects, actors, and overlapping objects that are present in sporting video footage make detecting and classifying different objects challenging. However, with the introduction of deep learning, researchers now have the available methods that can learn semantic, high level, and deeper features that can be used to solve problem areas within existing research. Cricket is a sporting domain that exhibits many of these challenges with multiple moving actors and objects. This research paper implements RetinaNet architecture to detect and classify multiple objects within a scene. Six different objects/classes are addressed: fielder, batsman, non-striker, bowler, umpire, ball, and wicket-keeper. Following the dataset preparation, using transfer learning, the images are trained on the RetinaNet architecture, and the architecture proved to be successful by producing a mean average precision score of 86.78%. The trained model manages class precision scores all above 98% except that of the ball class. Upon further investigation, the poor performance of the ball class is due to occlusion and the ball's small size relative to the overall frame. The proposed model can successfully detect and classify the different objects/classes within a cricket scene and serves as a promising foundation for further research within the cricketing domain.

1 Introduction

Within computer vision, object detection is an integral task toward finding instances of real-world objects [2]. Objects within an image allow the prediction of their location along with the class to which they belong. Attempting to determine the coordinates of an object within an image and the object's classification is a problem that is prevalent within various domains [16]. Unfortunately, there is no one fit-all solution. The different computer vision domains, such as image retrieval, security and surveillance, autonomous driving and many industrial applications, require different object detection approaches [2]. In many cases, object detection is seen as a prerequisite for activity recognition, which has evolved into an exciting area of research in the last decade [6]. Activity recognition can be challenging due to camera motion, camera switching, and variation in backgrounds [6].

© Springer Nature Switzerland AG 2021
J. M. R. S. Tavares et al. (Eds.): CIARP 2021, LNCS 12702, pp. 197–206, 2021.
https://doi.org/10.1007/978-3-030-93420-0_19

One such domain that exhibits these challenges is the cricketing domain. Cricket is a sport played by two teams with 11 players on each team [7]. Each side has batsmen, bowlers, fielders, a wicket-keeper, and an umpire that judges the match [14]. The game's objective is for the batsmen to score as many runs as possible whilst the bowler attempts to dismiss the batsman, in which the batsman is forced to leave the field of play. Due to the emphasis placed on accumulating runs, our previous research discusses the scene recognition of significant events within cricket game footage [11]. The significant events noted were when the ball is released by the bowler and the moment the batsman strikes the ball. Within these scenes, there are various objects and detecting these objects is needed for continued research.

This research paper discusses using the RetinaNet architecture to detect and classify objects within a scene. Section 2 will unpack the domain's challenges by discussing current similar works. Section 3 will discuss the experiment setup, which will illustrate the architecture implemented in Sect. 4. Finally, the discussion in Sect. 6 will highlight the results in Sect. 5 of the proposed architecture with an emphasis on the performance and challenges encountered.

2   Problem Background

Object detection remains an open research problem within various domains [1]. Difficulty in object detection arises due to the variations in images, which is especially problematic in sports. Object detection in sports has been noted for suffering from problems, such as high false positives, due to varying poses of the athlete and high-computational complexity due to the varying objects within the scene [12]. In sports such as football, there is a uniformly coloured field, which is addressed with static camera usage, image differencing, background subtraction, or a colour based elimination of the ground. However, these methods are prone to missed detections [12]. The environment in which cricket games are played is usually a large open field, with challenging illumination conditions, multiple moving actors and objects, and conditions that are not ideal. In the cricket domain's highly dynamic setting, the motion blur and shadows that actors cast over different lighting conditions are often apparent in videos and images. The positioning and shape of the actors who overlap with other actors vary greatly, making object detection within the domain more challenging.

2.1   Related Works

Buriśc, Pobar, and Ivasić-Kos propose an object detector that handles a different number of objects of varying sizes, partially occluded, lousy illumination, and cluttered scenes in order to detect objects within a handball scene [2]. Their works make use of current state-of-the-art detection methods that rely on convolutional neural networks. To achieve their outcome, the most prominent methods used are a mask regional convolutional neural network (R-CNN) and the "you only look once" (YOLO) object detector. Their findings show that mask R-CNN is better in team sports as it can inherently detect individual actors even when in cluttered groups [2]. The mask is also noted to provide, with more computation power, more significant features. The YOLO object detector was noted to test faster on real-time data that performs well but ultimately produces too many false positives. The R-CNN lacked the possibility of fast and on-field analysis. However, with additional hardware and algorithm development, their research problem would be addressed in the future [2]. One significant challenge was that both methods struggled to reliably detect the sports ball, where researchers noted it might be due to the fact that the texture and shape of the ball share similar features that are apparent in many other objects.

Research conducted by Liang, Mei, Wu, Sun, and Wang proposes automatic basketball detection in sports video based on Region-based Fully Convolutional Networks (R-FCN) built on the residual network with 50 layers [9]. Newer methods are then applied, Online Hard Example Mining (OHEM), Soft-NMS, and a multi-scale training strategy to achieve higher detection accuracy [9]. The OHEM method reduces the cost associated with training by calculating the loss of the region of interests. Using the Soft-NMS, the false positive rate was reduced by decreasing the object detection rate between the overlapping objects [9]. A mean Average Precision (mAP) value of 89.7% was achieved, proving that a deep-learning approach to basketball detection is effective and potentially in other sporting domains.

Zaveri, Merchant, and Desai propose an algorithm for detection and tracking fast and small moving objects, like a ping pong ball or a cricket ball [18]. Zaveri et al. note that detecting and tracking small and fast-moving objects is becoming increasingly important within the sporting domain. With the objective to develop algorithms that can further assist in making decisions and analysing sports video sequences [18]. Using a wavelet transform for temporal filtering enables one to detect and characterise the dynamical behaviour of the elements present in the scene. Their proposed method uses motion as a cue for detection, which allows for the detection of the object with shallow contrast and negligible texture content [18]. However, many challenges were encountered, most noteworthy that of the cricket ball's varying size, which made detection difficult due to the camera's different zooming.

Object detection has been applied in multiple sporting domains, where different approaches are adapted to cater to its inherent challenges encountered. Within a cricket scene, there is a vast number of objects that vary between each image. The same scene is used in each image to apply continued research within our domain, where the bowler is about to release the ball. Using the RetinaNet architecture, this research paper aims to detect and recognise different objects within a specific scene during a cricket game, which will subsequently further the research within the domain.

3   Dataset and Experiment Setup

To construct the dataset, a scene recognition model was implemented from previous research to find samples for the study [11]. Through a comprehensive search on YouTube, video footage of First-Class International Test Match highlights were captured and saved. The scene recognition model allows for video footage processing, capturing and saving the corresponding image of the desired scene. Each image is then manually inspected to remove any corresponding false positives that may have been produced by the scene recognition model, resulting in a total of 200 images ready for processing.

Once the images were correctly stored, using the Labelbox editor [8], the image annotations were created by drawing the bounding box and labelling each object within the scene. The following objects were labelled; fielder, batsman, non-striker, bowler, umpire, ball, and wicket-keeper. For each image, the number of objects varied, the count for each object's ground truth label is as follows; fielder: 222, batsman: 200, non-striker: 200, bowler: 199, umpire: 198, ball: 179, and wicket-keeper: 128. The ball had fewer occurrences due to the ball being a fast-moving object within the often blurred scene, making it difficult to detect. The wicket-keeper had the fewest number of occurrences, and earlier, it is highlighted that the cricket environment is filled with many overlapping actors. Whenever the wicket-keeper is directly behind the batsman, as per the Fig. 2, it causes uncertainty as to the area the actor spans.

Figure 2 illustrates the different positioning of the actors and objects within a cricket game for this particular dataset used. The batsman faces the delivery and is situated at one end on the pitch, where the non-striker is positioned on the opposite end of the pitch, waiting for the ball to be delivered. The wicket-keeper is positioned either right behind the wickets or further away, depending on the type of bowler. The bowler is the actor delivering the ball toward the opposite end of the pitch. Fielders are strategically positioned across the field to deny the batsmen from accumulating runs. The umpire is the match official who stands right behind the wickets on the non-striker's end. The second umpire is not considered as he/she is not in the frame of shot. Finally, the ball is a moving object usually in the bowler's hand right before releasing it toward the batsman.

Using Roboflow software [15], the training and testing datasets were created using an 80:20 data split, resulting in 160 images being used for training and the remaining 40 images for testing. The annotation files were created along with the data split, which contained each objects bounding box dimensions (top, left, height, and width), label, and the uniform resource location (URL) of each object instance paired with a file containing each class name. The mAP and precision scores for each class are computed to determine the performance of the architecture [17]. The mean Average Precision (mAP) is a common score used in measuring the accuracy scores in object detectors. mAP computes the average precision value for recall value over 0 to 1 for each class. The mean Average Precision score uses precision and recall, where precision refers to the number of observations that are, in fact, true. The recall is the number of correctly predicted positive observations [5]. The mAP of a set of queries is defined by:

$$AP = \sum_n (R_n - R_{n-1})P_n \tag{1}$$

where R_n represents precision and P_n represents recall at the nth threshold, and mAP is the mean overall the subsequent queries [17].

4   RetinaNet Architecture

Object detection relates to image and computer processing, facilitating the detection of semantic objects of a particular class within digital images and videos [3]. To achieve object detection for the proposed problem, a deep learning approach will be adapted. Deep learning forms part of machine learning based on artificial neural networks, in which multiple layers of processing are used to extract features from data. The idea behind deep learning is that we are given depth in multiple layers, which enables the computer to learn in a multi-step computer program [4]. Each layer in the given representation can be seen as the state of the computer's memory after executing another set of instructions in parallel. Within deep learning models, the function mapping from a set of pixels to an object can be complicated. Deep learning breaks down the mapping into a series of nested mappings, each of which is described by a different layer of the model.

The modern paradigm for object detection is based on one-stage and two-stage approaches. The two-stage approach is pioneered by selective search; the first stage generates the candidate proposals that should contain all the objects. The second stage classifies the proposals into their respective classes [10]. R-CNN was introduced, which saw large improvements in accuracy by upgrading the second-stage classifier to a CNN. One-stage treat object detection as a simple regression problem that learns the class probabilities and bounding boxes of input image detectors, which was later renewed with YOLO and Single Shot MultiBox Detector (SSD's), where these methods are tailored for speed [13]. RetinaNet shares many qualities as previous dense detectors, in particular the concept of anchors introduced by Region Proposal Network (RPN) that predicts the probability of an anchor being a background or foreground. Figure 1 depicts the RetinaNet architecture, a single, unified network composed of a backbone network and two task-specific subnetworks [10]. The backbone network is an off-the-shelf fully connected convolutional network responsible for computing a convolutional feature map over the input image. The Feature Pyramid Network (FPN) backbone is built on top of the ResNet18 architecture, where each level of the pyramid can be used for detecting objects at a different scale. The levels range from $p1$ to pi where i indicates the pyramid level. Translation-invariant anchor boxes are responsible for size and shape within a grid cell, the anchor box with the highest degree of overlap with an object is responsible for predicting the object's class and its location [10].

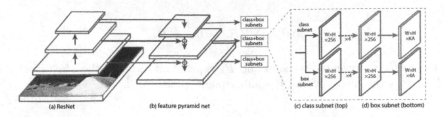

Fig. 1. The one-stage RetinaNet network architecture uses a FPN backbone on top of a feedforward ResNet architecture, that is attached by two subnetworks, one for classifying anchor boxes and one for regressing from anchor boxes to ground-truth object boxes [10].

The RetinaNet architecture was implemented using the RetinaNet model described by Lin, Goyal, Girshick, He, and Dollár [10]. Using transfer learning, whereby the RetinaNet architecture developed is reused, the following parameters were the only parameters altered through testing and validation to achieve the optimal results. The image dimensions were set to 1024 × 1024 pixels, epochs: 100, the number of steps: 40, batch size: 4, and patience level: 10. The patience level was altered to allow the model to improve over a larger number of epochs. The batch size and number of steps were altered to accommodate the number of samples within the dataset. The next section will unpack the results achieved by the model.

5   Results

A more in-depth investigation of the images is conducted to assess the architecture's performance on the given dataset. Figure 2 presents a sample image that illustrates the manner in which the bounding boxes and objects are annotated. The number of instances for each object in the scene is captured upon completion of training; fielder: 183, batsman: 160, non-striker: 159, bowler: 159, umpire: 159, ball: 141, and wicket-keeper: 102. Early in the research, we noted that the ball is in the bowler's hand before the ball is released. From the human eye, it is difficult to determine the difference between the ball and the bowler's hand, as seen in Fig. 2. The inability to view the ball supports the poor classification score, and further conclusions will be unpacked in Sect. 6.

The following metrics found in Table 1 illustrates the performance of the proposed solution. For each class the precision score is computed; batsman: 99.15%, wicket-keeper: 98.46%, fielder: 98.38%, bowler: 99.97%, non striker: 99.46%, umpire: 99.99%, and ball: 12.06%. Each class performs well apart from the ball object, which has a significantly lower score. Section 6 will further unpack this finding.

Fig. 2. A sample image depicting the ground truth labels and bounding boxes for each object, which was created using Labelbox editor [8].

Table 1. The table illustrates the precision scores for each object within the image and the weighted average precision (mAP) of 86.78% for all classes

Class	Precision
Batsman	99.15%
Wicket keeper	98.46%
Fielder	98.38%
Bowler	99.97%
Non striker	99.46%
Umpire	99.99%
Ball	12.06%
Weighted average precision mAP for all classes: 87.69%	
mAP score 86.78%	

6   Discussion and Critique

The RetinaNet architecture can detect and classify different objects within a cricket scene successfully. The most noteworthy finding was that of the ball object, which produced a low precision score of 12.06%. To determine the challenge related to the poor detection of the ball, Zaveri et al. [18] also noted that the ball is a small and fast-moving object, which is often subjected to low contrast within object tracking of the cricket ball and suffers from a shadow effect created by the changing environment [18]. However, the actors are easily segmented using an edge operator that exploits high contrast. The issue of low contrast, shadow effect coupled with the small size of the ball is one possible reason as to why the architecture struggles to detect and classify the ball. Another possible conclusion may be due to the bounding area size around the ball, which is significantly smaller, as seen in Fig. 2. Zaveri et al. also note the ball's varying size due to camera zooming that made detection difficult, which supports the issue relating to the ball's size. The small size of the object is noted as the primary reason for the mAP score.

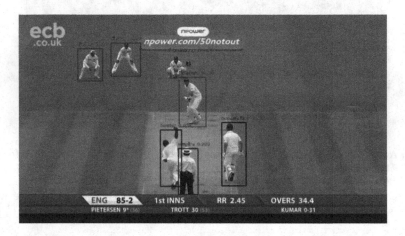

Fig. 3. A test image that was loaded into the model, where the model was able to successfully detect and classify the different objects within the scene.

When testing the model's performance, several new images were used to validate the architecture's performance. As seen in Fig. 3, a test image is shown to emphasise the success of the architecture while taking into account the ball class's poor performance. Figure 3 illustrates each object and the accuracy to which the model classifies that object. The batsman is correctly classified with a score of 98.87%, fielder 1: 100%, fielder 2: 99.8%, the non-striker: 98.90%, umpire 99.9%, and the bowler with 96.98%. The ball is not classified at all, which supports the metrics found in Table 1. From Fig. 4, the ball is difficult to see from the human eye as it can be argued that a human with little knowledge of the domain would not be able to reliably detect the ball object either. The ball in each image within the dataset is often in the bowler's hand. As seen in Fig. 1, there are apparent noise and anomalies, which requires further processing. A solution to capture a different frame after the ball is released may allow the architecture to detect and classify the object. Researchers suggest a possible

Fig. 4. An image representing different samples of the ball to illustrate the positioning of the ball, which supports the recommendation to select a frame where the ball is not in the bowler's hand.

solution to the inability to detect and classify balls within sport is to take into account the existence and position of the ball near the actor during the training phase [2].

7   Conclusion

Balaji, Karthikeyan, and Manikandan note the challenges around performing object detection within the sporting environment [1], detecting and classifying different objects is challenging with constant changes in the environment, moving objects and multiple moving actors. However, with the introduction of deep learning, researchers are now able to use more powerful methods to learn semantic, high-level, deeper features, which can solve problems existing in traditional architectures [19]. The architectures performance is highlighted in Table 1, where an mAP score of 86.78% is achieved. Each class apart from the ball has a precision score greater than 98%, which further supports the architecture's success. The most significant challenge encountered in this research was the architectures' inability to detect and classify the ball. A possible solution to the challenge encountered is to consider the ball's position relative to the bowler. Ensuring that the ball is isolated may serve as a plausible solution. This research serves as a promising foundation for further research within the cricketing domain. By selecting each object within the scene, further analysis is possible to potentially determine each object's impact within the scene.

References

1. Balaji, S.R., Karthikeyan, S., Manikandan, R.: Object detection using metaheuristic algorithm for volley ball sports application. J. Ambient Intell. Humanized Comput. **12**(1), 375–385 (2021)
2. Burić, M., Pobar, M., Ivašić-Kos, M.: Object detection in sports videos. In: 2018 41st International Convention on Information and Communication Technology, Electronics and Microelectronics (MIPRO), pp. 1034–1039. IEEE (2018)
3. Dasiopoulou, S., Mezaris, V., Kompatsiaris, I., Papastathis, V.K., Strintzis, M.G.: Knowledge-assisted semantic video object detection. IEEE Trans. Circ. Syst. Video Technol. **15**(10), 1210–1224 (2005)
4. Goodfellow, I., Bengio, Y., Courville, A., Bengio, Y.: Deep Learning, vol. 1. MIT Press, Cambridge (2016)
5. Goutte, C., Gaussier, E.: A Probabilistic Interpretation of Precision, Recall and F-Score, with Implication for Evaluation. In: Losada, D.E., Fernández-Luna, J.M. (eds.) ECIR 2005. LNCS, vol. 3408, pp. 345–359. Springer, Heidelberg (2005). https://doi.org/10.1007/978-3-540-31865-1_25
6. Gupta, A., Muthiah, S.B.: Viewpoint constrained and unconstrained cricket stroke localization from untrimmed videos. Image Vis. Comput. **100**, 103944 (2020)
7. Knight, J.: Cricket for Dummies. Wiley (2013)
8. Labelbox: Labelbox. Online (2021). https://labelbox.com
9. Liang, Q., Mei, L., Wu, W., Sun, W., Wang, Y., Zhang, D.: Automatic basketball detection in sport video based on R-FCN and Soft-NMS. In: Proceedings of the 2019 4th International Conference on Automation, Control and Robotics Engineering, pp. 1–6 (2019)

10. Lin, T., Goyal, P., Girshick, R., He, K., Dollár, P.: Focal loss for dense object detection. IEEE Trans. Pattern Anal. Mach. Intell. **42**(2), 318–327 (2020)
11. Moodley, T., van der Haar, D.: Scene recognition using AlexNet to recognize significant events within cricket game footage. In: Chmielewski, L.J., Kozera, R., Orłowski, A. (eds.) ICCVG 2020. LNCS, vol. 12334, pp. 98–109. Springer, Cham (2020). https://doi.org/10.1007/978-3-030-59006-2_9
12. Mustamo, P.: Object detection in sports: TensorFlow object detection API case study. University of Oulu (2018)
13. Pang, J., Chen, K., Shi, J., Feng, H., Ouyang, W., Lin, D.: Libra R-CNN: towards balanced learning for object detection. In: Proceedings of the IEEE/CVF Conference on Computer Vision and Pattern Recognition, pp. 821–830 (2019)
14. Pardiwala, D.N., Rao, N.N., Varshney, A.V.: Injuries in cricket. Sports Health **10**(3), 217–222 (2018). pMID: 28972820
15. roboflow: roboflow. Online, January 2020. https://roboflow.com/
16. Shaikh, S.H., Saeed, K., Chaki, N.: Moving object detection approaches, challenges and object tracking. In: Moving Object Detection Using Background Subtraction. SCS, pp. 5–14. Springer, Cham (2014). https://doi.org/10.1007/978-3-319-07386-6_2
17. Yue, Y., Finley, T., Radlinski, F., Joachims, T.: A support vector method for optimizing average precision. In: Proceedings of the 30th Annual International ACM SIGIR Conference on Research and Development in Information Retrieval, pp. 271–278. SIGIR 2007. Association for Computing Machinery (2007)
18. Zaveri, M.A., Merchant, S.N., Desai, U.B.: Small and fast moving object detection and tracking in sports video sequences. In: 2004 IEEE International Conference on Multimedia and Expo (ICME) (IEEE Cat. No.04TH8763), vol. 3, pp. 1539–1542 (2004)
19. Zhao, Z., Zheng, P., Xu, S., Wu, X.: Object detection with deep learning: a review. IEEE Trans. Neural Netw. Learn. Syst. **30**(11), 3212–3232 (2019)

Texture-Based Image Transformations for Improved Deep Learning Classification

Tomáš Majtner[1]($\boxtimes$), Buda Bajić[2], and Jürgen Herp[3]

[1] Central European Institute of Technology (CEITEC),
Masaryk University, Brno, Czech Republic
tomas.majtner@ceitec.muni.cz
[2] Faculty of Technical Sciences, University of Novi Sad, Novi Sad, Serbia
[3] The Maersk Mc-Kinney Moller Institute,
University of Southern Denmark, Odense, Denmark

Abstract. In this paper, we examine the effect of texture-based image transformation on classification performance. A novel combination of mathematical morphology operations and contrast-limited adaptive histogram equalization is proposed to enhance image textural features. The suggested operations are applied in HSV colour space, where the intensity component is separated from the colour information. Two publicly available, texture-oriented datasets are used for evaluation in this study. The KTH-TIPS2-b dataset is utilised to illustrate the general effectiveness and applicability of the proposed solution on standardized texture images. The Virus Texture dataset is subsequently used to demonstrate a statistically significant classification improvement in a particular biomedical image recognition task.

Keywords: Texture recognition · Image processing · Transfer learning · HSV colour model

1 Introduction

The texture analysis has been a topic of intensive research in the last decades and a variety of techniques for texture discrimination was introduced [15]. Because natural textures display different properties, like regularity and randomness, or uniformity and distortion, texture can hardly be described in a unified manner. Textural characteristics are important especially for image recognition and classification using convolutional neural networks (CNNs), as it was recently demonstrated by Geirhos et al. [5]. Their paper challenges the traditional intuition of shape-based CNN features and suggests that CNNs trained on ImageNet are biased towards texture. One of the key observations for our work is that the model performs better as the amount of texture on the object increases. Based on this statement, the modification or enhancement of the texture information in the image is potentially interesting for image classification tasks, when deep learning models pre-trained on ImageNet are considered.

Our paper explores this idea by incorporating the textural information enhancement methods to the recognition process. In particular, we employ the

© Springer Nature Switzerland AG 2021
J. M. R. S. Tavares et al. (Eds.): CIARP 2021, LNCS 12702, pp. 207–216, 2021.
https://doi.org/10.1007/978-3-030-93420-0_20

granulometry principle from mathematical morphology that uses a series of morphological opening operations to remove intensity peaks of a chosen size. The motivation for this approach comes from the earlier work on HEp-2 cells' classifier [20], where a hand-crafted granulometry-based descriptor was introduced. In this work, we use slightly modified approach, where we apply our methods in the HSV colour space.

In the next section, we briefly describe the recent related works and applications of texture-based analysis. Section 3 introduces our proposed method for texture improvement. Section 4 presents two datasets employed in this study, namely the KTH-TIPS2-b dataset [13] and the Virus Texture dataset v.1.0 [9] that will be further referred to as the *virus dataset*. At the end of the paper, we present our results together with a discussion, a conclusion, and future work.

2   Related Work

The recent work in texture-oriented deep learning is focused on medical or biomedical applications, where texture analysis is a common issue [1]. Deep learning techniques were utilised for texture analysis of chest and breast images [3] or the recognition of HEp-2 cell images [2]. Texture approaches were also used to analyse transmission electron microscopy (TEM) images [16]. All these works employed methods based, for example, on local binary patterns to support medical or biomedical image analysis applications.

For the virus particle detection, a method based on integration of the theory of rough sets and the merits of local texture descriptors was suggested by Kumar et al. [7]. Their method identifies the relevant local texture descriptor for representing the intrinsic properties of each pair of virus classes. Other approaches to improve the virus recognition include, for example, the reduction of trainable weights [14] or the usage of local texture descriptors [18]. Morphological layers in deep learning framework called MorphNet were introduced to perform atomic morphological operations like dilation and erosion [17]. Morphological granulometric features were used for a white blood cell classification [22], where four so-called granulometric moments were extracted from the image.

Our method employs morphological openings in a preprocessing stage. The algorithm changes the internal structure of textural area, which leads to enhanced image output. The modified outcome combined with the original image brings higher, state-of-the-art classification performance demonstrated in this paper on two different public datasets.

3   Proposed Method

In this section, we provide a detailed description of our proposed method. The visual illustration of our pipeline is shown in Fig. 1. In the beginning, we convert the input RGB image to HSV colour space that describes colours similarly to how the human eye tends to perceive them. For greyscale images, we copy the single intensity channel to all three colour components of RGB and continue

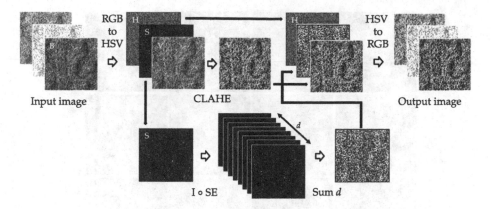

Fig. 1. Illustration of the pipeline of our method for image transformation.

with indicated conversion to HSV. In HSV colour space, hue (H) channel specifies the base colour, saturation (S) channel represents the vibrancy of the colour
and captures the amount of grey in a particular colour, and value (V) channel
in conjunction with S channel describes the intensity or brightness of the colour.
The motivation is to separate the intensity component from the colour information. This conversion is essential for our method and it was already employed
for textural image enhancement [6].

Since we are interested in texture transformation, we perform the transformation on S and V channels. First, we take the S channel that consists of greyscale
values. We define a disk structuring element SE of size d. The morphological
opening operation on the input image I is defined as an erosion followed by a dilation: $I \circ SE = (I \ominus SE) \oplus SE$. The erosion operation removes objects smaller than
SE and subsequent dilation operation restores the shape of remaining objects.
The opened value of a pixel is the maximum of the minimum value of the image
in the SE neighbourhood.

In the next step, we define $i = \{1, .., d\}$ and a morphological opening $I \circ SE$
with SE of size i is performed. For each i, we calculate the pixel-wise difference
between I and $I \circ SE$. We stack these differences into a volumetric 3D image,
where the third dimension has length corresponding to d. In the end, we transform the 3D image back to 2D by calculating the pixel-wise sum of intensities over
all d slices. This technique by its definition characterises the peak distribution
of grains in the image. The hyper-parameter d needs to be determined experimentally. Based on the parametric study presented in the evaluation section, the
optimal value determined for the KTH-TIPS2-b dataset is eight. We employed
the same value also for the *virus dataset*.

The second modification is performed on the V channel, where we apply
CLAHE [24]. This non-parametric equalization operates on small regions in the
image and computes histograms corresponding to distinct sections of the image.
Subsequently, it uses them to redistribute the lightness values of the image. The
application of CLAHE on the V channel was successfully used for retinal image

Fig. 2. Samples from KTH-TIPS2-b. *The top row*: images from the same class (wool) demonstrating high intra-class variance; *the middle row*: images from different classes (cork, linen, cotton, and wood); *the bottom row*: middle row images after application of our proposed method.

enhancement [6]. This variant of adaptive histogram equalization helps to stretch intensities by the contrast amplification.

After described two transformations, we convert the HSV image back to RGB colour space. The example of output images for the KTH-TIPS2-b dataset are in the bottom row of Fig. 2, while output images for the greyscale *virus dataset* are in the right part of Fig. 3. Please note that greyscale *virus images* have three channels after our transformation. However, they visually appear as greyscale samples. In both presented datasets, finer textural details with stretch intensities are visible after the proposed transformation.

4   Datasets

Since we study the texture transformation, the evaluation is conducted on textural datasets. The first one is a standard benchmark texture dataset known as the KTH-TIPS2-b dataset [13], where TIPS stands for *textures under varying illumination, pose, and scale*. The dataset contains four physical samples of eleven different materials (44 samples in total) including, cork, linen, cotton, wood, and others (see first two rows of Fig. 2 for illustration). Each image is obtained with three viewing angles, four illuminations, and nine different scales. Images were manually cropped to 200×200 pixels by their creators. In total, there are 108 images per each sample. The dataset has eleven image classes (classification categories) corresponding to eleven materials, where each category is formed by four samples, which leads to $4 \times 108 = 432$ images.

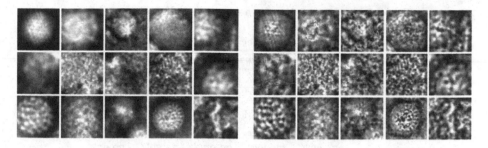

Fig. 3. *The left part*: original samples from *virus dataset*, each from a different class; *the right part*: the same images after application of our method.

There are two evaluation scenarios commonly used for KTH-TIPS2-b dataset. The first one employs training on three samples ($3 \times 11 \times 108 = 3564$ images) and evaluation on the remaining physical sample ($11 \times 108 = 1188$ images). We further refer to this scenario as *train3test1*. The second scenario utilises a single physical sample for training and the remaining three samples for evaluation. We refer to it as *train1test3*. The common practice is to present results for a single chosen scenario. However, we report both scenarios, which better illustrates the effect of the number of training samples on the classification performance.

The second dataset employed in our study is *virus dataset* [9], which is based on TEM images of 15 virus types (see the left part of Fig. 3 for illustration). It was chosen to demonstrate the practical application of our method. The texture samples have been automatically extracted from objects segmented using a specific method [8]. There are 100 unique greyscale texture patches per image class and each image has 41×41 pixels. For evaluation, we follow the 10-fold cross-validation procedure suggested by creators.

5   Evaluation and Results

The evaluation was conducted using a fine-tuning approach that is known as the *transfer learning*. For each CNN model pre-trained on ImageNet, we replaced its last three layers with the fully-connected layer, the softmax layer, and the classification layer, which classifies images directly to our desired categories. All images were resized to an appropriate input size for each network separately using bicubic interpolation and all tests were performed using MATLAB R2019b.

5.1   Results for KTH-TIPS2-b Dataset

For the evaluation on KTH-TIPS2-b dataset, we used two recent network architectures, namely Xception and NASNet. The fine-tuning utilised a stochastic gradient descent with momentum optimizer and an initial learning rate of 10^{-3}. For Xception, we used the mini-batch size of 16 images and the network was trained for 20 epochs. For NASNet, we employed the mini-batch size of eight

Table 1. Accuracy values for a pre-trained Xception model.

	$sample_a$	$sample_b$	$sample_c$	$sample_d$	average
	train1test3 scenario				
$orig$	77.95%	76.74%	75.08%	79.10%	77.22%
$gran_4$	74.38%	72.95%	**70.79%**	74.72%	73.21%
$gran_8$	**77.08%**	**78.23%**	69.92%	**76.80%**	**75.51%**
$gran_{12}$	77.02%	73.63%	68.55%	76.49%	73.92%
$gran_{16}$	76.68%	73.32%	66.27%	75.62%	72.97%
late fus. [23]	78.56%	74.97%	71.97%	**81.20%**	76.68%
$orig^*gran_8$	**79.38%**	**78.82%**	**73.48%**	79.91%	**77.90%**
	train3test1 scenario				
$orig$	93.35%	83.16%	90.74%	84.68%	87.98%
$gran_4$	91.33%	**81.99%**	87.21%	81.82%	85.59%
$gran_8$	**94.19%**	81.82%	**88.55%**	84.18%	**87.19%**
$gran_{12}$	93.52%	81.14%	83.92%	83.42%	85.50%
$gran_{16}$	92.34%	81.14%	86.20%	81.48%	85.29%
late fus. [23]	94.44%	81.06%	**89.90%**	84.34%	87.44%
$orig^*gran_8$	**94.78%**	**83.08%**	**89.90%**	**86.11%**	**88.47%**

images and it was trained for ten epochs. During the training, we controlled the convergence using the validation dataset that was separated from the training dataset using random 10% of files from each label. We tested also image augmentation techniques but for KTH-TIPS2-b dataset, no improvements were measured by augmenting training data. Results for Xception are in Table 1 and results for NASNet are in Table 2.

In both tables, we present accuracy values in columns $sample_{\{a,b,c,d\}}$. For the *train1test3* scenario, the column header specifies the physical sample used for training, while the images from the other three samples were used for testing. For the *train3test1* scenario, each column header specifies the physical sample used for the evaluation, while images from the other three samples were used for training. The original dataset without any transformation is referred to as *orig*. Subsequently, we present four entries corresponding to $gran_d$, where $d = \{4, 8, 12, 16\}$, which refer to the performance of our method for different parameter d described above. As we can see, the highest average values are consistently achieved for $d = 8$.

The stand-alone results of our method for KTH-TIPS2-b dataset are typically slightly lower when they are compared with results for original images. However, the key factor is that we bring complementary texture information with our approach. Therefore, an introduction of a fusion strategy for the network trained on original images and the network trained on modified images is highly relevant here. The last two rows in both results tables compare two fusion strategies. The first one comes from a comprehensive comparison of fusion strategies that

Table 2. Accuracy values for a pre-trained NASNet model.

	$sample_a$	$sample_b$	$sample_c$	$sample_d$	average
	train1test3 scenario				
orig	76.99%	83.08%	73.23%	77.50%	77.70%
$gran_4$	75.06%	78.54%	67.90%	74.16%	73.92%
$gran_8$	76.96%	**80.42%**	**71.75%**	**81.12%**	**77.56%**
$gran_{12}$	**77.72%**	76.96%	71.35%	79.94%	76.49%
$gran_{16}$	75.84%	78.73%	69.87%	76.49%	75.23%
late fus. [23]	77.67%	81.43%	73.29%	77.50%	77.47%
$orig^*gran_8$	**79.12%**	**86.98%**	**74.41%**	**83.92%**	**81.11%**
	train3test1 scenario				
orig	92.51%	81.48%	97.98%	85.19%	89.29%
$gran_4$	91.16%	79.21%	**96.04%**	76.77%	85.80%
$gran_8$	**94.78%**	**83.08%**	94.02%	**84.09%**	**88.99%**
$gran_{12}$	93.77%	81.14%	93.27%	83.67%	87.96%
$gran_{16}$	93.52%	77.10%	90.40%	79.29%	85.08%
late fus. [23]	94.36%	**83.50%**	93.86%	84.76%	89.12%
$orig^*gran_8$	**97.31%**	83.16%	**97.73%**	**88.38%**	**91.65%**

was done by Wetzer et al. [23]. Based on their findings, a so-called *late fusion* strategy was suggested, where the output probabilities of the softmax layer of both networks are concatenated and a linear SVM is trained to classify the data based on concatenated vectors.

In our experiments, we included a strategy, where we multiplied corresponding softmax probabilities for each original test image and its modified $gran_8$ version. This strategy is illustrated in Fig. 4 and results are presented as $orig^*gran_8$. The average performance of this fusion strategy over all four samples outperforms all other tested scenarios, including the late fusion strategy [23]. Moreover, our top-performing fused model outperforms state-of-the-art techniques on KTH-TIPS2-b dataset for *train3test1* scenario (see Table 3).

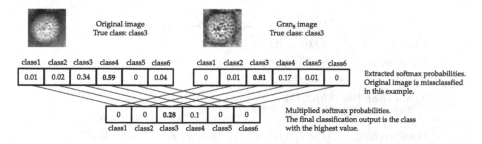

Fig. 4. Illustration of our proposed merging strategy, where softmax probabilities are multiplied. For simplicity, only six classes are used in this example.

Table 3. Comparison of our results on KTH-TIPS2-b dataset with other approaches.

	train1test3	train3test1
Li et al. [10]	–%	83.7%
Liu et al. [12]	–%	88.2%
Sulc et al. [21]	76.0%	–%
Song et al. [19]	79.3%	–%
Liu et al. [11]	**86.7%**	–%
Our NASNet	81.11%	**91.65%**

5.2 Results for Virus Dataset

The practical application of our method is demonstrated on the *virus dataset*, where we followed suggested evaluation using randomized 10-fold cross-validation. The fine-tuning employed the same optimizer but with the initial learning rate of 10^{-4}. For the augmentation of the training part, the image mirroring and rotation around the central pixel were used. We rotated each image by 90°, 180°, and 270°, which together with the mirroring operation resulted in seven unique training samples derived from each original one.

Since virus images are very small, we opted for pre-trained models with lower input image size, namely for VGG-16 and GoogLeNet. During the fine-tuning, they were both trained for 30 epochs with the mini-batch size of 32. The accuracy values are presented in Table 4. For comparison purposes, we added also results for a pre-trained Xception model but its higher input image size leads to slightly worse results in this particular task.

The VGG-16 results for original images and images after our modification are in line with most of the previously published methods. However, after our suggested multiplication of softmax probabilities described above, the accuracy was increased significantly ($p < 0.01$). This behaviour was measurable for all tested pre-trained models. The paired sample t-test was used to confirm statistical significance.

Table 4. Comparison of our results on *virus dataset* with current state-of-the-art approaches.

	Reported average on 10-fold cross-validation		
Kumar et al. [7]	82.07 ± 2.62%		
Faraki et al. [4]	82.5 ± 2.9%		
Sintorn et al. [18]	88.24%		
Nanni et al. [16]	**98.57%**		
	orig	*gran$_8$*	*orig*gran$_8$*
Our Xception	72.67 ± 0.03%	77.20 ± 0.02%	87.60 ± 0.03%
Our GoogLeNet	80.73 ± 0.03%	78.60 ± 0.04%	93.33 ± 0.02%
Our VGG-16	82.40 ± 0.02%	84.93 ± 0.04%	**95.40 ± 0.02%**

Since original images and our modified images are qualitatively different, their combined classification leads to a significant boost in the performance. Typically, an original image is classified incorrectly by a small margin and its modified version is classified correctly with high probability, as it is illustrated in Fig. 4. This scenario can also happen vice versa. Our results on the *virus dataset* were outperformed only by Nanni et al. [16], where authors fused a complex long list of features from multiple CNN models with handcrafted features.

6   Conclusion

In this paper, the texture transformation based on the granulometry principle from the mathematical morphology and the CLAHE algorithm was presented. As we have demonstrated on two independent public texture-oriented datasets, our suggested method brings complementary textural information to the classification process. Especially results on the *virus dataset* are very encouraging and promise a big potential for further practical applications. Our presented results support the observation of Geirhos et al. [5] that ImageNet-trained CNNs are biased towards texture. The modification of image texture has the potential to significantly improve the classification output. For the future work, we see the possibility for improvements by retraining CNNs from scratch on ImageNet samples that are modified by our approach.

Acknowledgement. The work was supported from European Regional Development Fund-Project "Postdoc2@MUNI" (No. CZ.02.2.69/0.0/0.0/18_053/0016952), and from Ministry of Education, Science, and Techn. Development of the Republic of Serbia project ON174008, and through the Project EFFICACY from the University of Southern Denmark.

References

1. Andrearczyk, V., Whelan, P.F.: Deep learning in texture analysis and its application to tissue image classification. In: Biomedical Texture Analysis, pp. 95–129. Elsevier (2017)
2. Bajić, B., Majtner, T., Lindblad, J., Sladoje, N.: Generalised deep learning framework for HEp-2 cell recognition using local binary pattern maps. IET Image Process. **14**(6), 1201–1208 (2020)
3. Cheng, J.Z., Chen, C.M., Shen, D.: Deep learning techniques on texture analysis of chest and breast images. In: Biomedical Texture Analysis, pp. 247–279. Elsevier (2017)
4. Faraki, M., Harandi, M.T., Porikli, F.: Approximate infinite-dimensional region covariance descriptors for image classification. In: IEEE International Conference on Acoustics, Speech and Signal Processing (ICASSP), pp. 1364–1368. IEEE (2015)
5. Geirhos, R., Rubisch, P., Michaelis, C., Bethge, M., Wichmann, F.A., Brendel, W.: ImageNet-trained CNNs are biased towards texture; increasing shape bias improves accuracy and robustness. arXiv preprint arXiv:1811.12231 (2018)
6. Jintasuttisak, T., Intajag, S.: Color retinal image enhancement by Rayleigh contrast-limited adaptive histogram equalization. In: 14th International Conference on Control, Automation and Systems, pp. 692–697. IEEE (2014)

7. Kumar, D., Maji, P.: An efficient method for automatic recognition of virus particles in TEM images. In: Deka, B., Maji, P., Mitra, S., Bhattacharyya, D.K., Bora, P.K., Pal, S.K. (eds.) PReMI 2019. LNCS, vol. 11942, pp. 21–31. Springer, Cham (2019). https://doi.org/10.1007/978-3-030-34872-4_3

8. Kylberg, G., Uppström, M., Hedlund, K.O., Borgefors, G., Sintorn, I.M.: Segmentation of virus particle candidates in transmission electron microscopy images. J. Microsc. **245**(2), 140–147 (2012)

9. Kylberg, G., Uppström, M., Sintorn, I.-M.: Virus texture analysis using local binary patterns and radial density profiles. In: San Martin, C., Kim, S.-W. (eds.) CIARP 2011. LNCS, vol. 7042, pp. 573–580. Springer, Heidelberg (2011). https://doi.org/10.1007/978-3-642-25085-9_68

10. Li, C., Guan, X., Yang, P., Huang, Y., Huang, W., Chen, H.: CDF space covariance matrix of Gabor wavelet with convolutional neural network for texture recognition. IEEE Access **7**, 30693–30701 (2019)

11. Liu, L., Chen, J., Zhao, G., Fieguth, P., Chen, X., Pietikäinen, M.: Texture classification in extreme scale variations using GANet. IEEE Trans. Image Process. **28**(8), 3910–3922 (2019)

12. Liu, L., Fieguth, P., Guo, Y., Wang, X., Pietikäinen, M.: Local binary features for texture classification: taxonomy and experimental study. Pattern Recognit. **62**, 135–160 (2017)

13. Mallikarjuna, P., Targhi, A.T., Fritz, M., Hayman, E., Caputo, B., Eklundh, J.O.: The KTH-TIPS2 Database (2006)

14. Matuszewski, D.J., Sintorn, I.M.: Reducing the U-Net size for practical scenarios: virus recognition in electron microscopy images. Comp. Meth. Program Biomed. **178**, 31–39 (2019)

15. Mirmehdi, M., Xie, X., Suri, J.S.: Handbook of Texture Analysis. World Scientific (2008)

16. Nanni, L., Ghidoni, S., Brahnam, S.: Handcrafted vs. non-handcrafted features for computer vision classification. Pattern Recognit. **71**, 158–172 (2017)

17. Shih, F., Shen, Y., Zhong, X.: Development of deep learning framework for mathematical morphology. Int. J. Pattern Recognit Artif Intell. **33**(6), 1954024 (2019)

18. Sintorn, I.M., Kylberg, G.: Virus recognition based on local texture. In: 22nd International Conference on Pattern Recognition, pp. 3227–3232. IEEE (2014)

19. Song, Y., Cai, W., Li, Q., Zhang, F., Dagan Feng, D., Huang, H.: Fusing subcategory probabilities for texture classification. In: IEEE Conference on Computer Vision and Pattern Recognition, pp. 4409–4417 (2015)

20. Stoklasa, R., Majtner, T., Svoboda, D.: Efficient k-NN based HEp-2 cells classifier. Pattern Recognit. **47**(7), 2409–2418 (2014)

21. Sulc, M., Matas, J.: Fast features invariant to rotation and scale of texture. In: Agapito, L., Bronstein, M.M., Rother, C. (eds.) ECCV 2014. LNCS, vol. 8926, pp. 47–62. Springer, Cham (2015). https://doi.org/10.1007/978-3-319-16181-5_4

22. Theera-Umpon, N., Dhompongsa, S.: Morphological granulometric features of nucleus in automatic bone marrow white blood cell classification. IEEE Trans. Inf. Technol. Biomed. **11**(3), 353–359 (2007)

23. Wetzer, E., Lindblad, J., Sintorn, I.-M., Hultenby, K., Sladoje, N.: Towards automated multiscale imaging and analysis in TEM: glomerulus detection by fusion of CNN and LBP maps. In: Leal-Taixé, L., Roth, S. (eds.) ECCV 2018. LNCS, vol. 11134, pp. 465–475. Springer, Cham (2019). https://doi.org/10.1007/978-3-030-11024-6_36

24. Zuiderveld, K.: Contrast limited adaptive histogram equalization. In: Graphics Gems IV, pp. 474–485. Academic Press Professional (1994)

Towards Precise Recognition of Pollen Bearing Bees by Convolutional Neural Networks

Fernando C. Monteiro[(✉)] [ID], Cristina M. Pinto, and José Rufino [ID]

Research Centre in Digitalization and Intelligent Robotics (CeDRI),
Instituto Politécnico de Bragança, Campus de Santa Apolónia,
5300-253 Bragança, Portugal
{monteiro,rufino}@ipb.pt
https://cedri.ipb.pt

Abstract. Automatic recognition of pollen bearing bees can provide important information both for pollination monitoring and for assessing the health and strength of bee colonies, with the consequent impact on people's lives, due to the role of bees in the pollination of many plant species. In this paper, we analyse some of the Convolutional Neural Networks (CNN) methods for detection of pollen bearing bees in images obtained at hive entrance. In order to show the influence of colour we pre-processed the dataset images. Studying the results of nine state-of-the-art CNNs, we provide a baseline for pollen bearing bees recognition based in deep learning. For some CNNs the best results were achieved with the original images. However, our experiments showed evidence that Dark-Net53 and VGG16 have superior performance against the other CNNs tested, with unsharp masking preprocessed images, achieving accuracy results of 99.1% and 98.6%, respectively.

Keywords: Pollen bearing bees · Convolutional neural network · Deep learning

1 Introduction

Honey bees and other insect pollinators are an essential component of ecosystems, being necessary for the successful reproduction of a wide variety of flowering plants, including agricultural crops [8]. The recognition of bee foraging behaviour brings an important input to identify the colony balance and health [4].

The scientific observation of honey bees activities within and outside their colonies began nearly a century ago [7]. In this seminal work, Lundie referred to the extreme difficulty to obtain accurate records by counts made at the entrance of the hive. Therefore, he claimed the use of some mechanical means to automatically register the exit and return of the bees over long periods of time. Despite his

This work has been supported by FCT – Fundação para a Ciência e Tecnologia within the Project Scope: UIDB/05757/2020.

ⓒ Springer Nature Switzerland AG 2021
J. M. R. S. Tavares et al. (Eds.): CIARP 2021, LNCS 12702, pp. 217–226, 2021.
https://doi.org/10.1007/978-3-030-93420-0_21

visionary proposals, the traditional technique still remains the human observation and manual annotation, as it is the only approach that enables the extraction of a wide range of behaviours available to bee specialists [1]. As such, developing image and video acquisition at the entrance of the hive, and recognizing bees bearing pollen, enables automatizing bee foraging behaviour analysis, which is of great interest for ecological and ethological studies [2,10,13].

Pattern recognition from images has a long and successful history. In recent years, deep learning, and in particular Convolutional Neural Networks (CNNs), has become the dominant machine learning approach in the field of computer vision and, to be specific, in image classification and recognition. While still being a classification based on a combination of key image features, this new approach involves a model determining and extracting the features itself, rather than these being predefined by human analysts. Since the number of labelled pollen bearing bees images in the publicly available datasets is too small to train a CNN from scratch, transfer learning can be employed.

This paper analyzes the performance of 9 state-of-the-art deep learning CNNs (VGG16, VGG19, ResNet50, ResNet101, InceptionV3, Inception-ResNetV2, Xception, DenseNet201 and DarkNet53) in classifying images of pollen bearing and non-bearing bees. As colour is *a priori* a relevant feature for the presence of pollen, CNNs were trained and tested using different colour image preprocessing techniques. Besides original images, grayscale images were used, as well as Contrast Limit Adaptive Histogram Equalization and Unsharp Masking techniques.

The rest of the paper is organized as follows: in the next section, related work is presented; Section 3 provides an overview of the CNNs used; Section 4 describes the proposed approach and the experimental setup; this is followed by the results analysis in Sect. 5, and conclusions and future work in Sect. 6.

2 Related Works

In this section, we review some papers that provide pollen bearing bees recognition systems and define computer vision techniques as baselines.

Babic [2] used background subtraction, colour segmentation and morphology methods for honey bees segmentation. Classification of pollen bearing bees and bees that do not have pollen load is done using a nearest mean classifier, with a simple descriptor consisting of colour variance and eccentricity features. This achieved a correct classification rate of 88.7% with 50 training images per class.

Stojnić [14] proposed an approach that starts by segmenting honey bees from images using two segmentation methods based on colour descriptors. Then, Support Vector Machines (SVM) are trained on few variations of VLAD and SIFT descriptors to classify the images into two classes (with or without pollen). Finally, the classification results are evaluated using Area Under a Curve (AUC), obtaining a score of 91.5% for a dataset of 1000 images.

Rodriguez [10] proposed a colour vision system for detecting pollen-bearing bees, using a Convolutional Neural Network (CNN) for detecting pollen on bees entering the hive. To highlight most of the pollen balls in the dataset, the colour information is filtered using a Gaussian model. Using three models of supervised classification (KNN, Naive Bayes and SVM), and three deep CNN architectures (VGG16, VGG19 and ResNet50), accuracy varies from 50% to 96%.

Sledevic [13] presented the classification of images with pollen bearing bees using a simple scratched CNN with a sufficient configuration required for future implementation on a low-cost FPGA. The three hidden layer CNN was selected as a trade-off between performance and number of required arithmetic operations, yielding an accuracy of 94%.

Yang and Collins [17] applied deep learning techniques to pollen sac detection and measurement on honeybee monitoring video. The pollen sac detection model is built using a faster RCNN architecture with VGG16 core network. This pollen detection model is then combined with a bee tracking model, so that each flying bee tracked on successive video frames is identified as bearing pollen or not. The classification score obtained was 93%.

3   Convolutional Neural Networks Architectures

Convolutional Neural Networks (CNN) is a type of deep learning model for processing data with a grid pattern (e.g., images), inspired by the organization of animal visual cortex and designed to automatically and adaptively learn spatial hierarchies of features through back-propagation by using multiple building blocks, like convolution layers, pooling layers, and fully connected layers from low to high level patterns. It is especially suited for image processing as it uses 2D hidden layers to convolve the features with the input data. The main strength of CNN is that it suppresses the need for feature extraction by automatically extracting the more discriminant features of a set of training images.

Next, we provides an overview of the main features of the CNNs used in this study. We choose nine popular architectures due to their performance on several classification tasks as well as on ImageNet dataset [11].

VGG16 and **VGG19** architectures [12] are 16 and 19 layers deep networks, respectively, on 224 × 224 RGB images input. They use 3 × 3 kernels in every convolution layer. They have five convolutional blocks where the first two blocks have two convolution layers and one max-pooling layer in each block. The remaining three blocks of the network have three fully-connected layers equipped with the rectification (ReLU) non-linearity and the final softmax layer. The main difference between VGG16 and VGG19 is the number of convolutional layers.

ResNet50 and **ResNet101** [5] also apply to 224 × 224 RGB image inputs, and have some design similarities with the VGG architectures with convolutional layers with 3 × 3 filters with convolution stride of one pixel, except the first convolutional block, whose filter is 7 × 7 with convolution stride of two pixels. The batch normalization is adopted right after each convolution layer and before activation. These architectures introduce the *residual block* whose aim is to address the degradation problem observed while training the networks. In the residual block the identity mapping is performed, adding the output from the previous layer to the non-linear layer ahead. The network finishes with a global average pooling layer and a fully-connected layer with the softmax function.

In **Inception-V3** [16], the first three convolutional layers have 3×3 filters with convolution strides, respectively, of two, one, and one pixel. These convolutional layers are followed by max-pooling and other three convolutional layers with 3×3 filters. This network has three inception modules where the resulting output of each module is the concatenation of the outputs of three convolutional filters with different sizes. The goal of these modules is to capture different visual patterns of different sizes and approximate the optimal sparse structure. Finally, before the final softmax layer, an auxiliary classifier acts as a regularization layer.

Inception-ResNet-V2 [15] combines residual connections and the Inception architecture. Since Inception networks tend to be very deep, they are hard to train because of the notorious vanishing gradient problem - as the gradient is back-propagated to earlier layers, repeated multiplication may make the gradient indefinitely small, so Inception-ResNet-V2 replaced the filter concatenation stage of the Inception architecture with residual connections as in ResNet.

The **Xception** architecture [3] has an initial convolutional layer with 3×3 filters with a convolution stride of 2 pixels on 299×229 RGB image input. The architecture has three blocks in a sequence where are carried out convolution, batch normalization, ReLU, and max pooling operations. Besides, at the output of each block, residual connections are made as in ResNet50.

DenseNet201 [6] is built from dense blocks and pooling operations, where each dense block is an iterative concatenation from all previous layers. Within those blocks, the layers are densely connected together: each layer gets the input from all preceding layers and passes on its own feature maps to all subsequent layers. This extreme reuse of residuals creates deep supervision because every layer receives more supervision from the previous layer and thus the loss function will react accordingly which makes it a more powerful network. Like on ResNet, the first block has 7×7 filters with a convolution stride of two pixels on 224×224 RGB images input; however, max-pooling in this convolutional block is carried out over a 3×3 pixel window.

Darknet53 [9] has 53 layers deep and acts as a backbone for the YOLOv3 object detection approach for a 256×256 image input. This network uses successive 3×3 and 1×1 convolutional layers with shortcut connections (introduced by ResNet to help the activations propagate through deeper layers without gradient vanishing) to improve the learning ability. Batch Normalization is used to stabilize training, speed up convergence, and regularize the model batch.

3.1 Transfer Learning

Constraints of practical problems, such as the limited size of training data, refrain the performance of deep CNNs, trained from scratch, to be satisfactory. Since there is so much work that has already been done on image recognition and classification, we can use transfer learning technique to solve the problem. With transfer learning, instead of starting the learning process from scratch, we can start from patterns that have been learned when solving a similar problem.

In deep learning, transfer learning is a technique whereby a CNN model is first trained on a large image dataset with a similar goal to the problem that is being solved. One or more layers from the trained model are then used in a new CNN, trained with sampled images for the current task. This way, the learned features in re-used layers may be the starting point for the training process and adapted to predict new classes of objects. Transfer learning has the benefit of decreasing the training time for a CNN model and can result in lower generalization error due to the small number of images used in the training process.

The weights in re-used layers may be used as a starting point for the training process and adapted in response to the new problem. This usage treats transfer learning as a type of weight initialization scheme. This may be useful when the first related problem has a lot more labelled data than the problem of interest and the similarity in the structure of the problem may be useful in both contexts.

4   Pollen Bearing Bees Dataset

The PollenDataset[1] [10] used in this work was created in 2018 for classifying pollen bearing honeybees obtained at hive entrance. This dataset features high resolution images (180×300) (as illustrated in Fig. 1) of segmented honey bees with a total of 714 labelled images (369 Pollen and 345 Non-Pollen bearing).

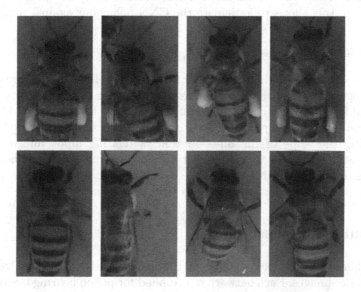

Fig. 1. PollenDataset samples: bees with (1st row) and without (2nd row) pollen bag.

As described in [10], the images were extracted and manually annotated from videos using a protocol defined to avoid near-duplicate samples, remove misaligned samples and ensure a balanced and representative dataset. The orientation of the bees was compensated to ensure in all image samples that the bee is facing upwards. The authors did not preprocessed the images.

---

[1] https://github.com/piperod/PollenDataset.

5   Experimental Setup

The nine CNN architectures adopted were set up to use the fine-tuning strategy as well as the stochastic gradient descent with momentum optimizer at their default values; also, dropout rate was set at 0.5, early-stopping was used to prevent over-fitting, and the learning rate was established at 0.0001. Additionally, the CNNs batch size was configured to be 12 and they were trained with 30 epochs. This specific batch size was chosen to consume less memory and train the CNNs faster since it allows to update the network weights more often. During training, all images go through a heavy data augmentation which includes horizontal and vertical flipping, 360° random rotation, rescaling factor between 70% and 130%, and horizontal and vertical translations between -20 and $+20$ pixels. All CNNs were trained using the MatConvNet MATLAB toolbox in a virtual machine with a pair of NVIDIA RTX 2080 Ti GPUs attached, hosted at the CeDRI cluster.

We adopted 5-Fold Cross-Validation on the dataset, such that in each fold the dataset was split on 70% (499 images) for training and 30% (215 images) for test, allowing the CNN models to be iteratively trained and tested on different sets. Since the testing set is composed of images not seen by the model during the training step, this allows to anticipate the CNN behaviour against new images.

5.1   Colour Preprocessing Techniques

As colour is *a priori* a relevant feature for the presence of pollen, we used different colour image preprocessing techniques, as exemplified in Fig. 2. Grayscale images where used to remove the colour information. Contrast limit adaptive histogram equalization (CLAHE) improved the low contrast issue, without producing noise, by changing the slope of the function that is relating input image intensity value to desired resultant image intensities. Unsharp masking is a linear image processing technique which sharpens the image. The sharp details are identified as a difference between an original image and its blurred version.

6   Results and Discussion

This section shows the performance of the deep learning networks trained on individual bee images. Each network was applied for pollen bearing bee detection on the 215 images validation set (111 pollen bee images and 104 non-pollen bee images). To study the influence of colour features in the recognition process, besides using the original images, CNNs were also trained and tested with the results produced by different colour image preprocessing techniques applied on those images: grayscale images, local contrast enhancing and unsharp masking.

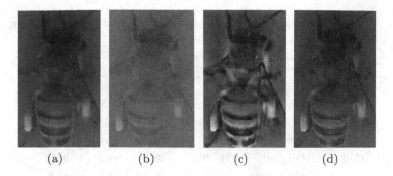

(a)   (b)   (c)   (d)

Fig. 2. Colour image preprocessing. (a) Original version. (b) Grayscale version. (c) After contrast limit adaptive histogram equalization. (d) After unsharp masking.

The training step produces a network model for pollen bearing bee detection images. The network is applied for pollen detection on the validation images producing correct classification (true positive or true negative) results and incorrect classification (false positive or false negative) results, as shown in Fig. 3.

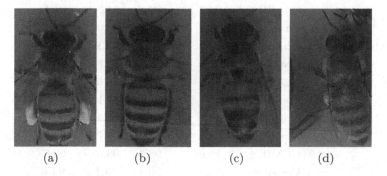

(a)   (b)   (c)   (d)

Fig. 3. Example of pollen recognition results. (a) A true positive result. (b) A true negative result. (c) A false positive result. (d) A false negative result.

An interesting observation is that the false negative samples have smaller pollen bags than the true positive samples, which confirms the intuition that the reduced size of pollen bags increases the difficulty of the classification.

The classification results for Precision, Recall, F1-score and Accuracy for the nine state-of-the-art CNN architectures considered, with different colour preprocessing techniques, are presented in Table 1. The numbers exhibited in bold indicate the best F1-score and Accuracy results obtained for each architecture.

The highest F1-score and Accuracy, both of 99.1%, were achieved by the DarkNet53 architecture with the unsharp masking preprocessing technique, showing that the rates of true positives, true negatives, false positives and false

Table 1. Classification results at percentage through 5-Fold Cross-Validation on the test set for different CNNs and preprocessing techniques.

CNN	Preprocessing	Precision	Recall	F1-score	Accuracy
VGG16	Original	94.8	99.1	96.9	96.7
	Grey-scale	86.4	97.3	91.5	90.7
	CLAHE	99.1	94.6	96.8	96.7
	Unsharp	100	97.3	**98.6**	**98.6**
VGG19	Original	98.2	98.2	**98.2**	**98.1**
	Grey-scale	82.7	99.1	90.2	88.8
	CLAHE	94.7	97.3	96.0	95.8
	Unsharp	94.8	98.2	96.5	96.3
ResNet50	Original	86.6	99.1	92.4	91.6
	Grey-scale	90.0	97.3	93.5	93.0
	CLAHE	99.0	87.4	92.8	93.0
	Unsharp	92.3	97.3	**94.7**	**94.4**
ResNet101	Original	94.7	96.4	**95.5**	**95.4**
	Grey-scale	92.3	97.3	94.7	94.4
	CLAHE	96.3	92.8	94.5	94.4
	Unsharp	89.9	96.4	93.0	92.6
Inception V2	Original	92.0	93.7	**92.9**	**92.6**
	Grey-scale	84.8	95.5	89.8	88.8
	CLAHE	90.2	91.0	90.6	90.2
	Unsharp	89.7	94.6	92.1	91.6
Inception V3	Original	98.0	87.4	92.4	92.6
	Grey-scale	94.2	87.4	90.7	90.7
	CLAHE	89.9	96.4	93.0	92.6
	Unsharp	93.9	96.4	**95.1**	**94.9**
Xception	Original	86.0	82.9	84.4	84.2
	Grey-scale	79.2	92.8	85.5	83.7
	CLAHE	85.2	88.3	86.7	86.1
	Unsharp	85.2	93.7	**89.3**	**88.4**
DenseNet201	Original	95.6	97.3	**96.4**	**96.3**
	Grey-scale	92.9	93.7	93.3	93.0
	CLAHE	95.4	93.7	94.5	94.4
	Unsharp	96.4	95.5	95.9	95.8
DarkNet53	Original	99.1	96.4	97.7	97.7
	Grey-scale	84.4	97.3	90.4	89.3
	CLAHE	98.1	93.7	95.9	95.8
	Unsharp	99.1	99.1	**99.1**	**99.1**

negatives did not present large distortions. On the opposite extreme, the Xception model achieved the lowest results of all CNNs, in all the experiments.

The generality of the CNNs used in this study considerably improve the results presented in [10] for the same dataset. In that work, researchers obtained the best results with an accuracy of 96%. However, they removed several images from the dataset without given any motives or information on that removal.

The validation score obtained with the DarkNet53 network for the images in the test dataset has a mean value of 96.7% with a variance of 1.2% for non-pollen images, and a mean of 98.8% with a variance of 0.4% for pollen images. This network produces only one false positive and one false negative with scores of 96.3% for the false positive and 68.7% for the false negative.

The grayscale preprocessing technique produced the worst evaluation results in all networks due to its lack of colour information, achieving the best result with ResNet101 with an Accuracy of 94.4%.

The CLAHE approach achieved good results, but its increasing of brightness enhanced the yellow regions of the bees body, which can be confused with pollen bags, thus biasing the training process.

As the unsharp masking technique enhances image details, this may help the training process, allowing the learning of different image characteristics. Thus, this technique achieved the best results for most of the networks, and good results even when the networks produced the best results for the original images.

Although the number of images in the dataset is still low, we foresee increasing the classification capacity of the tested approaches by expanding the number of images available via data augmentation with rotation, translation and scaling.

Also, the high values for the four evaluation metrics in all CNNs show that the number of correctly identified images is high when compared to the number of tested images. We believe that a 99.1% value for all the four evaluation metrics is enough to build an automatic classification system of pollen bearing bees, since the visual classification of that bees performed by humans is a hard task.

7   Conclusion

In this paper, an automated pollen bearing bee recognition approach is proposed. To promote research in the automation of pollen bearing bees classification, we report performance for nine pre-trained CNN topologies, applied in four different colour preprocessing techniques. We identify evidence that DarkNet53 has superior performance against other CNNs tested, specially with unsharp masking preprocessing technique, achieving an F1-score and Accuracy of 99.1%. VGG network produces the best evaluation results for original images with an Accuracy of 98.2% against the 97.7% of DarkNet53. This study proves that using the CNN architecture defined for the PollenDataset image collection, allows good classification results when used in a transfer learning approach. In the future, we plan to combine different CNNs in order to further improve the performance.

References

1. Abou-Shaara, H.: The foraging behaviour of honey bees, Apis mellifera: a review. Veterinarni Medicina **59**(1), 1–10 (2014)
2. Babic, Z., Pilipovic, R., Risojevic, V., Mirjanic, G.: Pollen bearing honey bee detection in hive entrance video recorded by remote embedded system for pollination monitoring. ISPRS Ann. Photogram. Remote Sens. Spat. Inf. Sci. **3**(7), 51–57 (2016)
3. Chollet, F.: Xception: deep learning with depthwise separable convolutions. In: 2017 IEEE Conference on Computer Vision and Pattern Recognition (CVPR), pp. 1800–1807 (2017)
4. Frias, B., Barbosa, C., Lourenço, A.: Pollen nutrition in honey bees (Apis mellifera): impact on adult health. Apidologie **47**, 15–25 (2016)
5. He, K., Zhang, X., Ren, S., Sun, J.: Deep residual learning for image recognition. In: 2016 IEEE Conference on Computer Vision and Pattern Recognition (CVPR), pp. 770–778 (2016)
6. Huang, G., Liu, Z., Van Der Maaten, L., Weinberger, K.Q.: Densely connected convolutional networks. In: IEEE Conference on Computer Vision and Pattern Recognition (CVPR), pp. 2261–2269 (2017)
7. Lundie, A.: The Flight Activities of the Honneybee, vol. 1328, pp. 1–38. United States Department of Agriculture (1925)
8. Madras Majewska, B., Majewski, J.: Importance of bees in pollination of crops in the European Union countries. In: International Conference Economic Science for Rural Development, pp. 114–119 (2016)
9. Redmon, J., Farhadi, A.: YOLOv3: an incremental improvement. ArXiv, p. 1804.02767 (2018)
10. Rodriguez, I.F., Megret, R., Acuna, E., Agosto-Rivera, J.L., Giray, T.: Recognition of pollen-bearing bees from video using convolutional neural network. In: 2018 IEEE Winter Conference on Applications of Computer Vision, pp. 314–322 (2018)
11. Russakovsky, O., et al.: ImageNet large scale visual recognition challenge. Int. J. Comput. Vis. **115**, 211–252 (2015)
12. Simonyan, K., Zisserman, A.: Very deep convolutional networks for large-scale image recognition. In: International Conference on Learning Representations, pp. 1–14 (2015)
13. Sledevič, T.: The application of convolutional neural network for pollen bearing bee classification. In: 2018 IEEE 6th Workshop on Advances in Information, Electronic and Electrical Engineering (AIEEE), pp. 1–4 (2018)
14. Stojnić, V., Risojević, V., Pilipović, R.: Detection of pollen bearing honey bees in hive entrance images. In: 17th International Symposium INFOTEH-JAHORINA (INFOTEH), pp. 1–4 (2018)
15. Szegedy, C., Ioffe, S., Vanhoucke, V., Alemi, A.: Inception-v4, InceptionResNet and the impact of residual connections on learning. In: Proceedings of the Thirty-First AAAI Conference on Artificial Intelligence, pp. 4278–4284 (2017)
16. Szegedy, C., Vanhoucke, V., Ioffe, S., Shlens, J., Wojna, Z.: Rethinking the inception architecture for computer vision. In: 2016 IEEE Conference on Computer Vision and Pattern Recognition (CVPR), pp. 2818–2826 (2016)
17. Yang, C., Collins, J.: Deep learning for pollen sac detection and measurement on honeybee monitoring video. In: 2019 International Conference on Image and Vision Computing New Zealand, pp. 1–6 (2019)

Web Application Attacks Detection Using Deep Learning

Nicolás Montes[3]([✉]), Gustavo Betarte[1]([✉]), Rodrigo Martínez[1]([✉]),
and Alvaro Pardo[2]([✉])

[1] Instituto de Computación, Facultad de Ingeniería, Universidad de la República,
Montevideo, Uruguay
{gustun,rodmart}@fing.edu.uy
[2] Departamento de Ingeniería, Facultad de Ingeniería y Tecnologías,
Universidad Católica del Uruguay, Montevideo, Uruguay
apardo@ucu.edu.uy
[3] Facultad de Ingeniería, Universidad de la República, Montevideo, Uruguay

Abstract. This work investigates the use of deep learning techniques to improve the performance of web application firewalls (WAFs), systems that are used to detect and prevent attacks to web applications. Typically, a WAF inspects the HTTP requests that are exchanged between client and server to spot attacks and block potential threats. We model the problem as a one-class supervised case and build a *feature extractor* using deep learning techniques. We treat the HTTP requests as text and train a deep language model with a transformer encoder architecture which is a self-attention based neural network. The use of pre-trained language models has yielded significant improvements on a diverse set of NLP tasks because they are capable of doing transfer learning. We use the pre-trained model as a *feature extractor* to map a HTTP request into a feature vector. These vectors are then used to train a one-class classifier. We also use a performance metric to automatically define an operational point for the one-class model. The experimental results show that the proposed approach outperforms the ones of the classic rule-based MOD-SECURITY configured with a vanilla OWASP CRS and does not require the participation of a security expert to define the features.

Keywords: Web application firewall · Anomaly detection · Deep learning

1 Introduction

It has become a regular security practice to deploy a Web Application Firewall (WAF) [9] to identify attacks that exploit vulnerabilities of web applications. A WAF is a piece of software that intercepts and inspects all the traffic between the

---

This research was partially supported by a grant given to Nicolás Montes from ANII (http://anii.org.uy) and was done in the context of projects FMV_1_2017_136337 (Fondo María Viñas, ANII) and WAFINTL from ICT4V center (http://ict4v.org).

© Springer Nature Switzerland AG 2021
J. M. R. S. Tavares et al. (Eds.): CIARP 2021, LNCS 12702, pp. 227–236, 2021.
https://doi.org/10.1007/978-3-030-93420-0_22

web server and its clients, searching for attacks inside the HTTP packet contents. An implementation of an open source WAF that has become a *de facto* standard is MODSECURITY [24]. The actions this WAF undertakes are driven by rules that specify, by means of regular expressions, the contents of the HTTP packets to be analyzed and eventually flagged as potential attacks. MODSECURITY comes equipped with a default set of rules, known as the OWASP Core Rule Set (OWASP CRS) [15], for handling the most usual vulnerabilities included in the OWASP Top Ten [16]. However, this rule-based approach has some drawbacks: rules are static and rigid by nature, so the OWASP CRS usually produces a rather high rate of false positives, which in some cases may be close to 40% [8] that would potentially lead to a denial of service of the application. The systematic review presented in [22] analyzes the available scientific literature focused on detecting web attacks using machine learning techniques. In [3,4,13] we have presented solutions where the rule-based detection approach of MODSECURITY is complemented with machine learning-based models to mitigate the rule-based approach's drawbacks.

In this work we present an approach that makes use of deep learning techniques to improve the performance of MODSECURITY. It consists of a two step learning framework: first we build a *feature extractor* using deep learning techniques; then we train a one-class supervised model. We treat the HTTP requests as raw text and pre-train a deep language model with the architecture proposed in [12]. These models can operate in huge amounts of text and are called self-supervised because the optimization of the network does not require labels (we will explain this in Sect. 3).

The structure of the rest of the paper is as follows: Sect. 2 describes the deep learning techniques and the related work. Section 3 presents the deep learning framework. The outcomes are described and discussed in Sect. 4. Further work and conclusions are presented in Sect. 5.

2   Background and Related Work

NLP techniques have been greatly improved by the advancements of deep learning [6,12,17,19]. These models rely on a two-step approach. First, they learn deep contextual word representation from raw text in a self-supervised way (stage referred as pre-training). Then, this pre-trained language model can be applied to downstream NLP tasks by choosing between two learning strategies: *feature-based* and *fine-tuning*.

Traditional NLP techniques represent words as atomic units and text is transformed into a numeric vector using one-hot encoding. There are two main problems with this approach. First, there is no notion of similarity between words, as they are represented as indices in a vocabulary [14]. Additionally, the size of the vector is as large as the size of the vocabulary, $|V|$, making machine learning methods prone to problems related with high dimensional feature spaces such as the curse of dimensionality. With the progress of machine learning techniques it has become possible to train more complex models. Probably one of the most successful concept is to use distributed representations of words, also know as *word embedding*. In this approach words are represented in a continuous vector space with much lower dimension than $|V|$. Additionally, it has been shown that words with semantic similarities tend to be nearby in the vector space [2]. In the last decade, word embeddings have established themselves as a core element of many NLP systems. However, as word embedding techniques are static they miss a crucial element for fully capturing local contexts, that is, the semantic and syntactic meaning of words. These methods actually learn to capture the general (most common) context of words in their representations, but they are not able to handle polysemy. Replacing static embeddings with deep contextualized representations has yielded significant improvements on a diverse set of NLP tasks. The idea is simple, a word is assigned a representation that is a function of the entire input sequence (the whole text sequence). The success of deep contextualized word representations suggests that despite being trained with only a language modelling goal, they learn highly transferable and task-agnostic properties of the language [7].

In this work we propose the use of deep contextualized representation of HTTP requests to extract feature vectors that then will be used to train a classifier to detect attacks to web applications. In a first step we create a deep pre-trained language model from scratch using a set of HTTP requests from the web application that we aim to protect. In a second step, we use the *feature-based* strategy to transform each HTTP request into a feature vector. That is, once we have obtained the pre-trained model, we convert each HTTP request into a numeric representation using the weights of the last layer of the network, also known as *feature extraction*. With these representations as input we build a One-Class Classification model (OCC).

Related Work. In [10] Kruegel and Vigna propose an anomaly detection approach where they model specific characteristics of the URL parameters, such as parameter length and input order to generate a probabilistic grammar of each parameter. In our one-class approach we work using the whole request, not only the URL parameters, capturing the normal behavior by modeling the occurrence of a specific set of tokens. This allows us to capture the behavior of the data sent in the normal use of the application. In our approach we also deal with attacks present in the body and header of the requests.

Several authors propose anomaly detection techniques that work over simplification of the application's parameter values. In [5], numbers and alphanumeric sequences are abstracted away, representing each category with a single symbol. In [23] Torrano-Giménez et al. present an anomaly detection technique that instead of using the tokens themselves uses a simplification that only considers the frequencies of three sets of symbols: characters, numbers and special symbol. In our approach the whole request is analyzed without any further simplification.

The work presented in [28] uses word embeddings to represent the URLs. This approach has three steps. First, an ensemble clustering model is applied to separate anomalies from normal samples. Then they use *word2vec* to get the semantic representations of anomalies. Finally, another multi-clustering approach clusters anomalies into specific types. In our model, static embeddings (word2vec) are replaced with deep contextualized representations. We use these representations to get the semantic representations of normal data and use it as input to build the one-class model. In [27] Yu et al. propose a method that uses Bidirectional

Long Short-Term Memory (Bi-LSTM) with an attention mechanism to model the HTTP traffic. This approach is supervised, as they train the Bi-LSTM network to predict whether a request is anomalous or not.

In [18] Qin et al. propose a model which learns the semantics of malicious segments in payload using a Recurrent Neural Network (RNN) with an attention mechanism. The payload is transformed into a hidden state sequence by a RNN and then an *attention mechanism* is used to weight the hidden states as the feature vector for further detection. Thus, they also can use the hidden state of the network as features for a second classifier. The difference with the learning technique that we propose is that they learn the weights of the RNN, the feature extractor model, using normal and abnormal instances. In our case, we build a self-supervised pre-trained model using only normal data.

The work [25] proposes a model that uses a stacked auto-encoder (SAE) and a deep belief network as feature learning methods in which only normal data is used in the learning phase. Subsequently, OCSVM, Isolation Forest and Elliptic Envelope are used as classifiers. In this work features of the HTTP are extracted using n-grams and then deep learning models are applied to reduce the dimensionality generated by the n-grams vectors. In our case we work directly with the HTTP request and avoid building the n-grams which require large amounts of data to correctly capture the statistics of each modelled field.

3   A Two-Step Learning Approach for Anomaly Detection

We propose a learning architecture composed of a two-step method to improve web application anomaly detection models. In a first step, we create a deep pre-trained language model using only normal HTTP requests to the web application. In a second step, we use this model as a feature extractor and train a one-class classifier. That is, each web application has its own model (both the pre-trained language model and the one-class classifier). In the following sections we describe the components of the proposed learning architecture depicted in Fig. 1.

3.1   Pre-training a HTTP Language Model

We train a language model for the HTTP requests in a self-supervised way. We use a Robustly Optimization Bidirectional Encoder Representations from Transformers architecture (RoBERTa) [12]. Using this model each HTTP request is transformed into a numeric vector that captures the contextual information of each token present in the request. The architecture of the network used to build the language model is a multi-layer bidirectional Transformer Encoder [26]. This is an attention-based architecture for modeling sequential data which is an alternative to recurrent neural networks (RNN) and is capable of capturing long range dependencies in sequential data.

The proposed model (see top Fig. 1), is composed of a stack of **L** identical transformer encoder layers, as detailed at the bottom of Fig. 1. Each encoder layer contains two types of sub-layers. The first one is a multi-head self-attention mechanism, which helps looking at other tokens in the sequence while encoding a specific token. The second sub-layer is simply a feed-forward network (FFN), which

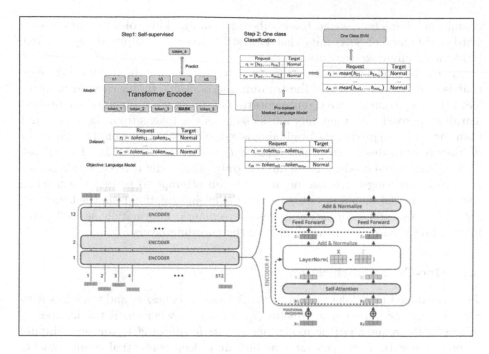

Fig. 1. Top: Proposed architecture. Left: transformer encoder used to extract the contextual representation of each token $token_{ij}$ in the request r_i. Right: each request r_i is represented as the mean of token deep contextualized representation h_{ij} and how they are used to train a one-class classifier. Bottom: Architecture of the Transformer Encoder (Jay Alammar, 2018 [1])

is applied to each position (token representation from previous layer) separately and identically. Because our implementation is almost identical to the original, for a detailed description of the model architecture, we refer readers to [12, 26]. We denote the number transformer encoder blocks as **L** (see top Fig. 1), the hidden size as **H** (the output of the transformer encoder denoted as h in Fig. 1), and the number of self attention heads as **A**. Our model uses the following set of parameters (L = 12, H = 768, A = 12, Total Parameters = 125M).

Token Encoding and Model Training. The input to the model composed of L blocks of transformers is a tokenized version of the HTTP request. For that we use a Byte-Pair Encoding (BPE)[21] tokenizer, a hybrid between character and token-level representations. It relies on sub-word units which are extracted by performing statistical analysis of a training corpus. We use the same tokenizer learned by [19][1], a clever implementation of BPE that uses *bytes* instead

---

[1] We could have chosen another pre-trained BPE tokenizer instead of the one proposed in [19]. The key point is to use a BPE tokenizer trained on huge corpus (40 GB of text) because they can tokenize any word (and any character) of any language without using the *unknown* token.

of unicode characters as the basic sub-words units. This tokenizer has a sub-word vocabulary of 50K units that can still encode any input string without introducing any *unknown* tokens.

Given the model architecture the next step is to define the training strategy, that is, the learning goal and the training mechanisms. In our case, in order to learn the deep contextualized representation of tokens we apply a self-supervised learning approach. We randomly masks some of the tokens from the input, and then the goal is to predict the original masked token based only on its context. In [6] they refer to this procedure as *Masked Language Model* (MLM). In contrast to denoising auto-encoders, these models only predict the masked words rather than reconstructing the entire input [6]. In an attempt to predict the masked tokens, the model should be able to extract some information from the language, not only structural information but some semantic information as well. This information is encoded in the weights of the encoding layers.

3.2   One-Class Classification

The pre-trained model detailed in Sect. 3.1 takes a request r_i and tokenizes it to obtain a representation $r_i = \{token_{i1}, ..., token_{in_i}\}$, where n_i is the number of tokens in the request r_i (the model has a *max length* of 2048 tokens as inputs to be processed). Then, generates as output a deep contextual representation of each token. This deep representation is obtained using the weights of the encoder's last layer. In this way, each request r_i is transformed into a numeric vector $\{h_{i1}, ..., h_{in_i}\}$. Each $h_{ij} \in R^H$ is the vector representation for the $token_{ij}$ in r_i (where H = 768 is the size of the encoder hidden layer).

In order to get a representation of the full request, the most common technique is to average token representations to produce a vector $\bar{r}_i \in R^H$ such that: $\bar{r}_i = \frac{1}{n_i} \sum_{j=1}^{n_i} h_{ij}$. With these procedure we transform a set of normal HTTP requests $D = \{r_1, .., r_m\}$ into a numeric form $\bar{D} = \{\bar{r}_1, .., \bar{r}_m\}$, each $\bar{r}_i \in R^H$. We perform a One-Class Classification model (OCC), with a One-Class Support Vector Machine (OCSVM), with these representations. We operate in a scenario in which only valid requests are known, and no requests tagged as attacks are necessary. We believe that is a realistic approach, where valid traffic could be collected, for instance, as the result of performing functional testing of the application.

Once we have the feature vector mentioned above, we apply the well-know OCSVM classifier introduced in [20] with a Radial Basis Function (RBF) Kernel. They develop an algorithm that returns a function f that takes the value +1 in a "small" region capturing most of the training data points and −1 elsewhere. The strategy is to map the data into the feature space corresponding to the kernel and to separate them from the origin with maximum margin. For a new point x, the value $f(x)$ is determined by evaluating which side of the hyper-plane it falls on in feature space. For each sample, we can calculate the signed distance to the separating hyper-plane. The distance is positive for an inlier (considered as normal) and negative for an outlier (a possible attack). We can set a threshold (θ)

and classify a sample as normal if the distance to the hyper-plane is greater than θ. Varying θ between -1 and $+1$ we obtain a ROC with different operational points. Below we will explain how to automatically obtain the best θ using a grid search approach.

Estimation of the Optimal Operational Point. We must set up two parameters to optimize the performance of the OCSVM: ν and γ. γ parameter is required by the RBF kernel to define a frontier. ν corresponds to the probability of finding a new, but normal, observation outside the frontier. To find the optimal parameters we use a traditional grid-search method. In the case of supervised classification, we can use performance metrics such as F-score or the overall accuracy to evaluate each configuration of parameters. However, these metrics rely on positive and negative samples so it is not possible to use them in an anomaly detection scenario. Nevertheless, [11] introduces a performance measure, $\hat{F}$, that can be estimated from normal and unlabeled examples. They show that $\hat{F}$ is proportional to the square of the geometric mean of precision and recall. Thus, they argue that has roughly the same behaviour as the F_1-score (the harmonic mean of the precision and recall). Therefore, we use a grid-search with the $\hat{F}$ metric for selecting the best parameters of the OCSVM classifier. We use a validation set with only normal and unlabeled examples. The best parameters for both datasets used in Sect. 4 are: $\nu = 0.05$ and $\gamma = 0.5$.

4   Results

The performance of the proposed method is analyzed in terms of True Positive Rate (TPR) and False Positive Rate (FPR). In our case, TPR and FPR indicate the ratio of requests correctly and incorrectly classified as attacks, respectively. To evaluate the proposed method we shall use the same datasets used in our previous work [3,13]. The CSIC2010 dataset embodies a collection of normal and abnormal HTTP requests for a web application that provides functionalities to perform an on-line shopping. The dataset contains 36.000 valid request for training, other 36.000 for testing and 25.000 request of anomalous traffic. We use the DRUPAL dataset in order to evaluate the model on a real life application. We crafted this dataset by registering three days of incoming traffic to the public website of a University. The dataset contains 65.000 valid request and 1.287 real attacks.

One of the main goals of this work is to reduce the amount of false positives generated by MODSECURITY without decreasing the TPR. For this reason we compare our results against MODSECURITY configured with the OWASP CRS version 3 out of the box with two paranoia levels. We also compare the results with the classic information retrieval approach based one-class model presented in [3] later improved in [13]. In this case, a security expert defines the features to be extracted and used to train a one-class classifier based on a Gaussian mixture model. One objective of this work is to compare the features automatically extracted with deep learning with the ones selected by the expert.

Table 1. TPR and FPR for each dataset. (*) One-class classifier using features manually selected by an expert (HTTP tokens). The operational point was manually selected by the authors. (+) The operational point was automatically selected (see Sect. 3.2).

Method	DRUPAL		CSIC 2010	
	TPR	FPR	TPR	FPR
ModSecurity OWASP CRS v3 -PL 1	29.55%	15.57%	26.62%	0.00%
ModSecurity OWASP CRS v3 -PL 2	77.89%	49.93%	29.48%	0.00%
One-class classifier from [13] (*)	94.43%	6.00%	39.63%	5.37%
RoBERTa + OCSVM (+)	95.00%	3.73%	47.10%	7.54%

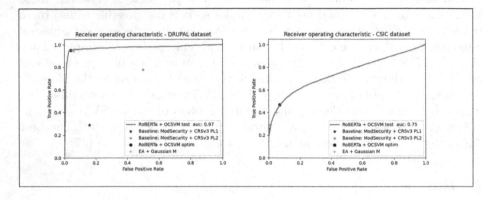

Fig. 2. One-class ROC curve varying the θ value

The evaluation was performed, on each of the datasets, using 70% of the valid requests for training and the rest of the dataset (30% of valid and 100% of attacks) for testing. The results of our proposal (RoBERTa + OCSVM) in Table 1 were obtained with parameters ν, γ and θ found automatically as explained in Sect. 3. In Fig. 2 we present the results in terms of a ROC curve (constructed in terms of TPR and FPR) The solid line represents the different operation points of the OCSVM model. The stars represents the performance of MODSECURITY. The plus symbol represents the operating point achieved by the EA+Gaussian Mixture model in [13]. The circle represents the operating point achieved by OCSVM with the estimated ν, γ and θ as explained above.

In the case of the DRUPAL dataset the ROC curve shows that there are several points that outperform all configurations of MODSECURITY. If we compare the results with our baseline, the best configuration MODSECURITY detects 75% of the attacks, whereas RoBERTa + OCSVM detects 95.00%. The FPR of MODSECURITY is 39.69% and RoBERTa + OCSVM is 3.37%. As to the dataset CSIC2010 the TPR is higher than all versions of MODSECURITY. CSIC2010 is a synthetic dataset constructed adding some attacks and anomalous requests to a set of normal ones. MODSECURITY with paranoia levels 1 and 2 does not produce any false positives at the expense of extremely low TPR. With a more strict paranoia level (PL 3) FPR is 13.95% and TPR is 52.61%. Our method produces a low FPR (7.54%) with a similar TPR (47.10%) for this dataset.

5 Conclusion and Further Work

To the best of our knowledge, the method we propose is a first attempt in using a deep transformer based language representation of HTTP requests to address the problem of web applications attack detection.

We used two different datasets to pre-train a deep language model for the HTTP requests without requiring a security expert to define the set of features. We have proposed a two-step learning approach consisting in first mapping a HTTP request into a continuous space using a transformer encoder and then applying a OCSVM to discriminate normal traffic from attacks. We have used a performance metric proposed by [11] to automatically obtain the parameters of the OCSVM.

We find that the results presented in Sect. 4 are quite promising. They outperform MODSECURITY using the most widely adopted rules and are slightly better than those presented in [3] later improved in [13] with the advantage of not requiring the participation of a security expert to define the features.

As future work, we intend to re-train the pre-trained language model with the dataset DRUPAL with more HTTP requests and check whether we can improve the language model. We also plan to use a set of attacks and explore the *fine-tuning* approach. That strategy consists in adding one additional layer and *fine-tuning* all the parameters of the whole network. The goal of this supervised downstream task is to check whether the model generalizes.

References

1. The Illustrated Transformer - Jay Alammar - Visualizing machine learning one concept at a time. jalammar.github.io/illustrated-transformer/. Accessed 14 Feb 2021
2. Bengio, Y., Ducharme, R., Vincent, P., Janvin, C.: A neural probabilistic language model. J. Mach. Learn. Res. **3**, 1137–1155 (2003)
3. Betarte, G., Giménez, E., Martinez, R., Pardo, Á.: Improving web application firewalls through anomaly detection. In: 2018 17th IEEE International Conference on Machine Learning and Applications (ICMLA), pp. 779–784. IEEE (2018)
4. Betarte, G., Martínez, R., Pardo, Á.: Web application attacks detection using machine learning techniques. In: 2018 17th IEEE International Conference on Machine Learning and Applications (ICMLA), pp. 1065–1072. IEEE (2018)
5. Corona, I., Ariu, D., Giacinto, G.: Hmm-web: a framework for the detection of attacks against web applications. In: Proceedings of ICC 2009, pp. 1–6 (2009)
6. Devlin, J., Chang, M.-W., Lee, K., Toutanova, K.: Bert: pre-training of deep bidirectional transformers for language understanding. arXiv preprint arXiv:1810.04805 (2018)
7. Ethayarajh, K.: How contextual are contextualized word representations? comparing the geometry of bert, elmo, and gpt-2 embeddings. arXiv preprint arXiv:1909.00512 (2019)

8. Folini, C.: Handling false positives with the owasp modsecurity core rule set (2016)
9. Hacker, A.J.: Importance of web application firewall technology for protecting web-based resources. ICSA Labs an Independent Verizon Business (2008)
10. Kruegel, C., Vigna, G.: Anomaly detection of web-based attacks. In: Proceedings of CCS 2003, pp. 251–261. ACM (2003)
11. Lee, W.S., Liu, B.: Learning with positive and unlabeled examples using weighted logistic regression. In: ICML, vol. 3, pp. 448–455 (2003)
12. Liu, Y., et al.: Roberta: a robustly optimized bert pretraining approach. arXiv preprint arXiv:1907.11692 (2019)
13. Martínez, R.: Enhancing web application attack detection using machine learning. Master thesis, Facultad de Ingeniería, UdelaR - Área Informática del Pedeciba, Uruguay (2019)
14. Mikolov, T., Chen, K., Corrado, G., Dean, J.: Efficient estimation of word representations in vector space. arXiv preprint arXiv:1301.3781 (2013)
15. OWASP. Owasp modsecurity core rule set project. coreruleset.org. Accessed 14 Feb 2021
16. OWASP. Owasp top ten project. https://www.owasp.org/index.php/Category:OWASP/Top/Ten/Project. Accessed 14 Feb 2021
17. Peters, M.E., et al.: Deep contextualized word representations. arXiv preprint arXiv:1802.05365 (2018)
18. Qin, Z.Q., Ma, X.K., Wang, Y.J.: Attentional payload anomaly detector for web applications. In: Cheng, L., Leung, A., Ozawa, S. (eds.) Neural Information Processing. ICONIP 2018. LNCS, vol. 11304. Springer, Cham (2018). https://doi.org/10.1007/978-3-030-04212-7_52
19. Radford, A., Wu, J., Child, R., Luan, D., Amodei, D., Sutskever, I.: Language models are unsupervised multitask learners. OpenAI blog 1(8), 9 (2019)
20. Schölkopf, B., Platt, J.C., Shawe-Taylor, J., Smola, A.J., Williamson, R.C.: Estimating the support of a high-dimensional distribution. Neural Comput. 13(7), 1443–1471 (2001)
21. Sennrich, R., Haddow, B., Birch, A.: Neural machine translation of rare words with subword units. arXiv preprint arXiv:1508.07909 (2015)
22. Sureda Riera, T., Bermejo Higuera, J.-R., Bermejo Higuera, J., Martínez Herraiz, J.-J., Sicilia Montalvo, J.-A.: Prevention and fighting against web attacks through anomaly detection technology. A systematic review. Sustainability, 12(12) (2020)
23. Torrano-Gimenez, C., Perez-Villegas, A., Marañón, G.Á., et al.: An anomaly-based approach for intrusion detection in web traffic. J. Inf. Assurance Secur. 5(4), 446–454 (2010)
24. Trustwave Holdings, I.: Modsecurity: open source web application firewall
25. Vartouni, A.M., Teshnehlab, M., Kashi, S.S.: Leveraging deep neural networks for anomaly-based web application firewall. IET Inf. Secur. 13(4), 352–361 (2019)
26. Vaswani, A., et al.: Attention is all you need. arXiv preprint arXiv:1706.03762 (2017)
27. Yu, Y., Yan, H., Guan, H., Zhou, H.: Deephttp: semantics-structure model with attention for anomalous http traffic detection and pattern mining. arXiv preprint arXiv:1810.12751 (2018)
28. Yuan, G., Li, B., Yao, Y., Zhang, S.: A deep learning enabled subspace spectral ensemble clustering approach for web anomaly detection. In: 2017 International Joint Conference on Neural Networks (IJCNN), pp. 3896–3903. IEEE (2017)

Less Is More: Accelerating Faster Neural Networks Straight from JPEG

Samuel Felipe dos Santos$^{(\boxtimes)}$ ⓘ and Jurandy Almeida ⓘ

Instituto de Ciência e Tecnologia, Universidade Federal de São Paulo – UNIFESP,
São José dos Campos, SP 12247-014, Brazil
{felipe.samuel,jurandy.almeida}@unifesp.br

Abstract. Most image data available are often stored in a compressed format, from which JPEG is the most widespread. To feed this data on a convolutional neural network (CNN), a preliminary decoding process is required to obtain RGB pixels, demanding a high computational load and memory usage. For this reason, the design of CNNs for processing JPEG compressed data has gained attention in recent years. In most existing works, typical CNN architectures are adapted to facilitate the learning with the DCT coefficients rather than RGB pixels. Although they are effective, their architectural changes either raise the computational costs or neglect relevant information from DCT inputs. In this paper, we examine different ways of speeding up CNNs designed for DCT inputs, exploiting learning strategies to reduce the computational complexity by taking full advantage of DCT inputs. Our experiments were conducted on the ImageNet dataset. Results show that learning how to combine all DCT inputs in a data-driven fashion is better than discarding them by hand, and its combination with a reduction of layers has proven to be effective for reducing the computational costs while retaining accuracy.

Keywords: Deep learning · Convolutional neural networks · JPEG · Discrete cosine transform · Frequency domain

1 Introduction

Convolutional neural networks (CNNs) have achieved state-of-the-art performance in several computer vision tasks, such as, classification, segmentation, object detection, image super resolution, denoising, medical images, autonomous driving, road surveillance, among others [2,8]. However, in order to achieve this performance, increasingly deeper architectures have been used, making computational cost one of the main problems faced by deep learning models [4].

For storage and transmission purposes, most image data available are often stored in a compressed format, like JPEG, PNG and GIF [2]. To use this data with a typical CNN, it would be required to decode it to obtain RGB images, a task demanding high memory and computational cost [2]. A possible alternative is to design CNNs capable of learning with DCT coefficients rather than RGB

© Springer Nature Switzerland AG 2021
J. M. R. S. Tavares et al. (Eds.): CIARP 2021, LNCS 12702, pp. 237–247, 2021.
https://doi.org/10.1007/978-3-030-93420-0_23

pixels [2,4,5,16]. These coefficients can be easily extracted by partial decoding JPEG compressed data, saving computational cost.

In this paper, we investigate strategies to accelerate CNNs designed for the JPEG compressed domain. The starting point for our study is a state-of-the-art CNN proposed by Gueguen et al. [5], which is a modified version of the ResNet-50 [6]. However, the changes introduced by Gueguen et al. [5] in the ResNet-50 raised its computational complexity and number of parameters. To alleviate these drawbacks, Santos et al. [16] proposed to feed the network with the lowest frequency DCT coefficients, thus losing image details irretrievably. Here, we explore smart strategies to reduce the network computation complexity without sacrificing rich information provided by the DCT coefficients.

Experiments were conducted on the ImageNet dataset, both in a subset and in the whole. Our reported results indicate that learning how to combine all DCT inputs in a data-driven fashion is better than discarding them by hand. We also found that skipping some stages of the network is beneficial, decreasing its computational costs, while also increasing the performance.

The remainder of this paper is organized as follows. Section 2 briefly reviews the JPEG standard. Section 3 discusses related work. Section 4 describes strategies to speed-up CNNs designed for DCT input. Section 5 presents the experimental setup and reports our results. Finally, Sect. 6 offer our conclusions.

2   JPEG Compression

The JPEG standard (ISO/IEC 10918) was created in 1992 and is currently the most widely-used image coding technology for lossy compression of digital images. The basic steps of the JPEG compression algorithm are described as follows. Initially, the representation of the colors in the image is converted from RGB to YCbCr, which is composed of one luminance component (Y), representing the brightness, and two chrominance components, Cb and Cr, representing the color. Also, the Cb and Cr components are down-sampled horizontally and vertically by a factor of 2, for human vision is more sensitive to brightness details than to color details. Then, each of the three components is partitioned into blocks of 8 × 8 pixels and 128 is subtracted from all the pixel values. Next, each block is converted to a frequency domain representation by the forward discrete cosine transform (DCT). The result is an 8 × 8 block of frequency coefficient values, each corresponding to the respective DCT basis functions, with the zero-frequency coefficient (DC term) in the upper left and increasing in frequency to the right and down. The amplitudes of the frequency coefficients are quantized by dividing each coefficient by a respective quantization value defined in quantization tables, followed by rounding the result to the the nearest integer. High-frequency coefficients are approximated more coarsely than low-frequency coefficients, for human vision is fairly good at seeing small variations in color or brightness over large areas, but fails to distinguish the exact strength of high-frequency brightness variations. The quality setting of the encoder affects the extent to which the resolution of each frequency component is reduced. If an excessively low-quality setting is used, most high-frequency coefficients are reduced to zero and thus discarded altogether. To improve the compression ratio, the quantized blocks are arranged into a zig-zag order and then coded by the run-length encoding (RLE) algorithm. Finally, the resulting data for all 8×8 blocks are further compressed with a lossless algorithm, a variant of Huffman encoding. For decompression, inverse transforms of the same steps are applied in reverse order. If the DCT computation is performed with sufficiently high precision, quantization and subsampling are the only lossy operations whereas the others are lossless, so they are reversible.

3 Related Work

The processing of JPEG compressed data has been widely-explored by many conventional image processing techniques as an alternative to speed up the computation performance in a variety of applications, such as face recognition [3], image indexing and retrieval [12], and many others. In deep learning era, the potential of the JPEG compressed domain for neural networks has received limited attention and a few works have emerged in the literature only recently.

To accelerate the training and inference speed, Ehrlich and Davis [4] reformulated the ResNet architecture to perform its operations directly on the JPEG compressed domain. Since the lossless operations used by the JPEG compression algorithm are linear, they can be composed along with other linear operations and then applied to the network weights. In this way, the basic operations used in the ResNet architecture, like convolution, batch normalization, etc., were adapted to operate in the JPEG compressed domain. For the ReLU activation, which is non-linear, an approximation function was developed.

In a different direction, Deguerre et al. [2] proposed a fast object detection method which takes advantage of the JPEG compressed domain. For this, the Single Shot multibox Detector (SSD) [10] architecture was adapted to accommodate block-wise DCT coefficients as input. To preserve the spatial information of the original image, the first three blocks of the SSD network were replaced by a convolutional layer with a filter size of 8×8 and a stride of 8. In this way, each 8×8 block from JPEG compressed data is mapped into a single position in the feature maps used as input for the next layer.

Similarly, the neural network introduced by Gueguen et al. [5] is a modified version of the ResNet-50 [6] capable of operating directly on DCT coefficients rather than RGB pixels. After the DCT coefficients are obtained by partial decoding JPEG compressed data, their Cb and Cr components are up-sampled to match the resolution of the Y component. Next, the Y, Cb, and Cr components are concatenated channel-wise, passed through a batch normalization layer, and fed to the convolution block of the second stage of the ResNet-50. Due to the smaller spatial resolution of the DCT inputs, the strides of the early blocks of the second stage were decreased, mimicking the increase in size of the receptive fields in the original ResNet-50. Also, the second and third stages were changed to accommodate the amount of input channels and to ensure that, at their end, they have the same number of output channels as the original ResNet-50.

However, these changes led to a significant increase in the computational complexity of the ResNet-50 network. To alleviate its network computation costs and number of parameters, Santos et al. [16] extended the modified ResNet-50 network of Gueguen et al. [5] to include a Frequency Band Selection (FBS) technique for selecting the most relevant DCT coefficients before feeding them to the network. The FBS technique relies on the idea that higher frequency information have little visual effect on the image, retaining the lowest frequency coefficients. Although this approach is efficient, image details are completely lost by discarding high frequency information, which may drop the model accuracy.

4   Speeding up CNN Models Designed for DCT Input

In this paper, we extend the work of Santos et al. [16], investigating smarter strategies to reduce the computational complexity and number of parameters of the ResNet-50 network proposed by Gueguen et al. [5]. In Sect. 4.1, we examine learning strategies to reduce the number of channels in the early stages of the ResNet-50 network but without sacrificing any information provided by the DCT coefficients. In a different direction, Sect. 4.2 investigates the possibility of reducing the computational complexity by decreasing the number of layers of the network, while attempting to keep the model accuracy.

4.1   Reducing the Number of Channels

To start, we evaluate the simple idea of reducing the number of channels in the early stages of the modified ResNet-50 of Gueguen et al. [5], which has been proven to be effective in reducing the computational costs of the network [16].

First, we reduce the number of input channels of the second stage to 64 but we kept the decreased strides at its early blocks, as proposed by Gueguen et al. [5]. To accommodate this amount of input channels, we change number of output channels of the second and third stages are changed to be the same as the original ResNet-50. Then, we evaluate different strategies to reduce the number of DCT inputs from 192 (i.e., 3 color components × 64 DCT coefficients) to 64.

Unlike the FBS of Santos et al. [16], where the DCT inputs are discarded by hand potentially losing image information, we take advantage of all DCT inputs and learn how to combine them in a data-driven fashion. For this, we evaluate three different approaches: (1) a linear projection (LP) of the DCT inputs (Sect. 4.1), (2) a local attention (LA) mechanism (Sect. 4.1), and (3) a cross channel parametric pooling (CCPP) (Sect. 4.1).

Linear Projection (LP). The ResNet-50 network [6] have residual learning applied to every block of few stacked layers, given by Eq. 1, where $F()$ is the residual mapping to be learned by the i-th block of stacked layers, W_i are its parameters, x are the input data, and y are the output feature maps.

$$y = F(x, W_i) + x \qquad (1)$$

The $F() + x$ operation is executed by a shortcut connection and a element-wise addition, but their dimensions must be equal. When they are not, a W_s linear projection can be applied in order to match the dimension. As can be seen in Eq. 2, assuming that x have n input features maps and W_s is a weight matrix of size $m \times n$, the product $W_s \cdot x$ will output in m feature maps, where each one is a linear combination of all the n inputs from x.

$$y = F(x, W_i) + W_s \cdot x \qquad (2)$$

We apply this linear projection to reduce the number of channels from 192 of the DCT inputs to 64 of the convolution block of the second stage. In this way, we consider the DCT inputs as a whole regardless the importance of each of their frequencies to the image content. Also, the skewness or kurtosis (shape) of their distribution is preserved by the linear transformation.

Local Attention (LA). The local attention proposed by Luong et al. [11] is a soft attention mechanism used on the machine translation task to analyze a word with a small context window of adjacent words, learning attention maps which focus on relevant parts of the input information.

We adapt this mechanism to be used in the DCT inputs in order to reduce the number of channels from 192 to 64. This is performed according to Eq. 3, where x is an input with n features maps, r is its reshaped version partitioning it into m groups of $\frac{n}{m}$ channels, W is a weight matrix of size $m \times (\frac{n}{m})$, y is an output with m feature maps, and $\odot$ is the Hadamard product.

$$r = reshape \left(x, \left[m, \frac{n}{m} \right] \right) \qquad (3)$$
$$s = W \odot r$$
$$a_i = softmax(s_i), \forall i \in \{1 \ldots m\}$$
$$y_i = a_i \cdot r_i, \forall i \in \{1 \ldots m\}$$

First, the input x is split into m partitions $r = \{r_1, \ldots, r_m\}$ with $\frac{n}{m}$ features maps. Then, alignment scores s are obtained by computing the Hadamard product between W and r. For each partition $i \in \{1 \ldots m\}$, alignment scores s_i are normalized by applying the softmax function, producing attention maps a_i which are used to amplify or attenuate the focus of the distribution of the input data r_i. Therefore, the feature map y_i outputted for the i-th partition is a linear combination of adjacent channels. In this way, we preserve information of the DCT spectrum for the entire range of frequencies.

Cross Channel Parametric Pooling (CCPP). In a cross channel parametric pooling layer, a weighted linear recombination of the input features maps is performed and then passed though a rectifier linear unit (ReLu) [9]. Min Lin et al. [9] proposed to use a cascade of such layers to replace the usual convolution layer of a CNN, since they have enhanced local modeling and the capability of being stacked over each other. Formally, a cascaded cross channel parametric pooling is performed according to Eq. 4 [9], where $f_{i,j,k}^l$ stands for the output of the l-th layer, $x_{i,j}$ is the input patch centered at the pixel (i,j), k is used to index the feature maps, $W_{l,k}$ and $b_{l,k}$ are, respectively, weights and biases of the l-th layer for the k-th filter, and N is the number of layers.

$$f_{i,j,k}^1 = max\left(0, W_{1,k}^T \cdot x_{i,j} + b_{1,k}\right) \tag{4}$$

$$\vdots$$

$$f_{i,j,k}^N = max\left(0, W_{N,k}^T \cdot f_{i,j}^{N-1} + b_{N,k}\right)$$

The cross channel parametric pooling is equivalent to a convolutional layer with a kernel size of 1×1 [9], which is also know as a pointwise convolution [1], being capable of projecting the input feature maps into a new channel space, increasing or decreasing the amount of channels.

We used a cross channel parametric pooling layer to reduce the number of channels from 192 to 64. Similar to the linear projection, the individual importance of each DCT coefficient for the image content is also not taken into account. On the other hand, the non-linear properties of the ReLU activation encourages the model to learn sparse feature maps, making it less prone to overfitting.

4.2   Reducing the Number of Layers

The modified version of the ResNet-50 introduced by Gueguen et al. [5] skips first stage of the original ResNet-50, feeding the DCT coefficients to the second stage, which is modified to accommodate the amount of input channels. In order to reduce the complexity of the network even further, we analyze the effects of skipping the second, third, and fourth stages of the original ResNet-50, but maintaining the stride reduction proposed by Gueguen et al. [5] at the early blocks of the initial stage in which the DCT coefficients are provided as input.

Different from Gueguen et al. [5], we do not increase the number of input channels at the initial stages, since it would lead to a great increase on the computational complexity of the network. Instead, we keep them the same as the original ResNet-50, whose the number of input channels in the second, third, fourth, and fifth stages are 64, 128, 256, and 512, respectively.

To accommodate such amount of channels, the strategies presented in the previous section were used to decrease or increase the DCT inputs from 192 (i.e., 3 color components × 64 DCT coefficients) to the amount of input channels of the initial stage in which they are provided as input. Notice that the number of DCT coefficients is close to the number of input channels of the third stage, requiring a less drastic reduction than the one needed to feed them on the second stage. On the other hand, the expected inputs for the fourth and fifth stages have a greater amount of channels than the DCT inputs and, for this reason, they need to be scaled up, however preserving the salient features as the original data.

5 Experiments and Results

Experiments were conducted on the ILSVRC12 [13] dataset, commonly known as ImageNet, and on a subset of it used by Santos et al. [16]. The ImageNet dataset has 1000 classes and is divided into a training set of 1,281,167 images and a test set of 50,000 images. The ImageNet subset has 211 of the 1000 classes, totaling 268,773 images that were split into a training set with 215,018 images and a test set of 53,755 images. Image classification tasks at two difficulty levels were considered for this subset: in the coarse granularity, the 211 classes were grouped according to their semantics into 12 categories, namely: ball, bear, bike, bird, bottle, cat, dog, fish, fruit, turtle, vehicle and sign; whereas in the fine granularity, all the 211 classes were used.

All the images were resized to 256 pixels on their shortest side and the crop size for all experiments was 224 × 224. In the experiments, the evaluated networks were trained for 120 epochs with batch size of 128, initial leaning rate of 0.05 reduced by a factor of 10 every 30 epochs, and momentum of 0.9. Data augmentation with random crop and horizontal flipping was applied on training phase, while on test, only center crop was used.

The experiments were implemented in PyTorch (version 1.2.0) and performed on a machine equipped with two 10-core Intel Xeon E5-2630v4 2.2 GHz processors, 64 GBytes of DDR4-memory, and 1 NVIDIA Titan Xp GPU. The machine runs Linux Mint 18.1 (kernel 4.4.0) and the ext4 file system.

Section 5.1 compares the performance of the different strategies used to reduce the number of channels from 192 to 64 before feeding them to the network. Section 5.2 shows the effects of reducing the number of layers of the network.

5.1 Effects of Reducing the Number of Channels

Table 1 presents a comparison of the computational costs and the accuracy for the coarse and fine granularity task for the ImageNet subset, and for the entire ImageNet. The computation complexity was measured by the amount of floating point operations (FLOPs) required for passing one image already loaded into the memory through the network and by its number of parameters. The value inside parentheses is the number of input channels at the initial stage of each network.

Table 1. Comparison of computational complexity (GFLOPS), number of parameters, and accuracy for the original ResNet-50 with RGB inputs and networks using DCT with different strategies to reduce number of input channels.

Approach	Computational cost		Accuracy		
	GFLOPs	Params	ImageNet Subset		ImageNet
			Fine	Coarse	
RGB (3×1) [6]	3.86	25.6M	76.28	96.49	73.46
DCT (3×64) [5]	5.40	28.4M	70.28	94.15	72.33
DCT + FBS (3×32) [16]	3.68	26.2M	69.79	94.53	70.22
DCT + FBS (3×16) [16]	3.18	25.6M	68.12	93.92	67.03
DCT + LP (1×64)	3.20	25.6M	70.08	93.17	69.62
DCT + LA (1×64)	3.20	25.6M	69.15	94.23	69.96
DCT + CCPP (1×64)	3.20	25.6M	70.09	94.85	69.73

For both tasks on the ImageNet subset, the RGB-based network performed better than the DCT-based ones. In the fine-grained task, the network of Gueguen et al. [5] achieved the highest accuracy among the DCT-based networks, but also have the highest number of parameters and computational complexity, even compared to the RGB-based network. Similar results were obtained by the networks of Santos et al. [16] and ours in terms of accuracy however greatly reducing the computational complexity and number of parameters. In the coarse-grained task, our LA and CCPP networks yielded better results than that of Gueguen et al. [5] and a similar performance to the DCT + FBS (3×32) of Santos et al. [16].

On the full ImageNet dataset, the network of Gueguen et al. [5] also achieved the highest accuracy, whereas the results obtained for those of Santos et al. [16] and ours were similar, with the advantage of reducing the network computation complexity. Among the strategies we proposed, LA performed slightly better than LP and CCPP. Compared to the networks of Santos et al. [16], our strategies yielded a similar accuracy to DCT + FBS (3×32), while having a computational complexity similar to DCT + FBS (3×16), showing that the use of smarter strategies to learn how to reduce the number of input channels is promising.

The computational complexity and number of parameters of all our strategies (LP, LA, and CCPP) are identical and better than the original ResNet-50, proving to be effective for accelerating computation without sacrificing accuracy.

5.2   Effects of Reducing the Number of Layers

When stages of the network are skipped, we need to decrease or increase the DCT inputs in order to match the amount of channels expected at the initial stage in which they are provided as input. For this, we use the CCPP method, since the results for all the strategies presented in Sect. 4.1 were similar. This strategy was chosen because it is commonly applied in CNNs in order to obtain channel-wise projections of the feature maps, like in depthwise separable convolutions [1].

Table 2 presents the computational complexity and number of parameters of our ResNet-50 using DCT as input when skipping different stages.

Table 2. Comparison of computational complexity (GFLOPS) and number of parameters for our ResNet-50 using DCT as input when skipping different stages and using the CCPP to accommodate the amount of channels expected at the initial stage.

Approach	GFLOPs	Params
Skip the first stage	3.20	25.6M
Skip the first and second stages	2.86	25.1M
Skip the first, second, and third stages	8.26	23.9M
Skip the first, second, third, and fourth stages	10.76	15.8M

As it can be seen, skipping the first and second stages was beneficial, reducing the computational complexity and number of parameters of the network. However, as more stages were skipped, although the number of parameters is decreased, the computational complexity is greatly increased. This is due to the decreased strides at the early blocks of the initial stage. For this reason, the skipping of the first and second stages is the only setting considered in the next experiments, since only it saves the computational costs of the network.

Table 3 compares the computational complexity, number of parameters, and accuracy between state-of-the-art methods and our proposed strategy, which skips the first and second stages and uses CCPP to accommodate DCT inputs. Skipping the first and second stages benefited not only computational costs of the network, but also its accuracy. In both tasks on the ImageNet subset, this modification led to the best performance among the DCT-based networks. On the full ImageNet dataset, it achieved the second best accuracy among the DCT-based networks, behind only the modified ResNet-50 of Gueguen et al. [5], whose computational complexity and number of parameters are considerably bigger.

Table 3. Comparison of computational complexity (GFLOPS), number of parameters, and accuracy for the original ResNet-50 with RGB, state-of-the-art networks designed for DCT, and our strategy for reducing the number of input channels and layers.

Approach	ImageNet Subset		ImageNet
	Fine	Coarse	
RGB (3×1) [6]	76.28	96.49	73.46
DCT (3×64) [5]	70.28	94.15	72.33
DCT + FBS (3×32) [16]	69.79	94.53	70.22
DCT + FBS (3×16) [16]	68.12	93.92	67.03
DCT + CCPP (1×64)	70.09	94.85	69.73
DCT + CCPP + skipping 1st and 2nd stages (1×128)	71.21	94.84	70.49

6   Conclusion

In this paper, we addressed the efficiency issues of CNNs designed for the JPEG compressed domain. More specifically, we speeded-up a modified version of the ResNet-50 proposed by Gueguen et al. [5] and improved by Santos et al. [16]. Although these proposals are effective, they introduced changes in the ResNet-50 [6] that either raised the computational costs or lost relevant information from the input. In contrast, we explored smart strategies to reduce the computational complexity without discarding useful information.

We conducted experiments on the ImageNet dataset, both in a subset and in the whole. Our results on both datasets showed that learning how to combine all DCT inputs in a data-driven fashion performs better than the FBS technique of Santos et al. [16], where the DCT inputs are discarded by hand. Also, we found that skipping some stages of the network is beneficial, proving to be effective for reducing the computational complexity while retaining accuracy.

As future work, we intend to evaluate other learning strategies for accelerating computation without sacrificing accuracy. Also, we want to evaluate the use of our strategies with other network architectures, like EfficientNet [17] and MobileNet [7]. In addition, we also plan to extend the ideas applied on networks designed for JPEG images to those devised for MPEG videos [14,15].

Acknowledgment. This research was supported by the FAPESP-Microsoft Research Virtual Institute (grant 2017/25908-6) and the Brazilian National Council for Scientific and Technological Development - CNPq (grant 314868/2020-8).

References

1. Chollet, F.: Xception: deep learning with depthwise separable convolutions. In: CVPR, pp. 1251–1258 (2017)
2. Deguerre, B., Chatelain, C., Gasso, G.: Fast object detection in compressed JPEG images. In: IEEE Intelligent Transportation Systems Conference (ITSC'19), pp. 333–338 (2019)
3. Delac, K., Grgic, M., Grgic, S.: Face recognition in JPEG and JPEG2000 compressed domain. Image Vision Comput. **27**(8), 1108–1120 (2009)
4. Ehrlich, M., Davis, L.S.: Deep residual learning in the JPEG transform domain. In: ICCV, pp. 3484–3493 (2019)
5. Gueguen, L., Sergeev, A., Kadlec, B., Liu, R., Yosinski, J.: Faster neural networks straight from JPEG. In: NIPS, pp. 3937–3948 (2018)
6. He, K., Zhang, X., Ren, S., Sun, J.: Deep residual learning for image recognition. In: CVPR, pp. 770–778 (2016)
7. Howard, A.G., et al.: Mobilenets: efficient convolutional neural networks for mobile vision applications (2017). arXiv preprint arXiv:1704.04861
8. Li, Y., Gu, S., Gool, L.V., Timofte, R.: Learning filter basis for convolutional neural network compression. In: ICCV, pp. 5623–5632 (2019)
9. Lin, M., Chen, Q., Yan, S.: Network in network (2013). arXiv preprint arXiv:1312.4400
10. Liu, W., et al.: SSD: single shot multibox detector. In: ECCV, pp. 21–37 (2016)

11. Luong, M.T., Pham, H., Manning, C.D.: Effective approaches to attention-based neural machine translation. In: Conference on Empirical Methods in Natural Language Processing (EMNLP'15), pp. 1412–1421 (2015)
12. Poursistani, P., Nezamabadi-pour, H., Moghadam, R.A., Saeed, M.: Image indexing and retrieval in JPEG compressed domain based on vector quantization. Math. Comput. Modell. **57**(5–6), 1005–1017 (2013)
13. Russakovsky, O., et al.: Imagenet large scale visual recognition challenge. Int. J. Comput. Vision **115**(3), 211–252 (2015)
14. Santos, S.F., Almeida, J.: Faster and accurate compressed video action recognition straight from the frequency domain. In: SIBGRAPI - Conference on Graphics, Patterns and Images (SIBGRAPI'20), pp. 62–68 (2020)
15. Santos, S.F., Sebe, N., Almeida, J.: CV-C3D: action recognition on compressed videos with convolutional 3d networks. In: SIBGRAPI - Conference on Graphics, Patterns and Images (SIBGRAPI'19), pp. 24–30 (2019)
16. Santos, S.F., Sebe, N., Almeida, J.: The good, the bad, and the ugly: neural networks straight from jpeg. In: ICIP, pp. 1896–1900 (2020)
17. Tan, M., Le, Q.V.: Efficientnet: rethinking model scaling for convolutional neural networks. In: ICML, pp. 6105–6114 (2019)

Optimizing Person Re-Identification Using Generated Attention Masks

Leonardo Capozzi[1,2]([⊠]) (ID), João Ribeiro Pinto[1,2](ID), Jaime S. Cardoso[1,2](ID), and Ana Rebelo[1](ID)

[1] INESC TEC, Porto, Portugal
{leonardo.g.capozzi,joao.t.pinto,jaime.cardoso,arebelo}@inesctec.pt
[2] Faculdade de Engenharia da Universidade do Porto, Porto, Portugal

Abstract. The task of person re-identification has important applications in security and surveillance systems. It is a challenging problem since there can be a lot of differences between pictures belonging to the same person, such as lighting, camera position, variation in poses and occlusions. The use of Deep Learning has contributed greatly towards more effective and accurate systems. Many works use attention mechanisms to force the models to focus on less distinctive areas, in order to improve performance in situations where important information may be missing. This paper proposes a new, more flexible method for calculating these masks, using a U-Net which receives a picture and outputs a mask representing the most distinctive areas of the picture. Results show that the method achieves an accuracy comparable or superior to those in state-of-the-art methods.

Keywords: Attention · Person re-identification · Feature extraction · Deep learning

1 Introduction

Person re-identification (re-ID) consists of matching a query image of an individual to other pictures of that same individual. The pictures are taken from a system of different cameras, each with a different view. Hence, this problem has important applications in security and surveillance systems. This task can be quite challenging, due to the differences in the position of the cameras, where the face of the person might not be visible, the possibility of occlusions, the large variation in poses, and differences in the background.

In the past, the problem of person re-ID was treated as a classification problem, and a common strategy was to use handcrafted features [11–13,15,26,32]. This was due to the small size of the datasets available in the past. With the rising interest in solutions for this problem, more complex datasets have been created. These new datasets include more identities and more pictures per identity. The approach to the problem has also changed with the use of Deep Learning, where instead of using handcrafted features, the networks encode the images by

© Springer Nature Switzerland AG 2021
J. M. R. S. Tavares et al. (Eds.): CIARP 2021, LNCS 12702, pp. 248–257, 2021.
https://doi.org/10.1007/978-3-030-93420-0_24

extracting relevant information into feature vectors that can be used to compare different images, using simple distance metrics [4,17,21,23,28].

The use of the same datasets, with the same train-test splits, allows to benchmark different methods, in order to compare their performance. Measures such as the Mean Average Precision (mAP) and the Cumulative Match Characteristic (CMC) curve are commonly used to compare the results.

One of the most commonly used loss functions is the Triplet Loss, which modifies the embedding space to pull together pictures belonging to the same person, and to push apart pictures of different identities [6]. This loss is also used with a batch mining process, which makes training more effective by choosing triplets where the error of the network is higher [8].

Many works also use attention mechanisms to force the network to focus on less distinctive areas, in order to improve performance in situations where important information may be missing [4,10,17,21,22].

In [4], the authors propose a method that randomly drops part of the input images in order to force the network to include less distinctive areas in the creation of the feature maps. This leads to a more robust network capable of performing better in situations where information may be missing or occlusions might occur.

In order to improve results, Quispe and Pedrini [17] proposed a method that instead of dropping random parts of the image, created an activation map. This allowed them to drop areas of the image with highly distinctive information, in order to force the network to perform better in situations where information was missing, by using less distinctive information. Although this method calculates which parts of the image it should drop, it can only drop horizontal stripes, which is limited and might lead to suboptimal results.

In this paper, we present a more flexible method, capable of calculating such masks using an auxiliary segmentation model, which allows it to define masks of any shape. This also removes the constraint of deriving the mask directly from the activations, which allows us to calculate the mask in a less predetermined manner.

2   Proposed Methodology

The proposed method is comprised of two networks. The feature extraction network uses a ResNet-50 backbone [7], pretrained on ImageNet [5]. This network receives as input an image and outputs a vector containing 512 features. This feature vector can be used to compare different pictures by computing their euclidean distance. After the network is trained, pictures containing the same individuals will have features vectors with a small distance and pictures with different individuals will have feature vectors with a large distance. These feature vectors allow us to sort a gallery of images. This is achieved by taking a query image and computing its distance to each image in the gallery.

This method also makes use of a mask generation network, to make the feature extraction network more robust. The mask generation network is trained to output a mask that highlights the parts of the image that the feature extraction network is using to generate the feature vector. To make the feature extraction network more robust we hide 30% of the image, by choosing the pixels where the mask has the highest values. This forces the feature extraction network to extract relevant information from other parts of the image. We set the drop ratio to 30% as it was the recommended value in [4].

2.1   Network Architecture

Feature Extractor. The backbone is comprised of a ResNet-50 Network, pretrained on ImageNet. Using a pretrained backbone helps our model to more easily extract relevant features in an image, by using previously learnt filters. The last layer of the ResNet-50 network was removed, therefore the output from the ResNet-50 model is a feature vector of shape 2048x1x1 (channels, height and width respectively). Then we added a convolutional layer with 512 filters, with a kernel size of 1×1 and stride 1×1, followed by batch normalization and ReLU activation.

The output of the network is a template that can be used to match individuals. This template is a unidimensional vector containing 512 features, that describes the picture of the person that was passed to the network. The picture is matched against a gallery of pictures containing a variety of different individuals, by computing the euclidean distance between the feature vectors and sorting the gallery based on the distance metric.

Mask Generator. The mask generation part of the model is performed by a U-Net [20], with an encoder that reduces the resolution at each level, and a decoder that increases the size back to its original resolution. The U-Net features skip connections between layers of the encoder and the decoder that have the same resolution. This allows the transmission of information between layers, which avoids information loss during the encoding and decoding process.

The encoder is composed of 9 Convolutional layers and 4 Max-Pooling layers distributed over 5 resolution levels. Convolutional layers have 32–512 filters, with a size of 3×3 and stride 1×1, each followed by a Batch Normalization layer and a ReLU activation function. Max-Pooling layers use a window size of 2×2 and a stride of 2×2, which reduces the resolution by a factor of 2 at each level.

The decoder mirrors the architecture of the encoder with convolutional and deconvolutional layers. Convolutional layers have 32 to 512 filters, with a size of 3×3 and stride 1×1, each followed by batch normalization and ReLU activation. Deconvolution Layers have 32–256 filters with a size of 2×2 and stride 2×2, which increases resolution by a factor of 2.

The final layer is a Convolutional layer and has one filter with size 1×1 and stride 1×1, followed by a sigmoid activation function to have pixel values between 0 and 1 for the mask.

2.2   Loss

Feature Extractor. The feature extraction model uses the triplet loss, the goal of this loss function is to reduce the distance between feature vectors of pictures belonging to the same individual, and to increase the distance between feature vectors of pictures belonging to different individuals. In the experimental settings section, we go over the selection process of the triplets.

The triplet loss can be written as:

$$\mathcal{L}_{triplet}(F) = \mathbb{E}_{a,p,n}[max(||F(a) - F(p)||_2 - ||F(a) - F(n)||_2 + m, 0)]. \quad (1)$$

where F is the feature extractor, a is the anchor image, p is the positive image, n is the negative image and m is the margin.

When both models are being trained simultaneously we take a weighted sum of the triplet loss of the regular images and the triplet loss of the images with the mask applied.

$$\mathcal{L}_{extractor}(F) = \lambda_1 \mathcal{L}_{triplet}(F) + (1 - \lambda_1)\mathcal{L}_{tripletwithmask}(F). \quad (2)$$

where $\mathcal{L}_{tripletwithmask}$ is the triplet loss applied to masked images (where some parts of the image has been removed to force the network to get relevant information from less distinctive areas).

Mask Model. In order to generate a mask that highlights the areas of the image that the feature extractor is using to calculate the feature vector, we use several losses combined.

The first component can be written as:

$$\mathcal{L}_{identity}(M) = \mathbb{E}_a[mse(F(a), F(M(a) \times a)]. \quad (3)$$

where mse is the mean squared error between two vectors, a is an image, M is the U-Net that generates the masks and F is the feature extractor. $M(a)$ is the mask generated for image a, and $M(a) \times a$ is the resulting image after applying the mask to image a.

The problem with this loss is that it will make the output of the mask model a mask filled with ones. This will make $M(a) \times a$ equal to a, minimizing the previous loss. This is not what we want, as our goal is to make the relevant pixels equal to one while making non-relevant pixels equal to zero. In order to achieve this we add a second loss component:

$$\mathcal{L}_{sparsity}(M) = \mathbb{E}_a[mean(M(a))]. \quad (4)$$

where $M(a)$ is the mask generated for image a. This loss function aims to reduce the mean value of the masks, in order to solve the previous problem.

In addition to both of these losses, we need to make the mask contiguous, in order to select an area of the image, and not single and separated pixels.

Hence, we add the third loss component:

$$\mathcal{L}_{contiguity}(M) = \mathbb{E}_a[\frac{1}{h \times w} \sum_{i,j} |M(a)_{i+1,j} - M(a)_{i,j}| \\ + |M(a)_{i,j+1} - M(a)_{i,j}|]. \tag{5}$$

where $M(a)_{i,j}$ is the value of the mask at index (i,j). This loss calculates the difference of consecutive pixels of the mask (vertically and horizontally), then it calculates the absolute value and finally the mean.

The final loss function becomes:

$$\mathcal{L}_{mask}(M) = \lambda_2 \mathcal{L}_{identity}(M) + (1 - \lambda_2)(\mathcal{L}_{sparsity}(M) + \mathcal{L}_{contiguity}(M)), \tag{6}$$

where $\mathcal{L}_{sparsity}$ and $\mathcal{L}_{contiguity}$ losses are adapted from [18].

3   Experimental Settings

3.1   Data

The proposed model was trained on three datasets that are commonly used in the person re-identification problem.

Market1501 dataset [30] contains data collected from six cameras, on an open environment, in Tsinghua University. It contains images of 1501 individuals. 751 individuals are used for training and 750 individuals for testing. There are a total of 12936 images in the training set, 19732 images in the test set and 3368 query images.

DukeMTMC-reID dataset [19,34] is a subset of the DukeMTMC dataset. The original dataset contains 85-minute videos of high-resolution captured from 8 different cameras. It contains images of 1404 individuals. 702 individuals are used for training and 702 individuals for testing. This dataset contains 16522 images in the training set, 17661 images in the testing set and 2228 query images.

CUHK03 dataset [9] is comprised of images collected from The Chinese University of Hong Kong (CUHK) campus. It contains images from 1467 identities collected from 5 different pairs of camera views. This dataset contains 7368 images for training, 5328 images for testing and 1400 query images. This dataset has two versions: labelled and detected. Labelled means that the bounding boxes were labelled by a human. Detected means that the bounding boxes were estimated by a pedestrian detector. This dataset is prone to missing body parts, misalignments and occlusions, especially on the "detected" version, which makes it more challenging.

3.2   Data Augmentation

There are a couple of pre-processing steps that are applied to the images during training. These improve training, as they make the model more robust to changes and avoid overfitting. During training, images are resized to a resolution of 234 × 117 pixels (height and width, respectively). Then a random section of the image is cropped, with a size of 224 × 112 pixels. The image is then randomly flipped horizontally, with a probability of 0.5. After all these steps are applied we normalize the images using the mean and standard deviation from ImageNet. This is needed because we use a pretrained ResNet-50 model, which was trained with normalized images.

3.3 Training

The proposed model was trained in three stages. The first stage consists of training the feature extraction model using $\mathcal{L}_{triplet}$. The second stage consists of training the mask model (while keeping the feature extraction model frozen) using $\mathcal{L}_{mask}$. The third stage consists of training both models simultaneously using $\mathcal{L}_{extractor}$ for the feature extractor model and $\mathcal{L}_{mask}$ for the mask generation model. After many experiments we set $\lambda_1 = 0.90$ and $\lambda_2 = 0.95$.

Since we are using the triplet loss we need a strategy to select the triplets that maximize the error of the network. Choosing triplets randomly is not a viable strategy, as the network can easily adapt itself to most pictures. This would result in a loss of progress, as the majority of the selected triplets would already have a difference in distances larger than the margin. To overcome this issue we used batch hard triplet mining [8].

Due to GPU memory constraints we used a batch size of 60 pictures (20 individuals and 3 pictures per individual).

We train the model for 50 epochs on the first stage, 25 epochs on the second stage and 100 epochs on the third stage.

For the feature extraction model we used the Adam optimizer with a learning rate of 2×10^{-4} as the default learning rate of 1×10^{-3} would make the pretrained ResNet-50 model unstable and led to a collapsed model, where every feature vector outputted by the model had the same values. The learning rate of 2×10^{-4} worked well for every dataset tested.

For the mask model we used the Adam optimizer with the default values.

4   Results and Discussion

To evaluate the performance of the model we used the mean average precision (mAP) and the rank-1 accuracy. We computed these values for the test sets of the Market1501, DukeMTMC-reID dataset and CUHK03 dataset. There are two versions of the CUHK03 dataset, labelled (L) and detected (D).

After training the model with the Market1501 dataset the model attained a mAP of 63.9% and a rank-1 accuracy of 97.4%. On the DukeMTMC-reID dataset, the model reached an mAP of 61.1% and a rank-1 accuracy of 90.2%. On the CUHK03(L) dataset the model attained an mAP of 69.0% and a rank-1 accuracy of 93.4%. On the CUHK03(D) dataset the model reached an mAP of 66.9% and a rank-1 accuracy of 92.7%.

Table 1. Comparison to state-of-the-art approaches

Method	Market1501 mAP	rank-1	DukeMTMC-ReID mAP	rank-1	CUHK03 (L) mAP	rank-1	CUHK03 (D) mAP	rank-1
IDE [31]	46.0	72.5	47.1	67.7	21.0	22.2	19.7	21.3
PAN [33]	63.4	82.8	51.5	71.6	35.0	36.9	34.0	36.3
DPFL [3]	73.1	88.9	60.0	79.2	40.5	43.0	37.0	40.7
HA-CNN [10]	75.7	91.2	63.8	80.5	41.0	44.4	38.6	41.7
PyrNet [14]	86.7	95.2	74.0	87.1	68.3	71.6	63.8	68.0
Auto-ReID [16]	85.1	94.5	–	–	73.0	77.9	69.3	73.3
MGN [24]	86.9	95.7	78.4	88.7	67.4	68.0	66.0	66.8
DenSem [27]	87.6	95.7	74.3	86.2	75.2	78.9	73.1	78.2
MHN [1]	85.0	95.1	77.2	89.1	72.4	77.2	65.4	71.7
ABDnet [2]	88.2	95.6	78.5	89.0	–	–	–	–
SONA [25]	**88.6**	95.6	78.0	89.2	**79.2**	81.8	**76.3**	79.1
OSNet [35]	84.9	94.8	73.5	88.6	–	–	67.8	72.3
Pyramid [29]	88.2	95.7	**79.0**	89.0	76.9	78.9	74.8	78.9
Top-DB-Net [17]	85.8	94.9	73.5	87.5	75.4	79.4	73.2	77.3
Proposed	63.9	**97.4**	61.1	**90.2**	69.0	**93.4**	66.9	**92.7**

Comparing the results to other state-of-the-art methods (Table 1) we can see that our model has the best results regarding the rank-1 accuracy metric.

Regarding the mAP score, our method is slightly inferior, but aligned, to the alternative methods. This might suggest that our method is better at making a single prediction (as it has a better rank-1 accuracy than other methods) but feels greater difficulty in sorting the whole gallery. To solve this problem we could add multiple streams (or networks), as in [4,17]. This would make the training more stable, since training with the masked images could have the opposite effect of what we want to achieve, which is to force the network to use less distinctive features to compute the feature vector, making the model more accurate.

Figure 1 shows the masks that were generated by our mask generation model. We select 30% of the pixels with the highest values and hide them, therefore black areas represent zones of high importance to our feature extraction model, while white areas represent zones of low importance. During training, we hide the high importance zones, in order to force our feature extraction model to use other parts of the image to generate the feature vectors. In Fig. 1 we can see that the mask generation model almost always tries to hide the person's face, as this is one of the most important attributes of the person for identification. It also hides information about clothing and other things that the feature extraction model finds useful.

Fig. 1. Masks generated by our method (the hidden parts are areas of high importance).

5   Conclusion

The proposed model generates masks that represent areas of the image with important information for the identification process. In order to improve the performance of our feature extraction algorithm, we train it with original images, and images with missing information. This forces the model to use less distinctive areas to compute the feature vector, which leads to better results.

Upon evaluation on three popular datasets, we show that the performance of the algorithm is superior to the alternatives regarding the rank-1 accuracy. However, when using the mAP metric the proposed algorithm was slightly inferior, but aligned, to the alternatives.

The process of removing information from the images could add noise and have the opposite effect that we want to achieve, making the training process unstable and decreasing the performance of the model. We tried to balance the losses to avoid this, but further efforts should be devoted to improving the architecture of the model, adding different streams to minimize the effects of removing information [4,17]. These changes could improve the performance of the model on the mAP metric.

Acknowledgement. This work is financed by National Funds through the Portuguese funding agency, FCT - Fundação para a Ciência e a Tecnologia, within project UIDB/50014/2020, within project UIDB/50014/2020, and within the PhD grant "SFRH/BD/137720/2018".

References

1. Chen, B., Deng, W., Hu, J.: Mixed high-order attention network for person re-identification. CoRR abs/1908.05819 (2019). arxiv.org/abs/1908.05819

2. Chen, T., et al.: Abd-net: Attentive but diverse person re-identification. CoRR abs/1908.01114 (2019). arxiv.org/abs/1908.01114
3. Chen, Y., Zhu, X., Gong, S.: Person re-identification by deep learning multi-scale representations. In: 2017 IEEE International Conference on Computer Vision Workshops (ICCVW), pp. 2590–2600 (2017). https://doi.org/10.1109/ICCVW. 2017.304
4. Dai, Z., Chen, M., Zhu, S., Tan, P.: Batch feature erasing for person re-identification and beyond. CoRR abs/1811.07130 (2018)
5. Deng, J., Dong, W., Socher, R., Li, L.J., Li, K., Fei-Fei, L.: ImageNet: a large-scale hierarchical image database. In: CVPR09 (2009)
6. Dong, X., Shen, J.: Triplet loss in siamese network for object tracking. In: Proceedings of the European Conference on Computer Vision (ECCV) (2018)
7. He, K., Zhang, X., Ren, S., Sun, J.: Deep residual learning for image recognition. CoRR abs/1512.03385 (2015)
8. Hermans, A., Beyer, L., Leibe, B.: In defense of the triplet loss for person re-identification. CoRR abs/1703.07737 (2017)
9. Li, W., Zhao, R., Xiao, T., Wang, X.: Deepreid: Deep filter pairing neural network for person re-identification. In: 2014 IEEE Conference on Computer Vision and Pattern Recognition, pp. 152–159 (2014). https://doi.org/10.1109/CVPR.2014.27
10. Li, W., Zhu, X., Gong, S.: Harmonious attention network for person re-identification. CoRR abs/1802.08122 (2018)
11. Li, Z., Chang, S., Liang, F., Huang, T.S., Cao, L., Smith, J.R.: Learning locally-adaptive decision functions for person verification. In: 2013 IEEE Conference on Computer Vision and Pattern Recognition, pp. 3610–3617 (2013). https://doi.org/ 10.1109/CVPR.2013.463
12. Liao, S., Hu, Y., Zhu, X., Li, S.Z.: Person re-identification by local maximal occurrence representation and metric learning. In: 2015 IEEE Conference on Computer Vision and Pattern Recognition (CVPR), pp. 2197–2206 (2015). https://doi.org/ 10.1109/CVPR.2015.7298832
13. Ma, A.J., Yuen, P.C., Li, J.: Domain transfer support vector ranking for person re-identification without target camera label information. In: 2013 IEEE International Conference on Computer Vision, pp. 3567–3574 (2013). https://doi.org/10.1109/ ICCV.2013.443
14. Martinel, N., Foresti, G.L., Micheloni, C.: Aggregating deep pyramidal representations for person re-identification. In: 2019 IEEE/CVF Conference on Computer Vision and Pattern Recognition Workshops (CVPRW), pp. 1544–1554 (2019). https://doi.org/10.1109/CVPRW.2019.00196
15. Pedagadi, S., Orwell, J., Velastin, S., Boghossian, B.: Local fisher discriminant analysis for pedestrian re-identification. In: 2013 IEEE Conference on Computer Vision and Pattern Recognition, pp. 3318–3325 (2013). https://doi.org/10.1109/ CVPR.2013.426
16. Quan, R., Dong, X., Wu, Y., Zhu, L., Yang, Y.: Auto-reid: searching for a part-aware convnet for person re-identification. CoRR abs/1903.09776 (2019). arxiv.org/abs/1903.09776
17. Quispe, R., Pedrini, H.: Top-db-net: top dropblock for activation enhancement in person re-identification (2020). arXiv preprint arXiv:2010.05435
18. Rio-Torto, I., Fernandes, K., Teixeira, L.F.: Understanding the decisions of cnns: an in-model approach. Pattern Recogn. Lett. **133**, 373–380 (2020)
19. Ristani, E., Solera, F., Zou, R.S., Cucchiara, R., Tomasi, C.: Performance measures and a data set for multi-target, multi-camera tracking. CoRR abs/1609.01775 (2016). arxiv.org/abs/1609.01775

20. Ronneberger, O., Fischer, P., Brox, T.: U-net: convolutional networks for biomedical image segmentation. CoRR abs/1505.04597 (2015)
21. Shen, Y., Li, H., Xiao, T., Yi, S., Chen, D., Wang, X.: Deep group-shuffling random walk for person re-identification. CoRR abs/1807.11178 (2018)
22. Si, J., et al.: Dual attention matching network for context-aware feature sequence based person re-identification. CoRR abs/1803.09937 (2018)
23. Sun, Y., Zheng, L., Deng, W., Wang, S.: Svdnet for pedestrian retrieval. CoRR abs/1703.05693 (2017)
24. Wang, G., Yuan, Y., Chen, X., Li, J., Zhou, X.: Learning discriminative features with multiple granularities for person re-identification. CoRR abs/1804.01438 (2018). arxiv.org/abs/1804.01438
25. Xia, B.N., Gong, Y., Zhang, Y., Poellabauer, C.: Second-order non-local attention networks for person re-identification. CoRR abs/1909.00295 (2019). arxiv.org/abs/1909.00295
26. Yang, Y., Yang, J., Yan, J., Liao, S., Yi, D., Li, S.Z.: Salient Color Names for Person Re-identification. In: Fleet, D., Pajdla, T., Schiele, B., Tuytelaars, T. (eds.) ECCV 2014. LNCS, vol. 8689, pp. 536–551. Springer, Cham (2014). https://doi.org/10. 1007/978-3-319-10590-1_35
27. Zhang, Z., Lan, C., Zeng, W., Chen, Z.: Densely semantically aligned person re-identification. CoRR abs/1812.08967 (2018). arxiv.org/abs/1812.08967
28. Zhao, L., Li, X., Wang, J., Zhuang, Y.: Deeply-learned part-aligned representations for person re-identification. CoRR abs/1707.07256 (2017)
29. Zheng, F., Sun, X., Jiang, X., Guo, X., Yu, Z., Huang, F.: A coarse-to-fine pyramidal model for person re-identification via multi-loss dynamic training. CoRR abs/1810.12193 (2018). arxiv.org/abs/1810.12193
30. Zheng, L., Shen, L., Tian, L., Wang, S., Wang, J., Tian, Q.: Scalable person re-identification: a benchmark. In: 2015 IEEE International Conference on Computer Vision (ICCV), pp. 1116–1124 (2015). https://doi.org/10.1109/ICCV.2015.133
31. Zheng, L., Yang, Y., Hauptmann, A.G.: Person re-identification: past, present and future. CoRR abs/1610.02984 (2016)
32. Zheng, W., Gong, S., Xiang, T.: Reidentification by relative distance comparison. IEEE Trans. Pattern Anal. Mach. Intell. **35**(3), 653–668 (2013). https://doi.org/ 10.1109/TPAMI.2012.138
33. Zheng, Z., Zheng, L., Yang, Y.: Pedestrian alignment network for large-scale person re-identification. CoRR abs/1707.00408 (2017)
34. Zheng, Z., Zheng, L., Yang, Y.: Unlabeled samples generated by GAN improve the person re-identification baseline in vitro. CoRR abs/1701.07717 (2017). arxiv.org/abs/1701.07717
35. Zhou, K., Yang, Y., Cavallaro, A., Xiang, T.: Omni-scale feature learning for person re-identification. CoRR abs/1905.00953 (2019). arxiv.org/abs/1905.00953

Self-supervised Bernoulli Autoencoders for Semi-supervised Hashing

Ricardo Ñanculef[1], Francisco Mena[1], Antonio Macaluso[2(✉)], Stefano Lodi[2], and Claudio Sartori[2]

[1] Department of Informatics, Federico Santa María Technical University, Valparaíso, Chile
{ricardo.nanculef,francisco.menat}@usm.cl
[2] Department of Computer Science and Engineering, University of Bologna, Bologna, Italy
{antonio.macaluso2,stefano.lodi,claudio.sartori}@unibo.it

Abstract. Semantic hashing is a technique to represent high-dimensional data using similarity-preserving binary codes for efficient indexing and search. Recently, variational autoencoders with Bernoulli latent representations achieved remarkable success in learning such codes in supervised and unsupervised scenarios, outperforming traditional methods thanks to their ability to handle the binary constraints architecturally.

In this paper, we propose a novel method for supervision (*self-supervised*) of variational autoencoders where the model uses its own predictions of the label distribution to implement the *pairwise* objective function. Also, we investigate the robustness of hashing methods based on variational autoencoders to the lack of supervision, focusing on two semi-supervised approaches currently in use. Our experiments on text and image retrieval tasks show that, as expected, both methods can significantly increase the quality of the hash codes as the number of labelled observations increases, but deteriorates when the amount of labelled samples decreases. In this scenario, the proposed *self-supervised* approach outperforms the classical approaches and yields similar performance in fully-supervised settings.

Keywords: Hashing · Variational autoencoder · Neural networks

1 Introduction

Given a dataset $D = \{x^{(1)}, x^{(2)}, \ldots, x^{(N)}\}$, with $x^{(\ell)} \in \mathbb{X} \; \forall \ell \in [N]$, *similarity search* is the problem of finding the elements of D that are *similar* to a query object $q \in \mathbb{X}$, not necessarily in D. If $\mathbb{X}$ is equipped with a similarity function[1] $s : \mathbb{X} \times \mathbb{X} \to \mathbb{R}$ and n is small, a simple approach to solve this problem is a *linear scan* which consists in comparing q with all the elements in D and returning $x^{(\ell)}$ if $s(x^{(\ell)}, q)$ is greater than some threshold θ. The value of θ is predetermined and can be computed either to return exactly k elements or chosen to maximize information retrieval metrics [1]. If $\mathbb{X} \subset \mathbb{R}^d$, with small d, tree-based indexing

---

[1] A high value in this function means that the objects are more similar.

© Springer Nature Switzerland AG 2021
J. M. R. S. Tavares et al. (Eds.): CIARP 2021, LNCS 12702, pp. 258–268, 2021.
https://doi.org/10.1007/978-3-030-93420-0_25

such as KD-trees are used to perform efficient scans when N is large. However, if d is large, the computational performance of these methods quickly degrades.

Semantic hashing aims to *learn* a similarity-preserving hash function $h(\boldsymbol{x}) \in \{0, 1\}^B$ that maps similar data to nearby positions in a hash table, preventing undesirable collisions. Similar items to a query $\boldsymbol{q}$ can be easily retrieved by simply accessing all the table cells that differ a few bits from $h(\boldsymbol{q})$. Since binary codes are storage-efficient, these operations can be performed in main memory even for very large datasets [15]. Although classical hashing algorithms based on randomized methods are able to preserve specific and well-known similarity functions (e.g. cosine) [8], the methods based on machine learning (ML) can significantly reduce the number (B) of bits required to preserve similarity by leveraging available real data.

In this context, two different supervision schemes are currently in use. The *pointwise* supervision augments the training objective to predict the label distribution of a training pattern [3]. The *pairwise* supervision leverages the labels to define an additional objective component in which pairs of items with the same label are required to have similar hash codes. This latter approach yields state-of-the-art performance [4] assuming as known the labels of all the training examples. However, in many real-world cases, obtaining labelled data is difficult and time-consuming, thus understanding the efficacy of the current methods when annotations are scarce is extremely important.

2   Related Works

The representation of high-dimensional data using binary codes that preserve the semantic content and support efficient indexing has been extensively studied.

Unsupervised methods rely purely on the properties of the points to indexing. For instance, *Iterative Quantization* (ITQ) [6] computes the codes by applying PCA followed by a rotation that minimizes the quantization error arising from thresholding. *Spectral Hashing* (SpH) [20] poses hashing as the problem of partitioning a graph that encodes information about the geometry of the dataset. Recently, more flexible unsupervised models based on autoencoders have been proposed. In [2], a classic (deterministic) autoencoder is trained to minimize the reconstruction error with an explicit constraint for handling the quantization error. The method proposed in [5] employs a deeper architecture for the encoder. Experimental results [2,5] suggest that hashing methods based on autoencoders can outperform other deep learning methods [13].

Supervised approaches leverage information about the semantic of the items in the dataset to improve the hash codes. Pointwise methods [19] use labels to implicitly enforce consistency between the codes and the annotations. Pairwise methods assume the *pairs* of objects to be annotated as similar and attempt to preserve these similarities explicitly. In [13] pairwise similarity relations are derived from class labels and integrated into the training objective of a neural network. Deep methods that leverage information from triplets or lists have also been proposed [12]. A different type of supervision that is often combined with deep learning techniques is known as *self-taught hashing* [22].

Hashing methods tailored to semi-supervised scenarios, in which labelled and unlabelled samples are available, have been studied in the last years. The method in [5] extends the objective function of a traditional (unsupervised) autoencoder with a term based on pairwise supervision. In [18], the authors present various methods based on linear projections, which combine pairwise supervision with an unsupervised learning goal inspired in information theory (max entropy). In [16], pointwise and pairwise supervision schemes are combined with spectral methods [20]. Building on the idea of self-training [17], the authors in [21] propose a classic approach to leverage partially labelled datasets where label representations and hash codes are learned together in the model.

In this work, we extend the approaches presented in [3,4,14] with a focus on semi-supervised scenarios. In particular, the proposed method can learn label representations and hash codes simultaneously, and while incorporating an explicit unsupervised mechanism, with the supervised one being complementary.

3   Methods

3.1   Generative Model and Bernoulli Autoencoders

As in [14], we pose hashing as an inference problem, where the goal is to learn a probability distribution $q_\phi(b|x)$ of the code $b \in \{0,1\}^B$ corresponding to an input pattern x. This framework is based on a generative process involving two steps: (i) choose an entry of the hash table according to some probability distribution $p_\theta(b)$, and (ii) sample an observation x indexed by that address according to a conditional distribution $p_\theta(x|b)$. The parameters of this random process are learnt in such a way that it approximates the real data distribution.

According to [10], the distribution $q_\phi(b|x)$ is called *encoder*, and the distribution $p_\theta(x|b)$ *decoder*. In the original formulation, $q_\phi(b|x)$ is a Gaussian distribution $\mathcal{N}(\mu_\phi(x), \sigma_\phi^2(x))$ and binary codes are obtained by thresholding $\mu_\phi(x)$ around its empirical median [3]. Instead, in Bernoulli variational autoencoders (B-VAEs), the encoder is a multi-variate Bernoulli distribution $\text{Ber}(\alpha(x))$ with activation probabilities $\alpha(x)$. This choice permits to handle the binary constraint underlying hashing architecturally, creating an inductive bias that can reduce the quantization loss incurred from thresholding Gaussian representations [14].

3.2   Parametrization by Neural Nets

To learn flexible non-linear mappings, the activation probabilities of the encoder $\alpha(x)$ can be represented using a neural net $f(x; \phi)$ whose architecture can be chosen according to the specific dataset. Usually, the model is composed by L fully-connected layers $f_1 \circ \ldots f_{L-1} \circ f_L$, where $f_1 : \mathbb{X} \to \mathbb{R}^{n_1}$ accommodates the input data (a feature vector) and $f_L : \mathbb{R}^{n_L} \to [0,1]^B$ produces the activation probabilities. The latter is usually obtained by using a layer of independent sigmoid neurons [7]. The architecture for the decoder depends also on the application. In text and image retrieval applications it is common to represent data using normalized features $x_i \in [0,1]$ (word frequencies or pixels). In this case $g(b; \theta)$ can be implemented using a net $g_1 \circ \ldots g_{L'-1} \circ g_{L'}$ that ends with a layer of sigmoid neurons (other layers can use other activations, e.g. ReLU).

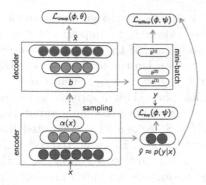

Fig. 1. Sketch of the architectures studied in this paper.

3.3 Unsupervised Training

As illustrated in Fig. 1, the composition an of an encoder $p_\theta(\boldsymbol{x}|\boldsymbol{b})$ and a decoder $q_\phi(\boldsymbol{b}|\boldsymbol{x})$ leads to a stochastic auto-encoder. This model can be trained in an unsupervised fashion using variational methods [10]. According to [14], if $S = \{\boldsymbol{x}^{(1)}, \boldsymbol{x}^{(2)}, \ldots, \boldsymbol{x}^{(n)}\}$ denotes the set of training examples, the negative log-likelihood corresponding to a single data point $\boldsymbol{x}^{(\ell)} \in S$, is upper bounded by the following loss function:

$$\mathcal{L}_{\text{unsup}}^{(\ell)} = \mathbb{E}_{q_\phi(\boldsymbol{b}|\boldsymbol{x}^{(\ell)})} \left[-\log p_\theta(\boldsymbol{x}^{(\ell)}, \boldsymbol{b}) + \log q_\phi(\boldsymbol{b}|\boldsymbol{x}^{(\ell)}) \right]$$

$$= \underbrace{\mathbb{E}_{q_\phi(\boldsymbol{b}|\boldsymbol{x}^{(\ell)})} \left[-\log p_\theta(\boldsymbol{x}^{(\ell)}|\boldsymbol{b}) \right]}_{\mathcal{L}_1} + \underbrace{\lambda \, D_{\text{KL}} \left(q_\phi(\boldsymbol{b}|\boldsymbol{x}^{(\ell)}) || p_\theta(\boldsymbol{b}) \right)}_{\mathcal{L}_2}. \qquad (1)$$

The term $\mathcal{L}_1$ measures the expected error in the reconstruction of $\boldsymbol{x}$ from the hash code $\boldsymbol{b}$. For instance, if the decoder uses a Gaussian distribution, $\mathcal{L}_1$ is proportional to the squared loss $\|\hat{\boldsymbol{x}} - \boldsymbol{x}\|^2$ between the decoder's output $\hat{\boldsymbol{x}}$ and the original observation $\boldsymbol{x}$. The term $\mathcal{L}_2$ measures the Kullback-Leibler divergence between the target distribution of the encoder $q_\phi(\boldsymbol{b}|\boldsymbol{x})$ and a prior distribution $p_\theta(\boldsymbol{b})$. In the context of hashing with Bernoulli autoencoders, $p_\theta(\boldsymbol{b}_i)$ is chosen as Ber(0.5) $\forall i \in [B]$, which expresses a preference for balanced hash tables. This choice allows computing $\mathcal{L}_2$ analytically [14].

3.4 Semi-supervised Training

In presence of training examples $S = \{\boldsymbol{x}^{(\ell)}\}$ annotated with class labels that describe the semantic content $\boldsymbol{y}^{(\ell)} \subset \mathbb{Y} = \{t_1, t_2, \ldots, t_K\}$, the unsupervised objective can be further expanded. Hereafter we assume that the annotations are one-hot encoded as probability distributions, i.e. $\boldsymbol{y}_j = 1$ if $\boldsymbol{x} \in S$ are annotated with a label t_j and $\boldsymbol{y}_j = 0$ otherwise. To accommodate the semi-supervised scenario, we assume that only the first $s \ll n$ examples from S are labelled.

Pointwise Supervision. A simple way to guide the model towards a more discriminative latent representation is to impose the learning $p(x)$ and $p(y|x)$ simultaneously. More specifically, if the neural net implementing the encoder is $f = f_1 \circ \ldots f_{L-1} \circ f_L$, we burden the model with the task of inferring $p(y|x)$ from the representation $z = f_1 \circ \ldots \circ f_{L-1}$ computed immediately before the bit-activation probabilities $\alpha(x)$. This approach is illustrated in Fig. 1. Intuitively, if two patterns $x^{(1)}, x^{(2)}$ have the same annotations, the model should learn $p(y|x^{(1)}) \approx p(y|x^{(2)})$. Since $p(y|x)$ is computed from z, we should have $p(y|x^{(1)}) \approx p(y|x^{(2)}) \Rightarrow z^{(1)} \approx z^{(2)}$. However, the codes also are computed from z, then we expect $z^{(1)} \approx z^{(2)} \Rightarrow b^{(1)} \approx b^{(2)}$. Therefore, the model learns those patterns with the same annotations allocated in nearby addresses of the hash table. This means that the model allows to preserve similarities between objects in the generated latent space. We can approximate the distribution $p(y|x)$, augmenting the autoencoder with an extra fully connected layer $\hat{y}(z; \psi)$. The parameters ψ of this layer can be jointly alongside the rest of the architecture to minimize the cross-entropy loss between the predicted label distribution for a labelled example $x^{(\ell)}$ and the ground-truth. In the simplest case (one-hot vectors representing mutually exclusive labels), the loss can be expressed as follows:

$$\mathcal{L}_{\text{sup}}^{(\ell)} = -\mathbb{E}\left[y^{(\ell)} \log p_\psi(y|x^{(\ell)})\right] = -\sum_k y_k^{(\ell)} \log \hat{y}_k^{(\ell)}. \tag{2}$$

Pairwise Supervision. A more straightforward way to preserve the similarities in the code space and the label space is by equipping the model with a pairwise loss function. If $b^{(\ell)}$ and $b^{(\ell')}$ denote the codes assigned to a pair of examples $x^{(\ell)}, x^{(\ell')}$ sampled from the labelled dataset, and $y^{(\ell)}, y^{(\ell')}$ denote their ground-truth labels, a loss that penalizes/rewards differences between the codes of similar/dissimilar pairs is

$$\begin{aligned} \mathcal{L}_{\text{pair}}^{(\ell,\ell')} &= I(y^{(\ell)} = y^{(\ell')}) D^+(b^{(\ell)}, b^{(\ell')}) \\ &\quad - I(y^{(\ell)} \neq y^{(\ell')}) D^-(b^{(\ell)}, b^{(\ell')}), \end{aligned} \tag{3}$$

where $D^\pm$ denote distance functions in the Hamming space. Usually, D^+ is chosen to be the standard Hamming distance $\|b^{(\ell)} - b^{(\ell')}\|_H$, but D^- is reduced according to $D^- = -(\rho - \|b^{(\ell)} - b^{(\ell')}\|_H)_+$ to avoid wasting efforts in separating dissimilar pairs beyond a margin ρ. This loss is also employed in a plethora of hashing algorithms and variational autoencoders [4].

Self-supervision. Both the pointwise and the pairwise supervision schemes suffer the lack of labelled examples, which is the most common case when dealing with real-world data. In particular, a hashing algorithm based on pairwise loss usually deteriorates even faster than a method using pointwise supervision because of relevant reduction in the number of pairs that can be generated reduces, which is the squared of the fraction of the training set whose label is known (ρ). We propose a self-supervised learning mechanism in which the ground-truth labels required in Eq. (3) are substituted by the predictions of the pointwisely supervised layer of the autoencoder ($\hat{y}(z; \psi)$).

To formally define the self-supervised loss function, we express Eq. (3) in matrix form. Since $\boldsymbol{y}^{(\ell)T}\boldsymbol{y}^{(\ell')} = 0$ if the points $\boldsymbol{x}^{(\ell)}, \boldsymbol{x}^{(\ell')}$ have different labels ($\boldsymbol{y}^{(\ell)T}\boldsymbol{y}^{(\ell')} = 1$ otherwise), the pairwise loss can be rewritten as

$$\mathcal{L}_{\text{pair}}^{(\ell,\ell')} = \boldsymbol{y}^{(\ell)T}\boldsymbol{y}^{(\ell')}D_{\ell,\ell'}^+ - (1 - \boldsymbol{y}^{(\ell)T}\boldsymbol{y}^{(\ell')})D_{\ell,\ell'}^- , \tag{4}$$

where $D_{\ell,\ell'}^\pm$ is used as short-hand for $D^\pm(\boldsymbol{b}^{(\ell)}, \boldsymbol{b}^{(\ell')})$. Then, the self-supervised loss is defined as

$$\mathcal{L}_{\text{selfsup}}^{(\ell,\ell')} = \hat{\boldsymbol{y}}^{(\ell)T}\hat{\boldsymbol{y}}^{(\ell')}D_{\ell,\ell'}^+ - (1 - \hat{\boldsymbol{y}}^{(\ell)T}\hat{\boldsymbol{y}}^{(\ell')})D_{\ell,\ell'}^- . \tag{5}$$

Intuitively, minimizing the pointwise loss in Eq. (2) requires fewer annotations than learning a label-consistent hash function. Thus, after some training rounds $\hat{\boldsymbol{y}}^{(\ell)}$ approximates $\boldsymbol{y}^{(\ell)}$ for many unlabelled observations. Notice that in this formulation the label distribution in Eq. (5) is trainable and the loss can be examined as a function of the labels $\hat{\boldsymbol{y}}$ assigned by the algorithm to the different points of the Hamming space. We can rewrite the self supervised loss as:

$$\mathcal{L}_{\text{selfsup}}^{(\ell,\ell')} = (D_{\ell,\ell'}^+ + D_{\ell,\ell'}^-)\hat{\boldsymbol{y}}^{(\ell)T}\hat{\boldsymbol{y}}^{(\ell')} - D_{\ell,\ell'}^- . \tag{6}$$

We can see that the loss function penalizes correlations between the label distributions proportionally to $D_{\ell,\ell'}^+ + D_{\ell,\ell'}^-$. The loss is 0 if and only if pairs (ℓ, ℓ') for which $D^\pm > 0$ get assigned orthogonal label distributions. Since there is a finite number of (normalized) distributions on $\mathbb{Y}$ which are mutually orthogonal, the proposed loss is minimized by reserving a different labelling to distant regions of the Hamming space ($D^\pm \gg 0$).

3.5 Efficient Implementation

The final objective function for training the autoencoder in semi-supervised scenarios based on self-supervised approach is the following:

$$\mathcal{L} = \sum_{\ell=1}^n \mathcal{L}_{\text{unsup}}^{(\ell)} + \beta \sum_{\ell=1}^s \mathcal{L}_{\text{sup}}^{(\ell)} + \alpha \sum_{\ell,\ell'=1}^n \mathcal{L}_{\text{selfsup}}^{(\ell,\ell')} \tag{7}$$

where $\beta, \alpha > 0$ are hyper-parameters. Note that only the supervised loss $\mathcal{L}_{\text{sup}}$ is computed on labelled instances. The unsupervised loss and the self-supervised loss are computed using all the available observations.

We optimize (7) using backpropagation. Indeed, thanks to the Gumbel-Softmax estimator [9], we can efficiently compute the gradients of $\mathcal{L}_{\text{unsup}}^{(\ell)}$ and $\mathcal{L}_{\text{selfsup}}^{(\ell,\ell')}$ w.r.t. all the model's parameters ϕ, θ, ψ. Being $\hat{\boldsymbol{y}}^{(\ell)}$ independent on the stochastic layer, the gradients of $\mathcal{L}_{\text{sup}}^{(\ell)}$ can be computed classically. However, the direct computation of the total loss has quadratic computational complexity in the number of examples. As sketched in Algorithm 1, we circumvent this problem by forming the pairs required for $\mathcal{L}_{\text{selfsup}}^{(\ell,\ell')}$ at a mini-batch level. In this way, the computational cost of the algorithm is only $\mathcal{O}(nM)$, where M is the mini-batch size which can be considered a (small) constant.

4   Experiments

We conduct experiments to compare different semi-supervised variational autoencoders in scenarios of label scarcity.

Data. The text retrieval tasks are defined on three annotated corpora: *20 Newsgroups*, containing 18000 newsgroup posts on 20 different topics; *TMC* containing 28000 air traffic reports annotated using 22 tags; and *Google Search Snippets*, with 12000 short documents organized in 8 classes (domains). We define an image retrieval task using the dataset *CIFAR-10*, containing 60000 32×32 RGB images of 10 different classes [11]. To facilitate comparisons, we represent the text using TD-IDF features [3,14]. *20 Newsgroups* (hereafter abbreviated *20News*) and *TMC* are splitted in train, validation and test (according to [3]). For *Snippets*, a random sampling of 2400 texts is performed (1200 for validation and 1200 for test) and the rest is used as training test. For *CIFAR-10*, we use a pre-defined test set [11] where the images are represented using deep *VGG* descriptors [5]. A validation set of the same size is randomly sampled from the training set.

Algorithm 1: SSB-VAE.

    Input: A set $S = \{x^{(1)}, \ldots, x^{(n)}\}$ and semantic labels $y^{(\ell)}$ for the first s.
    Output: Trained parameters ϕ, θ, ψ.

1   Initialize ϕ, θ, ψ;
2   **while** *not converged* **do**
3 | Randomly split S into n/M batches of size M;
4 | **foreach** *mini-batch* B_j **do**
5 | | Predict $\hat{y}^{(\ell)}$ for any $x^{(\ell)} \in B_j$;
6 | | Average the gradients of (1) and (5) w.r.t. ϕ, θ, ψ among all the examples in B_j;
7 | | Average the gradient of (2) w.r.t. ψ among the labelled examples in B_j;
8 | | Perform backpropagation updates for ϕ, θ, ψ;
9 | **end foreach**
10   **end while**

Methods. We compare three methods based on variational autoencoders: (1) **VDHS-S**, a variational autoencoder proposed in [3] employing Gaussian latent variables, that combine typical unsupervised learning approach and pointwise supervision; (ii) **PHS-GS**, a variational autoencoder proposed in [4] employing Bernoulli latent variables, that combines unsupervised learning, and both pointwise and pairwise supervision; (iii) **SSB-VAE**, our proposed method based on Bernoulli latent variable, unsupervised learning, pointwise and self-supervision.

Implementation. All models are implemented according to the architectures in [3]. We train the algorithms using 30 epochs, batch size $M = 100$, and the Adam learning rate scheduler [7]. The KL weight λ in Eq. (1) is fixed to the values reported in [14]. The parameters β and α required for **PHS-GS** and **SSB-VAE** are selected on the validation set, using a logarithmic search grid in the range $[10^{-6}, 10^6]$. For a fair comparison, we also allow **VDHS-S** to select the weight of the supervised loss in the objective function. The scores reported in figures and tables are obtained as an average over 5 runs. The experiments are implemented using Python 3.7 with TensorFlow 2.1 and executed on a GTX 1080Ti.[2]

Evaluation. To evaluate the effectiveness of the hash codes, each document/image in the test set is used as a query to search for similar items in the training set. Following previous works [4,5], a relevant search result is one which has the same ground-truth label (topic) as the query. To favour comparisons, the performance is measured using p@100, the precision within the first $k = 100$ retrieved documents/images, sorted according to the Hamming distances of their corresponding hash codes to the query. We also compute the mean average precision, the average of p@k varying k from 1 to the length of the retrieved list. This score penalizes missing relevant items among the first positions of the list. To assess the robustness of the algorithms in case of label scarcity, we train and evaluate the models at varying levels of supervision $\rho = s/n$, the ratio of labelled examples in the training set. Starting from $\rho = 1$ (fully supervised setting), we stress the algorithms reducing ρ till a 10% of supervision.

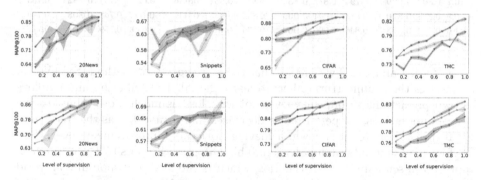

Fig. 2. Mean average precision (MAP@100) of the different algorithms for different levels of supervision. The first row shows the results with 16 hashing bits. The second row the results with 32 bits.

Results and Discussion. Table 1 shows the precision of the different methods on the four datasets using code lengths of $B = 16$ and $B = 32$ bits. When all the training instances are labelled ($\rho = 1$), the model PSH-GS often outperforms the VDSH-S. This result has been confirmed in other works [4]. However, as the supervision level decreases the advantage is reduced, specially when $\rho \leq 0.5$. For instance, with only 20% of training images labeled, VDSH-S provides a precision of 81.6% in CIFAR, 10% over PSH-GS. We see something similar in Snippets and 20News/32Bits. In this case, the performance of PSH-GS suffers significantly more the lack of supervision, with a precision loss over 20% in CIFAR, 25% in 20News and 20% in Snippets. To better illustrate this point, we display in Fig. 2 the mean average precision (MAP) of the algorithms as a function of ρ.

---

[2] Our code is made publicly available at https://github.com/amacaluso/SSB-VAE.

Table 1. P@100 of the different methods for different levels of supervision ρ. a) 32 hashing bits and b) 16 hashing bits. The algorithms PHS-GS [4] and VDHS-S [3] are abbreviated PSH and VDSH.

A) 20-News			CIFAR			SNIPPETS			TMC		
ρ	PSH	SSB-VAE	VDSH	PSH	SSB-VAE	VDSH	PSH	SSB-VAE	VDSH	PSH	SSB-VAE	VDSH
0.1	0.589	**0.734**	0.648	0.687	**0.825**	0.805	0.501	**0.565**	0.540	0.738	**0.750**	0.730
0.3	0.630	**0.787**	0.738	0.737	**0.847**	0.820	0.542	**0.620**	0.576	0.757	**0.759**	0.736
0.5	0.762	**0.824**	0.788	0.818	**0.879**	0.844	0.564	**0.641**	0.634	0.772	**0.778**	0.743
0.7	0.815	**0.841**	0.831	**0.889**	0.880	0.852	0.553	0.644	**0.648**	0.790	**0.795**	0.768
0.9	0.867	**0.880**	0.866	**0.903**	0.901	0.863	0.644	0.648	**0.656**	0.806	**0.813**	0.781
1.0	0.866	**0.878**	0.876	0.906	**0.910**	0.867	**0.696**	0.657	0.661	0.806	**0.818**	0.788

B) 20-News			CIFAR			SNIPPETS			TMC		
ρ	PSH	SSB-VAE	VDSH	PSH	SSB-VAE	VDSH	PSH	SSB-VAE	VDSH	PSH	SSB-VAE	VDSH
0.1	0.595	**0.711**	0.582	0.635	**0.816**	0.781	0.482	**0.621**	0.522	0.723	**0.725**	0.705
0.3	0.678	**0.762**	0.705	0.718	**0.849**	0.789	0.569	**0.612**	0.580	0.740	**0.751**	0.719
0.5	0.752	**0.770**	0.744	0.820	**0.870**	0.811	0.598	**0.614**	0.623	0.750	**0.763**	0.715
0.7	0.802	**0.829**	0.794	0.877	**0.884**	0.818	0.551	**0.634**	0.627	0.764	**0.782**	0.753
0.9	0.826	**0.870**	0.831	0.904	**0.906**	0.832	0.635	**0.647**	0.647	0.768	**0.803**	0.770
1.0	0.872	**0.873**	0.846	0.906	**0.909**	0.836	**0.666**	0.641	0.649	0.759	**0.808**	0.777

In Table 1 and Fig. 2 we can see that the *self-supervised* method (SSB-VAE), which uses the ground-truth labels to learn the label distribution and employs its own predictions to implement the pairwise loss, is much more robust to the lack of annotations. Its performance decreases more smoothly as the fraction of labelled instances reduces down, achieving noticeable improvements over PSH-GS for small amounts of supervision. This enforces the possibility to use such approach in semi-supervised scenarios, which is the most common in real-world data. In Snippets/16Bits, SSB-VAE provides a precision only 2% less than the precision achieved in the fully supervised case. In the cases of TMC dataset, the pairwise approach is able to maintain an advantage on VDSH-S almost uniformly as the supervision becomes lower. If this is the case, the proposed method is still competitive or better than the best baseline. Although the most significant improvements are obtained for smaller ρ, we also confirm that in scenarios of label abundance, using pairwise supervision based on the ground-truth distributions does not give a very significant advantage over the self-supervised approach. To test the robustness of the proposed method, two types of statistical tests are also conducted. We employ the Friedman's test to assess whether there is enough statistical evidence to reject the hypothesis that the three methods are statistically equivalent (in terms of p@100), when considering different levels of supervision. In this design, the method (SSB-VAE, PSH, VDSH) serves as the group variable, and the level of supervision serves as the blocking variable. In addition, when rejecting the null hypothesis of Friedman's test, we compare the proposed method against PSH and VDSH using the Nemenyi post-hoc test, to check for statistically significant differences. The p-values are reported in Table 2. In all but two cases we obtain values below 5%.

Table 2. P-values of the statistical tests

	16 *bits*			32 *bits*		
	Friedman test	PSH	VDSH	Friedman test	PSH	VDSH
20-NEWS	1.1×10^{-4}	6.3×10^{-5}	3.7×10^{-2}	5.0×10^{-4}	4.9×10^{-3}	1.0×10^{-3}
SNIPPETS	7.4×10^{-3}	1.0×10^{-2}	8.9×10^{-1}	7.4×10^{-3}	1.0×10^{-2}	8.9×10^{-1}
TMC	4.5×10^{-5}	6.5×10^{-2}	2.3×10^{-5}	2.2×10^{-4}	1.9×10^{-2}	1.6×10^{-4}
CIFAR	2.0×10^{-2}	1.1×10^{-1}	1.2×10^{-2}	1.8×10^{-3}	1.9×10^{-2}	2.2×10^{-3}

5   Conclusions

We studied the performance of semi-supervised hashing algorithms based on variational autoencoders in scenarios of label scarcity. As expected, the models that explicitly preserve pairwise similarities derived from the annotations, often yields better results than using pointwise supervision (only), confirming results of previous works. However, these methods tend to deteriorate more sharply when the number of labelled observations decreases. To overcome this problem, we proposed a new type of supervision in which the model uses its own beliefs about the class distribution to enforce a consistency between the similarities in the code space and the similarities in the label space. Experiments in text and image retrieval tasks confirmed that this method degrades much more gracefully when the models are stressed with scarcely annotated data, and very often outperforms the baselines by a significant margin. In scenarios of label abundance, the proposed method is competitive or even better than the best baseline. In future, we plan to equip the method with adaptive loss weights and extend the experiments to cross-domain information retrieval[3].

References

1. Baeza-Yates, R., Ribeiro-Neto, B.: Modern Information Retrieval. ACM, New York (1999)
2. Carreira-Perpinán, M.A., Raziperchikolaei, R.: Hashing with binary autoencoders. In: Proceedings of the CVPR, pp. 557–566 (2015)
3. Chaidaroon, S., Fang, Y.: Variational deep semantic hashing for text documents. In: SIGIR, pp. 75–84 (2017)
4. Dadaneh, S.Z., Boluki, S., Yin, M., Zhou, M., Qian, X.: Pairwise supervised hashing with Bernoulli variational auto-encoder and self-control gradient estimator. In: Proceedings of the UAI (2020)
5. Do, T.-T., Doan, A.-D., Cheung, N.-M.: Learning to hash with binary deep neural network. In: Leibe, B., Matas, J., Sebe, N., Welling, M. (eds.) ECCV 2016, Part V. LNCS, vol. 9909, pp. 219–234. Springer, Cham (2016). https://doi.org/10.1007/978-3-319-46454-1_14
6. Gong, Y., Lazebnik, S.: Iterative quantization: a procrustean approach to learning binary codes. In: Proceedings of the CVPR, pp. 817–824 (2011)

---

[3] A draft version of this work is available on arXiv:2007.08799.

7. Goodfellow, I., Bengio, Y., Courville, A.: Deep Learning. MIT Press, Cambridge (2016)
8. Indyk, P., Motwani, R.: Approximate nearest neighbors: towards removing the curse of dimensionality. In: Proceedings of the ACM STOC, pp. 604–613 (1998)
9. Jang, E., Gu, S., Poole, B.: Categorical reparameterization with Gumbel-softmax. In: Proceedings of the ICLR (2017)
10. Kingma, D.P., Welling, M.: Auto-encoding variational Bayes. In: Proceedings of the ICLR (2014)
11. Krizhevsky, A., Sutskever, I., Hinton, G.E.: Imagenet classification with deep convolutional neural networks. In: NIPS, pp. 1097–1105 (2012)
12. Lai, H., Pan, Y., Liu, Y., Yan, S.: Simultaneous feature learning and hash coding with deep neural networks. In: Proceedings of the CVPR, pp. 3270–3278 (2015)
13. Lu, J., Liong, V.E., Zhou, J.: Deep hashing for scalable image search. IEEE Trans. Image Process. **26**(5), 2352–2367 (2017)
14. Mena, F., Ñanculef, R.: A binary variational autoencoder for hashing. In: Nyström, I., Hernández Heredia, Y., Milián Núñez, V. (eds.) CIARP 2019. LNCS, vol. 11896, pp. 131–141. Springer, Cham (2019). https://doi.org/10.1007/978-3-030-33904-3_12
15. Norouzi, M., Punjani, A., Fleet, D.J.: Fast exact search in Hamming space with multi-index hashing. IEEE Pattern Anal. Mach. Intell. **36**(6), 1107–1119 (2014)
16. Song, T., Cai, J., Zhang, T., Gao, C., Meng, F., Wu, Q.: Semi-supervised manifold-embedded hashing with joint feature representation and classifier learning. Pattern Recognit. **68**, 99–110 (2017)
17. Triguero, I., García, S., Herrera, F.: Self-labeled techniques for semi-supervised learning: taxonomy, software and empirical study. Knowl. Inf. Syst. **42**(2), 245–284 (2015). https://doi.org/10.1007/s10115-013-0706-y
18. Wang, J., Kumar, S., Chang, S.F.: Semi-supervised hashing for large-scale search. IEEE Pattern Anal. Mach. Intell. **34**(12), 2393–2406 (2012)
19. Wang, Q., Zhang, D., Si, L.: Semantic hashing using tags and topic modeling. In: Proceedings of the SIGIR, pp. 213–222. ACM (2013)
20. Weiss, Y., Torralba, A., Fergus, R.: Spectral hashing. In: NIPS (2009)
21. Yang, H., Tu, C., Chen, C.: Adaptive labeling for hash code learning via neural networks. In: Proceedings of the ICIP, pp. 2244–2248 (2019)
22. Zhang, D., Wang, J., Cai, D., Lu, J.: Self-taught hashing for fast similarity search. In: Proceedings of the SIGIR, pp. 18–25 (2010)

Explainable Artificial Intelligence

Interpretable Concept Drift

João Guilherme Mattos[1](✉), Thuener Silva[1](✉), Hélio Lopes[1](✉),
and Alex Laier Bordignon[2](✉)

[1] Departamento de Informática, Pontifícia Universidade Católica do Rio de Janeiro,
Rio de Janeiro, Brazil
{jmattos,lopes}@inf.puc-rio.br
[2] Departamento de Geometria, Universidade Federal Fluminense,
Niterói, Brazil
alexb@id.uff.br

Abstract. In a dynamic environment, models tend to perform poorly once the underlying distribution shifts. This phenomenon is known as Concept Drift. In the last decade, considerable research effort has been directed towards developing methods capable of detecting such phenomenons early enough so that models can adapt. However, not so much consideration is given to explain the drift, and such information can completely change the handling and understanding of the underlying cause. This paper presents a novel approach, called *Interpretable Drift Detector*, that goes beyond identifying drifts in data. It harnesses decision trees' structure to provide a thorough understanding of a drift, i.e., its principal causes, the affected regions of a tree model, and its severity. Moreover, besides all information it provides, our method also outperforms benchmark drift detection methods in terms of false-positive rates and true-positive rates across several different datasets available in the literature.

Keywords: Data streams · Data stream mining · Concept drift · Drift detection · Drift understanding · Decision tree

1 Introduction

Data streams are a form of data where elements are produced sequentially and continuously. Although data streams can be stationary - i.e., their data is drawn from an unknown fixed probability distribution - for the majority of real-world scenarios that assumption is just not true, which means that a single probability distribution is unable to capture the behavior of the data at all the different periods encompassed by the stream. Streaming data normally present non-stationary characteristics, which could be due to seasonality, gradual or subtle changes in them. The main drawback of non-stationarity is that models once trained and fine-tuned to certain scenarios may perform sub-optimally once the data drifts to a new probability distribution, a phenomenon known as "Concept Drift". In supervised learning, a classification task can be defined as follows: y represents the target variable; and X, the feature variables. Their joint probability distribution at instant t is defined as $P_t(X, y)$ and is also known as "concept" at t.

© Springer Nature Switzerland AG 2021
J. M. R. S. Tavares et al. (Eds.): CIARP 2021, LNCS 12702, pp. 271–280, 2021.
https://doi.org/10.1007/978-3-030-93420-0_26

That way, there is a Concept Drift between two distinct timestamps t_0 and t_1 if $P_{t_0}(X, y) \neq P_{t_1}(X, y)$ [2].

The joint probability distribution can be stated as $P_t(X, y) = P_t(X)P_t(y|X)$. Thus a concept drift is either a mixture of changes in the evidence factor, $P(X)$, and in the posterior probability, $P(y|X)$, or a result of changes in one of these factors alone. Considering $P_{t_0}(y|X) = P_{t_1}(y|X)$ and $P_{t_0}(X) \neq P_{t_1}(X)$, we state that a *Virtual Drift* occurred between instants t_0 and t_1. It is called a virtual drift, because although the distribution of the feature variables, X, has changed, their relation to their corresponding target has not changed, and hence the drift does not affect the models previously trained. Nevertheless, they are still important for understanding the behaviour of the data and even coming up with a better model for it. On the other hand, if $P_{t_0}(X) = P_{t_1}(X)$ and $P_{t_0}(y|X) \neq P_{t_1}(y|X)$, a *Real Drift* is acknowledged between instants t_0 and t_1. This drift shows the opposite, i.e., the relation between target and feature variables has changed, while the underlying distribution of the feature variables has not changed. Therefore, they are highly correlated with the accuracy of the model.

Our framework aims at not only identifying drifts along the stream but also at providing an overall understanding of them. Recently, [1] presented a broader vision of Concept Drifts, which they named "Concept Drift Understanding". Instead of only focusing on the exact moment a drift happened (When), Concept Drift Understanding also aims at questioning other aspects of a drift, such as its severity (How) and the regions of the model most affected by it (Where). That way, three different questions are raised to drift detection methods: "When", "How", and "Where". In line with that, we propose another perspective for the area: "Concept Drift Interpretability", which is summarized by adding a new question for the drift: "Why did it happen?". Our method aims to answer all these four questions (When, How, Where, and Why) by leveraging decision trees' structure. That way, we frame our method as a drift interpretability method rather than a drift detection one.

2   Related Works

Concept Drift Detection is a topic of substantial research in academia. Many methods have been developed to identify the moment a drift occurs and provide appropriate windows for retraining the models. The two most common strategies are based on monitoring a classifier's performance (Error rate-based detectors) or monitoring changes in the data's underlying distribution (Data Distribution-based detectors).

Regarding the classifier's performance strategy, a popular approach is the DDM (Drift Detection Method) [3]. This method uses the overall classification error as the test statistic for a hypothesis test. The method runs the test on a landmark window. Although commonly used, it has some shortcomings which motivated the development of other methods. Early Drift Detection Method (EDDM) changed the DDM by considering the distance between two classification errors as the test statistic, which combined with other changes improved detection for gradual drifts [4]. DDM for Online Class Imbalanced (DDM-OCI)

[5] approaches the scenario where data is imbalanced by using the minority-class recall (True Positive Rate) as test statistic. Many other approaches have been developed based on the DDM algorithm. Apart from them, Linear Four Rates (LFR) [8], another drift detection method, proposes to monitor four different metrics of a confusion matrix (TPR, TNR, PPV, NPV), becoming more resilient to the different kinds of drift. As it also increases the rate of false positives, Hierarchical Linear Four Rates (HLFR) [9] proposes the addition of a second layer with a permutation test on the classifier's accuracy in order to double-check any drift raised by the LFR algorithm. There are also error rate methods based on sliding windows for detecting drifts: Adaptive Window (ADWIN) [7], which cuts a window W in all possible points, looking for two sub-windows with distinct enough averages in order to assume a drift happened and discard the older sub-window. Statistical Test of Equal Proportions Detection (STEPD) [6], which compares the overall time window with the most recent time window by applying a test of equal proportions on the accuracy of the classifier.

Data Distribution-based detectors use a dissimilarity measure to quantify drift between historical data and the new data. Normally, these methods use a fixed window to represent the historical data and a sliding window for the current data. They provide more information regarding the drift, since they act on their feature variables [1]. Dasu et al. [10] proposed the Information Theoretic Approach (ITA), which uses a *kdqTree*, to partition the historical window. This tree is then applied to the current sliding window, and the Kullback-Leibler divergence, $\hat{d}$, is calculated for each node between historical and current data. Then, by consecutively bootstrapping samples from both distribution, it calculates a KL-divergence interval for each node based on a parameter α. If $\hat{d}$ falls outside the interval, we assume a drift occurred for the node. There are still many other approaches for this sort of detector [11–13].

3   Visualizing Drift in Decision Trees

Most drift detection methods focus only on detecting the exact instant a drift occurred. Nevertheless, there are other aspects of a drift that are also essential, e.g. identifying affected regions of a model, assessing drift severity and finding their root causes. This knowledge can significantly impact businesses, since it enables them to handle the drift properly, understanding why it happened, what it affected, and correlating it to real-world situations. Deeply comprehending a drift is an important step into preventing it from happening again.

In order to assess drifting behaviors along the stream, our proposal is to exploit the structure of decision trees under two different perspective: (i) the frequency perspective, where nodes are analyzed according to how they segment their data through their child nodes, *Node Frequency Analysis*; (ii) and the accuracy perspective, where accuracies are computed for each node according to the prediction assertiveness of the observations passing through them, *Node Accuracy Analysis*.

Figure 1 illustrates a synthetic data set with 20000 instances. Note that features 0 and 2 were perturbed to simulate drifts according to a Gaussian noise for

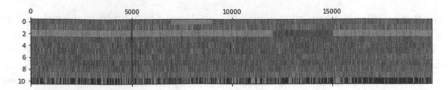

Fig. 1. Synthetic dataset with variables scaled to $[0,1]$ interval in order to visualize perturbations applied to them. First perturbation is applied to variable 0 for the interval $[7000, 9000]$; the second is applied to variable 2, from 12000 to 15000; and the last one is applied to the target variable (index 10) at the interval $[16000, 19000]$. The red line denotes the first 5000 instances which are used for training the decision tree. (Color figure online)

the periods $[7000, 9000]$ and $[12000, 15000]$ respectively. Furthermore, the binary target variable - denoted by the last row in the heat map (index 10) - was also perturbed using a set of Bernoulli trials with a success probability of 0.9 for the interval $[16000, 19000]$. The features were scaled to $[0,1]$ intervals for better visualization of their behavior. The first 5000 instances, denoted by the red vertical line, are selected for training a decision tree of maximum depth of 3 and the minimum amount of samples in a leaf of 5%. In the next subsections, we use this dataset to explain the analysis.

3.1 Node Frequency Analysis

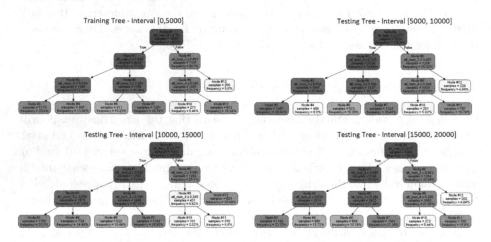

Fig. 2. Frequency visualization of the trained decision tree on distinct segments of the dataset: top left tree comprises the interval $[0, 5000]$ (training tree); top right tree, the next interval $[5000, 10000]$; bottom left tree, the interval $[10000, 15000]$; and bottom right tree, the interval $[15000, 20000]$. (Color figure online)

The Node Frequency Analysis aims at measuring how frequently each node is accessed on different periods. Variations on these frequencies give us a hint on the behavior of the feature variables along the stream. Hence, one can associate this analysis to the evidence factor, explained in Sect. 1. By monitoring how the evidence factor varies, we are actually monitoring where possible virtual drifts are taking place on the stream.

Figure 2 top left tree illustrates the trained decision tree with each node's frequency. A blue palette is used to denote their frequencies. The lighter the node's color, the less it is used by the tree. The trained decision tree was tuned, such that the leaves have a minimum access frequency of 5%, so the tree is initially balanced. Nevertheless, as the stream goes on, the tree may become imbalanced, with some nodes presenting drifting behaviors. We investigate this hypothesis on the other trees, each containing a fraction of the test set, which is segmented into three periods: [5000, 10000] (first period); [10000, 15000] (second period); and [15000, 20000] (third period).

By analyzing Fig. 2, it is possible to correlate drifts added to the stream to variations on the nodes' frequency. The first drift was placed in variable "att_num_0" at the interval [7000, 9000]. If we compare Node #2 for the top trees, we observe that their frequencies are almost the same (37,34% and 37,94%), but how they divide the data into their child nodes has changed. On the training set tree, nodes #3 and #4 have respectively 23,4% and 13,94%, while on the test set tree for the first period, the frequencies are 29,94% and 8,0%. Furthermore, when we move to the second period, another drift can be identified. Node #8, which is related to the variable "att_num_2", also perturbed in the data set, clearly shows a different segmentation of its data when compared to the training set tree. Identifying these drifts helps us better understand why the model might become outdated along the stream.

3.2   Node Accuracy Analysis

The node accuracy analysis is highly related to real drifts. A significant drop in accuracy for a node means that the hyperspace delimited by the set of rules that compose the path to this node is not reliable to approximate the target set's behavior passing through it anymore. Such variations in the distribution of a subset of the target variable is essentially the result of changes in the posterior probability, $P(y|X)$. That way, by identifying these nodes, where accuracy dropped significantly, we are actually identifying real drifts in the data as well.

Figure 3 top left tree represents the trained decision tree colored according to their node's accuracy. The colors used to illustrate accuracy are red and green. Nodes whose accuracy is below 60% are shown in a red palette, while the ones above 60% are shown in tones of green. The overall accuracy of the trained decision tree is expressed by the accuracy of its node #0, which is 82,68%.

By analyzing Fig. 3, we observe that the same nodes where virtual drifts were detected also presented significant drops in their accuracies, or real drifts as well. For instance, node #2 had 74,5% accuracy in training while in the first period of the test set it dropped to 63,15% (Top right tree). Observing the trees, it becomes obvious that it happened due to node #3, whose frequency became higher with the virtual drift, leading to a higher rate of wrong predictions. The same behavior can be observed for node #8 when comparing the second-period test set tree's training set. It presented a considerable drop in accuracy: from 86,67% to 51,87% (bottom left tree). Besides these drifts, the third period of the test set contains a drift on the target variable, which is not visible by just analyzing node frequencies. This analysis clearly shows drifting behavior in the last tree since many nodes (#2, #3, #8, #9, #10, and #11) are now painted in red, denoting a substantial drop in accuracy.

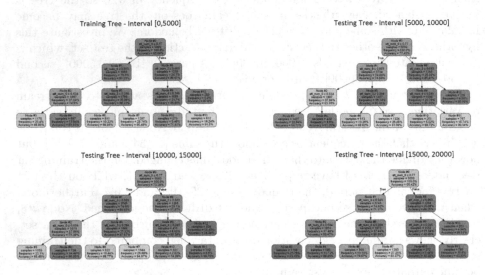

Fig. 3. Accuracy visualization of the trained decision tree on distinct segments of the data set: top left tree comprises the interval $[0, 5000]$ (training tree); top right tree, the next interval $[5000, 10000]$; bottom left tree, the interval $[10000, 15000]$; and bottom right tree, the interval $[15000, 20000]$. (Color figure online)

4   Interpretable Drift Detector

Visualizing the trees facilitates the detection of drifts along the data set. Nevertheless, when the drifts' position is not known beforehand, identifying the right intervals to detect them by plotting the trees can be quite overwhelming. Besides that, to develop a drift detector, a visual representation has to be translated into an automated behavior that either classifies as a drift or not. With that in mind, we propose to bootstrap samples of each node's training data, calculate the corresponding test statistics - the accuracy mean (Node Accuracy Analysis) and their child nodes frequencies (Node Frequency Analysis) - and compute the mean, μ_*, and standard deviation, σ_*, from their sampling distributions. A sliding window of a predefined size is then used for processing the whole data set in order to iteratively recalculate the test statistics, μ_{test}, for each node and update their z-scores, $z = \frac{\mu_{test} - \mu_*}{\sigma_*}$. Based on these z-scores and the minimum z-score threshold for considering a drift, z_{min} - which is based on a node's weight on the tree

- a drift grade is calculated by the following formula: $grade = \frac{1}{1+e^{-(z_{score}-z_{min})}}$. The grades help us assess the severity of a drift.

Since observations of streaming data are not necessarily i.i.d., we use a technique called *Block Boostrapping*, where the data is segmented into smaller blocks and the observations are selected from these blocks in order to generate samples that preserve the structure of the original data. By calculating the corresponding test statistic on each sample, we derive the appropriate sampling distribution. For the Node Accuracy Analysis, we are not interested in the upper tail of the distribution, since it means that accuracy actually improved, which is not a concern for our algorithm. On the other hand, for the Node Frequency Analysis, if the frequency of either one of the node's children, μ_{test}, lies anywhere beyond the interval: $[\mu_* - z_{min} * \sigma_*, \mu_* + z_{min} * \sigma_*)$, a higher drift grade is attributed to that node. In both analyses, nodes that are leaves in the tree are not considered by the algorithm and therefore receive drift grades equal to zero.

Fig. 4. Heat map of Node Frequency and Node Accuracy Grades for our example data set. The yellow color denotes the periods where grades are equal to 1, while the purple denotes periods of grades smaller or equal to 0.5 (not relevant for detecting drifts). (Color figure online)

Figure 4 illustrates drift grades calculated for Node Frequency and Node Accuracy Analysis. A sliding window of size 2500 is used for processing the data. Note that two specific nodes present distinct periods of high grades on the left chart: #2 and #8. These are clear indications of virtual drifts in the data. These nodes use the variables "att_num_0" and "att_num_2", which were respectively perturbed for the intervals [7000, 9000] and [12000, 15000]. Their grades start rising around 7300 (node #2) and 12300 (node #8) and falling around 11000 (node #2) and 17000 (node #8). The delay and size of the drifting periods can be controlled by tuning the sliding window size. A similar analysis could be performed for the node accuracy grades.

5   Experiments

Our algorithm was initially designed to detect drifts specifically on each node. Nevertheless, in order to provide a proper comparison to other benchmark algorithms, we adapted the *Interpretable Drift Detector* to be able to raise drifts every instant the mean z-scores of all nodes combined - excluding the leaves, which are left out of the equation - are higher than 3 (3-σ rule) for either one of both analyses: Node Accuracy and Node Frequency. Moreover, only when the mean drops below 2.5, we consider that the algorithm exited a drifting zone.

We compare our algorithm to all the benchmark methods available on *Scikit-Multiflow*: $ADWIN$, DDM, $EDDM$, $HDDM_{A_{test}}$, $HDDM_{W_{test}}$, $KSWIN$ and $Page-Hinkley$ [14]. We use those algorithms with their default parameters. Furthermore, we add Linear Four Rates to the list of benchmark methods with the following parameters: $\sigma_* = 1/100$, $\epsilon_* = 1/100K$ and $\eta_* = 0.99$ [8]. We opt to leave out $HLFR$, since it is essentially the LFR with a second layer permutation test, which could be adapted to most of the algorithms compared by our experiment. For our method, we train the decision tree with the first 10000 instances and set the sliding window to 1000 instances. The size of the sliding window is best determined through cross-validation. The bigger the size, the more the algorithm is likely to miss a drift, while the smaller the size, the more likely it is to detect false positives. So it is a clear trade-off between false positives and misses in our adaptation. The synthetic data sets used are also available on the *Scikit-Multiflow* library: SEA, $Sine$, $AGRAWAL$, $RandomRBF$, $Hyperplane$, $Mixed$, $RandomTree$ and $STAGGER$ [14]. They all have 40000 instances.

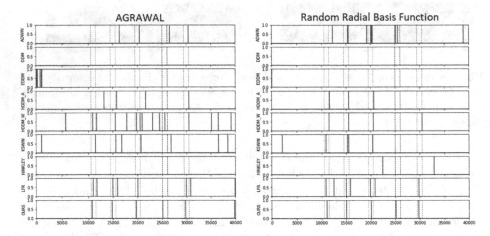

Fig. 5. Results of a single execution of the experiment for the data sets: $AGRAWAL$ and $RandomRBF$. The red vertical lines denote the intervals for the real drifts, while the green vertical lines, the interval for the virtual drift. The blue vertical lines correspond to the detection raised by each method. (Color figure online)

Figure 5 shows the results of one single execution of the experiment for data sets: $RandomRBF$ and $AGRAWAL$. On every data set, four distinct real drifts were added in the intervals: $[10500, 11500]$, $[14500, 15500]$, $[19500, 20500]$ and $[29500, 30500]$, denoted by the red dashed lines; and a single virtual drift was added to two random variables of the data set in the interval: $[25000, 26000]$, green dashed lines. Blue lines denote the drifts detected by each algorithm. Whenever they fall inside a drifting interval, we state that the algorithm detected it correctly (True Positive). If they fall outside these intervals, it is a false detection (False Positive). Furthermore, if the algorithm fails to detect a drift, it has

Table 1. Results of a hundred executions of the experiment. All data sets have 40000 instances and the drifts are set always on the same intervals. For each method and data set, we monitor three distinct metrics: True Positive (TP), False Positive (FP) and False Negative (FN). The best results in each one are marked in bold.

	AGRAWAL			RANDOM_RBF			SEA			SINE		
	TP	FP	FN	TP	FP	FN	TP	FP	FN	TP	FP	FN
ADWIN	2,51 ± 0,99	3,56 ± 1,49	2,49 ± 0,99	3,01 ± 0,93	6,67 ± 3,36	1,99 ± 0,93	2,20 ± 0,93	2,89 ± 1,63	2,80 ± 0,93	4,01 ± 0,10	36,03 ± 2,58	0,99 ± 0,10
DDM	0,04 ± 0,20	**0,40 ± 0,80**	4,96 ± 0,20	0,14 ± 0,35	0,64 ± 0,88	4,86 ± 0,35	0,02 ± 0,14	**0,21 ± 0,57**	4,98 ± 0,14	0,03 ± 0,17	3,70 ± 0,69	4,97 ± 0,17
EDDM	0,00 ± 0,00	4,86 ± 5,37	5,00 ± 0,00	0,00 ± 0,00	4,13 ± 4,38	5,00 ± 0,00	0,01 ± 0,10	5,20 ± 5,50	4,99 ± 0,10	0,19 ± 0,39	10,72 ± 6,51	4,81 ± 0,39
HDDM_A	0,03 ± 0,17	3,87 ± 0,42	4,97 ± 0,17	0,01 ± 0,10	3,81 ± 0,63	4,99 ± 0,10	0,00 ± 0,00	3,30 ± 0,89	5,00 ± 0,00	0,01 ± 0,10	4,00 ± 0,00	4,99 ± 0,10
HDDM_W	1,33 ± 0,88	15,59 ± 3,31	3,67 ± 0,88	0,46 ± 0,63	6,91 ± 2,27	4,54 ± 0,63	0,70 ± 0,70	8,34 ± 2,11	4,30 ± 0,70	0,00 ± 0,00	4,00 ± 0,00	5,00 ± 0,00
PAGE-HINKLEY	0,14 ± 0,35	2,21 ± 1,15	4,86 ± 0,35	0,14 ± 0,35	2,08 ± 0,98	4,86 ± 0,35	0,12 ± 0,33	1,96 ± 1,25	4,88 ± 0,33	0,02 ± 0,14	3,93 ± 0,26	4,98 ± 0,14
KSWIN	2,45 ± 1,10	14,68 ± 3,75	2,55 ± 1,10	1,69 ± 1,00	6,53 ± 2,66	3,31 ± 1,00	1,79 ± 1,12	7,00 ± 2,67	3,21 ± 1,12	1,79 ± 1,07	4,46 ± 0,67	3,21 ± 1,07
LFR	4,02 ± 0,14	5,88 ± 1,43	0,98 ± 0,14	4,01 ± 0,10	6,86 ± 5,79	0,99 ± 0,10	3,92 ± 0,37	5,87 ± 4,12	1,08 ± 0,37	4,16 ± 0,37	150,82 ± 63,96	0,84 ± 0,37
OURS	**4,55 ± 0,59**	2,06 ± 4,90	**0,45 ± 0,59**	**4,64 ± 0,50**	**0,52 ± 1,09**	**0,36 ± 0,50**	**4,72 ± 0,47**	0,49 ± 1,18	**0,28 ± 0,47**	**4,93 ± 0,36**	**2,75 ± 4,43**	**0,07 ± 0,36**

	HYPERPLANE			MIXED			RANDOMTREE			STAGGER		
	TP	FP	FN	TP	FP	FN	TP	FP	FN	TP	FP	FN
ADWIN	1,87 ± 1,25	2,78 ± 1,87	3,13 ± 1,25	4,00 ± 0,00	31,81 ± 3,74	1,00 ± 0,00	1,62 ± 1,44	2,91 ± 3,37	3,38 ± 1,44	**4,01 ± 0,10**	41,62 ± 3,57	**0,99 ± 0,10**
DDM	0,00 ± 0,00	**0,16 ± 0,47**	5,00 ± 0,00	0,01 ± 0,10	3,44 ± 1,04	4,99 ± 0,10	0,03 ± 0,17	**0,31 ± 0,71**	4,97 ± 0,17	0,00 ± 0,00	4,00 ± 0,00	5,00 ± 0,00
EDDM	0,00 ± 0,00	4,43 ± 4,59	5,00 ± 0,00	0,00 ± 0,00	3,82 ± 4,30	5,00 ± 0,00	0,00 ± 0,00	7,15 ± 7,54	5,00 ± 0,00	0,00 ± 0,00	**0,00 ± 0,00**	5,00 ± 0,00
HDDM_A	0,02 ± 0,20	3,12 ± 1,25	4,98 ± 0,20	0,01 ± 0,10	3,93 ± 0,26	4,99 ± 0,10	0,03 ± 0,17	2,92 ± 1,39	4,97 ± 0,17	0,02 ± 0,14	3,93 ± 0,26	4,98 ± 0,14
HDDM_W	1,00 ± 1,26	8,46 ± 3,68	4,00 ± 1,26	0,12 ± 0,36	4,00 ± 0,00	4,88 ± 0,36	1,50 ± 1,14	16,43 ± 6,35	3,50 ± 1,14	0,16 ± 0,44	3,99 ± 0,10	4,84 ± 0,44
PAGE-HINKLEY	0,34 ± 0,48	1,66 ± 1,00	4,66 ± 0,48	0,00 ± 0,00	4,00 ± 0,00	5,00 ± 0,00	0,12 ± 0,33	2,18 ± 1,21	4,88 ± 0,33	0,00 ± 0,00	3,01 ± 0,10	5,00 ± 0,00
KSWIN	1,74 ± 1,21	7,33 ± 3,58	3,26 ± 1,21	1,78 ± 0,95	4,63 ± 0,99	3,22 ± 0,95	2,35 ± 1,13	15,68 ± 6,58	2,65 ± 1,13	1,30 ± 1,10	4,20 ± 0,49	3,70 ± 1,10
LFR	3,63 ± 0,76	0,78 ± 1,14	1,37 ± 0,76	4,24 ± 0,43	110,24 ± 63,08	0,76 ± 0,43	3,72 ± 0,74	4,21 ± 13,05	1,28 ± 0,74	5,00 ± 0,00	981,05 ± 16,61	0,00 ± 0,00
OURS	**4,70 ± 0,72**	**0,34 ± 0,82**	**0,30 ± 0,72**	**4,45 ± 0,80**	**2,96 ± 2,61**	**0,55 ± 0,80**	**4,02 ± 1,23**	1,20 ± 3,10	**0,98 ± 1,23**	2,51 ± 1,94	1,36 ± 1,93	2,49 ± 1,94

missed the drift (False Negative). Observing Fig. 5, we state that our approach performs best for these instances of the data sets, followed closely by the *LFR* algorithm.

We executed the experiment a hundred times, always changing the seed for generating the data sets. That way, we have a more reliable estimation of the parameters: *TP*, *FP*, and *FN* for each kind of data. Table 1 shows each metric's mean and standard deviation per data set for each algorithm. The best results are marked in bold. Our method outperforms others in almost every data set, except for *STAGGER*, for which *ADWIN* had the best performance. We also point out that *DDM* and *EDDM* probably performed poorly, due to the standard parametrization of these algorithms in the library.

6 Conclusion

This paper presents the *Interpretable Drift Detector*, which aims at providing more insights into the detected drifts. Using our method, we can assess drift severity, identify the most impacted regions of a tree model, and visualize which variables are the drift's root causes. This knowledge leads to a deeper understanding of drifts. Furthermore, it can be easily adapted to run exactly as most drift detectors.

Our method also outperforms benchmark detectors in terms of false-positive and true-positive rates. It is important to emphasize that the method is computationally more expensive than most benchmark methods, since it navigates through the whole tree structure, analyzing nodes individually on their frequency and accuracy patterns. Moreover, it is also limited to the variables selected by the tree model.

Drift identification is a valuable information, however, with limited knowledge of the event, it is not possible to clearly understand the root causes and main impacts of a drift. A lot of applications can benefit considerably from a thorougher analysis on drifts. Our method is a step forward into exploring other aspects of concept drifts.

References

1. Lu, J., et al.: Learning under concept drift: a review. IEEE Trans. Knowl. Data Eng. **31**(12), 2346–2363 (2018)
2. Khamassi, I., Sayed-Mouchaweh, M., Hammami, M., Ghédira, K.: Discussion and review on evolving data streams and concept drift adapting. Evol. Syst. **9**(1), 1–23 (2016). https://doi.org/10.1007/s12530-016-9168-2
3. Gama, J., Medas, P., Castillo, G., Rodrigues, P.: Learning with drift detection. In: Bazzan, A.L.C., Labidi, S. (eds.) SBIA 2004. LNCS (LNAI), vol. 3171, pp. 286–295. Springer, Heidelberg (2004). https://doi.org/10.1007/978-3-540-28645-5_29
4. Baena-Garcıa, M., et al.: Early drift detection method. In: Fourth International Workshop on Knowledge Discovery from Data Streams, vol. 6 (2006)
5. Wang, S., et al.: Concept drift detection for online class imbalance learning. In: The 2013 International Joint Conference on Neural Networks (IJCNN). IEEE (2013)
6. Nishida, K., Yamauchi, K.: Detecting concept drift using statistical testing. In: Corruble, V., Takeda, M., Suzuki, E. (eds.) DS 2007. LNCS (LNAI), vol. 4755, pp. 264–269. Springer, Heidelberg (2007). https://doi.org/10.1007/978-3-540-75488-6_27
7. Bifet, A., Gavalda, R.: Learning from time-changing data with adaptive windowing. In: Proceedings of the 2007 SIAM International Conference on Data Mining. Society for Industrial and Applied Mathematics (2007)
8. Wang, H., Abraham, Z.: Concept drift detection for streaming data. In: 2015 International Joint Conference on Neural Networks (IJCNN). IEEE (2015)
9. Yu, S., Abraham, Z.: Concept drift detection with hierarchical hypothesis testing. In: Proceedings of the 2017 SIAM International Conference on Data Mining. Society for Industrial and Applied Mathematics (2017)
10. Dasu, T., et al.: An information-theoretic approach to detecting changes in multi-dimensional data streams. In: Proceedings of the Symposium on the Interface of Statistics, Computing Science, and Applications (2006)
11. Ditzler, G., Polikar, R.: Hellinger distance based drift detection for nonstationary environments. In: 2011 IEEE Symposium on Computational Intelligence in Dynamic and Uncertain Environments (CIDUE). IEEE (2011)
12. Qahtan, A.A., et al.: A PCA-based change detection framework for multidimensional data streams: change detection in multidimensional data streams. In: Proceedings of the 21th ACM SIGKDD International Conference on Knowledge Discovery and Data Mining (2015)
13. dos Reis, D.M., et al.: Fast unsupervised online drift detection using incremental Kolmogorov-Smirnov test. In: Proceedings of the 22nd ACM SIGKDD International Conference on Knowledge Discovery and Data Mining (2016)
14. Montiel, J., et al.: Scikit-multiflow: a multi-output streaming framework. J. Mach. Learn. Res. **19**(1), 2915–2924 (2018)

Interpreting a Conditional Generative Adversarial Network Model for Crime Prediction

Mateo Dulce[1]([⊠]), Óscar Gómez[1], Juan Sebastián Moreno[1],
Christian Urcuqui[1], and Álvaro J. Riascos Villegas[1,2]

[1] Quantil, Bogotá, Colombia
mateo.dulce@quantil.com.co , juansebastian.moreno@quantil.com.co
[2] Universidad de los Andes, Bogotá, Colombia

Abstract. Crime prediction models seek to assist policymakers and law enforcement agencies in the allocation of scarce resources intended to prevent crime occurrences. This paper proposes an extension and two interpretation methods for a novel conditional GANs architecture for crime (robberies) prediction in Bogota, Colombia. The model's performance on the area under the Hit Rate - Percentage Area Covered by Hotspots curve increases from 0.86 to 0.88 AUC when extended by conditioning on holidays. The proposed interpretability methods can help study the effect of crime occurring in a region on the likelihood of occurrence in other regions through the use of SHAP values. These interpretations can prove to be very useful for policymakers and law enforcement agencies in designing interventions and preventing future crime from inferring potential regions of displacement. (Results of the project "Diseño y validación de modelos de analítica predictiva para la toma de decisiones en Bogotá" funded by Colciencias with resources from the Sistema General de Regalías, BPIN 2016000100036. The opinions expressed are solely those of the authors.)

Keywords: Generative adversarial network · Interpretability · Crime prediction

1 Introduction

Crime prediction models have become useful in public policy to allocate law enforcement resources more efficiently. They have been increasingly used by police departments across the globe to aid in decision-making and optimizing resources. Their objective is to accurately provide the location of future crime events, but most of them are hard, if not impossible, to interpret. In previous research [19], conditional Generative Adversarial Networks (cGANs) were used to predict crime by using the spatial distribution of past crimes as conditional inputs. The model developed significantly outperforms other state-of-the-art crime prediction models, even when re-calibrated to correct for biases affecting vulnerable populations. However, due to its complexity and relatively

© Springer Nature Switzerland AG 2021
J. M. R. S. Tavares et al. (Eds.): CIARP 2021, LNCS 12702, pp. 281–290, 2021.
https://doi.org/10.1007/978-3-030-93420-0_27

abstract way to make predictions, it lacks interpretability. In an attempt to understand this black-box model, we extract knowledge to aid public policy-makers in the analysis of crime patterns to design future interventions. Particularly, we use SHAP (SHapley Additive exPlanations) on the conditions of the cGAN to study how the crime occurrence in a region affects the likelihood of occurrence in others.

Model interpretability has become a subject of paramount importance for machine learning applications. For complex machine learning solutions to be successfully deployed in highly sensitive domains such as healthcare and crime prediction, stakeholders need to gain trust in the models. In contrast to simple models, deep learning techniques such as the one we study here are not readily interpretable by humans and therefore are considered in many cases to act as black boxes. This lack of understanding of the model's internal behavior can lead to discriminatory results and other undesired outcomes. Furthermore, a proper understanding of these more complex models can serve to inform the development of public policy and resource allocation.

This paper focuses on the interpretation of the cGAN model developed in [19]. We first extend the model to include features related to the spatio-temporal behavior of crime phenomena. We then make use of existing interpretability techniques, specifically the SHAP framework, to gain a better understanding of the model's behavior. We demonstrate two methodologies that can help interpret the model and make it more useful and accessible to stakeholders.

The rest of this paper is organized as follows. In Sect. 2 we discuss the related literature. In Sect. 3 we discuss our methodology: data, conditional GANs network architecture, training and evaluation metrics, and model interpretation. Section 4 provides the results of the model validation and its interpretation. Section 5 concludes.

2   Related Work

The benchmark models in the crime prediction literature include [11] and [10] among others based on self-exciting Poisson processes. Extensions of them use past crime events and other variables that correlate spatially and temporally with crime (public infrastructure, police stations, weekdays, holidays, etc.) [2,9,14]. More recently deep learning architectures have been used to make next day crime predictions [18]. General approaches for spatio-temporal data prediction have been proposed based on GANs architectures. These models [6,17] have used external data (e.g. weather, weekday data) and conditioned data from previous time steps to make predictions. In the same spirit, [19] uses the spatial distribution of past crimes as conditional inputs to make next day crime predictions.

Machine learning interpretability techniques can generally be classified as global or local, and as model-agnostic or model-specific. Global techniques aim to explain the overall behavior of a model, while local ones explain a specific decision over a single instance. Model agnostic techniques, which often rely on perturbations of the inputs and analyzing the outputs of a model, see models as a black box, while model-specific techniques exploit the characteristics of a model's architecture to provide insights into their behavior.

Some of the most popular and widely adopted interpretability techniques are LIME (Local Interpretable Model-Agnostic Explanations) [15] and SHAP (SHapley Additive exPlanations) [8]. These are model agnostic local techniques that provide a measure of each feature's impact on a given prediction and have been widely adopted in practice to support decision-makers and extract knowledge in various domains [7,16]. When applied to classification for image data, they can operate either on individual pixels or on super-pixels of the image, assigning to each of them an attribution value related to the class the image was classified as, with larger values indicating a greater attribution in the prediction.

GAN-specific techniques have generally focused on finding structure and consistent effects of the noise vectors that serve as randomized inputs to the models. Examples of these include InfoGAN [3] which learns disentangled representations in an unsupervised manner through a latent input code and GANSpace [5] which create interpretable controls for image synthesis by performing Principal Component Analysis (PCA) on the latent or feature space. In [12], an overview of other methods for interpreting deep neural networks is provided.

3   Methodology

3.1   Dataset

We use the official statistics of crime incidents (SIEDCO) [1] that reports all crime incidents in the city of Bogotá. We restrict to the period 2016–2019 and focus only on robberies. We also focus on a locality, Chapinero, that has many crimes and heterogeneity in the income level and other characteristics of the different regions within.

For each day, crime incidents were grouped, counted, and assigned to a latitude-longitude pair corresponding to a cell in a uniform grid covering the city. This daily data is transformed into a matrix $X^{(t)}$ where the entries are the number of crimes, and the indexes are the cell index (i, j) in a day $t \in T$. That is, $x_{i,j}^{(t)}$ represents the number of crimes in cell (i, j) in a day $t \in T$. We split the dataset into a training set with data from 01/01/2016 to 01/07/2019 (1277 days), and a test set with data from 01/07/2019 to 30/09/2019 (92 days). The training set had 5,057 crime reports and the test set 498 reports. Figure 1 shows the total number of crimes in each cell in the data set for both training and testing periods, as well as the locations of police stations within the locality.

To understand the features related to the crime phenomena, especially their contribution to the spatio-temporal behavior behind the crime pattern, this study explores some covariants proposed by the literature as conditional inputs. The conditional features analyzed were holidays and the number of each day, these kinds of data were processed using a one-hot encoding to represent their categories as matrices.

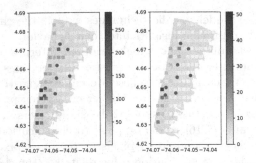

Fig. 1. Total number of crimes in each cell of the grid (red scale) and locations of police stations (green). Left: Crimes in train set. Right: Crimes in test set. (Color figure online)

3.2 Conditional Generative Adversarial Network

We use the same architecture used in [19] for the original model, and add a conditional input for processing the covariants. The generator (G) receives three inputs, first, the conditional time input is encoded using three convLSTM layers 128 (3 × 3), 128 (3 × 3), and 1 (3 × 3) filters with stride 1, the second parameter is the noise which is processed by two dense layers of 400 units and leaky ReLU [13] activation functions, this output is concatenated with the first result mentioned. The last input is the covariant that is processed like a conditional feature using a dense layer. Next, both outputs are concatenated and the result flows through two dense layers of 400 units using leaky ReLU, and a final sigmoid activation.

The discriminator (D) receives the conditional input and the predicted $\hat{X}^{(t)}$ or real reports in a day $X^{(t)}$, both inputs are concatenated and encoded using two ConvLSTM layers 68 (3 × 3), 68 (3 × 3) filters with stride 1, the result is integrated to the output of a dense layer oriented to process a covariant input (conditional variable), the flow is processed using one layer with 32 units and a leaky ReLU. Finally, the results pass through a final 1-unit sigmoid activation function. A dropout of 60% was applied on the ConvLSTM layer and the first dense layer. The neural network conditions spatio-temporal inputs with a lag of n days, $X^{(t-n)}$, that is, to make a prediction for day t it uses as conditional inputs the matrices (or images) of crime occurrences in days $t - n$ to $t - 1$ (see Fig. 2).

3.3 Training and Evaluation Metrics

We use the same architecture and parameters from the original model [19]. Each model was trained using adversarial learning (minmax game [4]) where both G and D implemented a criterion that measures binary cross-entropy between the target and the output. The batch size used for each training was 128 days (t), the learning rate was 0.0002, and the models ran for 400 epochs. We used $n = 42$ as the number of conditional days as this was found to produce the best results.

To evaluate the predictive capacity of the models we used a modification of the Hit Rate measure, which indicates the percentage of crimes that occurred

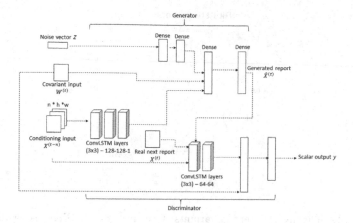

Fig. 2. cGAN architecture [19].

in the test set that were successfully predicted by the model. Since increasing the number of hotspots increases the number of crimes successfully predicted, we control for the percentage of the covered area labeled as hotspots. Therefore in our evaluation, we consider as a target metric the area under the Hit Rate - Percentage Area Covered by Hotspots curve.

$$\text{Hit Rate} = \frac{\text{\# of crimes that occurred on marked hotspots}}{\text{\# Crimes}} \tag{1}$$

3.4   SHAP Values and Model Interpretation

To interpret and extract knowledge from the model, we perform an analysis of the conditional inputs used by the generator with the SHAP framework [8]. SHAP values provide a measure of importance for each feature for any particular prediction of the model. This method constitutes the unique solution for an additive feature attribution method which satisfies three desirable properties: local accuracy (the sum of the feature attributions equals the model output), missingness (missing features get no attribution), and consistency (if a model changes and a feature's contribution increases its attribution will not decrease). The method is based on the traditional formulation of Shapley values from cooperative game theory. However, calculating these values exactly is computationally infeasible, so we use the Deep SHAP method which offers an accurate and fast approximation for deep learning models.

Recall that, to make a next-day prediction, the generator uses as conditional inputs the crime images from the previous 42 days. Its output is a generated image indicating the predicted probability for a crime to occur in each cell of the grid on that day. With the SHAP framework, we obtain a feature attribution for each cell in each of the conditional images. That is, given the output image for day t, $Y^{(t)}$, we obtain 42 SHAP values (one for each conditional image) for each of the (i, j) cells in the grid. This results in a matrix of SHAP values $S^{(t)}$,

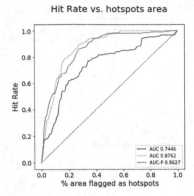

Fig. 3. Results of using different covariants. Green: original, Blue: using day of the week, Orange: using holidays. (Color figure online)

where $s_{a,b_{n,i,j}}^{(t)}$ is the contribution of cell (i,j) on day $t-n$ over the output of cell (a,b) on day t. These attributions allow us to study how the occurrence of crime in a cell affects the predictions made in other cells.

4   Results

4.1   cGAN

This section shows the settings and results using different covariants as conditional inputs to the cGANs. The first feature evaluated was the day in the week. The second variable allowed us to understand if special days, specifically, holidays impact the model's predictive power. Figure 3 shows the performance of the three models under consideration. Compared to the original model (0.86 AUC), we can see that the model that used the day in the week decreased its prediction performance (0.74 AUC), while the model that was processed using holidays improved the baseline performance (0.88 AUC).

4.2   Analysis of SHAP Values

Although the performance of the model improves when holidays are included as conditional inputs, the original model performs considerably well, in particular, it outperforms the others when the percentage of area flagged is below 15%. Given that in this setting the model learns only from past crime occurrences and their spatial location, we perform our SHAP value analysis on this model aiming to understand the spatio-temporal patterns the model captures.

As an approximation method for SHAP values, Deep SHAP relies on sampling a subset of images over which the average model prediction is calculated. To make the calculations computationally feasible, we randomly select 200 days

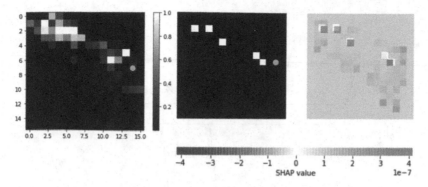

Fig. 4. Left: Example of generated image for a day. Right: SHAP values for the 8th conditional day on the output for cell (14,7) - green dot. (Color figure online)

from the training set and use the generated images corresponding to these days as a set of background samples for Deep SHAP.

Figure 4 shows an example of a generated image and the SHAP values attributed to each of the cells in the grid for the 8th conditional over the output of cell (14,7). Positive SHAP values (in red) indicate that the feature contributed towards predicting crime in the cell (14,7), while negative values (in blue) show the opposite effect. In this example, we see that three of the five cells where the crime occurred in the conditional (including the two closest to cell (14,7)) have a significant positive attribution, while the other two cells where the crime occurred, and the vast majority of cells where the crime did not occur, have negative attributions that are slightly smaller in magnitude. However, when added up they sum up to the final prediction for that day on cell (14,7) which is small compared to cells such as (4,2), (5,2), and (13,5).

Figure 5 shows an example of a time series of SHAP values. Each line represents the evolution of the SHAP values for a given cell. We plot the lines for the cell over which the effect is measured (dark blue) and the closest neighbors (gray). The color of the dots represents the number of crimes occurring in that cell in the given day. It can be seen that in these plots, the non-occurrence of crime in a cell is usually assigned a negative SHAP value with a small magnitude. However, when crime occurs, the SHAP value becomes positive and with a much larger magnitude. We also notice that there is no clear time-effect on the magnitude of the values, as they are comparable across the entire domain. Finally, even though the values of these cells in the past have a large effect on their prediction, neighboring cells can also have effects as large or even larger than the same cell.

Because the observations above are consistent across different cells and periods, we aggregate them to produce a single set of maps for each cell in the grid. For a given time series, we take the sum of the effects for each cell, then divide by the maximum absolute value, obtaining the average effect of each cell for a given period. We then average this effect over all the periods. The result is a map for each cell in the grid, indicating the average effect that each of the other

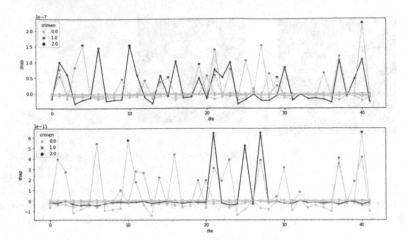

Fig. 5. Time series for the SHAP values of conditional inputs on two different cells on the same day. Top: cell (6,2). Bottom: cell (7,2). (Color figure online)

cells has on it. The inverse map can be obtained through an analogous exercise, where the result is a map for each cell indicating the average effect that the cell has on the other cells in the grid.

Figure 6 shows the resulting maps for two cells in the grid. The first one (4,2) is a cell with a high concentration of crime, while the second one has a low concentration of crime. The first one is affected mostly by cells in its vicinity, which also are cells with high crime concentration. In turn, this cell also has a negative effect on most of its vicinity, but its effect reaches further and extends

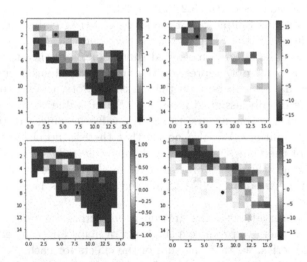

Fig. 6. Average effect of cells on the marked cell (left) and average effect of the marked cell on the other cells (right) for cells (4,2) (top) and (8,8) (bottom)

to the cell in the opposite corner of the locality. In contrast, for the second cell, only three cells have a negative average effect, while the average effect of cells in the area of high crime concentration is close to zero. Its effect over other cells is also widespread, however, it is overwhelmingly a positive effect.

5   Conclusions and Future Work

Crime prediction models seek to assist policymakers and law enforcement agencies in the allocation of scarce resources intended to prevent crime occurrences. In this paper, we extended and interpreted a novel cGAN model that achieves a very high performance of 0.86 AUC using 42 conditional days in the locality of Chapinero (Bogotá). The performance of the model improved to 0.88 AUC when using holidays as a conditional input.

Our interpretability analysis made use of the SHAP value framework to understand how the occurrence of crime in a cell affects the predictions made in other cells. By aggregating the attributions for each cell, we described two useful methodologies that shed light on the model's inner workings. The first is a time series analysis that can help examine the effect of each conditional input over a specific prediction. Such an analysis can be very useful for a deployed system to perform an in-depth analysis of the day-to-day crime predictions of a model, by understanding the cells with the most impact in the recent past over the new predictions. It could also help uncover new trends of crime patterns or could help preemptively detect if the model is focusing on irrelevant areas of the city to make its predictions.

The second methodology further aggregates the attributions, resulting in two types of maps for each cell. One of them allows the study of the effect of the occurrence of crime across all cells over a particular region. The second one allows the study of the effect of a particular region over the rest of the cells. Such maps can serve as valuable input for stake-holders to plan interventions through public policy on certain regions given their influence over other regions. They can also help identify potential areas where crime might be displaced in the face of such interventions.

Further work can be done to improve the performance and interpretation of the GAN model. Our results show how including temporal covariates such as holidays can improve the performance of the model. Further inclusion of other temporal and spatial covariates that characterize the city could be used to improve the model's performance. Furthermore, this could help in finding relationships between the cells that most affect each other and understand what type of interventions could prove most effective. Finally, SHAP values and other interpretability techniques could be used on these covariates as conditional inputs to understand which of them play a key role in the model's predictions.

References

1. Sistema de información estadística, delincuencial, contravencional y operativa (siedco) - policía nacional de colombia (2020)

2. Barreras, F., Díaz, C., Riascos, Á.J., Ribero, M.: Comparación de diferentes modelos para la predicción del crimen en bogotá. Economía y seguridad en el posconflicto, p. 209 (2018)
3. Chen, X., Duan, Y., Houthooft, R., Schulman, J., Sutskever, I., Abbeel, P.: InfoGAN: interpretable representation learning by information maximizing generative adversarial nets. In: Advances in Neural Information Processing Systems (2016)
4. Goodfellow, I., et al.: Generative adversarial nets. In: Advances in Neural Information Processing Systems, pp. 2672–2680 (2014)
5. Härkönen, E., Hertzmann, A., Lehtinen, J., Paris, S.: GANSpace: discovering interpretable GAN controls. arXiv preprint arXiv:2004.02546 (2020)
6. Jonietz, D., Kopp, M.: Towards modeling geographical processes with generative adversarial networks (GANs). In: 14th International Conference on Spatial Information Theory (COSIT 2019) (2019)
7. Lima, T., Santana, R., Teodoro, M., Nobre, C.: Knowledge extraction from vector machine support in the context of depression in children and adolescents. In: Nyström, I., Hernández Heredia, Y., Milián Núñez, V. (eds.) CIARP 2019. LNCS, vol. 11896, pp. 545–555. Springer, Cham (2019). https://doi.org/10.1007/978-3-030-33904-3_51
8. Lundberg, S.M., Lee, S.I.: A unified approach to interpreting model predictions. In: Advances in Neural Information Processing Systems, pp. 4765–4774 (2017)
9. Mohler, G., Raje, R., Carter, J., Valasik, M., Brantingham, J.: A penalized likelihood method for balancing accuracy and fairness in predictive policing. In: 2018 IEEE International Conference on Systems, Man, and Cybernetics (SMC), pp. 2454–2459. IEEE (2018)
10. Mohler, G.: Marked point process hotspot maps for homicide and gun crime prediction in Chicago. Int. J. Forecast. 30(3), 491–497 (2014)
11. Mohler, G.O., Short, M.B., Brantingham, P.J., Schoenberg, F.P., Tita, G.E.: Self-exciting point process modeling of crime. J. Am. Sta. Assoc. 106(493), 100–108 (2011)
12. Montavon, G., Samek, W., Müller, K.R.: Methods for interpreting and understanding deep neural networks. Digit. Signal Process. 73, 1–15 (2018)
13. Radford, A., Metz, L., Chintala, S.: Unsupervised representation learning with deep convolutional generative adversarial networks. arXiv:1511.06434 (2015)
14. Reinhart, A., Greenhouse, J.: Self-exciting point processes with spatial covariates: modeling the dynamics of crime. arXiv preprint arXiv:1708.03579 (2017)
15. Ribeiro, M.T., Singh, S., Guestrin, C.: "Why should I trust you?" Explaining the predictions of any classifier. In: Proceedings of the 22nd ACM SIGKDD International Conference on Knowledge Discovery and Data Mining, pp. 1135–1144 (2016)
16. Rodríguez-Pérez, R., Bajorath, J.: Interpretation of machine learning models using Shapley values: application to compound potency and multi-target activity predictions. J. Comput. Aided Mol. Des. 34(10), 1013–1026 (2020). https://doi.org/10.1007/s10822-020-00314-0
17. Saxena, D., Cao, J.: D-GAN: deep generative adversarial nets for spatio-temporal prediction. arXiv preprint arXiv:1907.08556 (2019)
18. Stec, A., Klabjan, D.: Forecasting crime with deep learning. arXiv preprint arXiv:1806.01486 (2018)
19. Urcuqui, C., Moreno, J., Montenegro, C., Riascos, A., Dulce, M.: Accuracy and fairness in a conditional generative adversarial model of crime prediction. In: 7th International Conference on Behavioural and Social Computing (BESC 2020), Bournemouth, UK (2020)

Interpreting Decision Patterns
in Financial Applications

Tiago Faria, Catarina Silva[✉], and Bernardete Ribeiro

Department of Informatics Engineering, Centre for Informatics and Systems
of the University of Coimbra, University of Coimbra, Coimbra, Portugal
tiagofaria@student.dei.uc.pt, {catarina,bribeiro}@dei.uc.pt

Abstract. Decisions in financial applications that directly impact citizens are often based on black-box intelligent methods. Given the growing interest in making these decisions more transparent, and the emergent legislation on interpretability and privacy, new solutions to give some insight on such black-boxes, presenting explanations on the decision patterns are being sought. In this paper we propose a method that transfers knowledge from black-box models to more interpretable models to understand the decision patterns in financial applications. Results on credit risk and stock market data show that it is possible to use white-box methods that work on black-box results to show the potential interpretation of the decision patterns.

Keywords: Pattern recognition · Distillation · Interpretability · Decision trees

1 Introduction

The rapid digitalization of our world has led us to great advances in services and activities we are involved in. Artificial Intelligence (AI) is now a big part of our lives even though not all of us are aware of it. From simple things like selecting the content we see online to partaking in critical decision making in our lives, AI has a strong presence across most activity sectors. The increasing awareness that these systems do in fact exist and the notion that there is not always a human supervising them has surfaced the need for explanations. From deciding if you get a loan to what is about to be shown on your Facebook feed AI is here, and it's not leaving. One of the sectors particularly impacted is the financial sector. As of the beginning of 2020 it was estimated that in the next two years there will be a mass adoption of AI in the financial sector with an impressive 77% expecting that AI will become essential to their business within the next two years [10], as can be gleaned from Fig. 1.

Intelligent systems are being applied in the financial sector in areas like:

- **Customer service**, e.g. operational cost savings from using chatbots in banking will reach $7.3 billion globally by 2023.[1]

---

[1] https://www.juniperresearch.com/press/press-releases/bank-cost-savings-via-chatbots-reach-7-3bn-2023.

© Springer Nature Switzerland AG 2021
J. M. R. S. Tavares et al. (Eds.): CIARP 2021, LNCS 12702, pp. 291–300, 2021.
https://doi.org/10.1007/978-3-030-93420-0_28

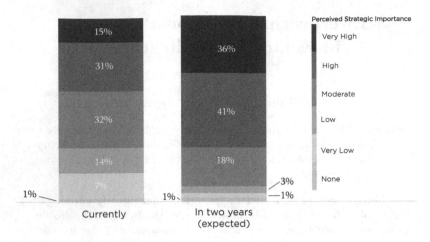

Fig. 1. Perceived strategic importance of AI services in the finance sector for 2022 [10].

- **Banking operations**, e.g., credit scoring, the use of AI technology enables more accurate scoring and allows for improved access to credit by reducing the risks and the number of false positives and false negatives.
- **Security purposes**, e.g., AI is providing great assistance in the detection of fraud [5] and other suspicious activities that are linked to financial crime generally.

The financial industry is highly regulated and in the case of loan issuers, laws around the world, e.g. the European Union General Data Protection Regulation (EU GDPR), start to determine that in a not far away future, financial institutions must effectively show that the decisions they take are fair. The systems implemented in the financial sector are usually black-box models, highly capable of achieving their goal with high performance. The problem with black-box models is, although they are usually very capable, their decision processes are not clear and also prone to bias. Thus, one significant challenge of using AI-based systems that, for instance predict credit scores, is that there is no underlying interpretability infrastructure that can provide *reason code* to borrowers, e.g., when a credit is denied. In this work we propose a method that transfers knowledge from deep models to decision-tree models to understand the decision patterns in financial applications. Results obtained in two distinct financial applications: credit risk and stock market show that it is possible to use white-box methods that work on black-box results to show the potential interpretation of the decision patterns. The rest of the paper is organized as follows. Section 2 introduces relevant background on interpretable AI in finance and describes previous works in this research area. Section 3 details the proposed approach and Sect. 4 presents the experimental setup. Section 5 discusses the results and finally Sect. 6 highlights the conclusions and proposes lines of future research.

2   Background - Interpretable AI in Finance

Machine learning critical decision-making is a relatively recent topic. As humans get assisted or even replaced by intelligent models, existing legislation becomes obsolete and data regulation is often ineffective. Hence, new regulation like the European Union General Data Protection Regulation (EU GDPR), appeared which includes article 22 (see Fig. 2) on automated decision making establishing the need for interpretability in the sector. Although still in debate, it has been said that the GDPR has introduced the *right to explanation* based on the paragraph 1 of article 22" "shall implement suitable measures to safeguard...at least the right to obtain human intervention on the part of the controller, to express his or her point of view and to contest the decision" otherwise a person has "the right not to be subject to a decision based solely on automated processing". As safeguard for the companies implementing these models, as well as for the subjects that are targeted, interpretability starts to become essential in the transition to fully digital automated services.

Article 22. Automated individual decision making, including profiling

1. The data subject shall have the right not to be subject to a decision based solely on automated processing, including profiling, which produces legal effects concerning him or her or similarly significantly affects him or her.

2. Paragraph 1 shall not apply if the decision:

   (a) is necessary for entering into, or performance of, a contract between the data subject and a data controller;

   (b) is authorised by Union or Member State law to which the controller is subject and which also lays down suitable measures to safeguard the data subject's rights and freedoms and legitimate interests; or

   (c) is based on the data subject's explicit consent.

3. In the cases referred to in points (a) and (c) of paragraph 2, the data controller shall implement suitable measures to safeguard the data subject's rights and freedoms and legitimate interests, at least the right to obtain human intervention on the part of the controller, to express his or her point of view and to contest the decision.

4. Decisions referred to in paragraph 2 shall not be based on special categories of personal data referred to in Article 9(1), unless point (a) or (g) of Article 9(2) apply and suitable measures to safeguard the data subject's rights and freedoms and legitimate interests are in place.

Fig. 2. Article 22 of the EU GDPR.

2.1   Interpretability Approaches

A lot of work has been done in the field of interpretability in the recent years. When it comes to the classification of techniques we can typically classify them in three categories:

- **Scope** - if the aim is to achieve global explanations in order to get an understanding of the decision making process of a model as whole we are talking about global interpretability techniques. On the other hand, if we are trying to understand how a model came up with a certain outcome for a specific observation we are talking about local explanations and therefore global interpretability. Works in this area include Baehrens et al. [1] method for explaining local decisions taken by arbitrary nonlinear classification algorithm, using the local gradients that characterize how a data point has to variate in order to change its predicted label. Another very famous work is Ribeiro's et al. [9] **LIME** for Local Interpretable Model-Agnostic Explanation. A model that can approximate a black-box model locally in the neighborhood of any prediction of interest.

- **Relation to the model**, methods dependent on the model they are being applied to, intrinsic methods are of these class like Caruana's et al. [3] the drawback of this practice is that it is limited in to a certain class of models, this is why there's a preference for model-agnostic methods like **knowledge distillation**.

- **Complexity**, the most basic way of having an interpretable model would be making it inherently and intrinsically interpretable, a common challenge, which often hinders the usability of such methods. This is the trade-off between interpretability and accuracy, more interpretable models tend to be less accurate and vice-versa [11].

Some of these works have also given birth to toolboxes for interpretability like **ELI5** which aims to give local explanations and has a strong connection to LIME [9] which has also been turned into a toolbox or **Shap** which makes use of shapley values from game theory to see how features are impacting a model outcome by giving them respective weights across their whole range of values.

2.2 Interpretability Models

Surrogate Models

Surrogate models can be classified in two types, global surrogate models and local surrogate models. A global surrogate model is a model that is trained to mimic a black-box model giving us a global overview of what the black-box model is trying to achieve. This model is usually interpretable and can be used to draw conclusions about the way the mimicked model is trying to make its predictions. Local surrogate models on the other hand are interpretable models that are used to explain individual predictions of black-box machine learning models. **LIME** [9] makes use of this type of surrogate models. Local surrogate models forget all the data and focus on a specific observation and how small perturbations on its features affect the outcome on the black-box model.

Knowledge Extraction Methods

As explained before one way of interpreting black-models would be making then interpretable in the first place, but that's not feasible as most models already in-place, usually deep nets, are black-box. That would mean we would have to replace those models with completely new interpretable ones, structured to solve very specific problems that might not be as accurate as the previous ones. This would make these models not only expensive to run but also limited to the problems they are solving. Knowledge extraction techniques try to extract explanations about the internal representation of complex models like deep neural nets. One of these methods is **model distillation**.

Model Distillation

Distillation is the process of transferring **dark knowledge** [7] from a deep neural net (usually denominated "the teacher") to less complex models ("the students"), these can be smaller deep nets or an interpretable model like a decision tree. Dark knowledge, also referred as hidden knowledge or latent knowledge in some literature, can be understood as information that is not seen with the "naked eye". In machine learning it refers to all the information contained in the hidden layers of a neural network model: the weights and ways each neuron connects to each other, inputs and inputs of each one or the way they jointly activate for a certain observation.

In the case of model distillation one is particularly interested in the last layer of a model, this can be seen as the layer where a decision has matured and is ready to be output.

Let's say we have a model m for classification of 3 classes which has a **softmax** layer l_s as the last layer, and, for a given observation o we know that the model outputs $m(o) = $ "$class1$". If we are interested in dark knowledge we need to look deeper into the model. We'll find that the result was given by $argmax(output(l_s))$ and $output(l_s) = [0.6, 0.3, 0.1]$.

We can understand why the model output was "class 1": it presented the highest probability, we can also understand that our model learned that for the specific observation it would be 3 times more likely to be classified in "class 2" than in "class 3". This type of information a.k.a as **dark knowledge** is the rationale on which model distillation operates. It can be particularly useful when classes are strongly related to each other and it has been proven that model distillation can produce smaller models that can be as accurate as more complex ones [2] through the usage of **dark knowledge**.

Model compression [2] is also referred many times in the literature as one of the first examples of model distillation originally proposed to reduce the computational cost of a model at run-time by reducing its complexity which was later explored for interpretability. Tan et al. [12] proposed that model distillation can be used to distill complex models into transparent models like generalized additive models and splines. Che et al. [4] introduced in their paper a knowledge-distillation approach called Interpretable Mimic Learning, to learn interpretable phenotype features for making robust prediction while mimicking the performance of deep learning models. A recent work by Xu et al. [14] presented **DarkSight**, a visualization method for interpreting the predictions of a black-box classifier on a data set in a way inspired by the notion of **dark knowledge**.

The premise in all of the methods mentioned above is to use the capabilities of deep neural nets and translate the processes they learned during training to another model. We are interested not only on the ability to make the same class of models more efficient [2], but also in the ability of possibly changing their class [4] to a more interpretable one.

3 Proposed Approach

Although knowledge distillation is not a new topic, we believe that there's more to do with it when it comes to interpretability, the interaction between classes can be a good resource to explain how a model came up with a certain decision. Some work in distilling knowledge to interpretable models has been done by Che et al. [4], but the models used were GAM's and splines, which don't have a great visualization, decision-trees are very easy to visualize and better at capturing feature interaction.

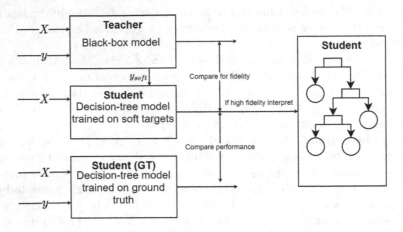

Fig. 3. Knowledge distillation process for the proposed approach

Methods to interpret trees can be more intuitively easy to come up with and explore. We believe that the tree structure is ideal for capturing inter-action between features in data, visualization of decision-trees is also human-friendly making them better for explanation and interpretation. Figure 3 depicts the knowledge distillation process for the proposed approach. We propose distilling knowledge from a deep neural net to a decision-tree by matching logits (scores before the last softmax layer), we do this by using these logits as targets to train a decision-tree for regression. This decision-tree should in theory mimic the way the deep neural net makes its decisions. It should capture not only the good parts but the bad parts. This tree can then be evaluated for interpretability. For each defined problem: 1) credit risk management, 2) stock movement prediction we define a supervised training dataset $D_{train} = \{X, y\}$. For each dataset we train a deep neural net model, which we will call **"Teacher"**. We then extract the **logits** (values of the last layer before the softmax), y_{soft}, and use them as soft targets to train a decision-tree based model, which we call **"Student"** using XGBoost toolkit. The results of these models are to be compared in order to check how closely the mimic model is following the deep net model, we do this by checking their performance scores on the assumption that for the same observations, a similar evaluation metric on both models indicates similar decision-making. If this holds true we can interpret a decision-tree assuming its decision process is similar to the neural net. Finally, we train a third model on ground-truth labels which we will call **"Student (GT)"**. This model serves as benchmark to validate the usage of a deep neural net on a problem in the first place. If the **"Student (GT)"** model proves to be more precise, than a neural net we have a problem that its not complex enough to justify the usage of deep neural nets, and so the usage of the distillation on said model.

4   Experimental Setup

4.1   Dataset Description

The German credit dataset was taken from UCI ML Repository [6] and is comprised of 1000 instances and classifies people described by a set of attributes as good or bad credit risks. The data have been contributed as part of a dataset collection created by the Statlog EU project[2] with Prof. Dr. Hans-Joachim Hofmann listed as the data donor.

There are 20 explanatory variables with seven being numerical and 13 being categorical, with 30% observations accounting for the positive class (having bad credit). Both stock price historical data was acquired using the Yahoo! finance API, raw data consists of a time series with the columns open Price, closing Price, adjusted closing price, volume, highest and lowest price of the day.

4.2   Evaluation Metrics

Table 1. Contingency table for binary classification

	Class positive	Class negative
Assigned positive	a	b
	(True positives)	(False positives)
Assigned negative	c	d
	(False negatives)	(True negatives)

Tests were done using basic metrics for model evaluation that tell us how well a model is performing, which are then to be compared between the "teacher" and "student" models. In order to evaluate the decision task, a contingency matrix can be defined to represent the possible outcomes of the classification, as shown in Table 1. In cases where the weight of false positives and false negatives have different cost or in unbalanced datasets its better to use metrics that difference

---

[2] https://cordis.europa.eu/project/rcn/8791/factsheet/en.

into account, as such for our problem we look to F1-score as being a more important metric than accuracy. If the F1-score of the student models is somewhat similar or better than the teacher model we can presume that its reliable to use this models as surrogate. In specific cases where the weight of the false negatives is greater, such is the case of credit risk classification, we give more importance to the recall score, while trying to maintain a good F1-score.

4.3 Models

Two neural network architectures were used for the teacher model. A feedforward neural network for the credit risk classification dataset with 2 layers of 256 and 128 hidden units respectively and a long short-term memory architecture for the stock movement prediction problem with 2 layers of 256 hidden units. The selected interpretable model a gradient boosted regression tree from XBGBoost's python library with the default parameters.

5 Experimental Results and Analysis

The models were evaluated based on their respective accuracy, precision, recall and F1-score, paying special attention to the F1-score and recall in the case of credit risk classification. For the German credit dataset, we have indication that the student is capturing the teacher's decisions very close by checking the that the scores are similar across all four metrics. We pay special attention to the recall metric, that is in fact exactly the same in the student and teacher models (see Table 2). This is particularly good in this context since the weight of having **false negatives** is far greater than the weight of **false positives**.

Table 2. German Credit Dataset credit default prediction results

Model	Accuracy	Precision	Recall	F1-score
Teacher	76.80%	60.61%	55.56%	57.97%
Student	77.20%	61.54%	55.56%	58.40%
Student (GT)	77.20%	63.64%	48.61%	55.12%

It is more important to not misclassify people with bad credit as having good credit than the inverse. We believe that the higher complexity of a neural net helps in better classifying a minority class. In the german credit dataset we have a minority class that represents only 30% of the total observations. Not only that but if we look at the F1-score across all tree models, we get the best performance on the student model, which tells us that training a model with the support of a neural net's **dark knowledge** might be beneficial to get better performance on less complex models. In the context of stock prediction we find a similar behaviour. Stock price history datasets are in constant change, being updated everyday. At time of acquisition the class balance was at around between both sets ranged from 45% to 50% for the minority class, meaning these were relatively balanced. As the problem is more complex than the credit risk classification and has a much larger scale, the teacher outperformed both the student and student (GT) models. Since in stock prediction its as important to know when a stock price is going up as well as when its going down we look to the F1-score for comparison (see Tables 3 and 4). We still see a slight improvement on the student when compared with the student GT, which enforces the belief that in general, less complex models can benefit from distillation. We also see a tendency for high recall scores that should represent a better classification of minorities which requires further investigation.

Table 3. Alphabet Inc. (GOOGL) stock movement prediction results

Model	Accuracy	Precision	Recall	F1-score
Teacher	64.31%	65.35%	76.10%	70.31%
Student	54.69%	56.07%	85.09%	67.60%
Student (GT)	56.15%	57.69%	78.95%	66.67%

Table 4. Tesla Inc. (TSLA) stock movement prediction results

Model	Accuracy	Precision	Recall	F1-score
Teacher	68.95%	71.03%	71.27%	71.15%
Student	55.05%	56.18%	74%	63.92%
Student (GT)	54.29%	56.18%	65.96%	60.78%

6   Conclusions and Future Work

The results obtained show that the method can be used to improve solutions in particular contexts, as the credit risk case. An exploration that can be interesting is working on the inner layers of neural net which have more complex interactions. If the method proves to be able to output a model that can closely mimic a cumbersome one in a consistent way, the same pipeline can be used for classification problems in other domains such as recidivism in the criminal justice sector or medical diagnosis in the health sector [8,13] for example.

The focus of the future work will be to optimize and tune the transfer process to better adapt it to decision-trees as well as define new metrics for fidelity. In order to better define the decision process we pretend to check how the decisions represented by the tree are different from the model by looking at specific cases and checking how differences in features change the outcome in both models.

Acknowledgements. This Research is developed under the EU COST Action: Fintech and Artificial Intelligence in Finance funded by Horizon 2020 Framework Programme of the European Union.

References

1. Baehrens, D., Schroeter, T., Harmeling, S., Kawanabe, M., Hansen, K., Müller, K.R.: How to explain individual classification decisions. J. Mach. Learn. Res. **11**(61), 1803–1831 (2010)
2. Buciluă, C., Caruana, R., Niculescu-Mizil, A.: Model compression. In: Proceedings of the 12th ACM SIGKDD International Conference on Knowledge Discovery and Data Mining, KDD'06, pp. 535–541. Association for Computing Machinery, New York (2006)
3. Caruana, R., Lou, Y., Gehrke, J., Koch, P., Sturm, M., Elhadad, N.: Intelligible models for HealthCare. In: Proceedings of the 21th ACM SIGKDD International Conference on Knowledge Discovery and Data Mining. ACM, August 2015
4. Che, Z., Purushotham, S., Khemani, R., Liu, Y.: Distilling knowledge from deep networks with applications to healthcare domain (2015)
5. Dhieb, N., Ghazzai, H., Besbes, H., Massoud, Y.: A secure AI-driven architecture for automated insurance systems: fraud detection and risk measurement. IEEE Access **8**, 58546–58558 (2020). https://doi.org/10.1109/ACCESS.2020.2983300
6. Dua, D., Graff, C.: UCI machine learning repository (2017). http://archive.ics.uci.edu/ml
7. Hinton, G., Vinyals, O., Dean, J.: Distilling the knowledge in a neural network. In: NIPS Deep Learning and Representation Learning Workshop (2015)
8. Ho, T.K.K., Gwak, J.: Utilizing knowledge distillation in deep learning for classification of chest X-ray abnormalities. IEEE Access **8**, 160749–160761 (2020)
9. Ribeiro, M.T., Singh, S., Guestrin, C.: Why should I trust you?: explaining the predictions of any classifier (2016)
10. Ryll, L., et al.: Transforming paradigms: a global AI in financial services survey. SSRN Electron. J. (2020). https://doi.org/10.2139/ssrn.3532038
11. Sarkar, S., Weyde, T., Garcez, A., Slabaugh, G., Dragicevic, S., Percy, C.: Accuracy and interpretability trade-offs in machine learning applied to safer gambling. In: CoCo@NIPS (2016)
12. Tan, S., Caruana, R., Hooker, G., Lou, Y.: Distill-and-compare. In: Proceedings of the 2018 AAAI/ACM Conference on AI, Ethics, and Society, December 2018
13. Vellido, A.: The importance of interpretability and visualization in machine learning for applications in medicine and health care. Neural Comput. Appl. **32**(24), 18069–18083 (2019). https://doi.org/10.1007/s00521-019-04051-w
14. Xu, K., Park, D.H., Yi, C., Sutton, C.: Interpreting deep classifier by visual distillation of dark knowledge. arXiv e-prints arXiv:1803.04042, March 2018

Image Processing

Metal Artifact Reduction Based on Color Mapping and Inpainting Techniques

Rafaela Souza Alcântara[1]($\boxtimes$), Antônio Lopes Apolinário Jr.[1]($\boxtimes$),
Perfilino Eugênio Ferreira Jr.[1]($\boxtimes$), Gilson Antonio Giraldi[2]($\boxtimes$),
Iêda Margarida Crusoé Rocha Rebello[3]($\boxtimes$),
and Joaquim de Almeida Dultra[3]($\boxtimes$)

[1] Mathematics and Statistics Institute, Federal University of Bahia, Salvador, Brazil
{rafaela.alcantara,antonio.apolinario,perfeuge}@ufba.br
[2] National Laboratory for Scientific Computing, Petrópolis, Brazil
gilson@lncc.br
[3] Faculty of Dentistry, Federal University of Bahia, Salvador, Brazil
ieda@radiologia.odo.br

Abstract. Data acquisition process in computed tomography (CT) can promote the generation of artifacts in image slice, making the patient diagnosis more difficult. Thus, metal artifact reduction (MAR) techniques must be applied to recover the damaged information from bone and tissue regions. In this paper, we propose a novel pipeline to reduce these artifacts in image domain. Specifically, this procedure computes the local tone mapping (TM) operator for each metallic artifact damaged slice. Thereby, it is possible to detect and classify these artifacts using morphological operations and its geometry features for restoration through Inpainting algorithm to fill them with image patterns visually plausible. The proposed pipeline is demonstrated in buco-maxillo facial CT images. Results show that restored slices enhance teeth structures allowing a better visualization of the reconstructed surface.

Keywords: Computed Tomography · Metallic artifacts · Tone mapping

1 Introduction

Computed Tomography (CT) has become one of the most important tool used in different health fields for diseases diagnosis since its development in the 1970s [3]. In dentistry, CT is widely used in surgical planning, allowing simulations in 3D reconstructed model before the procedure is performed on patients.

Filtered Back Projection (FBP) method performs image reconstruction from CT raw data (sinogram). Although it uses filtering techniques to remove noise, there are artifacts caused by the presence of metallic objects in scanned structures, such as implants and dental braces that are not eliminated after reconstruction step.

© Springer Nature Switzerland AG 2021
J. M. R. S. Tavares et al. (Eds.): CIARP 2021, LNCS 12702, pp. 303–312, 2021.
https://doi.org/10.1007/978-3-030-93420-0_29

To remove these structures, several metal artifact reduction (MAR) algorithms have been developed over the years [6,9,11,14]. Most of these solutions detect metallic objects in image domain and remove them from sinogram. However, depending on how damaged the image is, the detection of these structures becomes a challenging task.

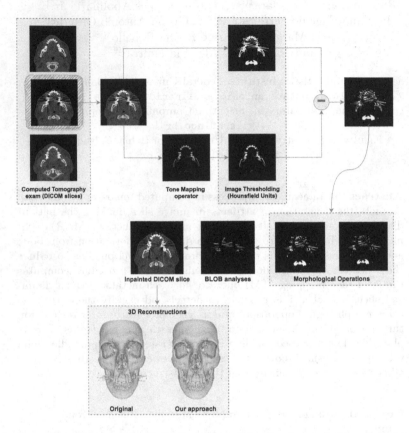

Fig. 1. Overview of our approach. At first, Tone Mapping is applied to original image. Next, segmentation step will be processed on both slices (original and tone mapped) to calculate subtraction operation and highlight metallic artifacts that will be evaluated in the next steps. Resulting artifacts will be restored through Inpainting and reconstructed on 3D domain.

In this work, we developed an image-based MAR technique to reduce artifacts on CT scans. The key idea is to transform pixels intensity through Tone Mapping (TM) to improve artifacts detection. Then, refinement steps are applied to classify and remove remaining bone structures, resulting in an image containing only metal artifacts structures that will be removed in the final step through Inpainting algorithm. The proposed pipeline was applied into buco-maxillo CT images and we achieved visually satisfactory results on severely damaged slices as will be described in the following sections.

2   MAR Based on Color Mapping and Inpainting Techniques

Figure 1 presents an overview of our approach. From manual selection of metallic artifacts damaged slices (Fig. 1 - right), we first apply a threshold segmentation step based on Hounsfield Units (HU) [5]. In parallel, we apply TM operation (Sect. 2.1) on the same original slice. As highlighted in Fig. 1, this technique reduces the rays generated by artifacts preserving only non-artifacts structures after segmentation step. Next, we calculate the subtraction between these two segmented image (Fig. 1 - left), preserving artifacts structures in the final image.

However, this result may still contain some non-metallic structures. Thus, two refinement steps are applied to reduce them. Morphological operations (Sect. 2.2) are used to detect artifacts regions while removing noise from bone and soft tissue structures. Then, each structure from resulting image will be evaluated based on its geometry (Sect. 2.3) to create an image mask (Fig. 1 - bottom) for Inpainting algorithm (Sect. 2.4) to restore damaged regions. The final step of our proposal consists on surface reconstruction (Sect. 2.5) to evaluate MAR effects on 3D domain.

To provide a better understanding of the following sections, we will adopt a generic notation for image representation, in which a gray level image with pixel resolution $X \times Y$ is represented through a real matrix $I \in \mathbb{R}^{X \times Y}$, where $I = I(x, y)$ means the intensity in the pixel on (x, y) position.

2.1   Tone Mapping

Metallic artifacts promotes x-ray beam reflection in a similar way of overexposed photographs. Tone Mapping algorithm aims to map image tones in order to reduce high (or low) intensities. This solution was first proposed by Reinhard et al. [13] and can be described using a scaled luminance formulation:

$$L(x, y) = a \frac{I(x, y)}{\exp\left(\frac{1}{Y \cdot X} \sum_{x,y} \log(\delta + I(x, y))\right)}, \tag{1}$$

where a is a parameter that represents the key of a scene. In our application, a local correction that applies dodging-and-burning [13] is used. This is implemented through convolutions:

$$V_i(x, y, s) = L(x, y) * G_i(x, y, \alpha_i s), \quad i = \{1, 2\} \tag{2}$$

with Gaussian kernel profiles $G_i(x, y, \alpha_i s)$ and $\alpha_i s$ as the smoothing scale. Finally, a normalized difference of Gaussians (DoG) is computed as follows:

$$V(x, y, s) = \frac{V_1(x, y, s) - V_2(x, y, s)}{(2^\phi a) / s^2 + V_1(x, y, s)}, \tag{3}$$

where ϕ is the sharpening parameter and the value of scale s will be changed according to convolution result.

Although Tone Mapping reduces considerably the presence of metallic artifacts on original image it also removes some important bone structures, compromising the final 3D reconstruction. Image segmentation is then applied to both
images (original and tone mapped) using the bone range [226, 3070] in Hounsfield
scale [15]. From these two segmented images, we can isolate only region of interest (artifacts) through subtraction operation (Fig. 1 - left), creating a prior mask
for final recovering step.

2.2   Metallic Artifacts Classification

Although results from previous step preserved artifacts structures, some tissue
and bone parts are still present in the final image. To reduce them, morphological
operations [4] will be applied (Fig. 1 - bottom), guaranteeing this first stage of
refinement. On image domain, mathematical morphology is frequently used to
identify and to extract relevant components in image processing applications [4].

According to Naranjo et al. [11], metallic artifacts have a geometry similar
to an ellipsoid, being easily enhanced through morphological operations. In this
work, we applied opening operation, formulated as:

$$A \circ B = (A \ominus B) \oplus B, \qquad (4)$$

where A is the original image, B is known as structuring element and the morphological operators $\ominus$ and $\oplus$ represent erosion and dilation [4], respectively.

Firstly, erosion operation will remove dental arch and spine bones structures
contours. However, this step also removes contours from artifacts structures that
will be restored by the dilation operator. At the end of this step, a clustering
process is applied to define a set of pixels to represent image structures, the
blobs. These blobs will be evaluated based on its geometric features to ensure
the number of non-artifacts structures will be further reduced.

2.3   Artifacts Geometry Evaluation

For this work, we calculate image moments from two shape characteristics [10]:
area and eccentricity. The discretization of raw image moment of order $(p + q)$
can be expressed as:

$$m_{pq} = \sum_{x=0}^{Xb-1} \sum_{y=0}^{Yb-1} x^p y^q I(x, y), \qquad (5)$$

where p and q are the indexes, defining the order of the moments and Xb and
Yb represents the blob region.

According to [12], low-order moments provide information about geometry
properties. For instance, zero order moment (m_{00}) will represent blob area. From
empirical tests, we set the blob area $m_{00} \geq 5$, to ensure that minor artifacts were
also eliminated. The second property evaluated is the eccentricity, that measures
blob roundness. For an ellipsoid structure it can be expressed as:

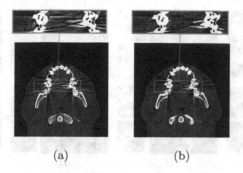

Fig. 2. Comparative image showing the application of Inpainting algorithm. (a) Original image with artifacts represented by white rays, (b) Inpainted image covered parts of artifacts.

$$\varepsilon = \sqrt{1 - \frac{\mu'_{20} + \mu'_{02} - \sqrt{(\mu'_{20} - \mu'_{02})^2 + 4(\mu'_{11})^2}}{\mu'_{20} + \mu'_{02} + \sqrt{(\mu'_{20} - \mu'_{02})^2 + 4(\mu'_{11})^2}}}, \tag{6}$$

where μ'_{20}, μ'_{02} and μ'_{11} are computed using Eq. 5. For further explanation, see the reference [7].

Ellipsoid eccentricity value should be in range $[0.0, 1.0]$, where the values closest to 1.0 should represent more elongated geometry. In our proposal, we used two values to evaluate the results: $\varepsilon \geq 0.7$ and $\varepsilon \geq 0.5$.

The main goal of these last two refinement steps is to remove the largest number of non-artifacts structures, creating a final binary mask containing only regions that we intent to restore with Inpainting. In this way, we can ensure that bone and soft-tissue pixels will not be modified by restoration step.

2.4 Inpainting

Inpainting algorithm is commonly used to restore damaged parts of original image based on pixel intensity collected from the neighboring regions. In this work, we applied the well-known Navier-Stokes solution proposed by Bertalmio et al. [1], available on OpenCV library [2]. This approach receives as input parameters the mask generated above and the damaged image, promoting the propagation of pixel intensities closest to the mask boundary.

Since we are working with CT, original image values are described in Hounsfield Units (i.e. signed 16-bits), thus to apply Inpainting algorithm we first convert these values into unsigned 16-bits pixels intensity:

$$pixel = \frac{HU - intercept}{slope}, \tag{7}$$

where HU represents the Hounsfield Units of pixel, *intercept* and *slope* corresponds to the linear transformation values. The last two values are available on CT metadata.

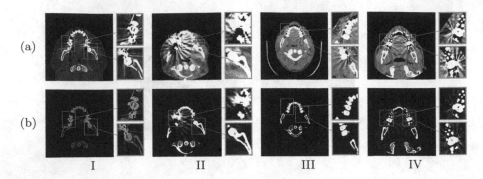

Fig. 3. Comparative scheme of the three selected cases to evaluate Tone Mapping algorithm on DICOM slices. Each column represents a case, where **row (a)** corresponds to the original image and the **row (b)** the tone mapped one.

Figure 2 presents a comparison between original image artifacts and its reduction after Inpainting step. In the Fig. 2b it is possible to notice that we were able to restore most of the damaged part, although some pixels closest to dental arch were not fully recovered. This problem occurs due to the size of Structuring Element (SE) described in Sect. 2.2. However, the selection of a larger SE could cause the removal of bone structures present in the slice. Further discussion will be provided in Sect. 3.

2.5 3D Reconstruction

Once the slice-by-slice restoration step is completed, we must evaluate results in the 3D domain. So, we applied Marching Cubes (MC) [8] algorithm for surface reconstruction. Considering that our pipeline was only applied to damaged slices, we first need to convert their pixels values from unsigned 16-bit pixel intensity to HU scale to ensure that all slices have values in the same domain. Next, we apply MC, using isovalues in HU range to generate the surface and compare original reconstruction with the processed one.

Table 1. Set of parameters in each selected case for Tone Mapping application

	α	ϕ	**s**	a
Case I	2.0	6.0	1.2	0.12
Case II	0.55	7.0	1.2	0.12
Case III	4.0	6.0	1.2	0.12
Case IV	0.55	8.0	1.2	0.12

3   Results

In this section, we will discuss the results achieved using the methodology proposed in Sect. 2. The developed algorithm was tested into 11 different cases provided by the Dentomaxilofacial Radiology Division of Federal University of Bahia under the approved ethical committee number 1.208.317. In this paper we selected four of them to discuss the results, focusing on cases with artifacts generated by dental braces and restorations.

All the images are in Digital Imaging and Communications in Medicine (DICOM) format, with 512×512 size and compressed into signed 16-bits. Images in Case I were acquired using the Siemens Spirit device while in Cases II, III and IV, were acquired by GE Optima CT660.

3.1   Tone Mapping Enhancement

Figure 3 presents a summary of TM results from selected cases. For each case, it is possible to notice that the artifacts rays were reduced drastically. This method is controlled by a set of parameters: a, α, ϕ and s described in Sect. 2.1. Based on empirical tests, we were able to determine the parametrization of Table 1.

In some cases, bone structures were also removed after TM application. Therefore, we used this technique associated with segmented original image, to ensure that these structures will remain intact in the final image.

3.2   Structuring Element Analysis

Another important step in our pipeline is the application of morphological operators. Results from this step are directly associated with Structuring Element type and size. As described in Naranjo et al. [11], SE must be orthogonally oriented in relation to the artifacts. Furthermore, for CT images, artifacts rays appears closest to bone structures, making it difficult to define an optimal size for the structuring element. Empirically, we were able to define a square SE with size 5×5 and horizontal 90° orientation.

3.3   Reconstruction

Figure 4 shows the results of our 3D reconstruction step after artifacts were reduced. As shown in Fig. 4 (c-d), to evaluate our approach, we calculated the distance between original mesh points and the final processed mesh points (after Inpainting application).

In Case I, we were able to present visually satisfactory results, considering that most part of external artifacts were removed. This is due to the fact that these artifacts presented a little variation in their geometry, promoting a better classification on blob evaluation step, described in Sect. 2.3.

In contrast, for Case II, it is possible to notice remaining artifacts in 3D reconstruction although results shows a significant metallic reduction on internal

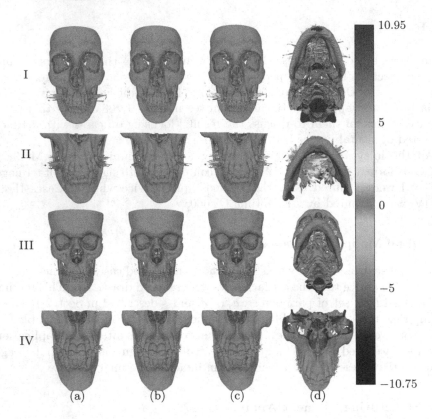

Fig. 4. 3D reconstruction comparison. Original reconstruction in **column (a)** and our approach result in **column (b)**. Columns (c) and (d) presents different views with color mapping scale representing the distance between original mesh points and processed one.

jaws structures (palate region). Despite presenting a shape more similar to an ellipse, these structures vary in size and consequently its eccentricity values are changed. In addition, as shown in Fig. 3 (**II-b**), some artifacts are so close to dental arch region, that after applying Tone Mapping they will not be segmented, considering that their values (in HU scale) will be the same as for bone structures.

Cases III and IV present an interesting characteristic to evaluate this work. In these scenarios, metallic artifacts appear due to the presence of dental braces. As shown in Fig. 3 (**III-a**) and (**IV-a**), these artifacts have a different geometry from the cases presented above. Thus, the eccentricity value used to classify the remaining blobs had to be reduced $\varepsilon \geq 0.5$ in order to remove most of the artifacts.

Despite the number of parameterizations, our algorithm guarantees the removal of severe artifacts and the preservation of important bone structures for evaluation by the dentist surgeon.

4   Discussion

Results presented in Sect. 3 achieved a satisfactory level in metallic artifact reduction on 2D domain as described on color mapping difference (Fig. 4 (**c-d**)). It is important to highlight that the chosen cases have regions that were severely damaged by these artifacts. As discussed in Sect. 3.3, selected cases present different geometries. One of the main factors observed was the variation in the shape of the artifacts and we can attribute this fact to the presence of different types of metallic materials.

In Case I, artifacts are generated by dental braces and restorations in both lower and upper teeth. The material of these restorations is composed with amalgam and therefore produces an artifact with a shape more similar to an ellipse, propagating through all cuts in a more consistently way. On the contrary, in Case II, although there is no dental braces, the number of amalgam restorations is higher when compared to Case I, creating more artifacts with sizes and shapes that vary along the slices. Thus, after the TM step, some artifacts closest to the dental arch are not eliminated and, consequently, morphological operators are unable to remove them without compromising bone structures.

It is possible to notice that in cases where the artifacts are very small (Cases III and IV) caused only by the presence of dental braces, the algorithm did not present such significant results. As already discussed in Sect. 2, the chosen morphological operator ensure that bone structures are not removed. In addition, area and eccentricity metrics also need to ensure that important regions are not identified as artifacts for removal. For these reasons, despite the values adjusts (Sect. 3.3), we were unable to eliminate all of these smaller artifacts from the dental braces.

5   Conclusion

In this paper, a novel method for reducing metallic artifacts was proposed based on Color Mapping and Morphological Operations on CT images domain. The main contribution of our approach is the possibility to apply the technique to restore more severe metal artifact degraded images with better results than similar works in the literature. The proposed methodology does not require large datasets for training, like deep learning counterparts.

Although it is not the main objective of this work, it is important to mention that our pipeline has a relatively low computational cost and can be improved to be used in clinical environments equipment. For future work, we intend to evaluate these performance metrics by optimizing this method using GPU parallel processing techniques.

We also plan to develop a novel approach with an automatic selection step for TM parameters based on image analysis and its artifacts. Another point to be evaluated is the application of morphological operations in the 3D domain after Inpainting step.

Acknowledgements. This work was financed by the FAPESB doctoral scholarship under the number BOL0098/2017. We also thank the Dentomaxilofacial Radiology Division of Federal University of Bahia for providing the CT scans used in this project.

References

1. Bertalmio, M., Bertozzi, A.L., Sapiro, G.: Navier-stokes, fluid dynamics, and image and video inpainting. In: IEEE Computer Society Conference on Computer Vision and Pattern Recognition, CVPR 2001, vol. 1, p. 1. IEEE (2001). https://doi.org/10.1109/CVPR.2001.990497
2. Bradski, G.: The OpenCV library. Dobb's J. Softw. Tools **25**, 120–123 (2000)
3. Gjesteby, L., et al.: Metal artifact reduction in CT: where are we after four decades? IEEE Access **4**, 5826–5849 (2016)
4. Gonzalez, R.C., Woods, R.E.: Digital Image Processing. Prentice Hall, Upper Saddle River, N.J. (2008)
5. Hounsfield, G.N.: Computerized transverse axial scanning (tomography): Part 1. Description of system. Brit. J. Radiol. **46**(552), 1016–1022 (1973)
6. Kalender, W.A., Hebel, R., Ebersberger, J.: Reduction of CT artifacts caused by metallic implants. Radiology **164**(2), 576–577 (1987)
7. Khan, Y.D., Khan, S.A., Ahmad, F., Islam, S.: Iris recognition using image moments and k-means algorithm. Sci. World J. **2014** (2014)
8. Lorensen, W.E., Cline, H.E.: Marching cubes: a high resolution 3D surface construction algorithm. In: Proceedings of the 14th Annual Conference on Computer Graphics and Interactive Techniques, SIGGRAPH '87, pp. 163–169. Association for Computing Machinery, New York (1987). https://doi.org/10.1145/37401.37422
9. Luzhbin, D., Wu, J.: Model image-based metal artifact reduction for computed tomography. J. Dig. Imaging **33**, 1–12 (2019)
10. Ming-Kuei, H.: Visual pattern recognition by moment invariants. IRE Trans. Inf. Theory **8**(2), 179–187 (1962)
11. Naranjo, V., Lloréns, R., Alcañiz, M., López-Mir, F.: Metal artifact reduction in dental CT images using polar mathematical morphology. Comput. Methods Prog. Biomed. **102**(1), 64–74 (2011)
12. Prokop, R.J., Reeves, A.P.: A survey of moment-based techniques for unoccluded object representation and recognition. CVGIP Graph. Models Image Process. **54**(5), 438–460 (1992)
13. Reinhard, E., Stark, M., Shirley, P., Ferwerda, J.: Photographic tone reproduction for digital images. In: Proceedings of the 29th Annual Conference on Computer Graphics and Interactive Techniques, pp. 267–276. Association for Computing Machinery (2002). https://doi.org/10.1145/566654.566575
14. Zhang, Y., Yu, H.: Convolutional neural network based metal artifact reduction in x-ray computed tomography. IEEE Trans. Med. Imaging **37**(6), 1370–1381 (2018)
15. Zheng, G., Li, S.: Computational Radiology for Orthopaedic Interventions. Springer, Heidelberg (2016). https://doi.org/10.1007/978-3-319-23482-3

New Improvement in Obtaining Monogenic Phase Congruency

Carlos A. Jacanamejoy[1] and Manuel G. Forero[2([⊠])]

[1] Semillero Lún, Grupo Naturatu, Faculty of Basic and Natural Sciences,
Universidad de Ibagué, Ibagué, Colombia
carlos.jacanamejoy@unibague.edu.co
[2] Semillero Lún, Grupo D+Tec, Faculty of Engineering, Universidad de Ibagué,
Ibagué, Colombia
manuel.forero@unibague.edu.co

Abstract. Phase congruency is an advanced technique for edge detection in images. However, in the original technique, edge detection errors can occur when at least one side of the image is not a power of two. In this paper, this problem, not reported before, and its origin are exposed and two ways of correction are proposed to reduce this problem. The proposed solutions allow to overcome the presented problem by obtaining more accurate and uniform contours than the original technique.

Keywords: Monogenic Phase congruency · Tile-mirror · Edge detection · Filter bank · Image segmentation

1 Introduction

Automatic edge detection in images is an area of great interest to industry and the scientific community. A problem usually experienced is that edge detectors are sensitive to the magnitude of changes in brightness. However, this disadvantage disappears when employing the technique known as phase congruency (PC), which allows edge detection in an image regardless of its illumination level [5]. This technique is based on phase alignment of frequency components. This principle states that the edges of an image occur when the phases of the Fourier components coincide. By using phase, the direct dependence on brightness intensity in edge detection is avoided.

Generally speaking, phase congruency implementations can be classified into two different groups according to the way the frequency components of the image to be analyzed are calculated. The first one consists in using wavelet filters, so that for each component different directional Log-Gabor filters are applied [6]. The second consists in using monogenic filters, so that only one filter is required for each component, substantially reducing the computational cost [9]. Using directional filters increases the computational cost, but has applications in corner detection [7], which has evolved to point of interest detection [1,12,13]. Regarding phase congruency using monogenic filters (MPC), progress has also

© Springer Nature Switzerland AG 2021
J. M. R. S. Tavares et al. (Eds.): CIARP 2021, LNCS 12702, pp. 313–323, 2021.
https://doi.org/10.1007/978-3-030-93420-0_30

been made to improve detection of contiguous edges [3], estimate noise with greater versatility [2], and extend it to color images [11].

A difficulty with MPC, not previously mentioned, is that edge detection, when these are very close, is affected by the image dimensions. Thus, for example, if the PC of an image of 256×256 pixels is obtained, the result is accurate, but if its width and height are reduced by only one pixel, to 255×255 pixels, the edge detection is highly affected, losing many of the edges, as seen in the example in Fig. 1. To overcome this problem, this paper presents the causes of the error and its solution by increasing the image size or properly adjusting the filter bank.

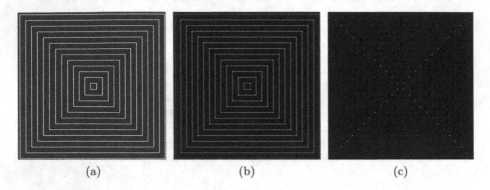

(a)   (b)   (c)

Fig. 1. Problem presented in edge detection in MPC. (a) Original image of 256×256 pixels. (b) MPC obtained from the original image. (c) MPC obtained from the original image cropped to 255×255 pixels.

2   Monogenic Phase Congruency

Figure 2 shows the approximation of a square and a triangular signal using four Fourier components. As can be seen, the phases of all the components coincide at the edges of the signals, i.e. in the image of the square signal when there is a sudden change in the signal, and in the triangle at the maximum or minimum. Thus, a phase congruency is present at all edges of an image [6].

PC is defined by Eq. (1), where $\overline{\phi}(x)$ is the phase that maximizes it, defined in terms of the Fourier components A_n of a signal at position x [10].

$$PC(x) = \max_{\overline{\phi}(x) \in [0,2\pi]} \frac{\sum_{n=1}^{N} A_n \cos\left(\phi_n(x) - \overline{\phi}(x)\right)}{\sum_{n=1}^{N} A_n}, \tag{1}$$

By using monogenic filters it is possible to implement phase congruency [9, 11]. Thus, Kovesi proposed a PC implementation using these filters through the Equation:

$$PC(\vec{x}) = W(\vec{x}) . \lfloor 1 - \alpha \lvert \delta(\vec{x}) \rvert \rfloor . \frac{\lfloor E(\vec{x}) - T \rfloor}{E(\vec{x}) + \varepsilon} \tag{2}$$

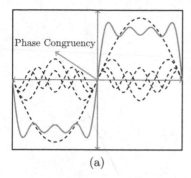
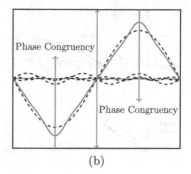

(a)   (b)

Fig. 2. Approximate signals with four Fourier components. (a) Square signal. (b) Triangular signal.

where the noise threshold T allows, for lower energy values E, the PC to become zero, avoiding producing false edges due to noise. This noise compensation is used by involving the positive part function, denoted as $\lfloor \cdot \rfloor$, in Eq. (3).

$$\lfloor f(x) \rfloor = \max(f(x), 0) = \begin{cases} f(x) & \text{if } f(x) > 0 \\ 0 & \text{if } f(x) \leq 0 \end{cases} \tag{3}$$

$W(x)$ is the phase weighting function [6]. To calculate it, a sigmoid function is used according to Eq. (4), where s is the quantification of the frequency distribution given in Eq. (5). The most important variables used for the calculation of the PC, i.e. the energy, the average phase deviation and the noise threshold can be represented in a geometrical scheme, as illustrated in Fig. 3,

$$W(x) = \frac{1}{1 + e^{\gamma(c - s(x))}}, \tag{4}$$

$$s(x) = \frac{1}{N} \left(\frac{\sum_{n=1}^{N} A_n(x)}{\varepsilon + A_{max}(x)} \right), \tag{5}$$

3   Incorrect Edge Detection in MPC

Edge detection, when borders are very close, is affected by the dimensions of the image. In Fig. 4 the result and the vertical profile of the MPC on a square image of side 256, the edges are detected correctly. On the contrary, when the same image, but with 255×255 dimensions, the result is drastically deteriorated, as seen in Fig. 5, where, as can be seen in the profile plotted on the right, the value of the detected phase congruency is reduced almost twenty times, from about 0.1 to 0.005. This problem arises from the way the FFT interprets the periodicity of the incoming signal and the distribution and shape of the filter bank.

Thus, on the one hand, because the FFT is a discrete approximation of the Fourier transform, it does not produce results identical to the continuous one.

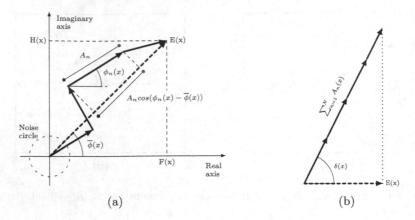

Fig. 3. Geometric scheme of Phase Congruency. (a) Relationship between phase congruency, local energy, and the sum of the Fourier amplitudes, adapted from [6]. (b) Representation of the PC by a triangle inequality where the $\sum_n A_n(x)$ is always greater than or equal to $E(x)$.

Therefore, if the period of the fundamental signal is considered equal to the total of acquired samples, the FFT result will be given directly in frequency units. In this sense, if pure frequency components are desired, the quotient between the number of samples and the period of the signal measured in samples must be an integer. Therefore, for 256 samples there are more integer divisors than for 255, thus offering a wider range of signals that can be approximated with pure components, as illustrated in the example of the Fig. 6, which shows the FFT result obtained from a sinusoidal signal of frequency 64, sampled 256 Hz, taking 256 and 255 samples. As can be seen, when the number of samples is 255, the discrete approximation of a non-integer frequency, between 63 and 64 and real and imaginary components, is obtained as transformed, while if the number of samples is 256, a signal with a single frequency component, equal to 64, is obtained as transformed. On the other hand, due to the fact that the joint response of the filter bank is not completely uniform and the phase congruency quantification is non linear, incorrect responses are generated, as shown in Fig. 1c.

Thus, since to obtain the PC of an image it is required to convert it to frequency space by using the FFT, and subsequently obtain the A_n components through a bank of filters whose joint response is not linear, as shown in Fig. 7, when performing the inverse process to obtain the resulting image is affected by its dimension, as illustrated in Fig. 6. In other words, if the processing in the frequency space were a linear process, this problem would not occur.

4   Materials

For this work, a synthetic image was created, which allows to clearly observe the studied problem, and six images of different specialties were used to illustrate

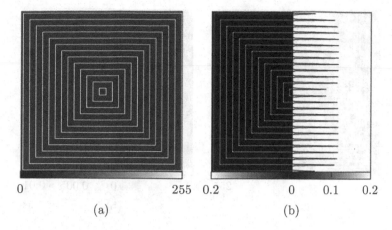

0   255 0.2   0   0.1   0.2
                    (a)   (b)

Fig. 4. MPC of a synthetic square image of side 256. (a) Original image. (b) MPC and vertical profile.

the obtained results. The MPC method, in which the proposed corrections are included, was developed in java as a plugin of the open source software imageJ and is available for free use in [4]. The study of the Fourier transform and MPC profiles was performed in Octave, using the code developed by Kovesi [8].

5   Problem Solution

A solution usually employed in 1d signal processing to solve the problem explained above, when the signal does not fit the desired size q, it is resized and padded with zeros to become equal to q. This solution is adequate when the signal at the cutoff point is zero or close to it, since it does not generate noticeable discontinuities in the signal that can generate high frequency spurious components. In the case of MPC, this solution is not adequate due to its high sensitivity to discontinuities, detecting edges independently of their amplitude, accentuating the nonlinear response of the filters, as shown in Fig. 8, which shows the profile obtained by traversing the image in the middle from top to bottom. As can be seen, the response at the edges located at the upper and left ends of the image are much greater, due to the fact that they are located next to the padded zone, which is also observed in the profile, which shows that the PC value of the upper edge is more than twice as large as that of the others. To avoid affecting the MPC values and resize the image to the desired dimensions, the signal continuity must be preserved. For this purpose, the signal is mirrored on each side of the image until the desired size is reached. Thus, it is now possible to correctly detect edges within the image preserving the PC value for all edges. In the case of the synthetic image in Fig. 5, the same MPC will be obtained, since the mirror image results equal to the original one.

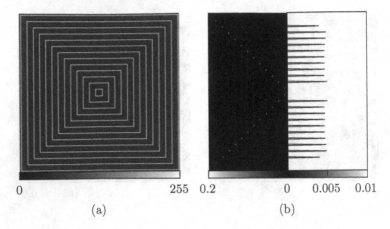

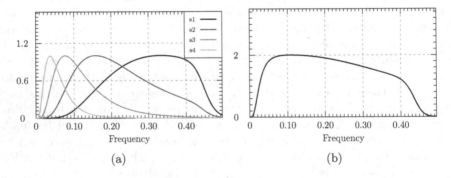

Fig. 5. MPC of a synthetic square image of side 255. (a) Original image. (b) MPC and vertical profile.

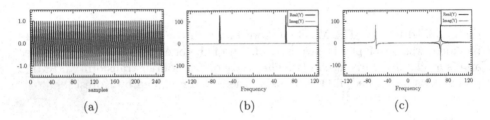

Fig. 6. FFT of 64 Hz cosine signal sampled at a frequency 256 Hz. (a) Original signal. (b) FFT of (a). (c) FFT of (a) taking only the first 255 samples, i.e., excluding the last one.

Fig. 7. Filter bank spectra of phase congruency with default parameters. (a) Scales spectra. (b) Sum of scales spectra.

Figure 8 shows the result of the phase congruency of the image in Fig. 1a cropped to 255 × 255 pixels, after resizing it to 256 × 256 pixels, by padding it with zeros and mirror extension.

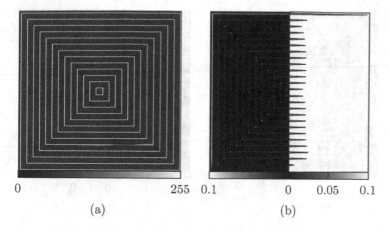

Fig. 8. MPC of a synthetic square image of side 255 and fitted with zeros. (a) Synthetic image adjusted with tile mirror. (b) MPC of (a) and a vertical profile.

Since the nonlinearity of the joint response of the filter bank is, as mentioned above, one of the causes of the problem encountered, another possible solution, although more complex, consists of experimentally adjusting the response of the filter bank until the MPC identifies the edges, which can be achieved by modifying the separation between them, looking for a better response. As shown in Fig. 9, in this example the response of the filter bank is increased in the low frequencies, allowing to reduce the effect introduced by the high ones, obtaining an improvement in edge detection as shown in Fig. 10.

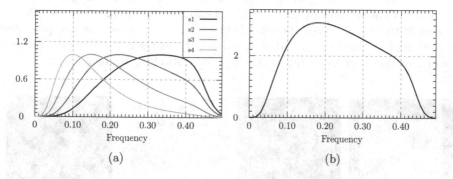

Fig. 9. Bank filter spectra of MPC with factor between filter at 1.5. (a) Scales spectra. (b) Sum of scales spectra.

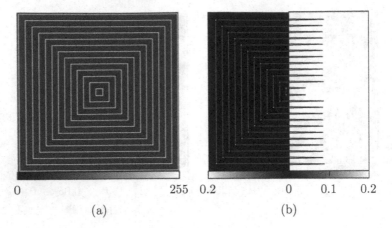

0   255 0.2   0   0.1   0.2

(a)   (b)

Fig. 10. MPC of a synthetic square image of side 255 with factor between filter at 1.5. (a) Original image. (b) MPC and vertical profile.

6   Experimental Results and Analysis

The example shown in Fig. 1, illustrates a critical case presented in the MPC, when the sides of the image do not have a value equal to a power of two. In practice, this deficiency is reflected in the incorrect detection of edges at the extremes of the image, as shown in Fig. 11b. As can be seen, when comparing the image obtained by mirror reflection of the edges, in Fig. 11d, with the one obtained by padding with zero, the first image allows obtaining a better edge detection at the extremes of the image, while in the second one a false edge is produced, as could be expected due to the abrupt change in the gray levels in that place. Other results are illustrated in the images in Fig. 12, where details can be seen in the red and green boxes, where it can be observed how the proposed correction allows to obtain a better edge detection at the extremes of the images. This correction is of great importance in image stitching in microscopy.

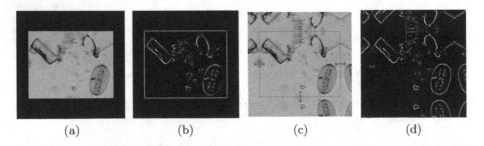

(a)   (b)   (c)   (d)

Fig. 11. MPC of diatom image. (a) Image resized using zero padding. (b) MPC of (a). (c) Image resized using tile-mirror. (d) MPC of (c).

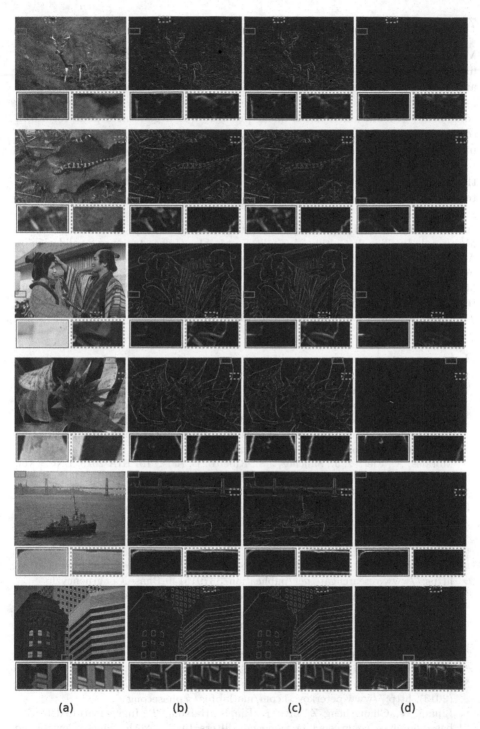

(a)   (b)   (c)   (d)

Fig. 12. Results. (a) Sample image. (b) MPC. (c) MPC with tile-mirror. (d) Difference between (b) and (c).

7   Conclusions

MPC is a more efficient technique than the one based on the use of directional filters. However, it presents deficiencies such as incorrect edge detection when any of its dimensions is not a power of two. This deficiency, not mentioned in previous works, is due to the fact that the filter bank and its mode of use for the congruency calculation is not a linear process, which generates poor results in some cases. Two solutions are proposed. The first one consists in enlarging the image by tile-mirror and the second one in readjusting the behavior of the filter bank, being the first one the simplest option and producing optimal results in the whole image, being of interest in cases where image stitching is required and to obtain a homogeneous result of the phase congruency.

Acknowledgment. This work was funded by the OMICAS program: Optimización Multiescala In-silico de Cultivos Agrícolas Sostenibles (Infraestructura y validación en Arroz y Caña de Azúcar), anchored at the Pontificia Universidad Javeriana in Cali and funded within the Colombian Scientific Ecosystem by The World Bank, the Colombian Ministry of Science, Technology and Innovation, the Colombian Ministry of Education, the Colombian Ministry of Industry and Tourism, and ICETEX, under grant FP44842-217-2018 and OMICAS Award ID: 792-61187.

References

1. Fu, Z., Qin, Q., Luo, B., Sun, H., Wu, C.: Hompc: a local feature descriptor based on the combination of magnitude and phase congruency information for multi-sensor remote sensing images. Remote Sens. **10**(8), 1234 (2018)
2. Jacanamejoy Jamioy, C., Meneses-Casas, N., Forero, M.G.: Image feature detection based on phase congruency by monogenic filters with new noise estimation. In: Morales, A., Fierrez, J., Sánchez, J.S., Ribeiro, B. (eds.) IbPRIA 2019. LNCS, vol. 11867, pp. 577–588. Springer, Cham (2019). https://doi.org/10.1007/978-3-030-31332-6_50
3. Jacanamejoy, C.A., Forero, M.G.: A note on the phase congruence method in image analysis. In: Vera-Rodriguez, R., Fierrez, J., Morales, A. (eds.) CIARP 2018. LNCS, vol. 11401, pp. 384–391. Springer, Cham (2019). https://doi.org/10.1007/978-3-030-13469-3_45
4. Jacanamejoy, C.A., Forero, M.G.: Phase congruency with monogenic filters (2018). https://www.researchgate.net/publication/337915149_Phase_Congruencyzip
5. Kovesi, P.: Invariant measures of image features from phase information. Ph.D. thesis, University of Western Australia (1996)
6. Kovesi, P.: Image features from phase congruency. Videre J. Comput. Vision Res. **1**(3), 1–26 (1999)
7. Kovesi, P.: Phase congruency detects corners and edges. In: The Australian Pattern Recognition Society Conference: DICTA, vol. 2003 (2003)
8. Kovesi, P.: Matlab and octave functions for computer vision and image processing (2013). http://www.peterkovesi.com/matlabfns/#phasecong
9. Lijuan, W., Changsheng, Z., Ziyu, L., Bin, S., Haiyong, T.: Image feature detection based on phase congruency by monogenic filters. In: The 26th Chinese Control and Decision Conference (2014 CCDC), pp. 2033–2038. IEEE (2014)

10. Morrone, M.C., Owens, R.A.: Feature detection from local energy. Pattern Recogn. Lett. **6**(5), 303–313 (1987)
11. Shi, M., Zhao, X., Qiao, D., Xu, B., Li, C.: Conformal monogenic phase congruency model-based edge detection in color images. Multimedia Tools Appl. **78**(8), 10701–10716 (2018). https://doi.org/10.1007/s11042-018-6617-x
12. Wang, L., Sun, M., Liu, J., Cao, L., Ma, G.: A robust algorithm based on phase congruency for optical and SAR image registration in suburban areas. Remote Sens. **12**(20), 3339 (2020)
13. Zhang, L., Li, B., Tian, L., Zhu, W.: LPPCO: a novel multimodal medical image registration using new feature descriptor based on the local phase and phase congruency of different orientations. IEEE Access **6**, 71976–71987 (2018)

Machine Learning

Evaluating the Construction of Feature Descriptors in the Performance of the Image Data Stream Classification

Mateus Curcino de Lima[✉], Alex J. S. de Abreu, Elaine R. Faria,
and Maria Camila N. Barioni

Faculdade de Computação, Universidade Federal de Uberlândia (UFU),
Uberlândia, MG, Brazil
{mateuscurcino,alex.abreu,elaine,camila.barioni}@ufu.br

Abstract. The image data stream classification presents a number challenges, such as the evolution of previously known classes and the emergence of new classes. A large part of current studies for image data stream classification uses classifiers that are able to evolve. However, these do not take into consideration that the image descriptors also need to evolve, in order that these improve the representation of features on new images available in the stream. Nonetheless, studies that explore the evolution of image feature descriptors do not analyze a number of the scenarios that may occur in these applications. This work presents an experimental study on the construction of image feature descriptors for the image data stream classification, while considering different aspects, as in the fact that only part of the image instances can be made available or only part of the image classes are known at the moment of constructing the descriptors. Experiments were performed on 4 image datasets, considering the state-of-art descriptors, BoVW and CNN, as well as with an algorithm that considers the evolution of the image feature descriptor. The obtained results show that the performance of a classifier may degrade its performance when submitted to scenarios that were not explored in previous studies.

Keywords: Feature evolution · Image classification · Image data stream

1 Introduction

Currently, with the high availability of image acquisition devices across different applications, the need has arisen for the development of strategies that deal with image data stream. Data stream consist of massive quantities of data, generated over a short period of time, in a continuous and potentially infinite way [8]. Among the examples of applications for image data stream, one finds the real-time object classification, face recognition and movement detection. Although

---

This study was financed in part by the Coordenação de Aperfeiçoamento de Pessoal de Nível Superior - Brasil (CAPES) - Finance Code 001.

© Springer Nature Switzerland AG 2021
J. M. R. S. Tavares et al. (Eds.): CIARP 2021, LNCS 12702, pp. 327–339, 2021.
https://doi.org/10.1007/978-3-030-93420-0_31

the aforementioned research works, there are still questions regarding tasks dealing with the image data stream.

Among machine learning tasks, related to the generation and data mining in image data stream, image classification stands out. This task is faced with a series of challenges: i) the constant arrival of new images, which imposes memory requirements; ii) constant learning with the emergence of new classes (*openset*); iii) detection and reaction to change in data distribution (*concept drift*); iv) updating of the feature descriptor to represent the appearance of new features (*feature evolution*) [8]. These challenges have been covered in only few studies in the literature [5,11]. One such challenge, as in *feature evolution*, was most recently explored in [3,10]. The general idea of these studies is to create a classifier and update it incrementally to the degree that the labeled data windows become available. Therefore, the phenomena of *concept drift* and *open-set* are explored with by updating the model with recent data from the stream. However, a majority of these studies deals with a fixed feature descriptor, that is, it does not evolve as images with distinct features appear along the stream.

The evolution of feature descriptors is an important question in image data stream environments. In these environments, when considering descriptors based on BoVW (*Bag-of-Visual Words*) and CNN (*Convolutional Neural Network*) [4], the construction of visual vocabulary on BoVW and the process of adjusting the weight of a CNN is performed based on a data sample. This sample may not be adequate for representing all the images of the stream, especially as new classes can emerge with features different to those already seen. This question was initially covered in studies described in [3,10]. However, although promising results were obtained, a comprehensive analysis was not performed on the parameters for the construction of the descriptor, such as the quantity of images per class and the number of image classes, which can affect the performance of the classifier. In addition, the BoVW feature descriptor, considered as one of the state-of-art in images, was not considered.

The study presented herein has as its objective to perform an experimental study for evaluating how the construction of the image feature descriptor can affect the performance of an image data stream classifier. Another objective of this present study is to evaluate algorithms that have already been developed for *feature evolution*, while highlighting the main questions that still have not been dealt with by these studies. The main contributions of the study described in this paper are:

- An experimental study with two state-of-art image descriptors, CNN and BoVW, with the objective to verify those questions that still have not been explored in previous studies on image data stream classifiers;
- Experiments with an algorithm that considers *feature evolution* for the image data stream classification;
- Experiments using 4 image datasets that possess different quantities of images and classes, and which are commonly used in scientific literature studies in the area of image classification.

The remainder of the paper is organized as described in the following. Section 2 presents some of the fundamental concepts. Correlated studies are discussed in Sect. 3. The experimental method is presented in Sect. 4. The discussion on the obtained experimental results is presented in Sect. 5. Finally, the conclusions and future works are described in Sect. 6.

2   Background

The image classification task receives as input an image set, where each one is represented by a feature set (or attributes). The process of transforming images on a numerical vector representation is known as feature extraction. Examples of feature descriptors commonly used in image classification are BoVW and CNN [4]. In general, these studies use these descriptors in a fixed manner, without updates, which consider the availability of all the images in the construction phase of the feature descriptor.

In the scenario of the image data stream, the classifiers possess some restrictions: restricted processing time, limited memory and a single scan of the images [13]. In addition, the classifiers need to evolve in an incremental way, as the known image concepts can change *(concept drift)* and new classes can emerge *(open-set)*. Traditionally, studies on the image data stream classification consider the incremental classification algorithm denominated as NCM *(Nearest Class Mean)* [5]. This strategy uses a centroid to represent each image class. For the incremental update of the classifier, one should recalculate the centroid of the classes when newly labelled images are available for training the classifier. In addition, when new classes emerge, one defines a new centroid on the decision model to represent the new class based on the labelled images from this class. When dealing with the classification of a new test image, one considers the centroid label with the shortest distance for that respective image.

3   Related Works

A number of recent studies have investigated the image data stream classification [3,5,10,11]. The study described in [5] considers the NCM algorithm for classifying personal image stream, while the study [11] described an approach that combines NCM and *Random Forest* algorithms for classifying large image datasets. The focus of these studies is related to improving incremental classifiers, in order to deal with *concept drift* and *open-set*. However, these algorithms ignore an important question related to the *feature evolution* [8].

The feature evolution question has been covered in a more increasing manner in recent studies on incremental classification of images [3,10]. The study described in [10] used CNN for the description of images, through the proposal of an approach that allows for the updating of the image descriptors from new images that emerge in the image data stream. This study concluded that the NCM is not adequate for classification in *feature evolution* contexts, due to the difficulty in updating the features of the centroid that represent each image class. A variation of the NCM algorithm, called iCaRL (*Incremental Classifier and Representation Learning*), was proposed with the aim of dealing with this question [10]. This variation considers multiple centroids for each image class. These centroids refer to a parametrized quantity of images to be selected from within the respective classes. In this way, through the storing of images the possibility is open for verifying whether their features change in regards to new images, which emerge in the image data stream.

The study presented in [3] proposes a single CNN architecture for representing images and classification, i.e., the same CNN was used for the representation and updating of feature descriptors as it was for the image classification. Despite the important contributions of the studies realized in [3,10], these studies had as their focus the descriptors based on CNN and do not investigate other descriptors, such as BoVW. Another question that was not evaluated in these studies is the influence exerted by the number of images and the classes used to construct the image features. In [10], an evaluation was made only of the reduction in the number of classes in the representation of the images, but variations were not performed on the number of images of each class. As such, an extensive analysis of the parameters, considered in the construction of image descriptors, can contribute to a greater understanding into the influence of each parameter in the quality of the representation of images for classifying image data stream.

4   Experimental Method

Formally, the extraction process of features from an image i generates a vector $\vec{w}$ of d dimensions for representing the image i, where each dimension of $\vec{w}$ corresponds to a feature of the image i obtained by the adopted feature extraction method. This process is performed by a feature descriptor method, which can be understood as a function ε, constructed from an image set $Y = \{i_1, i_2, ..., i_m\}$.

The image set Y, hereby called the initial set, is a finite set, which possesses a quantity c of image classes. However, considering the context of image data stream, new image classes can emerge in addition to the initial c classes. Formally, an image data stream X, is a very large, potentially infinite, sequence of images $i_1, i_2, ..., i_n$, which arrives at the timestamps $t_1, t_2, ... , t_n$. As such, after the construction steps of the feature descriptor, each image $i_i \in X$ is described by ε, thus resulting in feature vector $\vec{w}_i$. Noteworthy here is that over the data stream X new image classes can emerge, where X possesses a quantity of image classes $c' > c$. Moreover, the d of image features may not be sufficiently representative for representing the new images.

With the emergence of new classes and new instances with different features at each timestamp of X, there may arise the need for the evolution of the feature descriptor ε in order to consider the new features in the images. Therefore, the objective of the experimental method herein is to perform numerous variations concerning the quantity of image classes and in the number of images of the initial set (Y) for assessing the influence of these parameters in the representation of the image data stream X. From these variations, one also notes the influence from the image description on the quality of the result for image classification. The steps in the proposed experimental method are illustrated in Fig. 1.

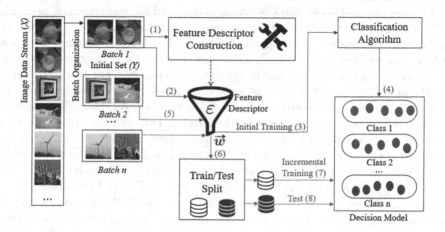

Fig. 1. Experimental Method - Steps.

Figure 1 demonstrates that the image data stream (X) is organized into batches. The number of batches, the quantity of image classes and the number of images that belong to each batch are parameterized options. This organization of X uses the same strategy as that described in [10]. This strategy considers the availability of all labels from images of X. The first batch represents the initial set (Y), which is used as input for Step (1) from the experimental method. In this stage, the feature descriptor (ε) is constructed. In Step (2), the extraction of the features of the images is performed for Y from ε. The training of the classification algorithm is performed after the extraction of features in Step (3), using as input the feature vectors $(\vec{w})$ obtained in Step (2), which resulted in the decision model in Step (4).

After the processing of the first batch, the remaining batches are considered. In Step (5), the feature extraction of the images from each batch is performed, considering ε. Following this, the vectors $\vec{w}$ of the images from each batch are splitted into training and test portions in Step (6). After this step, the incremental training of the decision model is performed in Step (7). Finally, the decision model is tested in Step (8). Noted here is that Steps (5) to (8) should be performed for each batch.

5   Experimental Evaluation

In order to perform the experiments described herein, we selected four real datasets, which are commonly used in studies of image classification. These datasets possess different quantities of images and classes. The dataset *Scenes15* [7] has 4,485 images with 15 classes. The dataset *UIUC Sports* [6] possesses 8 classes and 1,574 images. The dataset *Flowers17* [1] possesses 1,360 images and 17 classes. The largest image dataset considered is the *Pascal VOC 2007* [2], with 9,963 images and 20 classes.

The descriptors BoVW and CNN were considered for feature extraction. The CNN considered for the performing of experiments was VGG16. The feature vector, with 25,088 dimensions, produced by the last layer of pooling, known as flatten, was selected to represent the images. For the implementation, VGG16 was considered pre-trained in the image dataset *ImageNet*. For the adjusting of weights from the network, 10 epochs were considered, along with a learning rate equal to 0.1. In order to describe images considering BoVW, the SIFT descriptor was used for identifying *key points* of the images. Different quantities of visual words were considered for each image dataset, based on studies in the scientific literature of the area. For the datasets *Flowers17* and *UIUC Sports*, 512 visual words were considered [1,6]. In the dataset *Scenes15*, 800 visual words were considered [7]. Finally, we used 8,192 visual words in *Pascal VOC 2007* [2].

The algorithms considered for classification were the NCM [5] and iCaRL [10]. The iCaRL algorithm possesses a feature description step, where this method allows for the evolution of the feature descriptor with new images. The implementation considered and the configurations used for the iCaRL were the same as those published in [10]. The efficiency of the feature descriptors for the classification was evaluated considering two metrics - *accuracy* and *F-Measure* [9].

5.1   Experimental Results

Through the proposed experimental method, described in Sect. 4, the experiments were performed, while considering 3 different scenarios related to a Baseline. The scenarios of the proposed experiments have as their objective to analyze different parametrizations, thus promoting combinations that still have not yet been explored in previous studies, for the construction of the feature descriptors BoVW and CNN. These parametrizations are related to the composition of the image datasets used for the construction of the descriptors. In addition, the iCaRL algorithm, which feature descriptor is CNN based, was considered in order to allow for the analysis of the results from one method that considers the evolution of the image feature descriptor.

Baseline In the Baseline, the image descriptors, BoVW and CNN, consider all the images from the dataset, in order to compose the initial set (Y) for the construction of the visual vocabulary and for training the CNN, respectively. The objective is to obtain a Baseline for the proposed analyses, while considering a scenario in which all the classes of images are known and all the images are accessible. Likewise, in the iCaRL algorithm, all the images were used in the feature descriptor construction. Therefore, the approach of feature evolution of the iCaRL algorithm was not considered in this scenario. In regards to the classification step of the iCaRL and NCM, the images for each batch were randomly selected, at 70% of the images for training and 30% for the classifier tests. Each batch possesses images from 2 image classes. The results from the experiments with the Baseline are described on Table 1.

Table 1. *Accuracy* (A) and *F-Measure* (M) of iCaRL and NCM (BoVW and CNN) algorithms. All images are used for the construction of the feature descriptors.

| Image dataset | NCM | | | | iCaRL | |
| | BoVW | | CNN | | | |
	(A)	(M)	(A)	(M)	(A)	(M)
Flowers17	36.51	37.67	83.08	84.03	82.49	82.61
Scenes15	49.92	48.99	84.76	84.73	83.22	83.15
Sports	37.81	37.71	90.75	90.52	90.47	90.04
Pascal VOC 2007	27.36	28.3	67.19	65.45	66.74	65.12

Through an analysis of the results encountered on Table 1, one notes that the images dataset *Pascal* is that which presented itself as the most challenging, for BoVW as well as for CNN. This image dataset is that which possesses the highest quantity of images and classes. The results for the iCaRL algorithm were close to the results of the NCM with the descriptor VGG16. In both cases, the descriptors are based on CNN. Subsequently, under these circumstances, the similarity of the results between NCM and iCaRL is compatible with the results published in [10]. Under the intent of analyzing the influence of different quantities of images used in the construction of the descriptors, the three scenarios of experiments were used, as described in the following.

Scenario 1 - Does Reducing the Number of Images During the Construction of the Descriptor Have an Impact on the Representation Capacity over the Whole Image Data Stream? In this scenario of experiments, the objective was to assess the impact from reducing the number of images of Y, when generating the visual vocabulary of the BoVW and CNN training. The number of image classes was preserved, i.e., all the classes were considered. The number of images of each class was reduced to 70%, 50% and 30% in relation to the Baseline. The images were randomly selected, where each image set included the lower sets. According to that described in Sect. 4, the images used for the training of the classifier correspond to the same quantity considered by the descriptors. In terms of the iCaRL, the selected images were used in the first batch. The results from this scenario are described on Table 2(A).

In general, one notes a small degradation in the classification results with BoVW, when we compare the scenarios that consider 70% and 30% of the images. The maximum degradation noted was of 0.69% for the *F-Measure* from the image dataset *Scenes15*. For the CNN, degradations were noted that arrived at 2.48% on the *F-Measure* of the image dataset *Pascal VOC 2007*. For the algorithm iCaRL, the highest degradation noted was of 0.26% for the *F-Measure* of the image dataset *Pascal VOC 2007*. Note that the use of a method that considers the evolution of features can mitigate the degradation of the results.

When comparing the results of Table 2(A) to the Baseline, present on Table 1, one notes higher degradation of the *F-Measure*. For example, by analyzing the results obtained for the image dataset *Scenes15*, there is a noted degradation of up to 3.37% on the CNN descriptor. For the BoVW, the highest degradation noted was of 1.43% in the image dataset *Flowers17*. This same set also presented the highest degradation in the iCaRL algorithm, with a value of 0.36%. These results show that the reduction in the number of images when constructing descriptors undermined the representation capacity of the whole data image stream.

Table 2. *Accuracy* (A) and *F-Measure* (M) of iCaRL and NCM (BoVW and CNN) algorithms. Different numbers of images (2 (A)) and classes (2 (B)) are used for the construction of the feature descriptors.

Image		NCM				iCaRL			NCM				iCaRL	
		BoVW		CNN					BoVW		CNN			
dataset	Images	(A)	(M)	(A)	(M)	(A)	(M)	Classes	(A)	(M)	(A)	(M)	(A)	(M)
Flowers17	70%	35.78	36.73	80.63	82.02	82.42	82.45	70%	35.53	35.91	80.63	81.29	82.27	82.31
	50%	35.78	36.59	80.14	81.51	82.35	82.31	50%	34.8	36	80.14	81.28	82.13	82.18
	30%	35.29	36.24	79.9	81.46	82.2	82.25	30%	34.55	35.69	79.65	80.95	81.98	82.01
Scenes15	70%	50.22	48.96	82.98	82.53	83.16	83.11	70%	49.48	48.22	81.13	81.01	83.12	83.09
	50%	49.55	48.32	82.17	82.72	83.12	82.99	50%	49.33	48.24	80.99	81.18	83.07	82.88
	30%	49.48	48.27	81.65	81.36	83.09	82.81	30%	49.4	48.15	80.99	80.76	83.03	82.75
UIUC Sports	70%	37.81	37.56	90.75	89.89	90.33	90.21	70%	37.39	37.32	90.12	89.68	90.21	90.07
	50%	37.6	37.52	90.33	90.04	90.27	90.15	50%	37.39	37.26	89.91	89.55	90.15	89.97
	30%	37.39	37.42	90.33	89.67	90.27	90.11	30%	37.18	37.06	89.7	89.41	90.02	89.56
Pascal VOC 2007	70%	27.33	27.83	66.72	64.62	66.73	65.78	70%	27.23	27.57	63.83	61.75	66.67	65.51
	50%	27.09	28.02	64.5	63.42	66.69	65.56	50%	27.13	27.54	62.94	61.5	66.55	65.38
	30%	27.23	27.58	64.57	62.14	66.66	65.52	30%	26.93	27.57	62.82	61.17	66.51	65.25
		(2(A))							(2(B))					

Scenario 2 - Does Reducing the Number of Classes During the Construction of Descriptors Have an Impact on the Representation Capacity over the Whole Image Data Stream? The objective behind this scenario was to evaluate the effect from the reduction in images when generating the visual vocabulary for BoVW and for the training of the CNN. Regarding the Baseline, the number of classes was reduced to 70%, 50% and 30%. The considered classes were selected randomly. All the images from the selected classes were considered. Through such, in the classification step of the NCM, the same training and test sets from the Baseline were considered. A random selection was made of 70% of the images for training and 30% for the test, including the images of classes not used for the construction of the descriptors. In the same form as in the previous scenario, for the iCaRL, the selected images were used in the first batch and the remaining images were considered for the incremental training of the classifier, as well as for the evolution of the feature descriptor. The results from the experiments with this scenario are described on Table 2(B).

From the analysis on Tables 2(B), one notes that there exists a degradation at each reduction in the number of classes. As seen in the previous scenario, there exists a small degradation in the results of the *F-Measure* with the BoVW descriptor, where there was found at most 0.26% of degradation on the dataset *UIUC Sports*, between the cases of 70% and 30% of image classes. For the CNN, one notes a maximum degradation of 0.58% on the dataset *Pascal VOC 2007*. On the iCaRL, the highest degradation was noted on the dataset *UIUC Sports*, with the value of 0.51%.

By comparing the results presented on Table 2(B) and the results of the Baseline, described on Table 1, one notes higher degradation of the *F-Measure*. For the BoVW descriptor, the highest degradation seen was of 1.98% on the dataset *Flowers17*. However, for the descriptor CNN, the highest degradation was of 4.27% on the dataset *Pascal VOC 2007*. On the iCaRL, the highest degradation was noted using the dataset *UIUC Sports*, with a value of 0.41%. In general, it was also noted that the reduction in the number of classes influences the quality of the representation of the image data stream, than that of merely reducing the number of images, covered in Scenario 1 of experiments.

Scenario 3 - Does Reducing the Number of Images and the Quantity of Image Classes During the Construction of Descriptors Have an Impact on the Representation Capacity over the Whole Image Data Stream? In the last scenario of experiments, the objective was to assess the junction of the previous scenarios, i.e., the simultaneous reduction in the number of image classes and quantity of images. The number of classes used for generating the visual vocabulary of BoVW and for training the CNN was reduced to 70%, 50% and 30%. The number of images selected also suffered a reduction. The same percentages were considered (70%, 50% and 30%). The classes taken into consideration were selected with the same criteria as Scenario 2. In order to select the images, the same strategy of Scenario 1 was considered. In the classification step of the NCM algorithm, the same sets as those used in the descriptor construction were considered. However, it was necessary to perform incremental training with the image classes not recognized by the first set of images. For the incremental training, the same percentages of images were selected from the considered respective sets; the remaining images were used in the classifier test. For the iCaRL algorithm, the selected images were presented in the first batch and the remaining images in the following batches for the incremental training of the classifier, as well as for the evolution of the feature descriptor. The results of the experiments are described on Table 3.

Table 3. *Accuracy* (A) and *F-Measure* (M) of iCaRL and NCM (BoVW and CNN) algorithms. Different numbers of classes and images are used for the construction of the feature descriptors.

Image dataset	Classes	Images	NCM BoVW (A)	(M)	CNN (A)	(M)	iCaRL (A)	(M)	Image dataset	NCM BoVW (A)	(M)	CNN (A)	(M)	iCaRL (A)	(M)
Flowers17	70%	70%	35.29	35.46	79.41	80.89	81.98	82.01	*UIUC Sports*	37.39	37.18	89.7	89.33	90.08	89.97
		50%	34.55	35.57	79.65	80.61	81.91	81.94		37.18	36.98	89.28	88.79	90.02	89.88
		30%	34.8	35.49	79.16	80.66	81.69	81.63		37.18	36.91	89.07	88.47	89.89	89.78
	50%	70%	35.29	35.46	79.65	80.51	81.76	81.68		36.97	36.77	88.86	88.25	89.96	89.81
		50%	34.55	35.35	78.18	79.86	81.25	81.47		36.97	36.73	88.02	87.64	89.83	89.61
		30%	34.31	35.37	78.67	79.55	81.25	81.43		36.97	36.7	87.81	87.42	89.77	89.58
	30%	70%	34.06	35.28	78.67	76.13	80.88	81.15		36.97	36.69	86.76	85.96	89.64	89.51
		50%	34.31	34.94	73.52	76.13	80.8	81.07		36.97	36.61	86.13	85.27	89.45	89.33
		30%	34.06	34.92	72.54	73.6	80.73	80.98		36.55	36.28	85.92	85.25	89.39	89.31
Scenes15	70%	70%	49.11	47.76	80.47	80.07	82.94	82.6	*Pascal VOC 2007*	27.23	27.57	62.79	61.15	66.49	65.18
		50%	48.66	47.68	79.95	80.14	82.92	80.56		27.13	27.54	63.12	60.77	66.47	65.15
		30%	48.74	47.38	79.51	79.79	82.78	82.19		27.13	27.5	62.35	60.61	66.43	65.1
	50%	70%	48.74	47.37	79.06	79.07	82.83	82.33		26.99	27.5	62.52	60.43	66.45	65.13
		50%	48.59	47.25	77.95	77.97	82.76	82.17		26.62	27.5	60.37	59.08	66.34	65.08
		30%	48.44	47.19	76.77	76.67	82.72	82.08		26.56	27.2	60.24	58.96	66.24	65.03
	30%	70%	48.37	47.1	74.18	74.38	82.63	82.06		26.19	27.08	60.64	58.83	65.74	64.99
		50%	48.15	46.93	70.26	69.56	82.58	81.99		26.29	26.9	60.57	57.78	65.64	64.97
		30%	47.92	46.6	65.9	64.18	82.54	81.94		25.92	26.83	59.1	57.14	65.6	64.91

As seen from Table 3, one notes that there exists a gradual degradation of the results at each reduction of class as well as of images. By considering only the reduction in the quantity of images, one already notes degradations in the results. However, the lowest results noted were obtained for the cases of 30% of classes and 30% of images. Through a comparison of the *F-Measure* for the extreme cases on Table 3, i.e., 70% classes with 70% images and 30% classes with 30% images, the highest degradations noted were of 1.16% for the BoVW and 15.89% for the CNN. These degradations were seen in the image dataset *Scenes15*. For the algorithm iCaRL, the lowest degradations were seen, with a maximum value noted of 1.03% in the dataset *Flowers17*.

In regards to the Baseline, described on Table 1, here the highest degradations were seen for the *F-Measure* across all the scenarios of experiments. The degradation of 2.75% was seen for the BoVW descriptor in the dataset *Flowers17*. For the descriptor CNN, the highest degradation was of 20.55% in the dataset *Scenes15*. In the iCaRL, the highest degradation noted was of 1.63% in the dataset *Flowers17*.

From the undertaken analysis, the conclusion was reached that the simultaneous reduction in the number of image classes and in the quantity of images caused an even greater degradation in the results when compared to scenarios from previous experiments. This degradation was especially noted for the CNN descriptor. Across all image datasets, the worst results were noted with the lowest quantity of classes as well as images. In order to complement the analysis of the results, the statistical test was performed, as described in Sect. 5.2.

5.2   Statistical Analysis

The statistical analysis performed herein, compared the values for *accuracy* and *F-Measure* obtained in the different scenarios of the performed experiments. To this end, the *Wilcoxon signed-rank test* [12] was applied. This is a nonparametric statistical test used to compare two samples of results and to verify whether there exists a significant difference between the samples. The two hypotheses formulated on the *Wilcoxon signed-rank test* are H_0 — the results obtained from two scenarios of experiments are not statistically different at a confidence level of 95%, H_1 — the results obtained from two scenarios of experiments are statistically different at a confidence level of 95%.

The analyzes were performed in relation to the Baseline. For the CNN descriptor, in all scenarios it was possible to observe a statistical difference between the results. However, for the other descriptors only in Scenario 3 it was possible to observe this fact, thus H_0 was rejected in all cases only in the comparison between Scenario 3 and the Baseline. Therefore, from the statistical analysis, one concludes that the simultaneous reduction of the number of images and in the quantity of classes present the highest degradations in the results. Under such conditions, even for iCaRL, which is capable of updating the feature descriptor, presented statistically different degradations in the results. However, one notes that these degradations are eased in relation to methods that do not consider the updating of the image feature descriptor.

6   Conclusion

The description of images in environments of image data stream is an important research question, which still leaves some points open to discussion. The majority of existing correlated studies consider the evolution of classifiers and do not focus on *feature evolution*, while those studies that consider the evolution of descriptors are limited in terms of the experimental scenarios considered.

The study presented in this paper performed an experimental analysis, based on questions still unexplored in the image data stream classification, related to the construction of feature descriptors of BoVW and CNN images. Through the analysis of the experimental results obtained, it was found that the lower the number of images and classes of images used in the construction of the descriptors, the greater the deterioration of the results. Furthermore, from the statistical analysis the simultaneous reduction for the number of image classes and for the quantity of images is understood as being the most critical scenario for the experiments under consideration, which is able to affect even those methods that consider the evolution of the feature descriptor. Therefore, the result from the study presented herein has the potential of supporting the description of images and classification in the context of image data stream.

As future work, we intend to explore approaches for updating feature descriptors and assess other image classification algorithms, other CNN architectures, along with other image datasets. In addition, the evaluation method used in the study described herein, as in correlated studies, considers the immediate avail-ability of labels of new image classes for the incremental training of the classifier. This assumption is incompatible with several application scenarios. Therefore, an important objective to be explored consists of the exploration of an experi-mental method that allows for incremental updating of the classifier and of the feature descriptor of the images, while considering the restrictions that exist in real applications.

References

1. Avila, S., Thome, N., Cord, M., Valle, E., Araújo, A.D.A.: Bossa: Extended bow formalism for image classification. In: ICIP, pp. 2909–2912. IEEE, Brussels, Bel-gium (2011). https://doi.org/10.1109/ICIP.2011.6116268
2. Avila, S., Thome, N., Cord, M., Valle, E., Araújo, A.D.A.: Pooling in image repre-sentation: the visual codeword point of view. Comput. Vis. Image Underst. **117**(5), 453–465 (2013). https://doi.org/10.1016/j.cviu.2012.09.007
3. Castro, F.M., Marín-Jiménez, M.J., Guil, N., Schmid, C., Alahari, K.: End-to-end incremental learning. In: Ferrari, V., Hebert, M., Sminchisescu, C., Weiss, Y. (eds.) ECCV 2018. LNCS, vol. 11216, pp. 241–257. Springer, Cham (2018). https://doi.org/10.1007/978-3-030-01258-8_15
4. Kumar, M.D., Babaie, M., Zhu, S., Kalra, S., Tizhoosh, H.R.: A comparative study of CNN, BoVW and LBP for classification of histopathological images. In: SSCI, pp. 1–7. IEEE, Honolulu, Hawaii (2017). https://doi.org/10.1109/SSCI.2017.8285162
5. Hu, J., Sun, Z., Li, B., Yang, K., Li, D.: Online user modeling for interactive streaming image classification. In: Amsaleg, L., Guðmundsson, G.Þ, Gurrin, C., Jónsson, B.Þ, Satoh, S. (eds.) MMM 2017. LNCS, vol. 10133, pp. 293–305. Springer, Cham (2017). https://doi.org/10.1007/978-3-319-51814-5_25
6. Kwitt, R., Vasconcelos, N., Rasiwasia, N.: Scene recognition on the semantic mani-fold. In: Fitzgibbon, A., Lazebnik, S., Perona, P., Sato, Y., Schmid, C. (eds.) ECCV 2012. LNCS, vol. 7575, pp. 359–372. Springer, Heidelberg (2012). https://doi.org/10.1007/978-3-642-33765-9_26
7. Lazebnik, S., Schmid, C., Ponce, J.: Beyond bags of features: spatial pyramid matching for recognizing natural scene categories. In: CVPR, vol. 2, pp. 2169–2178. IEEE, New York, NY, USA (2006). https://doi.org/10.1109/CVPR.2006.68
8. Masud, M.M., et al.: Classification and adaptive novel class detection of feature-evolving data streams. IEEE Trans. Knowl. Data Eng. **25**(7), 1484–1497 (2013). https://doi.org/10.1109/TKDE.2012.109
9. Powers, D.: Ailab: Evaluation: from precision, recall and f-measure to roc, informedness, markedness correlation. J. Mach. Learn. Technol **2**, 2229–3981 (2011). https://doi.org/10.9735/2229-3981
10. Rebuffi, S.A., Kolesnikov, A., Sperl, G., Lampert, C.H.: iCaRL: incremental classi-fier and representation learning. In: CVPR, pp. 5533–5542. IEEE, Honolulu, Hawaii (2017). https://doi.org/10.1109/CVPR.2017.587
11. Ristin, M., Guillaumin, M., Gall, J., Gool, L.V.: Incremental learning of NCM forests for large-scale image classification. In: CVPR, pp. 3654–3661. IEEE, Colum-bus, Ohio (2014). https://doi.org/10.1109/CVPR.2014.467

12. Rosner, B., Glynn, R., Lee, M.L.: The wilcoxon signed rank test for paired comparisons of clustered data. Biometrics **62**, 185–92 (2006). https://doi.org/10.1111/j.1541-0420.2005.00389.x
13. Stefanowski, J., Brzezinski, D.: Stream classification. In: Sammut, C., Webb, G.I. (eds.) Encyclopedia of Machine Learning and Data Mining. Springer, Boston (2017). https://doi.org/10.1007/978-1-4899-7687-1_908

Clustering-Based Partitioning of Water Distribution Networks for Leak Zone Location

Marlon J. Ares-Milián[1], Marcos Quiñones-Grueiro[1(✉)], Carlos Cruz Corona[2], and Orestes Llanes-Santiago[1]

[1] Department of Automation and Computing, Universidad Tecnológica de La Habana "José Antonio Echeverría", Cujae, Havana, Cuba
[2] Department of Computer Science and Artificial Intelligence, University of Granada, Granada, Spain

Abstract. In recent years, there has been an increase in leak zone identification strategies in water distribution networks. This paper presents an analysis of the effect network partitioning techniques have on the performance of leak zone location methodologies. An SVM classifier is used to identify the leak zone location. The effect of the following clustering methods for network partitioning is analyzed: k-medoids, agglomerative clustering, DBSCAN, and Girvan-Newman algorithm. Both topological and hydraulic variables are considered when performing the clustering with three different sensor configurations. The results obtained demonstrate that the effect of each clustering method on the leak location performance is similar for both types of variables.

Keywords: Clustering · Leakage zone identification · Water distribution networks

1 Introduction

Efficient diagnosis of leakages in water distribution networks (WDNs) has been a topic of interest over the last years [8,15]. In particular, the accurate location of leakages is of significant economical and environmental importance. Research on leak location in WDNs follows two main lines: hardware-based and software-based methods, being the latter the main interest in this paper.

Software-based approaches comprehend data analysis [8], model-based [15], and hybrid strategies [16], most of which try to identify a node in the network as the leak location. However, these exact-node strategies present accuracy limitations due to the uncertainty in the data, the variability in consumer demands and the number of sensors installed, among other factors. Therefore, several software-based strategies in recent years have approached the leak location task as a leak zone identification problem aiming to narrow the possible leak location down to a group of nodes [2,4,16,20]. Once a leak zone has been proposed, the exact location of the leak can be pinpointed using hardware-based methods. This

© Springer Nature Switzerland AG 2021
J. M. R. S. Tavares et al. (Eds.): CIARP 2021, LNCS 12702, pp. 340–350, 2021.
https://doi.org/10.1007/978-3-030-93420-0_32

approach has generated an increase in leak location reliability while sacrificing leak location resolution. Leak location reliability measures the performance of the method in terms of identifying the correct zone where the leak is located while leak location resolution is associated with the size of the zone where the leak is located, i.e. the more resolution the smaller the size of the zone.

Quiñones-Grueiro [4] proposes to partition the network into zones by means of a k-medoids algorithm with topological variables. Data for each zone is artificially generated to form classes. The leak zone location problem is then solved by using random forests (RFs) and support vector machines (SVM) classifiers trained with the generated data. A similar strategy is followed by Zhang [20], who performs network partitioning using k-means clustering with hydraulic variables.

Iterative zone divisions have also been proposed by Shekofteh [16] and Chen [2] using Girvan-Newman algorithm and hydraulics-based k-means, respectively. Both propose the division of the network into two zones to then identify one as the leak location. The identified location is then divided again into two subzones repeating the process in an iterative manner until a previously determined division iteration is reached.

Despite the increasing number of leak zone location approaches presented recently, none of them has conducted an analysis of the influence of the network partitioning methods on the leak location reliability. Therefore, the main objective of this paper is to analyze the effect of different zone division strategies on the overall performance of the leak zone location task. Four clustering algorithms are studied: DBSCAN, k-medoids, Girvan-Newman algorithm and agglomerative clustering. A secondary objective is to analyze the effect of the type of variable used when performing the clustering. The clustering methods are tested considering both topological and hydraulic variables and the leak location task is solved by using an SVM classifier. The paper is structured as follows. In Sect. 2, a theoretical base is given for the methods used for clustering the nodes. Section 3 presents a summary of the methodology followed to locate a leakage. Section 4 introduces the Modena WDN which is used in this paper as the case study. The classification results corresponding to each clustering strategy are presented and compared in Sect. 5. Finally, conclusions are issued and recommendations for future studies are made.

2   Materials

2.1   WDN Partitioning Strategies

The partitioning of the network has been approached through clustering methods. The task of clustering a data set $X = \{x_1, x_2, \ldots x_n\}$ of n objects consists in identifying a number of disjoint subsets in X formed by objects with similar characteristics while the similarities among subsets are minimum [9]. These subsets or groups are called clusters. Several approaches have been proposed for solving the clustering task [3,10,12], differing in the way the clusters are shaped, the type of objects being clustered and the measures of object similarity, among other characteristics. In this paper four clustering strategies are used. They were selected due to their simplicity, their effectiveness and the fact they are widely known.

K-medoids Clustering. K-medoids is a partitioning clustering method [10] in which a set $M = \{m_1, m_2, ...m_k\}$ of k representative objects from a data set $X = \{x_1, x_2, ... x_n\}$ are selected in order to build k $X_c \subset X, c = 1, 2 ... k$ subsets (clusters) of X. Each representative object m_c is referred to as the *medoid* of cluster X_c. Once the medoids have been determined, the clusters are built by assigning each object from data set X to the cluster formed by the *nearest* medoid [9]. In order to do so, a similarity or distance measure $d_o(x, x')$ must be defined for every two objects $x, x' \in X$. The k-medoids clustering problem is then reduced to finding a set of medoids M such that the total distance from every object in X to its corresponding medoid is minimum. This can be formulated as the following optimization problem:

$$\min_M \sum_{i=1}^n \sum_{j=1}^n d_o(x_i, x_j) z_{ij} \quad s.t.$$

$$z_{ij} = \begin{cases} 1 & if \ x_i = m_c, m_c \in M \quad and \quad x_j \in X_c \\ 0 & otherwise \end{cases} \tag{1}$$

where m_c is the medoid representing cluster X_c and $c = 1, 2, ... k$.

Agglomerative Clustering. Agglomerative clustering is a hierarchical clustering method that, given a data set $X = \{x_1, x_2, ... x_n\}$, yields a hierarchical tree or dendrogram of clusters. Each junction in the dendrogram represents the merging of two clusters and the bottom of the tree is formed by n one-object clusters, while the top has only one cluster with all n objects [9].

In order to build a cluster dendrogram, the following iterative methodology is defined:

1. Create a starting set $C^0 = \{C_1^0, C_2^0, \ldots, C_n^0\}$ of one-object clusters which goes at the bottom of the dendrogram.
2. Merge the two nearest clusters to form a new cluster set $C^1 = \{C_1^1, C_2^1, \ldots, C_{n-1}^1\}$. In order to do so, a similarity or distance measure $d_c(C, C')$ between any two C, C' clusters must be defined; which, by itself, implies the need for a similarity or distance measure $d_o(x, x')$ for every two objects $x, x' \in X$.
3. Repeat step 2 for every cluster set $C^i = \{C_1^i, C_2^i, \ldots, C_{n-i}^i\}$ until a single-cluster set $C^{n-1} = \{C_1^{n-1}\}$ is generated.

The agglomerative clustering problem is reduced to finding a level in the hierarchy which renders a cluster set most suitable for the task at hand. Several suitability criteria might be followed, such as a previously determined optimal number of clusters or an analysis of the inconsistency coefficient, among others [7].

Girvan-Newman Community Algorithm. Girvan-Newman clustering algorithm is based on the determination of communities within a graph network. A community is a group of network nodes within which connections are dense while connections with other groups are sparse [12].

This is also a hierarchical clustering algorithm, however, in this case, it is a divisive algorithm; meaning it starts with a single cluster C^0 containing all network nodes and proceeds to iteratively split it until n one-node clusters $C^{n-1} = \{C_1^{n-1}, C_2^{n-1}, \ldots, C_n^{n-1}\}$ are generated.

A set of clusters $C^i = \{C_1^i, C_2^i, \ldots, C_{i+1}^i\}$ at any level of the hierarchy ($i = 0, 1, \ldots n - 2$) is divided to form a cluster set $C^{i+1} = \{C_1^{i+1}, C_2^{i+1}, \ldots, C_{i+2}^{i+1}\}$ by removing edges from the network with the highest *edge betwenness* score until one of the clusters in C^i is no longer connected [1]. Edge betwenness is a measure of how many paths in the network go through a given edge and its calculated in several ways. However, the particular edge betwenness calculation method does not have much impact on the overall performance of the clustering task.

Density Based Spatial Clustering Algorithm of Applications with Noise (DBSCAN). DBSCAN is a density based clustering algorithm which produces a set of clusters from a data set X while identifying objects in the data set that don't belong to any cluster and are, therefore, considered noise. It is particularly useful when the clusters to be formed might have arbitrary shapes [13].

Cluster formation in DBSCAN takes place by selecting a set of core objects and their neighborhoods. A core object x_p is defined as an object in X for which there are at least *MinP* other objects in X within an *Eps* radius neighborhood from x_p [3]. In order to construct an *Eps* neighborhood for each object, a similarity or distance measure $d_o(x, x')$ must be defined for every two objects $x, x' \in X$. In other words, an object x_p is considered a core object when the following is true:

$$| N_{eps}(x_p) | \geq MinP \quad s.t. \quad N_{eps}(x_p) = \{x_q \in X \mid d_o(x_p, x_q) \leq Eps\} \quad (2)$$

where $N_{eps}(x_p)$ is x_p's Eps-neighborhood and $| N_{eps}(x_p) |$ is the neighborhood cardinality (i.e. number of objects).

Once all core objects have been identified, clusters are formed by merging sets of core objects which are in each other's Eps-neighborhoods together with their respective neighborhoods. All objects that have no core objects within their Eps-neighborhoods are considered noise and are not assigned to any cluster. The *MinP* parameter defines a minimum cluster size while the combination of both *MinP* and *Eps* determines a minimum cluster density, (or a maximum noise density). These parameters can be adjusted for a known cluster density, or in search of a desired number of clusters. They can also be determined with the help of a *k-dist graph* defined in [3].

3   Methodology

The leak localization task is addressed as a classification problem using SVM classifiers. Given a sample vector of WDN operational variables, a possible leak zone location is represented by a class. The construction of these classes consists in the clustering of groups of nodes forming k zones within the network. In this section, the methodologies used for network partitioning are presented, as well as the methodology used in the classification step.

3.1   Class Formation

In order to form the classes, the four clustering methods presented in Sect. 2.1 were implemented. The Girvan-Newman method is graph-based and, therefore, topological in nature, since it works directly with the network's nodes and connections (pipes). However, the other three methods (k-medoids, agglomerative clustering and DBSCAN) are defined for any set of objects. With this in mind, two types of variables are considered for each of these three methods: one based on the network's structure (topology-based), and the other based on patterns generated by leaks in each node (hydraulics-based).

On one hand, in the case of topology-based clustering, the set of objects X is defined as the node tags for the WDN. The similarity measure between objects $d_o(x, x')$ is defined as the topological distance between two nodes, i.e. the total pipe distance traversed on the shortest path between both nodes. Both the topological object set X and the distance value between all its components can be easily determined from the structure of the WDN, making it an easily accessible strategy. Also, considering topological distance as a similarity measure, the formation of connected zones [1] is guaranteed, which should make the process of subsequently narrowing down the location of the leak more effective.

On the other hand, the use of a sensitivity matrix for zone generation is often found in leak diagnosis applications [2,20]. In this case, for the hydraulics-based clustering, the set of objects X is defined as the rows of the sensitivity matrix generated by a set of leaks simulated across the network. Since the objects in X are now feature vectors, several similarity measures $d_o(x, x')$ can be defined between objects. A downside of this approach is the need for a hydraulic model of the network in order to generate the sensitivity matrix, as well as the fact that the matrix is conditioned by the simulation parameters. Also, not having a set of objects or a similarity measure directly related to the network structure, the zones can be disconnected.

3.2   a DBSCAN Variation

As presented in Sect. 2.1, DBSCAN generates a number of clusters and a noise set with all the objects that don't belong to any clusters. However, in this case, a leak can occur in any node of the network, therefore, every object in the data set X must be assigned to a cluster. To solve this problem, an extra step was added to the DBSCAN clustering algorithm in which every object in the noise set is assigned to the nearest cluster, using single linkage [9] as a similarity measure.

4   Case Study

4.1   Modena WDN

The WDN of the city of Modena, in Italy was selected as case study to test the performance of the strategies proposed. This network (Fig. 1), has 268 junctions, 317 pipes, and 4 reservoirs, which make it a large-scaled network. The network is completely gravity-fed [19], therefore, it has no pumps.

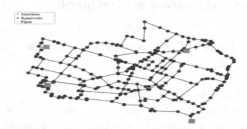

Fig. 1. Modena WDN topological representation

4.2   Data Generation

The hydraulic model of the Modena WDN was used in order to generate synthetic leakage data by means of EPANET 2.0 [14] hydraulic simulation software. The following considerations are regarded during simulation:

- A regime of minimum night flow (from 2AM to 6AM) is considered, where the variations caused by consumer demands is minimum, making the pattern recognition task simpler.
- A sample time of 15 min is set, for a total of 4 samples per hour and 4 h per day (due to minimum nigh flow regime). Samples are filtered by averaging the four samples in an hour. The 4 filtered samples in a day are considered a scenario.
- Leaks of random sizes were generated within the following interval: [2.7; 6.2]lps in every node of the network. This represents 1.6 to 3.6% of the network's total demand and 1 to 3 times a node's nominal demand.
- Aiming to simulate the variations caused in node demands by the consumers' water usage patterns and, therefore, generate more realistic data samples, a certain level of uncertainty was considered when simulating the consumer demands by sampling from a Gaussian distribution defined as follows: $\aleph \sim \{d_n, 0.05d_n\}$ where d_n is the nominal demand of the node.
- Also aiming towards realistic simulation, measurement noise (with mean 0 and standard deviation of $0.025mH_2O$) is added to the pressure values simulated.

4.3   Sensor Configuration

Being this such a large network, monitoring pressure and flow in all of its junctions and pipes would be an economically strenuous task; not to mention unnecessary, since several works [2,18] have achieved satisfactory results by selecting only a subset of locations within the network to position sensors. A set of pressure sensors, being economically more accesible, have been considered to be installed in this paper. The optimal sensor positions were selected aiming to maximize leak detection performance by means of a genetic optimization algorithm [6]. Three pressure sensor configurations were proposed:

Table 1. Pressure sensor configurations

No. Sensors	Configuration
10	85 23 54 79 120 113 187 202 225 232
15	14 23 33 70 93 111 133 156 161 183 184 201 206 251 263
20	3 33 43 62 69 70 76 99 133 149 155 157 161 162 190 201 206 208 213 220

5   Results and Discussion

A comparison was first considered among clustering methods based on topological variables and, afterwards, methods based on hydraulic variables were compared. Finally, an overall comparison between all methods was developed.

In order to test each method, the following methodology was developed:

1. The WDN was partitioned into k clusters, each representing a class.
2. A balanced training set was generated, with 400 leak scenarios per class.
3. An SVM classifier was trained to identify the leak location class. SVM optimal hyperparameters were determined using grid search [4].
4. The classifier performance was evaluated by means of a completely new generated test set with 50 scenarios per node.
5. Bayesian temporal reasoning was applied for all 4 samples in each scenario in order to improve classification performance [17].
6. Steps 3 to 5 were repeated 10 times with different training sets to assess the variability in performance given by the uncertainty.

The effect of each clustering method on the classification performance was tested for all three sensor configurations. A fixed number of classes $k = 25$ was selected for all experiments. A performance measure was defined in order to effect the comparison. Leak zone location performance was defined as: $LZP = 100\frac{CL}{TS}$, where TS is the total number of scenarios in the test set and CL is the number of scenarios for which the leak zone location was properly estimated.

5.1   Topology-Based Clustering Methods

As expected, topology-based clustering algorithms produced connected zones (classes) in every case. Figure 2 presents a performance comparison of the four clustering methods tested.

The obtained results can be interpreted as a two-factor full factorial experiment design with sensor configuration and clustering method as the two factors with three and four levels respectively [11]. Therefore, Friedman's nonparametric statistical test [5] was executed in order to find significant differences among clustering methods. No significant difference was found between DBSCAN and k-medoids. However, Girvan-Newman algorithm and agglomerative clustering present statistical differences. In general, lower performance is attained for the 10 sensor configuration, while the 15 sensor configuration shows a slightly higher performance than its 20 sensor peer. Regarding clustering methods, Girvan-Newman algorithm achieves the worst performance overall and agglomerative clustering renders the highest.

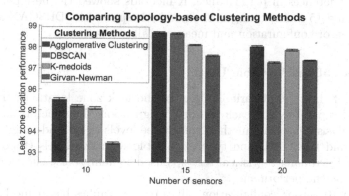

Fig. 2. Comparing Topology-based Clustering Methods

5.2   Hydraulics-Based Clustering Methods

Hydraulics-based clustering algorithms however, in some cases, did not generate connected zones (classes), which mostly depends on the number of classes and the distance measures used. By using hydraulic variables, the clustering methods k-medoids clustering, agglomerative clustering and DBSCAN were tested; and a performance comparison is presented in Fig. 3.

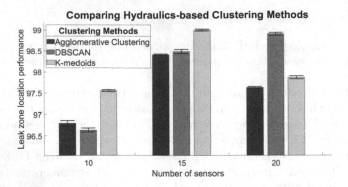

Fig. 3. Comparing Hydraulics-based Clustering Methods

This can also be interpreted as a full factorial experiment in the same way. Friedman's test showed similar results, with no method presenting significant statistical differences in performance. K-medoids showed the best performance for the 10 and 15 sensor configurations to be surpassed by DBSCAN algorithm in the 20 sensor configuration and mean overall performance.

5.3   Effect of the Variable Used

The effect of the type of variable used on the leak zone location performance is analyzed separately for each sensor configuration. Two factors are defined for this analysis: clustering method with three levels (k-medoids, agglomerative clustering and DBSCAN) and type of variable with two levels (topological or hydraulic). Friedman's statistical test was used to analyze the effect of the type of variable on the performance.

For the 10 sensor configuration, all three hydraulics-based methods show significantly better classification performance, with hydraulics-based k-medoids being the best clustering method with a mean performance of 97.56%.

For the 15 sensor configuration, both topology-based and hydraulics-based methods show similar performances, with hydraulics-based k-medoids yielding, again, the best results with a mean performance of 99%, closely followed by topology-based agglomerative clustering and DBSCAN with a mean performance of 98.68% and 98.64% respectively. Finally, for the 20 sensor case, there's a clear superiority shown by the hydraulics-based DBSCAN algorithm with a mean performance of 98.91%.

6   Conclusions

A comparison among 7 zone partitioning procedures for WDNs was developed for three different (10, 15 and 20) sensor configurations, aiming to study the effect of the clustering method used on the zone-based classification performance, as well as the effect of the type of variable used when performing the clustering.

Agglomerative clustering presented the best results among topology-based clustering methods while Girvan-Newman algorithm showed a poor effect on classification performance for all sensor configurations. However, no significant difference was found among hydraulics-based methods. For the 10 sensor configuration, hydraulics-based methods considerably outperform topology-based ones, however, for the 15 sensors and 20 sensors cases, both groups present strategies with similarly commendable results. For future studies, a study of the effect of combining different sensor positioning strategies with these clustering methodologies will be conducted.

References

1. Chartrand, G., Zhang, P.: A First Course in Graph Theory. Courier Corporation (2012)
2. Chen, J., Xin, F., Xiao, S.: An iterative method for leakage zone identification in water distribution networks based on machine learning. Struct. Health Monit. (2020). https://doi.org/10.1177/1475921720950470
3. Ester, M., Kriegel, H.P., Sander, J., Xu, X.: A density-based algorithm for discovering clusters in large spatial databases with noise. In: Proceedings of 2nd International Conference on Knowledge Discovery and Data Mining (1997)
4. Quiñones Grueiro, M., Verde, C., Llanes-Santiago, O.: Novel leak location approach in water distribution networks with zone clustering and classification. In: Carrasco-Ochoa, J.A., Martínez-Trinidad, J.F., Olvera-López, J.A., Salas, J. (eds.) Pattern Recogn., pp. 37–46. Springer International Publishing, Cham (2019)
5. Hollander, M., Wolfe, D.A., Chicken, E.: Nonparametric statistical methods. John Wiley & Sons, 2nd Edition (2013)
6. Jennings, P.C., Lysgaard, S., Hummelshoj, J.S., Vegge, T., Bligaard, T.: Genetic algorithms for computational materials discovery accelerated by machine learning. NPJ Computational Materials 5(46) (2019)
7. Jung, Y., Park, H., Du, D.Z., Drake, B.L.: A decision criterion for the optimal number of clusters in hierarchical clustering. J. Global Optim. **25**(1), 91–111 (2002)
8. Kang, J., Park, Y.J., Lee, J., Wang, S.H.: Novel leakage detection by ensemble CNN-SVM and graph-based localization in water distribution systems. IEEE Trans. Ind. Electron. **65**(5), 4279–4289 (2017)
9. Kaufman, L., Rousseeuw, P.J.: Finding Groups in Data. An Introduction to Cluster Analysis. John Wiley & Sons (2005)
10. Kaufman, L., Rousseeuw, P.J.: Clustering by means of medoids. In: Statistical Data Analysis Based on the L1-Norm Conference, pp. 405–416. Elsevier Science, Neuchatel (1987)
11. Montgomery, D.C.: Design and Analysis of Experiments. John Wiley & Sons, 8th Edition (2013)
12. Newman, M.E.J., Girvan, M.: Finding and evaluating community structure in networks. Phys. Rev. E **69**, 026113 (2004)
13. Rheman, S.U., Khan, K., Aziz, K., Fong, S., Sarasvady, S.: Dbscan: Past, present and future. In: 2014 Fifth International Conference on the Applications of Digital Information and Web Technologies (ICADIWT 2014) (2014)
14. Rossman, L.A.: Epanet 2 Users Manual (2000)

15. Sanz, G., Meseguer, J., Pérez, R.: Model calibration for leak localization, a real application. In: CCWI 2017: 15th Computing and Control for the Water Industry Conference 2017, pp. 1–9. Sheffield (UK) (September 2017)
16. Shekofteh, M., Jalili-Ghazizadeh, M., Yazdi, J.: A methodology for leak detection in water distribution networks using graph theory and artificial neural network. Urban Water J. **17**(6), 525–533 (2020)
17. Soldevila, A., Fernández-canti, R., Blesa, J., Tornil-sin, S., Puig, V.: Leak localization in water distribution networks using bayesian classifiers. J. Process Control **55**, 1–9 (2017)
18. Sun, C., Parellada, B., Puig, V., Cembrano, G.: Leak localization in water distribution networks using pressure and data-driven classifier approach. Water **12**, 54 (2020)
19. Wang, Q., Guidolin, M., Savic, D., Kapelan, Z.: Two-objective design of benchmark problems of a water distribution system via MOEAs?: towards the best-known approximation of the true pareto front. J. Water Resour. Plan. Manage. **141**(3), 1–14 (2015)
20. Zhang, Q., Wu, Z.Y., Zhao, M., Qi, J.: Leakage zone identification in large-scale water distribution systems using multiclass support vector machines. J. Water Resour. Plan. Manage. **142**(11), 04016042 (2016)

Bias Quantification for Protected Features in Pattern Classification Problems

Lisa Koutsoviti Koumeri[✉] and Gonzalo Nápoles

Department of Cognitive Science and Artificial Intelligence, Tilburg University,
Tilburg, The Netherlands
g.r.napoles@uvt.nl

Abstract. The need to measure and mitigate bias in machine learning data sets has gained wide recognition in the field of Artificial Intelligence (AI) during the past decade. The academic and business communities call for new general-purpose measures to quantify bias. In this paper, we propose a new measure that relies on the fuzzy-rough set theory. The intuition of our measure is that protected features should not change the fuzzy-rough set boundary regions significantly. The extent to which this happens can be understood as a proxy for bias quantification. Our measure can be categorized as an individual fairness measure since the fuzzy-rough regions are computed using instance-based information pieces. The main advantage of our measure is that it does not depend on any prediction model but on a distance function. At the same time, our measure offers an intuitive rationale for the bias concept. The results using a proof-of-concept show that our measure can capture the bias issues better than other state-of-the-art measures.

Keywords: Fuzzy-rough sets · Fairness-aware AI · Bias

1 Introduction

Artificial Intelligence (AI) systems are widely used in decision-making for tasks like predictive crime mapping, loan granting or resource allocation in healthcare. However, as historical data often involves implicit biases [10], these systems have been found to correlate their predictions with sensitive characteristics such as race or gender, leading to discriminatory decisions. In the literature, more than 20 definitions of fairness [9] and respective bias metrics have been proposed. However, existing metrics express different and often contradictory notions of fairness [4,5,11]. Deciding which definition is most appropriate for the task at hand is difficult [15] as several parameters need to be considered such as causal influences among features, mis-representation of groups and different modalities of data [9], among other factors.

Speicher et al. [16] distinguish between group and individual fairness metrics. The former ones explore disparities between groups within a sensitive feature to protect them from being discriminated against [16] and work well in settings

© Springer Nature Switzerland AG 2021
J. M. R. S. Tavares et al. (Eds.): CIARP 2021, LNCS 12702, pp. 351–360, 2021.
https://doi.org/10.1007/978-3-030-93420-0_33

where equal proportions across demographic groups are desirable. Some widely used group fairness metrics include statistical parity, disparate impact, equalized odds, and equal opportunity [3]. Individual fairness metrics test whether similar individuals are treated similarly [14]. They are suitable when bias needs to be considered from the perspective of the subject that is targeted in the decision [8]. Moreover, these measures offer strong semantics [1]. The individual fairness metrics explored in this paper (as state-of-the-art measures) are *Consistency* [13] and *Generalized Entropy Index*(GEI) [16].

Group fairness metrics have been criticized for, among others, leading to inverse discrimination [12] and being oblivious to features other than the sensitive feature [6]. Additionally, they require discretization when dealing with numeric sensitive features such as age. Such preprocessing steps can have an impact on accuracy and the outputs of fairness measures [4]. Individual fairness metrics require strong assumptions such as the availability of an agreed-upon similarity metric, or knowledge of the underlying data generating process [7]. Furthermore, both groups of measures are machine learning model-dependent meaning that they depend on the base-rate ratio among groups [2], are not intuitively explainable due to the black-box manner that machine learning models operate and tend to be sensitive to variations in the input arising from variations in training-test splits [4]. Moreover, state-of-art measures act as bias proxies as they do not measure bias directly: *Consistency* measures consistency in classification and *GEI* can be understood as a measure of redundancy in data.

In this paper, we propose a bias quantification measure, called *fuzzy-rough uncertainty*, that acts as an individual fairness metric as it focuses on instance-wise inconsistencies. It quantifies the relevance of a protected feature in classification problems as a proxy for measuring fairness. To that end, we use the advantages of rough sets [20] for analyzing inconsistency in decision systems. This means that our measure does not rely on any prediction model. To cope with the issue of defining similarity thresholds when handling problems with continuous features, we use fuzzy-rough sets as defined by Inuiguchi et al. [22]. This allows us to compute membership values that express the extent to which instances belong to each information granule [19]. The intuition behind our measure is that, in fair decision-making scenarios, the removal of a protected feature should not cause big changes in the decision boundaries. The extent to which that happens can be understood as a bias quantification. In practice, this means that we should quantify the changes to the membership functions attached to granular regions after removing the protected feature.

We tested our measure on the *German Credit* data set [18] which is widely used in the context of AI Fairness and classifies loan applicants in terms of creditworthiness. In our experiment, removing the protected feature *sex* causes more uncertainty (26%) than removing the feature *age* (11%). We contrasted our measure to state-of-the-art metrics as well. The results show that our measure captures a different trend than that of the state-of-the-art metrics showing (1) the risk of focusing on a particular feature-category pair instead of analyzing the feature as a whole and (2) that our FRS-based measure can capture differences that other individual fairness measures cannot.

The next section introduces the mathematical formalism behind the computation of the membership values. Section 3 describes the similarity function we deployed and the proposed bias quantification measure. Section 4 presents the experimental setup and analyzes the measures' outputs. Finally, Sect. 5 discusses possible implications to the field.

2   Fuzzy-Rough Set Theory

The section introduces the theoretical formalism behind the FRS theory [21]. This mathematical theory is used to transform tabular data into information granules describing each decision class. This process is referred to as *universe granulation*. The outputs of this fuzzy granulation process are membership values that we input to the proposed bias quantification measures.

Fuzzy-rough sets combine two important notions in computing [26]: soft computing and granular computing. The first deals with imperfect data or knowledge and the latter semantically transforms data to information granules which can be compared and grouped with each other in terms of similarity [19]. Objects can belong to a concept to varying degrees and can be similar to a certain extent which is expressed by a fuzzy relation. Based on their similarity to one another, objects are categorized into classes, or granules, with soft boundaries, in this case, a positive, negative and boundary fuzzy-rough region.

In this paper, we will use the FRS formalization proposed by Inuiguchi et al. [22]. Let us assume that we have a universe of discourse U, a fuzzy set $X \in U$ and a fuzzy binary relation $R \in Q(U \times U)$ where $\mu_X(x)$ and $\mu_R(y, x)$ are their respective membership functions. The membership function $\mu_R : U \to [0, 1]$ determines the degree to which $x \in U$ is a member of X, whereas $\mu_R : U \times U \to [0, 1]$ denotes the degree to which y is presumed to be a member of X from the fact that x is a member of the fuzzy set X. Whenever opportune, $R(x)$ is denoted with its membership function $\mu_{R(x)}(y) = \mu_R(y, x)$.

Firstly, let us build a partition of U according to the decision classes. The X_k set contains all objects associated with the k-th decision class. In previous research works, the membership degree of $x \in U$ to a subset X_k was computed using the following hard membership function,

$$\mu_{X_k}(x) = \begin{cases} 1, & x \in X_k \\ 0, & x \notin X_k \end{cases} \tag{1}$$

as we do not argue about the labeling accuracy. In other words, we will assume that all problem instances are correctly labeled.

Then, we need to determine the similarity between instances x and y. Such a function, denoted by $\mu_R(y, x)$, can be constructed by combining the previously described membership degree $\mu_{X_k}(x)$ with the similarity degree $\phi(x, y)$ between two objects $x, y \in U$. Equation (2) shows how to compute this fuzzy binary relation $\mu_R(y, x)$ for two given instances,

$$\mu_R(y, x) = \mu_{X_k}(x)\phi(x, y). \tag{2}$$

In the next section, we will give more details about the similarity function, which is expressed in terms of a distance function.

Aiming at defining the lower approximations, we use the degree of x being a member of X_k under the knowledge R. This can be measured by the truth value of the statement '$y \in R(x)$ implies $y \in X_k$' under fuzzy sets $R(X)$ and X_k. We use a necessity measure $inf_{y \in U} \mathcal{I}(\mu_R(y, x), \mu_{X_k}(y))$ with a fuzzy implication function $\mathcal{I} : [0, 1] \times [0, 1] \rightarrow [0, 1]$ such that $\mathcal{I}(0, 0) = \mathcal{I}(0, 1) = \mathcal{I}(1, 1) = 0$ and $\mathcal{I}(1, 0) = 1$. It also holds that $\mathcal{I}(., a)$ decreases and $\mathcal{I}(a, .)$ increases, $\forall a \in [0, 1]$. Recall that X_k is the set of objects labeled with the k-th decision class. Equation (3) displays the membership function for the lower approximation $R_*(X_k)$ associated with the k-th decision class,

$$\mu_{R_*(X_k)}(x) = min\{\mu_{X_k}(x), inf_{y \in U} \mathcal{I}(\mu_R(y, x), \mu_{X_k}(y))\} \tag{3}$$

where $\mathcal{I}(a, b) = min(1 - a + b, 1)$ is the Łukasiewicz implication operator. Of course, other fuzzy operators can be adopted as well.

To derive the upper approximations, we measure the truth value of the statement '$\exists y \in U$ such that $x \in R(y)$' under fuzzy sets $R(x)$ and X_k. The true value of this statement can be obtained by a possibility measure $sup_{y \in U} \mathcal{T}(\mu_R(x, y), \mu_{X_k}(y))$ with a conjunction function $\mathcal{T} : [0, 1] \times [0, 1] \rightarrow [0, 1]$ such that $\mathcal{T}(0, 0) = \mathcal{T}(0, 1) = \mathcal{T}(1, 1) = 1$ and $\mathcal{T}(1, 0) = 0$, where both $\mathcal{T}(., a)$ and $\mathcal{T}(a, .)$ increase, $\forall a \in [0, 1]$. Equation (4) displays the membership function for the upper approximation $R^*(X_k)$ associated with the k-th decision class,

$$\mu_{R^*(X_k)}(x) = max\{\mu_{X_k}(x), sup_{y \in U} \mathcal{T}(\mu_R(x, y), \mu_{X_k}(y))\} \tag{4}$$

where $\mathcal{T}(a, b) = max(a + b - 1, 0)$ is the Łukasiewicz conjunction operator.

This model does not assume that $\mu_R(x, x) = 1, \forall x \in U$. Instead, we compute the minimum between $\mu_{X_k}(x)$ and $inf_{y \in U} \mathcal{I}(\mu_R(y, x), \mu_{X_k}(y))$ when calculating $\mu_{R_*(X_k)}(x)$, and the maximum between $\mu_{X_k}(x)$ and $sup_{y \in U} \mathcal{T}(\mu_R(x, y), \mu_{X_k}(y))$ when calculating $\mu_{R^*(X_k)}(x)$. This allows preserving the inclusiveness of $R_*(X_k)$ in the fuzzy set X_k and the inclusiveness of X_k in $R^*(X_k)$.

Finally, we define fuzzy-rough regions using the upper and lower approximations. The membership functions for the fuzzy-rough positive, negative and boundary regions can be defined as $\mu_{POS(X_k)}(x) = \mu_{R_*(X_k)}(x)$, $\mu_{NEG(X_k)}(x) = 1 - \mu_{R^*(X_k)}(x)$ and $\mu_{BND(X_k)}(x) = \mu_{R^*(X_k)}(x) - \mu_{R_*(X_k)}(x)$, respectively. Membership values to positive regions indicate the extent to which the instances belong to a decision class, membership values to negative regions indicate the extent to which the instances do not belong to a decision class, and finally, membership values to boundary regions indicate the extent to which the instances are uncertain to the decision problem at hand.

3   Similarity Function and Bias Quantification Measure

This section introduces the proposed granular measure to quantify bias in tabular data sets used for pattern classification. This measure starts from the assumption that the experts determine the set of protected features (i.e., those likely related to bias). The intuition of our measure is as follows. The removal of a protected feature should not impact the decision boundary regions heavily. For example, let us assume that we have a problem described by several features where *gender* is deemed a protected feature. If we remove that feature and the modified decision boundary causes a significant increase in the misclassifications, then one can conclude that such a feature is important to separate the decision classes. The extent to which the decision boundaries become less separate can be understood as a bias indicator. Of course, we are assuming that a protected feature is not expected to be a pivotal feature in the classification.

Before presenting our measure, we have to describe the similarity function used to compare the instances. Vluymans et al. [27] state that the construction of appropriate similarity measures is dependent on the application. In this paper, the similarity measure will be derived from a distance function. In particular, we will use the Heterogeneous Manhattan-Overlap Metric (HMOM) [24] because of its ability to deal with instances having mixed-type features. The HMOM distance function can be formalized as follows:

$$d(x,y) = \sum_{j=1}^{|F|} \rho_j(x,y) \tag{5}$$

such that

$$\rho_j(x,y) = \begin{cases} 0 & if \ f_j \in F \ is \ nominal \wedge x(j) = y(j) \\ 1 & if \ f_j \in F \ is \ nominal \wedge x(j) \neq y(j) \\ |x(j) - y(j)| & if \ f_j \in F \ is \ numerical \end{cases}$$

where F is the set of features, while $x(j)$ and $y(j)$ denote the values of the j-th feature according to instances x and y, respectively.

The distance function above is the cornerstone of the similarity function. Equation (6) portrays the similarity function, which produces values in the $(0,1)$ interval, derived from the previous distance function,

$$\phi(x,y) = e^{-\lambda\left(d(x,y)\right)} \tag{6}$$

where $\lambda > 0$ is a user-specified smoothing parameter that prevents passing very large values to the exponential function. We can tune the λ parameter such that we can have more separable granular regions, which will likely translate into lower uncertainty. Such a tuning might be necessary to prevent the similarity function to report only values close to zero or one. Overall, membership values are sensitive to the distance metric within the similarity function, so selecting the proper configuration depends on the problem at hand.

The proposed measure, termed *fuzzy-rough uncertainty*, aims at quantifying change between the fuzzy rough boundary regions and is the main contribution of this work. Measuring the distance or the change between the regions of fuzzy-rough sets has been examined in the literature [28,29], but not in the context of bias quantification, as far as we know. The measure quantifies how much the absence of the f_i feature modifies the fuzzy-rough boundary regions of the same decision class. If the difference is positive, we can conclude that the boundary regions became bigger after removing the protected feature, so there is more uncertainty (the feature is important for the classification). If the difference is negative, we can conclude that the boundary regions became smaller after removing the protected feature, so there is less uncertainty (the feature was causing uncertainty and its removal was convenient).

To quantify these differences, we use the membership values of instances in U to the boundary regions. Such granules are computed using (i) the full set of features, and (ii) a set of features where the protected feature is removed (denoted by $\neg f_i$). Recall that we are only interested in positive differences (instances for which the uncertainty increased when compared with the membership values to the boundary regions). Equation (7) shows how to compute the *fuzzy-rough uncertainty* value for the k-th decision class after removing the i-th feature from the feature set describing the problem at hand,

$$\Omega_k(f_i) = \frac{\sqrt{\Sigma_{x \in U}(\Delta^+_{B_k \neg f_i}(x))^2}}{\sqrt{\Sigma_{x \in U}(\mu_{B_k}(x))^2}} \tag{7}$$

such that $\Delta^+_{B_k \neg f_i}(x) = \mu_{B_k}(x) - \mu_{B_k \neg f_i}(x)$ when the removal of the i-th feature increased the uncertainty. Otherwise, we will assume that $\Delta^+_{B_k \neg f_i}(x) = 0$, which means that the uncertainty attached to that instance did not increase after suppressing the protected feature. Moreover, to lighten the notation, we denote the k-th region $\mu_{BND(X_k)}(x)$ with $\mu_{B_k}(x)$. The reader can notice that this granular measure is similar to computing the relevance of the protected feature for preserving the decision boundaries attached to the problem.

4   Experiments, Results and Discussion

In our experiment, we attempt to quantify the extent to which suppressing the protected features *age* and *gender* [17] of the *German Credit* data set modifies the membership values to the fuzzy-rough boundary regions of each decision class. The target feature is binary denoting credit-worthy and non-creditworthy applicants. During preprocessing, the nominal feature *sex&marital status* was re-coded to include only gender-related information. Data preprocessing included (1) normalization of numeric features such that their minimum and maximum values are 0.0 and 1.0 respectively, (2) encoding target classes as integers starting at zero, and (3) encoding nominal variables as integers ranging between zero and the number of categories in the feature.

Next, membership values to the positive, negative and boundary fuzzy-rough regions per decision class are calculated following the approach in Sect. 2. In these simulations, we used $\lambda = 0.5$ arbitrarily. First, membership values are calculated using all features and their trend is visualized in Fig. (1). The graphs show that the fuzzy-rough regions are relatively distinct from one another with only a few instances belonging to the boundary values by more than 50%. In other words, instances are classified with relative certainty. Then, membership values are computed again using two different feature combinations where each of the protected features is removed. Figure 2 shows the change in the fuzzy-rough boundary regions before and after protected features are suppressed.

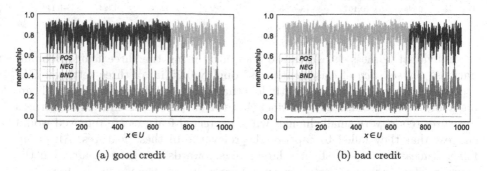

(a) good credit   (b) bad credit

Fig. 1. Membership values computed using the complete feature set. The x axis represents the instances and y axis the number the membership values. (Color figure online)

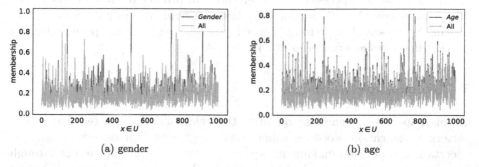

(a) gender   (b) age

Fig. 2. Membership values to the boundary regions. The x axis corresponds to the instances and y axis represents the range of the membership values. (Color figure online)

The state-of-the-art measures are computed using aif360.sklearn package [17] using our preprocessed data set. To compute the literature metrics, a logistic regression is used with a prediction accuracy of 78%. Prior to calculating the group fairness metrics, *age* was discretized into people younger and older than 25 years old [17]. The literature measures' outputs along with the results of our proposed measure are shown in Table 1.

Table 1. Results of proposed and state-of-the-art measures. The ideal value of *Consistency* and *Disparate Impact* is 1 and of the rest of the measures is 0.

Individual fairness metrics				Group fairness metrics				
Feature set	Consist	GEI	**FR Uncertainty**	Protected group	Statistical parity	Disparate Impact	Equal Opportunity	Average odds
All	0.746	0.094	–	–	–	–	–	–
Excl. gender	0.743	0.094	**0.226**	gender/female	−0.135	0.834	−0.056	−0.132
Excl. age	0.746	0.093	**0.107**	age/young	−0.202	0.752	−0.124	−0.149

Individual fairness measures are computed using the three feature combinations. *Consistency* of classification is roughly 0.75 in all three settings. *GEI* reports relatively low values (slightly less than 0.1) in all three situations indicating a low level of redundancy. The fact that the outputs of the individual fairness measures report very small changes when protected features are removed would suggest that they failed to capture the relevance of these features. All group fairness measures report slightly larger bias towards *age* than *gender*. On the contrary, *fuzzy-rough uncertainty* quantifies a greater change in the boundary regions when protected feature *gender* is excluded compared to *age*. In other words, when *gender* is removed from the data set greater uncertainty is reported in classification which serves as an indication of greater bias towards *gender*. Recall that bias is expressed as the relevance of protected features when classifying instances. Overall, our measure behaves fundamentally differently from existing metrics and, therefore, is worth introducing to avoid proliferation of existing measures as Friedler et al. [4] suggest.

5   Concluding Remarks

In this paper, we attempt to quantify the relevance of protected features in pattern classification problems using fuzzy-rough sets. Our measure quantifies uncertainty in decision making as expressed by the changes in the fuzzy-rough boundary regions before and after the removal of a protected feature. In our experiment, although the state-of-the-art individual fairness measures are not completely sensitive to the exclusion of protective features, our measure shows that ignoring *gender* causes more uncertainty in the decision-making process than ignoring *age*. Moreover, even if group fairness measures show that *gender-female* seems to be less affected than *age-young*, our measure captures the exact opposite trend. Overall, our study shows that (i) the fuzzy information granulation approach can express the relevance of protected features as a whole in decision-making scenarios and that (ii) focusing on a particular feature-category pair instead of analyzing the protected feature as a whole might give rise to misleading results.

There are several advantages to our approach. First, no user intervention is needed to detect the discriminated subgroup. Our measure takes into account, first, all feature categories as a whole, and, second, all features and categories at once. Second, no discretization is needed to handle protected numeric features, such as *age*. Third, our measure does not depend on a machine learning model to compute its outcomes but on a solid mathematical foundation. Limitations of our measure are that it does not account for (1) non-sensitive features that might contain implicit bias due to social factors [15] and (2) that its performance depends on the distance function and fuzzy operators.

References

1. Kearns, M., Neel, S., Roth, A., Wu, Z.: Preventing fairness gerrymandering: auditing and learning for subgroup fairness. In: ICML (2018)
2. Kleinberg, J., Mullainathan, S., Raghavan, M.: Inherent trade-offs in the fair determination of risk scores. In: 8th Innovations in Theoretical Computer Science Conference, pp. 43:1–43:23 (2017)
3. Hardt, M., Price, E., Srebro, N.: Equality of opportunity in supervised learning. In: NIPS (2016)
4. Friedler, S.A., Scheidegger, C., Venkatasubramanian, S., Choudhary, S., Hamilton, E., Roth, D.: A comparative study of fairness-enhancing interventions in machine learning. In: Proceedings of the Conference on Fairness, Accountability, and Transparency (2019)
5. Corbett-Davies, S., Pierson, E., Feller, A., Goel, S., Huq, A.: Algorithmic decision making and the cost of fairness. In: Proceedings of the 23rd ACM SIGKDD International Conference on Knowledge Discovery and Data Mining (2017)
6. Choi, Y., Farnadi, G., Babaki, B., Van den Broeck, G.: Learning fair Naive Bayes classifiers by discovering and eliminating discrimination patterns. In: Proceedings of the AAAI Conference on Artificial Intelligence, 34(06) (2020)
7. Kehrenberg, T., Chen, Z., Quadrianto, N.: Tuning fairness by balancing target labels. Front. Artif. Intell. **3**, 33 (2020)
8. Varona, D., Lizama-Mue, Y., Suárez, J.L.: Machine learning's limitations in avoiding automation of bias. Artif. Intell. Soc. **36**(1), 197–203 (2020). https://doi.org/10.1007/s00146-020-00996-y
9. Ntoutsi, E., et al.: Bias in data-driven artificial intelligence systems-An introductory survey WIREs. Data Min. Knowl. Disc. **10**(3), e1356 (2020)
10. Fuchs, D.: The Dangers of Human-Like Bias in Machine-Learning Algorithms, Missouri S&T's Peer to Peer2, (1) (2018)
11. Verma, S., Rubin, J.: Fairness definitions explained. In: Proceedings of the International Workshop on Software Fairness , pp. 1–7 (2019)
12. Chouldechova, A.: Fair prediction with disparate impact: a study of bias in recidivism prediction instruments. Big Data **5**(2), 153–163 (2017)
13. Zemel, R., Yu Wu, Y., Swersky, K., Pitassi, T., Dwork, C.: Learning fair representations. In: Proceedings of the 30th International Conference on Machine Learning, PMLR 28(3), 325–333 (2013)
14. Dwork, C., Hardt, M., Pitassi, T., Reingold, O., Zemel, R.: Fairness through awareness. In: Proceedings of the 3rd Innovations in Theoretical Computer Science Conference (ITCS), pp. 214–226 (2012)
15. Kusner, M.J., Russell, C., Loftus, J., Silva, R.: Counterfactual fairness. In: Proceedings of the 31st International Conference on Neural Information Processing Systems, pp. 4069–4079 (2017)
16. Speicher, T., et al.: A unified approach to quantifying algorithmic unfairness: measuring individual & group unfairness via inequality indices. In: CoRR (2018)

17. Bellamy, R., et al.: AI fairness 360: an extensible toolkit for detecting, understanding, and mitigating unwanted algorithmic bias (2018)
18. Dua, D., Graff, C.: UCI machine learning repository. University of California, School of Information and Computer Science, Irvine, CA (2019)
19. Pedrycz, W., Vukovich, G.: Feature analysis through information granulation and fuzzy sets. Pattern Recogn. **35**, 825–834 (2002)
20. Pawlak, Z.: Rough sets. Int. J. Comput. Inf. Sci. **11**, 341–356 (1982)
21. Dubois, D., Prade, H.: Rough fuzzy sets and fuzzy rough sets. Int. J. Gen. Syst. **17**, 191–209 (1990)
22. Inuiguchi, M., Wu, W., Cornelis, C., Verbiest, N.: Fuzzy-rough hybridization. Handbook of Computational Intelligence, pp. 425-451. Springer, Berlin (2015)
23. Jensen, R., Cornelis, C.: Fuzzy-rough nearest neighbour classification and prediction. Theoret. Comput. Sci. Rough Sets Fuzzy Sets Nat. Comput. **412**(42), 5871–5884 (2011)
24. Wilson, D.R., Martinez, T.R.: Improved heterogeneous distance functions. J. Artif. Intell. Res. (JAIR) **6**, 1–34 (1997)
25. Nápoles, G., Mosquera, C., Falcon, R., Grau, I., Bello, R., Vanhoof, K.: Fuzzy-rough cognitive networks. Neural Netw.: Official J. Int. Neural Netw. Soc. **97**, 19–27 (2017)
26. Cornelis, C., De Cock, M., Radzikowska, A.: Fuzzy rough sets: from theory into practice. Handbook of Granular Computing. Wiley, pp. 533–553 (2008)
27. Vluymans, S., D'eer, L., Saeys, Y., Cornelis, C.: Applications of fuzzy rough set theory in machine learning: a survey. Fundam. Inf. **142**(1–4), 53–86 (2015)
28. Yang, J., Xu, T., Zhao, F.: Modified uncertainty measure of rough fuzzy sets from the perspective of fuzzy distance. Math. Problems Eng. 1–11 (2018)
29. Bello, M., Nápoles, G., Morera, R., Vanhoof, K., Bello, R.: Outliers detection in multi-label datasets. Advances in Soft Computing, pp. 65–75. Springer Nature Switzerland AG (2020)

Regional Commodities Price Volatility Assessment Using Self-driven Recurrent Networks

Pablo Negri[1,3]([✉]), Priscila Ramos[2,4], and Martin Breitkopf[4]

[1] Departamento de Computacion, UBA-FCEyN, Buenos Aires, Argentina
pnegri@dc.uba.ar
[2] Departamento de Economia, UBA-FCE, Buenos Aires, Argentina
[3] Instituto de Investigacion en Ciencias de la Computacion (ICC), CONICET-UBA, Buenos Aires, Argentina
[4] Instituto Interdisciplinario de Economia Politica (IIEP), CONICET-UBA, Buenos Aires, Argentina

Abstract. The high volatility of the agricultural and energy commodity prices in the international market is a concern due to their transmission to regional prices, increasing instability in domestic markets. This paper evaluates the performance of recurrent networks (RNN and LSTM) to predict regional prices reactions under international shock simulations. Experiments are run to soybean and corn regional prices in Argentine by considering exogenous changes of the international oil price - both agricultural commodities are inputs for biofuels' production - and also of their international prices. Results are in line with the econometric literature and consistent with the dynamic of regional prices in Argentina's markets. Thus, the RNNs could be a useful tool for timely economic policy decisions that cushion external price shocks in domestic markets.

Keywords: Recurrent Neural Networks · Regional commodities prices · Shock simulations

1 Introduction

The definition of new trade policy instruments for monitoring and stabilizing agricultural commodities prices at borders must meet specific domestic socio-economic objectives. Thus, it is essential to understand how changes in international and internal prices propagate geographically within a country. Without an accurate measurement of these effects, any quantitative analysis would be flawed, and the calibration of contingency measures distorted. For example, assuming perfect price transmission would be a risky simplification and would lead to an overestimation of the corrective power of trade policy instruments (e.g. export duties or subsidies).

The literature on price volatility focuses mainly on the cases of large exporters (e.g. United States) and more recently on the case of countries with a high food

This work was funded by the University of Buenos Aires PDE Nro. 24-2019.

© Springer Nature Switzerland AG 2021

J. M. R. S. Tavares et al. (Eds.): CIARP 2021, LNCS 12702, pp. 361–370, 2021.
https://doi.org/10.1007/978-3-030-93420-0_34

dependence on agricultural imports (e.g. Sub-Saharan African countries). The related economic and econometric literature evidences the inter-dependencies between the different agricultural products [8,12–14], and between agricultural and energy markets [8,15], explaining the dynamics of price volatility between markets. Most of these works use GARCH or MGARCH models [8,15] to assess agricultural price volatility as a function of its history. The first one captures the effects on short-term but also long-term price volatility between markets, and the second analyzes the interdependence between them (e.g. spillover effects).

While econometric methods of Vector Auto-regressive (VAR) models remain as the benchmark for price forecast, many research works are pointing to neural networks as a more precise method. Wang et al. [17] use a Back Propagation Neural Network (BPNN) to predict prices of agricultural commodities such as wheat, soy, or corn, and conclude that their predictions are more accurate than an econometric method used for comparison. Fang et al. [7] arrive at similar conclusions using a traditional Neural Network (NN).

Most of the existing research uses NN static models to predict future prices; however, they only use the state of the network in one period to predict values for the next, losing all memory of the network for the next step [11]. For time series, where each value is related to previous and next values, using static models does not properly capture the dynamics. This is particularly true for series with sudden movements or "shocks", where predictions for static models tend to detach rapidly from real values. Conversely, a dynamic model could accurately learn from shocks and consider their information for prediction.

Recurrent Neural Networks (RNN) are a potential accurate prediction model for agricultural prices. RNNs are Neural Networks that link actual variables on their prior states, giving them a "dynamic memory" [6]. This is extremely useful to predict within a time series, where each element fed to the model is related to the previous and next values. Wang [20] uses an Echo State RNN to predict stock prices from the S&P 500, while Boyko et al. [4] use Long-Short Term Memory (LSTM), to predict upon the same database. Both papers arrive at satisfying conclusions. Moreover, Wang and Wang [18] use an Elman RNN, similar to the one used in our experiments, with a successful prediction to estimate future oil price. It is worth noting that data harmonization before applying any Machine or Deep Learning method can improve these RNN performance [7,17,19]. Furthermore, this RNN literature makes predictions based only on one single input (i.e., time lags of the same price). Nevertheless, a dynamic network could learn and forecast based also on other elements (e.g., international oil price) strongly related to the variable target.

This work implements RNN and LSTM architectures to simulate the dynamics of a closed system of prices (i.e., international prices of oil, soybean, and corn and Argentina's regional -Bahia Blanca, Rosario and Quequen - prices the same agricultural products). We focus on the training and evaluation of these models to estimate inter-dependencies between the inputs, and predict the dynamics of the regional prices. In our experiments, each international commodity is stressed under a strong shock (i.e. international price of oil), and the evolution of the regional prices on each recurrent model is evaluated as a self-driven dynamic. Recurrent models' results show good performance compared to econometric analysis, validating the use of the RNN and LSTM as a realistic engine for this application.

The paper is organized as follows. The next section states the problem, depicts the recurrent models, and details the training procedure. Experiments and analysis are detailed in Sect. 3. Section 4 concludes the paper and propose future works.

2   Commodities Prices Prediction Models

2.1   Problem Formulation

The prediction models will work with temporal sequences corresponding to commodities prices. We define three kinds of series:

- $e^{(t)}$ an exogenous price sequence dependent to $i^{(t)}$.
- $i^{(t)}$ a price sequence that it is related with $e^{(t)}$.
- $r^{(t)}$ a price sequence dependent to $i^{(t)}$ and $e^{(t)}$.

were (t) indicates the value of the price at time t. In our experiments, $e^{(t)}$ is the international price of oil. The sequences $i^{(t)}$ are international prices of agricultural commodities associated with bio-diesel (soybean) and bio-ethanol (corn). Because these bio-fuels (partially) replace gasoline, we can state that $e^{(t)}$ and $i^{(t)}$ are interdependent variables. Finally, $r^{(t)}$ corresponds to agricultural commodities prices in different regions of Argentina. The dynamic of these prices involves local factors, and (what we expect to prove) external ones such as the $i^{(t)}$ sequences.

The model, which simulates the behavior of the closed price system, could capture variables' inter-dependencies from the data at the learning process. This dynamic can be evaluated using *shocks*. A shock is an abrupt change in the price of one of the products in the system that could affect other products' prices. For instance, we are interested in evaluating prices' inter-dependence when applying an oil price shock. This kind of behavior happens in real life, due to political changes, wars, pandemics, and more lastly, environmental concerns.

We choose the RNN model to learn the dynamics of the closed system and predict the stationary values after the shock. Static models could not produce this kind of results as it is needed a system that receives as inputs their precedent outputs. The next section introduces the RNN models.

2.2   Recurrent Neural Network Architecture

Temporal series denoted as $(\mathbf{x}^{(1)}, \mathbf{x}^{(2)}, ..., \mathbf{x}^{(T)})$ are usually the inputs of RNN models. In our case, $\mathbf{x}^{(t)}$ is a vector containing the commodity prices at week t including prices data from the three series $(e^{(t)}, i^{(t)}, r^{(t)})$. Equivalently, the target sequences corresponding to the expected commodity prices is stated as $(\mathbf{y}^{(1)}, \mathbf{y}^{(2)}, ..., \mathbf{y}^{(T)})$. The predictions produced by the recurrent model are denoted as $\hat{\mathbf{y}}^{(t)}$ (Fig. 1).

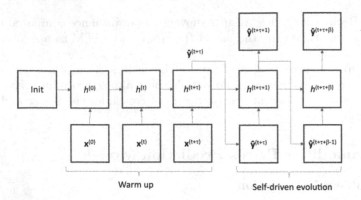

Fig. 1. System architecture and evolution.

The forward pass of a simple recurrent network model [11] introduces $h^{(t)}$, the hidden state of the network at time t and is defined by two equations:

$$\mathbf{h}^{(t)} = \sigma(W^{hx}\mathbf{x}^{(t)} + W^{hh}h^{(t-1)} + b_h) \tag{1}$$

$$\hat{\mathbf{y}}^{(t)} = \sigma(W^{yh}h^{(t)} + b_y) \tag{2}$$

Equation 1 obtains $h^{(t)}$ as the combination of the input $\mathbf{x}^{(t)}$ at time t and $h^{(t-1)}$, which corresponds to the hidden previous state. These recurrent connections are what give the model memory [6]. We express the estimation of target $\mathbf{y}$ of Eqs. 1 and 2 at time t as a dependent function R with internal parameters $\{W^{hx}, W^{hh}, W^{yh}, b_h\}$:

$$\hat{\mathbf{y}}^{(t)} = R(\mathbf{x}^{(t)}|h = h^{(t-1)}) \tag{3}$$

Modern RNN architectures introduce several improvements overcoming traditional training problems. Long-Short Term Memory model [9] (LSTM) is one of the most successful networks widely employed on several applications, such as natural language processing. LSTM deals with long-term dependencies incorporating gates to the recurrent cell.

This work implements recurrent neural networks with both RNN-Elman and LSTM cells with a forget gate. Also, we'll deploy a stacked RNN and LSTM network [21]. In practice, an easy way to increase the depth of the recurrent network is to stack the cells into L layers. This architecture has proved to improve efficiency and performance in problems like vehicle-to-vehicle communication [5] and French-English translation [16].

2.3 Training Procedure

The training follows a mini-sequences batch procedure. We split the training sequence into mini-sequences of τ length $(\mathbf{x}^{(t)}, ..., \mathbf{x}^{(t+\tau-1)})$, referred as $\mathbf{X}^{(t,\tau)}$. The target $\mathbf{Y}^{(t,\beta)}$ is also a sequence that consists of the price values of interest from τ to β: $(\mathbf{x}^{(t+\tau)}, ..., \mathbf{x}^{(t+\tau+\beta-1)})$. They are the "future" prices that the model should predict in a self-driven way. More precisely, the inputs always correspond to all the agricultural prices $\mathbf{x}^{(t)} = (\mathbf{e}^{(t)}, \mathbf{i}^{(t)}, \mathbf{r}^{(t)})$, while outputs are subject to which variable receives the exogenous shock. For example, if the shock is applied on the international oil price, the output becomes $\mathbf{y}^{(t)} = (\mathbf{i}^{(t)}, \mathbf{r}^{(t)})$. If another variable is selected to be shocked, it should be excluded from the target. The τ inputs feed the RNN model, updating the internal hidden states. This step can be thought as a *warm-up* of the internal variables from a (always the same) initial value. Then, for the next β time steps, Eq. 3 is modified by:

$$\hat{\mathbf{y}}^{(t)} = R(\hat{\mathbf{y}}^{(t-1)}|h = h^{(t-1)}) \tag{4}$$

this outputs are then reserved as sequence target $\hat{\mathbf{Y}}^{(t,\beta)} = (\hat{\mathbf{y}}^{(t+\tau)}, ..., \hat{\mathbf{y}}^{(t+\tau+\beta-1)})$. The loss function is defined as a mean squared error on the output sequence $\hat{\mathbf{Y}}$:

$$\mathcal{L} = \sum_{\beta} \frac{1}{\beta} ||\hat{\mathbf{Y}} - \mathbf{Y}|| \tag{5}$$

3   Experiments

3.1   Data

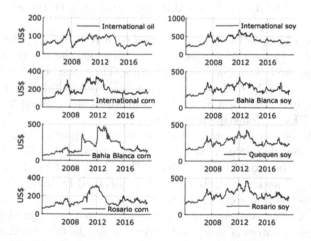

Fig. 2. International and regional commodities prices data series from 2005 to 2019.

We have built a database of weekly prices in US dollars between January 2005 and August 2019, leading to a sample of 772 observations for each price (Fig. 2).

Prices considered in the database are: Soybean and corn prices per ton in three regional markets in Argentina (Bahia Blanca - BB, Rosario - Ros, Quequén, QQ, the latter only for soybean) from GRANAR [2]; Soybean and corn international prices per ton from FAOSTAT [1]; Oil international price per barrel from the Western Texas Intermediate, WTI.

Before testing the RNN models, we have analysed the data in order to evaluate the presence of a stable long-term relationship between regional, international prices of each agricultural commodity and the international price of oil. We follow the Johansen's approach [10] for an appropriate cointegration analysis, so we evaluate multivariate stationarity of price variables in each system (i.e., each system is composed of regional and international prices of one of the agricultural commodities and the international oil price) [3,10]. Soybean markets (regional and international) and international oil market are integrated, being the Vector Error Correction (VEC) model most appropriate for the regional soybean price estimation. Cointegration between corn prices (regional and international) and oil is not verified, being regional corn prices estimated through a VAR model. Impulse-Response functions have been run in both regional price systems by shocking (own commodity and oil) international prices to know the convergence path for regional prices. These econometric estimations provide a reference for regional price behaviors under the recurrent network architectures.

3.2 Hyperparameters Selection

Four recurrent architectures are implemented: RNN-1c, RNN-2c, LSTM-1c and LSTM-2c. Two of them consist of a single RNN and an LSTM cell. The hidden states h for RNN and (h, c) for LSTM, have H hidden units. The other architectures stack a second recurrent cell to the network with the same number of hidden units H.

We run a K-fold cross validation training, with $K = 5$, using the following set of values for $H = [4, 8, 12, 16, 20, 24, 28, 32]$. Moreover, the training is controlled by τ (warm-up) and β (self-driven) variables. Thus, the set of values for each variable are $\tau = [6, 7, 8, 9, 10]$ and $\beta = [1, 2, 3, 4]$. Note that $\beta = 1$ corresponds to a classical single prediction of the $t + 1$ output value, while $\beta > 1$ applies the loss function of Eq. 5 to a sequence of targets.

Each K-fold is evaluated by two means squared error indices on the target prices of the validation split: a $MSE^{(t+1)}$ prediction, and a $MSE^{(t+N)}$ prediction. Let be $\mathbf{x}^{(t)}$ the model input, $MSE^{(t+1)}$ is computed by the mean squared error between $\hat{y}^{(t)}$ and $\mathbf{x}^{(t+1)}$. $MSE^{(t+N)}$ is obtained by using Eq. 4 for a self-driven estimation for N steps. Then, the error is computed between prediction $\hat{y}^{(t+N-1)}$ and $\mathbf{x}^{(t+N)}$, and measures how well the recurrent model adjusts the self-driven dynamic after N steps to the real values. In this work, we fix $N = 4$ which means a month of self-driven evolution. We employ an SGD optimizer with an initial learning rate of $1e-2$. After 20 epochs, the learning rate is reduced by half. Table 1 shows the best results of each architecture sorted by the $MSE^{(t+N)}$ index. As can be seen, recurrent cells with a high number of hidden units H get the lowest errors. In the case of τ, warm-up phase seems more important for RNN cells. LSTM cells incorporate additional gates, then, this is a normal conclusion. This is expected for models like LSTM having several gates to remember/forget input data. Increasing τ also increases the temporal drift of the system itself. In the case of β parameter, the best results for RNN are obtained using values greater than one. On the other hand, LSTM prefers lower values of β.

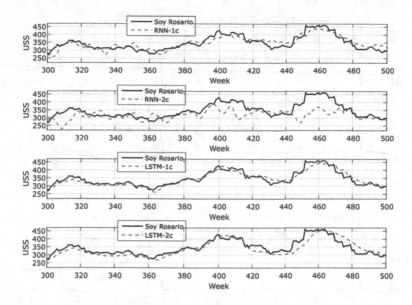

Fig. 3. System architecture and evolution.

Figure 3 samples the $t + 1$ predictions of the four models on a portion of the soybean times series prices from Rosario port. We can appreciate different behaviors for each model. RNN-1c model predicts the series values with a low error but a rapid dynamic. RNN-2c, on the other hand, seems to have a sinusoidal dynamic near the series values, but sometimes the error is high, which is consistent with the high value of their $MSE^{(t+1)}$ index on Table 1. LSTM-1c and LSTM-2c predict accurately the average of the series values but have a very low dynamic. This soothing effect is more remarkable on the LSTM-2c predictions.

Table 1. Hyperparameters with the best results of the K-Fold Cross Validation.

Architecture	H	τ	β	$MSE^{(t+1)}$	$MSE^{(t+N)}$
RNN-1c	32	10	2	0.159 ± 0.093	0.195 ± 0.156
RNN-2c	32	10	4	0.426 ± 0.182	0.207 ± 0.156
LSTM-1c	32	8	2	0.124 ± 0.083	0.247 ± 0.135
LSTM-2c	32	6	1	0.196 ± 0.182	0.288 ± 0.214

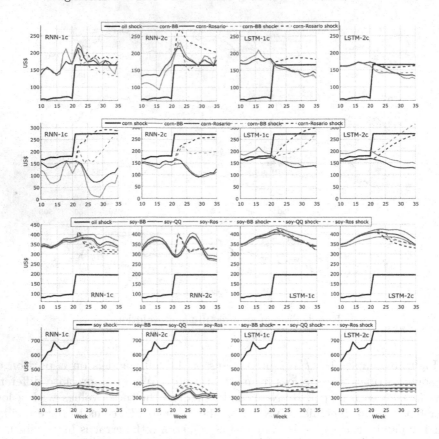

Fig. 4. Shock prediction results. (Color figure online)

3.3   International Shock Simulation

The experiments seek to validate the self-driven evolution of the recurrent networks when a permanent exogenous change (an increase of 100 US$) is introduced in each of international prices (own commodity and oil).

The tests are conducted as follows. For example, to test soybean exogenous change shock, we train the four models with all the commodities prices as inputs and a target that does not predict international soybean. Thus, we split the data sequences into temporal frames of $T = 35$ weeks. The first $\tau = 20$ weeks are employed as *warm-up*, and at $t = 20$, the value of the international soybean price is increased by 100 US$, keeping this value until the end of the test. At this point, the system uses both the new value of the international soybean price and the self-prediction of other prices as input.

For example, in Fig. 4, we depict in black line the international variable we employ to perform the shock and in colors (red, blue, green) the evolution of the regional prices. In solid lines, the picture draws the regional prices without the shock, and the dashed lines depict the self-driven dynamic of the system.

The chosen time-frames in Fig. 4 are in line with the regional prices behaviors in the Impulse-Response function based on econometric models considered as reference. It is worth mentioning that we need to train a different model each time we change the international price to perform the shock.

According to the results in Fig. 4, when considering an exogenous increase of the international oil price, soybean prices in regional markets of Argentina are immediately impacted, but the reaction depends on the model considered, e.g. the RNN-2c displays greater volatility. Nevertheless, the decreasing convergence paths of all models (consistent with econometric estimations) lead to the same new stationary state.

While the regional soybean prices in Argentina recover stability near to the path without shock, the regional corn prices show greater volatility facing the same exogenous shock. Except for the LSTM-2c, regional corn prices display a great difficulty to recover the path without shock, and Bahia Blanca and Rosario corn markets show different behaviors between them and across models. Their different paths of convergence increase the price-gap between regions (supported by the econometric estimations).

Finally, when assuming an exogenous increase in the international price of their agricultural commodity, regional markets prices display greater positive reactions (particularly for corn) and convergence towards higher values compared to their values without shock. Regional soybean prices converge to a higher price in the new stationary state, except under the LSTM-2c, which brings the price back to the path without shock. Reactions of regional corn prices to their international price increase are greater than in the case of soybean and tend to converge close to the new level of the international price of corn.

The difference between the reactions of soybean and corn regional prices to their own international prices is due to Argentina's soybean and corn markets particularities. These results are in line with the role of Argentina as a big soybean producer in the international market, so it is considered as a price maker. Conversely, in the international corn market Argentina is a relatively small player being a price-taker, so a change in the international price of corn is strongly transmitted to regional prices.

4   Conclusions

In this paper, we have trained four recurrent networks to forecast the reaction of regional commodity prices when an exogenous variable (i.e., an international price) is shocked. Results have been validated since they are in line with estimations from econometric auto-regressive models. The self-driven dynamic of recurrent networks has been demonstrated to be consistent with the behavior of Argentina's soybean and corn markets. To reduce regional price volatility, RNNs become a new tool to predict domestic prices' reactions to international changes and provide relevant insights for policy-makers decisions.

Further works should consider more complex recurrent networks, including other variables related to these agricultural and energy prices (e.g., bio-ethanol and bio-diesel prices) and also other regional variables that condition regional price path-through (e.g., transport costs).

References

1. Food and agriculture organization. http://www.fao.org/faostat/en/
2. Granar. http://www.granar.com.ar/
3. Ahumada, H., Cornejo, M.: Long-run effects of commodity prices on the real exchange rate: evidence from Argentina. Económica **61**, 3–33 (2015)
4. Boyko, N., Ivanets, A., Bosik, M.: Forecasting economic and financial indicators by supply of deep and recovery neural networks. ECONTECHMOD Int. Q. J. Econ. Technol. Model. Process. **7**, 3–8 (2018)
5. Du, X., Zhang, H., Van Nguyen, H., Han, Z.: Stacked LSTM deep learning model for traffic prediction in vehicle-to-vehicle communication. In: 2017 IEEE 86th Vehicular Technology Conference (VTC-Fall), pp. 1–5. IEEE (2017)
6. Elman, J.L.: Finding structure in time. Cogn. Sci. **14**(2), 179–211 (1990)
7. Fang, Y., Guan, B., Wu, S., Heravi, S.: Optimal forecast combination based on ensemble empirical mode decomposition for agricultural commodity futures prices. J. Forecast. **39**(6), 877–886 (2020)
8. Gardebroek, C., Hernandez, M.A., Robles, M.: Market interdependence and volatility transmission among major crops. Agric. Econ. **47**(2), 141–155 (2016)
9. Hochreiter, S., Schmidhuber, J.: Long short-term memory. Neural Comput. **9**(8), 1735–1780 (1997)
10. Johansen, S., et al.: Likelihood-based inference in cointegrated vector autoregressive models. Econ. Theor. **14**, 517–524 (1995)
11. Lipton, Z.C., Berkowitz, J., Elkan, C.: A critical review of recurrent neural networks for sequence learning. arXiv preprint arXiv:1506.00019 (2015)
12. Minot, N.: Food price volatility in Sub-Saharan Africa: has it really increased? Food Policy **45**, 45–56 (2014)
13. Minot, N., et al.: Transmission of world food price changes to markets in Sub-Saharan Africa. Citeseer (2010)
14. Pietola, K., Liu, X., Robles, M., et al.: Price, inventories, and volatility in the global wheat market. Technical report, International Food Policy Research Institute (IFPRI) (2010)
15. Serra, T., Gil, J.: Price volatility in food markets: can stock building mitigate price fluctuations? Eur. Rev. Agric. Econ. **40**(3), 507–528 (2013)
16. Sutskever, I., Vinyals, O., Le, Q.V.: Sequence to sequence learning with neural networks. arXiv preprint arXiv:1409.3215 (2014)
17. Wang, D., Yue, C., Wei, S., Lv, J.: Performance analysis of four decomposition-ensemble models for one-day-ahead agricultural commodity futures price forecasting. Algorithms **10**(3), 108 (2017)
18. Wang, J., Wang, J.: Forecasting energy market indices with recurrent neural networks: case study of crude oil price fluctuations. Energy **102**, 365–374 (2016)
19. Wang, J., Li, X.: A combined neural network model for commodity price forecasting with SSA. Soft. Comput. **22**(16), 5323–5333 (2018)
20. Wang, Y.: Applications of recurrent neural network on financial time series. M.Sc. thesis, Imperial College London (2017)
21. Yu, Y., Si, X., Hu, C., Zhang, J.: A review of recurrent neural networks: LSTM cells and network architectures. Neural Comput. **31**(7), 1235–1270 (2019)

Semi-supervised Deep Learning Based on Label Propagation in a 2D Embedded Space

Bárbara C. Benato[1(✉)], Jancarlo F. Gomes[1,2], Alexandru C. Telea[3], and Alexandre Xavier Falcão[1]

[1] Laboratory of Image Data Science, University of Campinas, Campinas, Brazil
{barbara.benato,jgomes,afalcao}@ic.unicamp.br
[2] Faculty of Medical Sciences, University of Campinas, Campinas, Brazil
[3] Department of Information and Computing Science, Utrecht University, Utrecht, The Netherlands
a.c.telea@uu.nl

Abstract. Expert human supervision of the large labeled training sets needed by convolutional neural networks is expensive. To obtain sufficient labeled samples to train a model, one can propagate labels from a small set of supervised samples to a large unsupervised set. Yet, such methods need many supervised samples for validation. We present a method that iteratively trains a deep neural network (VGG-16) from labeled samples created by projecting the features of VGG-16's last max-pooling layer in 2D with t-SNE and propagating labels with the Optimum-Path Forest semi-supervised classifier. As the labeled set improves along iterations, it improves the network's features. We show how this significantly improves classification results on test data (using only 1% to 5% of supervised samples) of three private challenging datasets and two public ones.

Keywords: Data annotation · Label propagation · Iterative feature learning

1 Introduction

Convolutional neural networks (CNNs) usually need large training sets (labeled images) [14,19]. While regularization, fine-tuning, transfer learning, and data augmentation [24] can help this, manually annotating enough images (human supervision) by expert uses, as in Biology and Medicine, remains expensive.

To build a large enough training set, Lee [12] propagated labels from a small set of supervised images to a large set of unsupervised ones, as an alternative to entropy regularization. In detail, Lee trained a neural network with 100 to 3000 supervised images, assigned the class with maximum predicted probability to the unsupervised ones, and then fine-tuned a neural network with the true-plus-artificially labeled (*pseudo* labeled) samples, showing advantages over other semi-supervised learning methods. Still, this required validation sets of over 1000

© Springer Nature Switzerland AG 2021
J. M. R. S. Tavares et al. (Eds.): CIARP 2021, LNCS 12702, pp. 371–381, 2021.
https://doi.org/10.1007/978-3-030-93420-0_35

supervised images for the optimization of hyperparameters; used a network with a single hidden layer; and was shown on a single dataset (MNIST).

Label propagation from supervised to unsupervised samples was recently used to build larger training sets [8,9,13,26]. Amorim et al. [2] used the semi-supervised Optimum Path Forest (OPFSemi) classifier [1] for this, outperforming several existing semi-supervised techniques when training CNNs. Yet, they did not explore CNNs pre-trained with large supervised datasets for transfer learning, and still needed many supervised samples (10% of the dataset) for validation.

Graph-based semi-supervised learning has recently received increasing attention [1,3,6,28]. By modeling training samples as nodes of a graph whose arcs connect adjacent samples in the feature space, one can propagate labels from supervised samples to their most strongly connected unsupervised neighbors. Benato et al. [4,5] showed the advantages of OPFSemi for label propagation in a 2D *embedded* space created by t-SNE [15] from the latent space of an autoencoder trained with unsupervised images – which differs from [2] where propagation is done in the *feature* space. Their supervised classifiers achieved higher performance on unseen test sets when trained with large sets of truly-and-artificially labeled samples, with OPFSemi surpassing LapSVM [22] for label propagation. Yet, they have not used this strategy to train deep neural networks.

We fill the above gaps by proposing a loop (Fig. 1) that trains a deep neural network (VGG-16, [21]) with truly-and-artificially labeled samples along iterations. At each iteration, we create a 2D embedded space by the t-SNE projection (like [4], different from [2]) of VGG-16's features at the last max-pooling layer (before the MLP) and propagate labels by OPFSemi, so that the labeled set *jointly* improves with the CNN's feature space over iterations. Our method can improve classification on unseen test data of challenging datasets.

2   Proposed Pipeline

After the user supervises a small set of training images, we execute a three-step loop (*deep feature learning, feature space projection*, and *label propagation*; Fig. 1).

2.1   Deep Feature Learning

To minimize user effort for annotation, we use the ability of pre-trained CNNs to transfer knowledge [27] between scenarios – e.g., from natural to medical images – using few supervised samples and few epochs. We use VGG-16, pre-trained on ImageNet [19], and fine tuned with the supervised images. In the next iterations of our loop, we train VGG-16 with all true-and-artificially labeled images.

2.2   Feature Space Projection

We project the features of the last max-pooling layer of VGG-16 by t-SNE [15] in a 2D embedded space. One may conceptually divide a deep neural network into (a) layers for feature extraction, (b) fully connected layers for feature space reduction, and (c) the decision layer (a MLP classifier). We explored features that

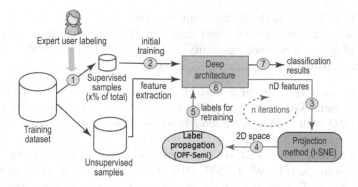

Fig. 1. Pipeline of our method. The user supervises a small fraction x of images (1). These are used to train a deep neural network (2), which extracts features from the unsupervised images (3). Features are projected in a 2D embedded space (4). A semi-supervised classifier propagates labels to the unsupervised images (5). The model is retrained by all images and their assigned labels (6), creating a new and improved feature space along iterations. Finally, the trained model is used for classification (7).

result from (a), where the feature space is still high and sparse; a comparison with the output of the last hidden layer is left for future work. Rauber et al. [18] showed that high classification accuracy relates to a good separation of classes in a 2D projection. Hence, if a 2D projection presents good class separation, then a good class separation can also be found in the data space.

Benato et al. [4] showed that label propagation (using two semi-supervised classifiers) in a 2D projection space leads to better classification results than using the latent feature space of an autoencoder. We also opt to investigate label propagation in a 2D projected space (created by t-SNE [15], as in [18] and [4]) to create larger training sets for deep learning.

2.3 Label Propagation

We used OPFSemi in both the 2D t-SNE projection [4] and the original feature space [2]. OPFSemi sees each sample as a node of a complete graph, setting the cost of a path between two nodes to the maximum arc weight (Euclidean distance between samples) along it. The supervised nodes seed the computing of a minimum-cost path forest – each seed propagates its label to the most closely connected unsupervised nodes of its tree.

3   Experiments and Results

3.1   Experimental Set-Up

We randomly divide each dataset into supervised training samples S, unsupervised training samples U, and testing samples T. To measure the impact of annotated samples on classification, we let $S \cup U$ have 70% of samples, while T

has 30%. To minimize user effort for supervision, we set $|S|$ to 1% up to 5% of the entire dataset, thus much smaller than $|U|$ For statistics, we generate three partitions of each experiment randomly and in a stratified manner. We validate our method, called *DeepFA looping*, by three experiments:

1. *Baseline*: train VGG-16 on S, test on T, ignore U (Fig. 1 steps 1, 2, 6, 7).
2. *DeepFA*: train VGG-16 on S; extract $S \cup U$ features from VGG-16 and project them in $2D$ with t-SNE; OPFSemi label estimation in U; train VGG-16 on $S \cup U$ and test on T (all of Fig. 1 for $n = 1$).
3. *DeepFA looping*: train VGG-16 on S; extract $S \cup U$ features from VGG-16; project them in $2D$ with t-SNE; OPFSemi label propagation on U; train VGG-16 on $S \cup U$; repeat from the projection step $n = 5$ times; test on T (all of Fig. 1 for $n > 1$).

To compare effectiveness, we compute accuracy from VGG-16's final probability. As we have unbalanced datasets, we also compute Cohen's κ coefficient, $\kappa \in [-1, 1]$, where $\kappa \leq 0$ means no possibility and $\kappa = 1$ means full possibility of agreement occurring by chance, respectively. For the experiments where OPFSemi propagates labeled samples, we also compute the label propagation accuracy in U, *i.e.*, the number of correct labels assigned in U over the size of U.

3.2   Datasets

We first use two public datasets: MNIST [11] contains handwritten digits from 0 to 9 as 28×28 grayscale images. We use a random subset of 5K samples from MNIST's total of 60K. CIFAR-10 [10] contains color images (32×32 pixels) in 10 classes: airplane, automobile, bird, cat, deer, dog, frog, horse, ship, and truck. We use a random subset of 5K images from CIFAR-10's total of 60K.

We also used three private datasets from a real-world problem (Fig. 2). These datasets contain color microscopy images (200×200 pixels) of the most common species of human intestinal parasites in Brazil, responsible for public health problems in most tropical countries [25]. These datasets are challenging, since they are unbalanced and contain an impurity class for the large majority of the samples, having samples very similar to parasites, which makes classification hard (Fig. 2). We explored two (out of three) datasets with and without the impurity class, yielding thus five total datasets: (i) *Helminth larvae*, (ii) *Helminth eggs* without impurities, (iii) *Helminth eggs* with impurities, (iv) *Protozoan cysts* without impurities, and (v) *Protozoan cysts* with impurities. The *Helminth larvae* dataset presents larvae and impurities (2 classes, 3514 images); the *Helminth eggs* dataset has several categories: *H.nana*, *H.diminuta*, *Ancilostomideo*, *E.vermicularis*, *A.lumbricoides*, *T.trichiura*, *S.mansoni*, *Taenia*, and impurities (9 classes, 5112 images); and the *Protozoan cysts* dataset has the categories *E.coli*, *E.histolytica*, *E.nana*, *Giardia*, *I.butschlii*, *B.hominis*, and impurities (7 classes, 9568 images). For more details, we refer to [17]. Table 1 presents the experimental set-up described in Sect. 3.1 for these 7 datasets.

Fig. 2. Datasets: (a) MNIST (b) CIFAR-10 and (c) H.eggs, with parasites (green box) and similar impurities (red box). (Color figure online)

3.3 Implementation Details

We implemented VGG-16 in Python using Keras [7]. We load the pre-trained weights from ImageNet [19] and fine-tuned this model using the supervised S first, and subsequently labeled sets $(S \cup U)$ for each chosen dataset. To guarantee convergence, we used 100 epochs with stochastic gradient descent with a linearly decaying learning-rate from 10^{-4} to zero over 100 epochs and momentum of 0.9.

3.4 Experimental Results

We use our results to address three joint questions:

Q1: How do more supervised samples improve the process? Per dataset, Table 2 first shows accuracy (mean, standard deviation) and κ for VGG-16 trained on S with 1%–5% supervised samples and tested on T (*baseline*). Accuracy and κ increase with the supervised sample count. For *H.larvae* and *H.eggs*, this trend cannot be seen for the given training-data fractions (3% to 4%). Still, VGG-16 performs better when the supervised training sample count increases.

Table 1. Number of samples in each set S, U, and T considering $|S|$ for five sample percentages $x = 1, 2, \ldots, 5\%$ of supervised images in each dataset.

Dataset	H.larvae					H.eggs					P.cysts				
x	1%	2%	3%	4%	5%	1%	2%	3%	4%	5%	1%	2%	3%	4%	5%
$|S|$	35	70	105	140	175	17	35	53	70	88	38	77	115	154	192
$|U|$	2424	2389	2354	2319	2284	1220	1202	1184	1167	1149	2658	2619	2581	2542	2504
$|T|$	1055	1055	1055	1055	1055	531	531	531	531	531	1156	1156	1156	1156	1156
Total	3514	3514	3514	3514	3514	1768	1768	1768	1768	1768	3852	3852	3852	3852	3852
Dataset	H.eggs imp					P.cysts imp					MNIST/CIFAR-10				
x	1%	2%	3%	4%	5%	1%	2%	3%	4%	5%	1%	2%	3%	4%	5%
$|S|$	51	102	153	204	255	95	191	287	382	478	50	100	150	200	250
$|U|$	3527	3476	3425	3374	3323	6602	6506	6410	6315	6219	3450	3400	3350	3300	3250
$|T|$	1534	1534	1534	1534	1534	2871	2871	2871	2871	2871	1500	1500	1500	1500	1500
Total	5112	5112	5112	5112	5112	9568	9568	9568	9568	9568	5000	5000	5000	5000	5000

Table 2. Results of the *baseline*, *DeepFA*, and *DeepFA looping* experiments, all datasets, five supervised sample percentages x, color-coded using a white-to-green colormap.

	Methods	Metrics	$x = 1\%$	$x = 2\%$	$x = 3\%$	$x = 4\%$	$x = 5\%$
H.larvae	baseline	accuracy	0.930806 ± 0.0266	0.958925 ± 0.0038	0.961138 ± 0.0066	0.960821 ± 0.0070	0.971564 ± 0.0068
		kappa	0.613432 ± 0.2334	0.824394 ± 0.0280	0.818082 ± 0.0492	0.808397 ± 0.0416	0.868460 ± 0.0361
	DeepFA	accuracy	0.962085 ± 0.0148	0.968721 ± 0.0053	0.967773 ± 0.0047	0.974092 ± 0.0112	0.977567 ± 0.0043
		kappa	0.819799 ± 0.0767	0.864692 ± 0.0251	0.854607 ± 0.0308	0.878935 ± 0.0624	0.900290 ± 0.0197
		propagation	0.961366 ± 0.0132	0.969364 ± 0.0070	0.963806 ± 0.0108	0.975193 ± 0.0135	0.979531 ± 0.0048
	DeepFA looping	accuracy	0.956398 ± 0.0202	0.974407 ± 0.0049	0.974092 ± 0.0071	0.978831 ± 0.0074	0.978515 ± 0.0031
		kappa	0.783412 ± 0.1150	0.886325 ± 0.0299	0.888211 ± 0.0298	0.904963 ± 0.0353	0.905292 ± 0.0138
		propagation	0.954860 ± 0.0153	0.976278 ± 0.0013	0.970991 ± 0.0049	0.981158 ± 0.0056	0.980887 ± 0.0015
H.eggs	baseline	accuracy	0.812932 ± 0.0599	0.925926 ± 0.0189	0.934714 ± 0.0313	0.929065 ± 0.0132	0.966101 ± 0.0032
		kappa	0.775954 ± 0.0737	0.912299 ± 0.0224	0.923002 ± 0.0367	0.916335 ± 0.0154	0.959807 ± 0.0039
	DeepFA	accuracy	0.949780 ± 0.0104	0.967985 ± 0.0082	0.962963 ± 0.0120	0.974263 ± 0.0203	0.978656 ± 0.0141
		kappa	0.940671 ± 0.0122	0.962146 ± 0.0097	0.956316 ± 0.0234	0.969631 ± 0.0239	0.974742 ± 0.0167
		propagation	0.957154 ± 0.0127	0.974400 ± 0.0055	0.971436 ± 0.0084	0.978442 ± 0.0102	0.987065 ± 0.0037
	DeepFA looping	accuracy	0.970496 ± 0.0039	0.969868 ± 0.0086	0.964846 ± 0.0237	0.974262 ± 0.0193	0.983679 ± 0.0078
		kappa	0.965144 ± 0.0046	0.964405 ± 0.0102	0.958557 ± 0.0278	0.969634 ± 0.0227	0.980694 ± 0.0093
		propagation	0.981676 ± 0.0031	0.979251 ± 0.0042	0.979520 ± 0.0140	0.983563 ± 0.0115	0.991107 ± 0.0035
P.cysts	baseline	accuracy	0.757209 ± 0.0158	0.881776 ± 0.0113	0.913783 ± 0.0079	0.914648 ± 0.0077	0.934545 ± 0.0133
		kappa	0.651933 ± 0.0232	0.837102 ± 0.0163	0.882551 ± 0.0108	0.884303 ± 0.0108	0.912294 ± 0.0177
	DeepFA	accuracy	0.847751 ± 0.0106	0.912341 ± 0.0154	0.914072 ± 0.0269	0.937428 ± 0.0018	0.950404 ± 0.0178
		kappa	0.794713 ± 0.0124	0.882537 ± 0.0213	0.885475 ± 0.0348	0.916035 ± 0.0028	0.933649 ± 0.0235
		propagation	0.832715 ± 0.0153	0.897997 ± 0.0231	0.893793 ± 0.0303	0.931627 ± 0.0042	0.937685 ± 0.0161
	DeepFA looping	accuracy	0.889562 ± 0.0030	0.925894 ± 0.0234	0.938870 ± 0.0126	0.964533 ± 0.0120	0.959919 ± 0.0088
		kappa	0.853264 ± 0.0021	0.900792 ± 0.0318	0.918280 ± 0.0163	0.952664 ± 0.0159	0.946442 ± 0.0117
		propagation	0.881800 ± 0.0130	0.925321 ± 0.0121	0.928783 ± 0.0110	0.959570 ± 0.0047	0.956726 ± 0.0044
H.eggs imp	baseline	accuracy	0.862234 ± 0.0157	0.900696 ± 0.0087	0.910256 ± 0.0167	0.931986 ± 0.0057	0.937419 ± 0.0086
		kappa	0.740861 ± 0.0287	0.815160 ± 0.0138	0.833168 ± 0.0301	0.876969 ± 0.0090	0.886231 ± 0.0159
	DeepFA	accuracy	0.928509 ± 0.0032	0.941113 ± 0.0010	0.935246 ± 0.0053	0.948501 ± 0.0111	0.956758 ± 0.0046
		kappa	0.873674 ± 0.0068	0.895627 ± 0.0014	0.885487 ± 0.0104	0.908733 ± 0.0190	0.923366 ± 0.0079
		propagation	0.913639 ± 0.0068	0.931433 ± 0.0058	0.920626 ± 0.0146	0.939631 ± 0.0129	0.945314 ± 0.0018
	DeepFA looping	accuracy	0.935680 ± 0.0014	0.949370 ± 0.0063	0.944372 ± 0.0037	0.956758 ± 0.0076	0.957844 ± 0.0046
		kappa	0.885645 ± 0.0033	0.910179 ± 0.0111	0.901216 ± 0.0078	0.922875 ± 0.0130	0.925353 ± 0.0080
		propagation	0.926681 ± 0.0022	0.939352 ± 0.0066	0.937675 ± 0.0077	0.951369 ± 0.0045	0.950252 ± 0.0024
P.cysts imp	baseline	accuracy	0.850691 ± 0.0189	0.865320 ± 0.0018	0.900383 ± 0.0072	0.903634 ± 0.0129	0.916173 ± 0.0045
		kappa	0.751667 ± 0.0280	0.776938 ± 0.0031	0.832300 ± 0.0106	0.840126 ± 0.0216	0.860640 ± 0.0076
	DeepFA	accuracy	0.852084 ± 0.0066	0.848717 ± 0.0090	0.884709 ± 0.0152	0.892140 ± 0.0144	0.916405 ± 0.0074
		kappa	0.755127 ± 0.0132	0.756045 ± 0.0138	0.811239 ± 0.0231	0.823333 ± 0.0223	0.862764 ± 0.0100
		propagation	0.845055 ± 0.0065	0.840924 ± 0.0056	0.880294 ± 0.0193	0.879200 ± 0.0181	0.901996 ± 0.0094
	DeepFA looping	accuracy	0.854522 ± 0.0013	0.860908 ± 0.0292	0.900035 ± 0.0140	0.892488 ± 0.0282	0.920933 ± 0.0032
		kappa	0.763711 ± 0.0026	0.774361 ± 0.0434	0.836986 ± 0.0217	0.825604 ± 0.0450	0.869900 ± 0.0051
		propagation	0.845652 ± 0.0072	0.853915 ± 0.0256	0.897367 ± 0.0071	0.882684 ± 0.0242	0.915335 ± 0.0106
MNIST	baseline	accuracy	0.661111 ± 0.0523	0.782222 ± 0.0269	0.870445 ± 0.0050	0.876444 ± 0.0132	0.909778 ± 0.0143
		kappa	0.623148 ± 0.0582	0.757848 ± 0.0298	0.855944 ± 0.0056	0.862635 ± 0.0147	0.899686 ± 0.0159
	DeepFA	accuracy	0.766222 ± 0.0252	0.852667 ± 0.0397	0.901556 ± 0.0170	0.899778 ± 0.0096	0.932889 ± 0.0136
		kappa	0.740028 ± 0.0280	0.836263 ± 0.0440	0.890529 ± 0.0119	0.888552 ± 0.0107	0.925403 ± 0.0151
		propagation	0.750571 ± 0.0320	0.833524 ± 0.0362	0.893524 ± 0.0064	0.895333 ± 0.0114	0.923619 ± 0.0148
	DeepFA looping	accuracy	0.815778 ± 0.0212	0.862222 ± 0.0396	0.908444 ± 0.0103	0.918000 ± 0.0077	0.936444 ± 0.0224
		kappa	0.795079 ± 0.0236	0.846885 ± 0.0439	0.898190 ± 0.0114	0.908816 ± 0.0086	0.929355 ± 0.0249
		propagation	0.806000 ± 0.0230	0.856667 ± 0.0353	0.905905 ± 0.0054	0.923238 ± 0.0068	0.939905 ± 0.0156
CIFAR10	baseline	accuracy	0.266000 ± 0.0264	0.321555 ± 0.0151	0.372889 ± 0.0341	0.417111 ± 0.0413	0.455333 ± 0.0263
		kappa	0.183681 ± 0.0301	0.245770 ± 0.0166	0.303050 ± 0.0377	0.352095 ± 0.0461	0.394558 ± 0.0291
	DeepFA	accuracy	0.228445 ± 0.0435	0.310000 ± 0.0790	0.365555 ± 0.0205	0.407778 ± 0.0136	0.424889 ± 0.0093
		kappa	0.142149 ± 0.0492	0.232875 ± 0.0880	0.295078 ± 0.0230	0.341907 ± 0.0148	0.360883 ± 0.0102
		propagation	0.219048 ± 0.0428	0.288952 ± 0.0790	0.356095 ± 0.0340	0.389619 ± 0.0190	0.421143 ± 0.0126
	DeepFA looping	accuracy	0.324000 ± 0.0418	0.375333 ± 0.0436	0.402444 ± 0.0125	0.448445 ± 0.0177	0.461555 ± 0.0211
		kappa	0.248837 ± 0.0463	0.305496 ± 0.0483	0.335927 ± 0.0138	0.387059 ± 0.0199	0.401490 ± 0.0236
		propagation	0.314286 ± 0.0277	0.369334 ± 0.0235	0.411238 ± 0.0253	0.446667 ± 0.0152	0.466857 ± 0.0301

Q2: What is OPFSemi's Effect? Table 2 next shows mean and standard deviation for accuracy, κ, and propagation accuracy for VGG-16 trained with $S \cup U$, with U labeled by OPFSemi in the 2D projection (*DeepFA*). As for *baseline*, accuracy and κ increase with the fraction of supervised training samples. The propagation accuracy of OPFSemi is related to the number of supervised samples used to train VGG-16. The labeling performance of OPFSemi in the 2D projected space can be verified by the propagation accuracy. For the parasites dataset, this accuracy is over 80% even when VGG-16 was trained with *just* 1% of the data.

Q3: What is Looping OPFSemi's Effect? Finally, Table 2 shows mean and standard deviation of 5 iterations of *DeepFA looping* for accuracy, Cohen's κ, and propagation accuracy for VGG-16 trained with $S \cup U$, with U labeled by OPF-Semi in the 2D projection. As for *baseline* and *DeepFA*, we see an increase of accuracy and κ with the fractions (1% to 5%) of supervised training samples. We see the same for propagation accuracy, which reflects the effect of 5 iterations of OPFSemi for labeling samples in the 2D projected space. For all datasets, except CIFAR-10, propagation accuracy is over 80% even when the VGG-16 feature space was trained with only 1% of data.

4   Discussion

Added-Value of DeepFA Looping: Figure 3 plots the average κ for our *baseline*, *DeepFA*, and *DeepFA looping* experiments, for all 7 studied datasets. *DeepFA looping* consistently obtains the best results, except for the *P.cysts with impurity* dataset. *DeepFA* shows an improvement over the *baseline* experiment, while the first one was improved by a looping addition in the method. The gain of *DeepFA looping* is even higher when using a low number of supervised samples – relevant when one cannot, or does not want to put effort, to supervise new ones. This gain is lower for CIFAR-10 and almost zero for *P.cysts with impurities*, as these datasets are more challenging, as their lowest κ scores show.

Effectiveness of OPFSemi Labeling: The positive results for VGG-16 rely on OPFSemi propagating labels accurately. Figure 3 shows this by the average propagation accuracy of OPFSemi for *DeepFA* and *DeepFA looping*, which is high for all datasets, being worst-case 50% for CIFAR-10. For CIFAR-10, the propagation accuracy *gain* of *DeepFA looping* is higher than for the other datasets. We see also the impurity class impact for H.eggs and P.cysts in propagation accuracy (roughly 5%). Propagation accuracy is high as long as the sample count increases. The *DeepFA looping* curve is on top of *DeepFA* curve for all datasets, so the effectiveness of OPFSemi label propagation consistently improves by the looping addition.

Feature Space Improvement: Figure 3 showed that OPFSemi improved VGG-16's effectiveness and also accurately propagated labels to unsupervised samples. The OPFSemi labeled samples also improve the VGG-16 feature space. Figure 4 shows this space projected with t-SNE for the studied datasets. Projections are colored by (i) labels (supervised samples colored by the true-label; unsupervised samples black), and (ii) OPFSemi's confidence in classifying a sample (red = low confidence, green = high confidence) [16,20,23]. For all datasets, we see a clear reduction of red zones from *baseline* to *DeepFA* and a good cluster formation in the projection for same-color (*i.e.*, same-class) supervised samples (Fig. 4a). From *DeepFA* to *DeepFA looping*, there is no further reduction of red zones. Yet, different-color groups get more clustered and better separated. This

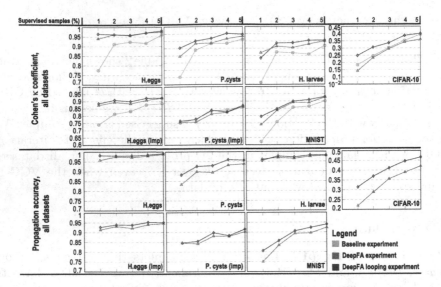

Fig. 3. Cohen's κ (top) and propagation accuracy (bottom), all datasets, for 1% to 5% supervised samples, *DeepFA* (red) *vs DeepFA looping* last iteration (blue). (Color figure online)

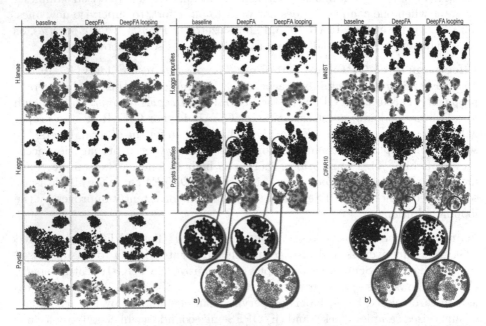

Fig. 4. 2D feature-space projections of training samples $(S \cup U)$ – *baseline, DeepFA,* and *DeepFA looping*, 1% supervised samples. Top row per experiment: Supervised samples colored by true labels, unsupervised ones are black. Bottom row per experiment: Samples colored by OPFSemi's confidence (red = low, green = high). Insets (a, b) show details. (Color figure online)

is clearer for CIFAR-10, which does not show good cluster separation for *DeepFA* (Fig. 4b). We conclude that OPFSemi's label propagation and the looping strategy improve VGG-16's feature space when this space is fed by those samples.

Figure 5 shows the projected space colored by class labels (unsupervised samples in black) and the OPFSemi's confidence values for 5 iterations of *DeepFA looping* on the *P. cysts* dataset with impurities, 1% supervised samples. Class separation and confidence values increase with the iterations. The red-class samples are well separated from samples of the other classes in the first iteration; some brown supervised samples get attached to them in iteration 2, creating a low-confidence region. From iteration 3 on, the problem is solved.

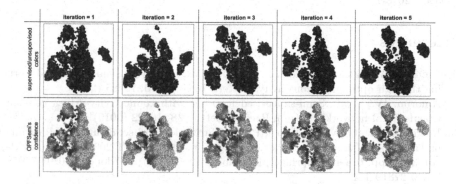

Fig. 5. 2D projections of training samples ($S \cup U$) for *DeepFA looping*, 1% supervised samples, *P. cysts impurities* dataset. Top row: Color shows class labels; unsupervised samples in black. Bottom row: Color shows OPFSemi's confidence (red = low, green = high). Class separation and confidence increase with iterations. (Color figure online)

Limitations: Our validation used only seven datasets, one deep-learning approach (VGG-16), one semi-supervised classifier (OPFSemi), and one projection method (t-SNE). Exploring more (combinations of) such techniques would be valuable. Also, using more than 5 iterations could help understand how OPF-Semi labels low-confidence regions and how it affects VGG-16's feature space.

5   Conclusion

We proposed an approach for increasing the quality of image classification and of extracted feature spaces when lacking large supervised datasets. From a few supervised samples, we create a feature space by a pre-trained VGG-16 model and use the OPFSemi technique to label unsupervised samples on a 2D t-SNE projection of that feature space. We iteratively improve labels (and the feature space) using labeled samples as input for the VGG-16 training.

OPFSemi shows low label-propagation errors and leads VGG-16 to good classification results for several tested datasets, thereby improving the VGG-16 training and hence the feature space. The small gain yielded by looping tells that OPFSemi can stagnate, its label-propagation errors lowering classification quality. To help OPFSemi during label propagation, we plan next a bootstrapping strategy to avoid propagation in low certainty regions. We also aim to include user knowledge to support OPFSemi's label propagation and to understand the VGG-16 training process and feature space generation. This co-training approach involving a bootstrapping strategy and two classifiers (OPFSemi and VGG-16) can lead to higher quality, and more explainable, deep-learning methods.

Acknowledgments. The authors are grateful to FAPESP grants #2014/12236-1, #2019/10705-8, CAPES grants with Finance Code 001, and CNPq grants #303808/2018-7.

References

1. Amorim, W., Falcão, A., Papa, J., Carvalho, M.: Improving semi-supervised learning through optimum connectivity. Pattern Recogn. **60**, 72–85 (2016)
2. Amorim, W., et al.: Semi-supervised learning with connectivity-driven convolutional neural networks. Pattern Recogn. Lett. **128**, 16–22 (2019)
3. Belkin, M., Matveeva, I., Niyogi, P.: Regularization and semi-supervised learning on large graphs. In: Shawe-Taylor, J., Singer, Y. (eds.) COLT 2004. LNCS (LNAI), vol. 3120, pp. 624–638. Springer, Heidelberg (2004). https://doi.org/10.1007/978-3-540-27819-1_43
4. Benato, B.C., Telea, A.C., Falcão, A.X.: Semi-supervised learning with interactive label propagation guided by feature space projections. In: Proceedings of the SIBGRAPI, pp. 392–399 (2018)
5. Benato, B.C., Gomes, J.F., Telea, A.C., Falcão, A.X.: Semi-automatic data annotation guided by feature space projection. Pattern Recogn. **109**, 107612 (2021)
6. Chapelle, O., Scholkopf, B., Zien, A.: Semi-supervised learning. IEEE TNN **20**(3), 542 (2009)
7. Chollet, F., et al.: Keras (2015). https://keras.io
8. Gong, M., Yang, H., Zhang, P.: Feature learning and change feature classification based on deep learning for ternary change detection in SAR images. J. Photogram. Remote Sens. **129**, 212–225 (2017)
9. Iscen, A., Tolias, G., Avrithis, Y., Chum, O.: Label propagation for deep semi-supervised learning. In: Proceedings of the IEEE CVPR (2019)
10. Krizhevsky, A., Nair, V., Hinton, G.: CIFAR-10 dataset. www.cs.toronto.edu/~kriz/cifar.html
11. LeCun, Y., Cortes, C.: MNIST handwritten digit database (2010). http://yann.lecun.com/exdb/mnist
12. Lee, D.H.: Pseudo-label: the simple and efficient semi-supervised learning method for deep neural networks. In: Proceedings of the ICML-WREPL (2013)
13. Li, Z., Ko, B.S., Choi, H.-J.: Naive semi-supervised deep learning using pseudo-label. Peer-to-Peer Netw. Appl. **12**(5), 1358–1368 (2018). https://doi.org/10.1007/s12083-018-0702-9
14. Lin, T.-Y., et al.: Microsoft COCO: common objects in context. In: Fleet, D., Pajdla, T., Schiele, B., Tuytelaars, T. (eds.) ECCV 2014. LNCS, vol. 8693, pp. 740–755. Springer, Cham (2014). https://doi.org/10.1007/978-3-319-10602-1_48

15. van der Maaten, L.: Accelerating t-SNE using tree-based algorithms. JMLR **15**(1), 3221–3245 (2014)
16. Miranda, P.A.V., Falcão, A.X.: Links between image segmentation based on optimum-path forest and minimum cut in graph. JMIV **35**(2), 128–142 (2009)
17. Peixinho, A.Z.: Learning image features by convolutional networks under supervised data constraint. Master's thesis, University of Campinas (2017)
18. Rauber, P., Falcão, A., Telea, A.: Projections as visual aids for classification system design. Inf. Vis. **17**, 282–305 (2017)
19. Russakovsky, O., et al.: ImageNet large scale visual recognition challenge. IJCV **115**(3), 211–252 (2015)
20. Silva, A.T., Santos, J.A., Falcão, A.X., Torres, R.S., Magalhães, L.P.: Incorporating multiple distance spaces in optimum-path forest classification to improve feedback-based learning. CVIU **116**(4), 510–523 (2012)
21. Simonyan, K., Zisserman, A.: Very deep convolutional networks for large-scale image recognition (2014). arxiv.org/abs/1409.1556
22. Sindhwani, V., Niyogi, P., Belkin, M.: Beyond the point cloud: from transductive to semi-supervised learning. In: Proceedings of the ICML, pp. 824–831 (2005)
23. Spina, T., Miranda, P., Falcão, A.: Intelligent understanding of user interaction in image segmentation. IJPRAI **26**(02), 126–001 (2012)
24. Sun, C., Shrivastava, A., Singh, S., Gupta, A.: Revisiting unreasonable effectiveness of data in deep learning era. In: Proceedings of the ICCV, pp. 843–852 (2017)
25. Suzuki, C., Gomes, J., Falcão, A., Shimizu, S., Papa, J.: Automated diagnosis of human intestinal parasites using optical microscopy images. In: Proceedings of the Symposium on Biomedical Imaging, pp. 460–463 (April 2013)
26. Wu, H., Prasad, S.: Semi-supervised deep learning using pseudo labels for hyperspectral image classification. IEEE TIP **27**(3), 1259–1270 (2018)
27. Yosinski, J., Clune, J., Bengio, Y., Lipson, H.: How transferable are features in deep neural networks? In: Proceedings of the NIPS, pp. 3320–3328 (2014)
28. Zhu, X.: Semi-supervised learning literature survey. Computer Science, University of Wisconsin-Madison (July 2008)

Iterative Creation of Matching-Graphs – Finding Relevant Substructures in Graph Sets

Mathias Fuchs[1]([✉])[iD] and Kaspar Riesen[1,2][iD]

[1] Institute of Computer Science, University of Bern, 3012 Bern, Switzerland
mathias.fuchs@inf.unibe.ch
[2] Institute for Informations Systems, University of Applied Sciences Northwestern Switzerland, 4600 Olten, Switzerland
kaspar.riesen@fhnw.ch

Abstract. Both the amount of data available and the rate at which it is acquired increases rapidly. The underlying data is often complex, making it difficult (or somehow unnatural) to represent it by vectorial data structures. Hence, graphs are a promising alternative for formalizing the data. Actually a large amount of graph-based methods for pattern recognition have been proposed. The vast amount of these methods rely on graph matching procedures. In a recent paper a novel encoding of graph matching information has been proposed. The idea of this encoding is to formalize the stable cores of specific classes by means of graphs (called matching-graphs). In the present paper we aim to further improve the relevance of these matching-graphs by using an iterative creation algorithm. In an empirical evaluation we show that these novel matching-graphs offer a more stable and significant representation of their respective class than the previous version.

Keywords: Graph matching · Matching-graphs · Graph edit distance

1 Introduction and Related Work

Pattern recognition emerged to a major field of research which aims at solving diverse problems like signature verification[1], situation recognition [2], or breast cancer detection [3], to name just a few prominent examples. Roughly speaking there are two main approaches for pattern recognition. *Statistical approaches*, which use data structures like *vectors* for data representation and *structural approaches*, which use *strings*, *trees*, or *graphs* for the same task. Graphs provide a powerful alternative to feature vectors and thus, they are widely used in various pattern recognition applications, ranging from protein function/structure prediction [4], over inferring the privacy risk of an image on social media [5], to the detection of Alzheimer's Disease [6]. The main drawback of graphs is, however, the computational complexity of basic operations, which in turn makes graph based algorithms often slower than their statistical counterparts.

Supported by Swiss National Science Foundation (SNSF) Project Nr. 200021_188496.

© Springer Nature Switzerland AG 2021

J. M. R. S. Tavares et al. (Eds.): CIARP 2021, LNCS 12702, pp. 382–391, 2021.
https://doi.org/10.1007/978-3-030-93420-0_36

A large amount of graph based methods for pattern recognition have been proposed from which many rely on *graph matching* [7,8]. Graph matching is typically used for quantifying graph proximity. *Graph edit distance* [9,10], introduced about 40 years ago, is recognized as one of the most flexible graph distance models available. In contrast with many other distance measures (e.g. *graph kernels* [11]), graph edit distance generally offers more information than merely a dissimilarity score, viz. the information which subparts of the underlying graphs actually match with each other (known as *edit path*).

In a recent paper [12], the authors of the present paper propose to explicitly exploit the matching information of graph edit distance. Formally, we encode the matching information derived from graph edit distance into a data structure, called *matching-graph*. The main contribution of the present paper is to further improve the quality of these matching-graphs by means of an iterative process, which selects the best matching-graphs of each iteration and continues to create new matching-graphs from these selected parent graphs. The proposed algorithm is remotely inspired by the idea of genetic algorithms that also aim at emulating the process of natural selection and improvement of a population [13].

Morerover, our approach is similar in spirit to approaches from graph transaction based *Frequent Subgraph Mining (FSM)* [14]. This field also focuses on the identification of frequent subgraphs within a set of graphs (extract all subgraphs that occur more often than a specified threshold). We observe two main categories in FSM, viz. *Apriori-based approaches* and *Pattern-growth approaches* [14]. The apriori-based methods proceed to grow subgraphs by using a *Breadth First Search (BFS)* strategy. Before they continue to graphs of size $k + 1$ it first searches for all frequent graphs of size k. Pattern-growth approaches, on the other hand, work by using a *Depth First Search (DFS)* strategy, where one graph is extended until all frequent supergrahs of this graphs are found.

Though the goal of our approach is comparable to that of FSM, our procedure is quite unique. We identify common subgraphs of pairs of graphs by using a graph matching procedure rather than using an algorithm stemming from one of the two main categories discussed above (Apriori or Pattern Growth). We do also not focus on finding comprehensive lists of frequent and large subgraphs but rather we aim at improving our initial set of matching-graphs by iteratively matching these matching-graphs with each other in order to extract more stable and robust graph representatives for each class.

In the present paper we conduct both a quantitative and qualitative experimental evaluation. First, we measure the frequencies of the found matching-graphs in their correct class and second, we visualize and inspect the most frequent subgraphs which in turn enables novel insights into the question which graph substructures actually make up a class of patterns.

The remainder of this paper is organized as follows. Section 2 makes the paper self-contained by providing basic definitions and terms used throughout this paper. Next, in Sect. 3 the general procedure for creating a matching-graph is explained together with a description of the novel algorithm that we propose. Eventually, in Sect. 4, we empirically confirm that our algorithm produces indeed a set of highly relevant graph structures. Finally, in Sect. 5, we conclude the paper and discuss some ideas for future work.

2   Graphs and Graph Edit Distance - Basic Definitions

Let L_V and L_E be finite or infinite label sets for nodes and edges, respectively. A *graph* g is a four-tuple $g = (V, E, \mu, \nu)$, where

- V is the finite set of nodes,
- $E \subseteq V \times V$ is the set of edges,
- $\mu : V \rightarrow L_V$ is the node labeling function, and
- $\nu : E \rightarrow L_E$ is the edge labeling function.

In the present paper we employ *graph edit distance* as basic dissimilarity model for graphs. One of the main advantages of graph edit distance is its high degree of flexibility, which makes it applicable to virtually any kind of graphs.

Given two graphs g_1 and g_2, the basic idea of graph edit distance is to transform g_1 into g_2 using some *edit operations*. A standard set of edit operations is given by *insertions*, *deletions*, and *substitutions* of both nodes and edges. We denote the substitution of two nodes $u \in V_1$ and $v \in V_2$ by $(u \rightarrow v)$, the deletion of node $u \in V_1$ by $(u \rightarrow \varepsilon)$, and the insertion of node $v \in V_2$ by $(\varepsilon \rightarrow v)$, where ε refers to the empty node. For edge edit operations we use a similar notation.

A set $\{e_1, \ldots, e_t\}$ of t edit operations e_i that transform a source graph g_1 completely into a target graph g_2 is called an *edit path* $\lambda(g_1, g_2)$ between g_1 and g_2. Let $\Upsilon(g_1, g_2)$ denote the set of all edit paths transforming g_1 into g_2 while c denotes the cost function measuring the strength $c(e_i)$ of edit operation e_i. The graph edit distance can now be defined as follows.

Let $g_1 = (V_1, E_1, \mu_1, \nu_1)$ be the source and $g_2 = (V_2, E_2, \mu_2, \nu_2)$ the target graph. The *graph edit distance* between g_1 and g_2 is defined by

$$d_{\lambda_{\min}}(g_1, g_2) = \min_{\lambda \in \Upsilon(g_1, g_2)} \sum_{e_i \in \lambda} c(e_i) \quad , \tag{1}$$

Optimal algorithms for computing the edit distance of two graphs are typically based on combinatorial search procedures. A major drawback of those procedures is their computational complexity, which is exponential in the number of nodes. To render graph edit distance computation less computationally demanding, we employ the often used approximation algorithm BP [15].

3   Matching-Graphs

3.1   Creating Matching-Graphs

Our novel approach is based on matching-graphs originally proposed in [12]. The general idea of the matching-graphs is to extract information on the matching of pairs of graphs in a new data structure that formalizes and encodes the matching parts of the two graphs.

Formally, we assume k sets of training graphs $G_{\omega_1}, \ldots, G_{\omega_k}$ stemming from k different classes $\omega_1, \ldots, \omega_k$. For all pairs of graphs g_i, g_j stemming from the same class ω_l, the graph edit distance is computed. Hence, a (suboptimal) edit path $\lambda(g_i, g_j)$ is obtained for each pair of graphs $g_i, g_j \in G_{\omega_l} \times G_{\omega_l}$. For each edit path $\lambda(g_i, g_j)$, two matching-graphs $m_{g_i \times g_j}$ and $m_{g_j \times g_i}$ are eventually built (for the source and the target graph g_i and g_j, respectively). To this end, all nodes of g_i and g_j that are actually substituted in edit path $\lambda(g_i, g_j)$ are added to $m_{g_i \times g_j}$ and $m_{g_j \times g_i}$, respectively. Vice versa, all nodes that are deleted in g_i or inserted in g_j are neither considered in the two matching-graphs.

Note that this procedure can result in matching-graphs with isolated nodes, which are eventually removed. If a node is not included in the matching-graph (since it was either deleted or inserted in the underlying edit path), the incident edges of this node are not included in the resulting matching-graph. Edges that connect two substituted nodes, however, are included in the matching-graphs. That is, if two nodes $u_1, u_2 \in V_i$ of a source graph g_i are substituted with nodes $v_1, v_2 \in V_j$ in a target graph g_j and there is an edge $(u_1, u_2) \in E_i$ available, (u_1, u_2) is actually included in the matching-graph $m_{g_i \times g_j}$ (whether or not edge (v_1, v_2) is available in E_j).

As shown in [12], the complete process leads to graph structures that can be interpreted as denoised core structures of their respective class. The present paper is built upon these matching-graphs by pursuing the goal of gradually improving the matching-graphs of the first iteration.

3.2   Iterative Building of Matching-Graphs

Using the described procedure for creating matching-graphs out of two input graphs, we now propose an algorithm that iteratively creates sets of matching-graphs out of existing sets of matching-graphs. The proposed procedure is formalized in Algorithm 1.

Algorithm 1: Algorithm for iterative matching-graph creation.

input : sets of graphs from k different classes $\mathcal{G} = \{G_{\omega_1}, \ldots, G_{\omega_k}\}$, number of matching-graphs c
output: sets of matching-graphs for each of the k different classes $\mathcal{M} = \{M_{\omega_1}, \ldots, M_{\omega_k}\}$

1 Initialize $\mathcal{M}$ as the empty set: $\mathcal{M} = \{\}$
2 **foreach** *set of graphs* $G \in \mathcal{G}$ **do**
3 | Initialize M as the empty set: $M = \{\}$
4 | **foreach** *pair of graphs* $g_i, g_j \in G \times G$ *with* $j > i$ **do**
5 | | $M = M \cup \{m_{g_j \times g_i}, m_{g_i \times g_j}\}$
6 | **end**
7 | reduce M to the c matching-graphs with highest quality q
8 | **do**
9 | | **foreach** *pair of graphs* $m_i, m_j \in M \times M$ *with* $j > i$ **do**
10 | | | $M = M \cup \{m_{m_j \times m_i}, m_{m_i \times m_j}\}$
11 | | **end**
12 | | reduce M to the c matching-graphs with highest quality q
13 | **while** *M has changed in the last iteration*
14 | $\mathcal{M} = \mathcal{M} \cup M$
15 **end**

The input of the algorithm is a set $\mathcal{G}$ that contains several sets of graphs $\{G_{\omega_1}, \ldots, G_{\omega_k}\}$ each representing members of a certain class ω_k. Additionally, the number of matching-graphs per class is fixed to a user-defined value c. The output is a set $\mathcal{M}$ which consists of k different sets $M_{\omega_1}, \ldots, M_{\omega_k}$ each containing c matching-graphs that represent one of the given classes.

First $\mathcal{M}$ is initialized to the empty set. The algorithm then actually starts on line 2 by iterating over each set of graphs $G \in \mathcal{G}$. For each of these sets the corresponding result M is initialized as the empty set.

As seen on line 4 to 6, before beginning the main iterative process, we first loop through all possible combinations of graphs $g_i, g_j \in G \times G$, where $j > i$. As mentioned in Sect. 3.1, from one edit path $\lambda(g_i, g_j)$ two matching-graphs are inferred, viz. $m_{g_i \times g_j}$ and $m_{g_j \times g_i}$, where g_i and g_j is the source and target graph, respectively (line 5). Hence, this process yields $n(n-1)$ matching-graphs, where n is the number of graphs in the current set G^1.

On line 7 we proceed to select the c graphs from M with the highest quality q by calculating the relative frequency of occurrence in their own class with respect to the occurrence in other classes. Formally, for a matching-graph $m \in M$ derived from graphs stemming from class ω_l, we verify for all graphs $g \in G_{\omega_l}$ whether or not m is a subgraph of g and store the number of positive matches in f_1. Likewise, we count all graphs $g' \in G_{\omega_i}$, where $\omega_i \neq \omega_l$, that contain m as subgraph and store this number in f_2.

Clearly, the higher f_1 and simultaneously the lower f_2 for a given matching-graph m, the better the quality of m. With $f_2 = max(1, f_2)$ (in order to avoid divisions by zero), we formalize the quality q of a matching-graph m by means of

$$q(m) = \frac{f_1}{f_2} \quad . \tag{2}$$

Given this initial set M of matching-graphs, the whole process is eventually repeated (lines 9 to 12). Yet, instead of creating the matching-graphs from the training set G, we produce matching-graphs from pairs of existing matching-graphs. This process is repeated as long as the c graphs in M have altered in the last iteration (line 13). Once the algorithm terminates, we obtain k sets of c matching-graphs for each class which are stored in $\mathcal{M}$.

4   Experimental Evaluation

4.1   Experimental Setup

The main question – from the experimental point of view – is whether or not our novel procedure is able to create more representative matching-graphs than

---

[1] Note that edit path $\lambda(g_i, g_j)$ is not necessarily the same as $\lambda(g_j, g_i)$ and thus, it could actually happen that the resulting matching-graphs stemming from these edit paths also differ. Yet, due to computational reasons we omit the computations of the edit paths and matching-graphs in both directions and assume two matching-graphs per graph pair.

the initial procedure (proposed in [12]). In order to answer this question, we count the occurrences of the matching-graphs found in a given test set (via subgraph isomorphism verification from graph-tool[2] that is based on the VF2 algorithm [16]). That is, we create the matching-graphs using the aforementioned algorithm, on the training sets and then count the actual occurrences of the created graphs as subgraphs in the corresponding test sets.

The proposed approach is evaluated on two different data sets from the IAM graph repository both providing graphs from two classes [17][3]:

- AIDS (active vs. inactive)
- Mutagenicity (mutagen vs. nonmutagen).

The single parameter of our algorithm – namely, the number of matching-graphs being generated – is set to $c = 15$ in our experiments for the sake of convenience.

4.2   Test Results and Discussion

First, we aim at researching whether or not the quality of the matching-graphs actually improves from iteration to iteration. To this end, we plot the qualities (according to Eq. 2) of the top c matching-graphs from the first to the last iteration (see Fig. 1). It is clearly observable that the quality of the matching-graphs increases by each iteration. For instance, for the AIDS data set and class active the initial matching-graphs offer quality values between 20 and 38, while the qualities of the final matching-graphs are between 39 and 45, which means that the final matching-graphs occur about 39 to 45 times more often in their own class than in the other class.

One could assume that this increase is mainly due to the fact that the matching-graphs become smaller from iteration to iteration (and are therefore found more often in the correct class). In fact, we observe only a marginal reduction of the average graph size (if any). This can be seen in Table 1 where we show the number of iterations per data set and class as well as the average number of nodes of the matching-graphs in the first and last iteration.

Table 1. Development of the average number of nodes from the top matching-graphs between the first and last iteration.

		# iterations	Avg. # nodes first iteration	Avg. # nodes last iteration
MUTA	Nonmutagen	2	17.9	18.1
	Mutagen	2	14.4	14.3
AIDS	Inactive	4	6.6	5.5
	Active	3	14.5	12.1

---

[2] https://graph-tool.skewed.de/static/doc/topology.html#graph_tool.topology. subgraph_isomorphism.

[3] www.iam.unibe.ch/fki/databases/iam-graph-database.

Next, we analyze the absolute frequencies of the resulting matching-graphs in the correct and false classes (see Fig. 2). It can be clearly observed that the resulting matching-graphs occur significantly more often in their correct classes than in the wrong class for the AIDS active, Mutagenicity mutagen and nonmutagen classes.

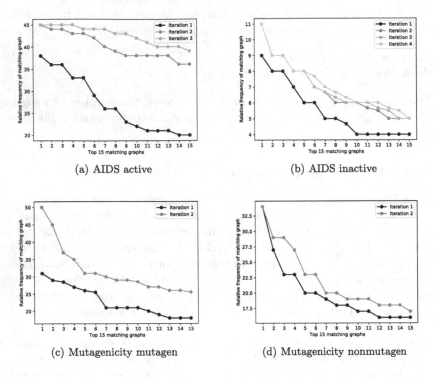

(a) AIDS active

(b) AIDS inactive

(c) Mutagenicity mutagen

(d) Mutagenicity nonmutagen

Fig. 1. Evolution of the relative frequencies of which a matching-graph occurs in the correct and incorrect class during the iterations.

In particular for AIDS active (Fig. 2(a)) we can report exciting results, where most of the matching-graphs occur in about 80% of the test graphs of the correct class, and only about 1% in the other class. However, for the AIDS inactive (Fig. 2(b)), the resulting matching-graphs do not seem to be representative.

Finally, we conduct a qualitative evaluation. In Fig. 3 we visualize the three matching-graphs with the best quality (according to Eq. 2) of each class for both data sets. Interestingly, as seen in Fig. 3(a), the matching-graphs for the AIDS active class consist of carbon atoms only (in very specific combinations, that seems to be exclusive for this class). The matching-graphs of the inactive class on the other hand consist of chains of various atoms, that seem to be less common overall and not very specific to the inactive class (as seen in the quantitative analysis in Fig. 2).

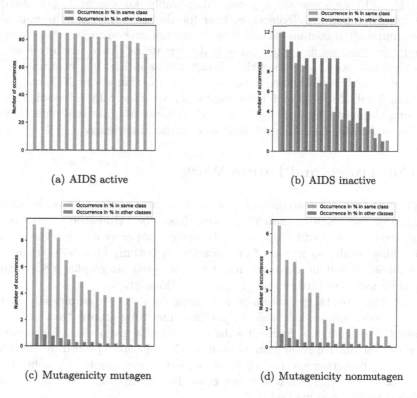

(a) AIDS active

(b) AIDS inactive

(c) Mutagenicity mutagen

(d) Mutagenicity nonmutagen

Fig. 2. Percentage of frequency of the final c matching-graphs in the test set. Bars in light gray show the frequency in the correct class while the darker bars show the frequency in the incorrect class.

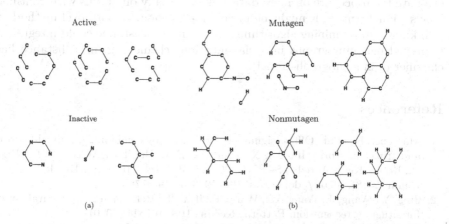

Fig. 3. The three matching-graphs with the best quality for the AIDS data set (a) and the Mutagenicity data set (b).

In Fig. 3(b) we show the top matching-graphs for the Mutagenicity data set. One of the major differences is, that for the mutagen class the matching-graphs found often contain carbon rings or partial carbon rings, whereas in the nonmutagen class we find much more hydrogen atoms. Also very interesting to see is that the second sample of the Mutagen class in Fig. 3(b) contains a NO_2 compound, which is well known to be mutagenic [18]. Overall the NO_2 compound occurs in 5 out of the 15 matching-graphs. This is especially interesting as the matching-graphs are automatically created on the basis of the edit path between training and matching-graphs without any further knowledge.

5   Conclusions and Future Work

We propose to build matching-graphs on the basis of the edit path between two graphs. The resulting matching-graphs basically include the nodes that are substituted via graph edit distance. In the present paper we advance the creation of matching-graphs by means of an iterative algorithm. That is, starting with an initial set of matching-graphs, novel sets of matching-graphs are iteratively created by means of further matchings of matching-graphs.

In an experimental evaluation on two graph data sets, we empirically confirm that our novel approach is able to produce matching-graphs that accurately represent significant and frequent substructures of a given class. Moreover, by means of a qualitative evaluation we confirm that our novel procedure offers high potential for detecting novel and relevant substructures in sets of graphs. To the best of our knowledge this is the first time that a graph matching algorithm is employed for this specific task.

There are several promising paths to be pursued in future work. First we feel that the classification accuracy of the framework presented in [12] can further be increased by using the novel improved matching-graphs. Moreover, one could evaluate the procedure on more data sets (especially on graphs with continuous labels). Furthermore, it might be interesting to compare our novel method with well known graph mining algorithms. Last but not least, one could integrate the improved matching-graphs in a classification scheme (e.g. in a distance based classifier or in a subgraph-kernel).

References

1. Maergner, P., et al.: Offline signature verification by combining graph edit distance and triplet networks. In: Bai, X., Hancock, E.R., Ho, T.K., Wilson, R.C., Biggio, B., Robles-Kelly, A. (eds.) S+SSPR 2018. LNCS, vol. 11004, pp. 470–480. Springer, Cham (2018). https://doi.org/10.1007/978-3-319-97785-0_45
2. Jing, Y., Wang, J., Wang, W., Wang, L., Tan, T.: Relational graph neural network for situation recognition. Pattern Recogn. **108**, 107544 (2020)
3. Khan, S., Islam, N., Jan, Z., Din, I.U., Rodrigues, J.J.C.: A novel deep learning based framework for the detection and classification of breast cancer using transfer learning. Pattern Recogn. Lett. **125**, 1–6 (2019)

4. Rieck, B., Bock, C., Borgwardt, K.: A persistent Weisfeiler-Lehman procedure for graph classification. In: International Conference on Machine Learning, pp. 5448–5458. PMLR (May 2019)
5. Yang, G., Cao, J., Chen, Z., Guo, J., Li, J.: Graph-based neural networks for explainable image privacy inference. Pattern Recogn. **105**, 107360 (2020)
6. Curado, M., Escolano, F., Lozano, M.A., Hancock, E.R.: Early detection of Alzheimer's disease: detecting asymmetries with a return random walk link predictor. Entropy **22**(4), 465 (2020)
7. Conte, D., Foggia, P., Sansone, C., Vento, M.: Thirty years of graph matching in pattern recognition. Int. J. Pattern Recogn. Artif. Intell. **18**(03), 265–298 (2004)
8. Foggia, P., Percannella, G., Vento, M.: Graph matching and learning in pattern recognition in the last 10 years. Int. J. Pattern Recognit Artif Intell. **28**(01), 1450001 (2014)
9. Bunke, H., Allermann, G.: Inexact graph matching for structural pattern recognition. Pattern Recogn. Lett. **1**(4), 245–253 (1983)
10. Sanfeliu, A., Fu, K.S.: A distance measure between attributed relational graphs for pattern recognition. IEEE Trans. Syst. Man Cybern. **3**, 353–362 (1983)
11. Vishwanathan, S.V.N., Schraudolph, N.N., Kondor, R., Borgwardt, K.M.: Graph kernels. J. Mach. Learn. Res. **11**, 1201–1242 (2010)
12. Fuchs, M., Riesen, K.: Matching of matching-graphs - a novel approach for graph classification. In: Proceedings of the 25th International Conference on Pattern Recognition, ICPR 2020, Milano, Italy, 10–15 January 2021 (2021)
13. Mitchell, M.: An Introduction to Genetic Algorithms. MIT Press, Cambridge (1996)
14. Jiang, C., Coenen, F., Zito, M.: A survey of frequent subgraph mining algorithms. Knowl. Eng. Rev. **28**(1), 75–105 (2013)
15. Riesen, K., Bunke, H.: Approximate graph edit distance computation by means of bipartite graph matching. Image Vis. Comput. **27**(4), 950–959 (2009)
16. Cordella, L.P., Foggia, P., Sansone, C., Vento, M.: A (sub)graph isomorphism algorithm for matching large graphs. IEEE Trans. Pattern Anal. Mach. Intell. **26**(10), 1367–1372 (2004)
17. Riesen, K., Bunke, H.: IAM graph database repository for graph based pattern recognition and machine learning. In: da Vitoria Lobo, N., et al. (eds.) SSPR /SPR 2008. LNCS, vol. 5342, pp. 287–297. Springer, Heidelberg (2008). https://doi.org/10.1007/978-3-540-89689-0_33
18. Luo, D., et al.: Parameterized explainer for graph neural network. arXiv preprint arXiv:2011.04573 (2020)

Semi-Autogeonous (SAG) Mill Overload Forecasting

R. Hermosilla[1]($\boxtimes$), C. Valle[2], H. Allende[1], E. Lucic[3], and P. Espinoza[4]

[1] Universidad Técnica Federico Santa María, Valparaíso, Chile
`rodrigo.hermosilla@sansano.usm.cl`, `hector.allende@inf.utfsm.cl`
[2] Universidad de Playa Ancha, Valparaíso, Chile
`carlos.valle@upla.cl`
[3] DeepCopper, Viña del Mar, Chile
`erich.lucic@deepcopper.com`
[4] BHP Billiton Minera, Escondida, Antofagasta, Chile
`pablo.pa.espinoza@bhpbilliton.com`

Abstract. In mining, the detection of overload conditions in SAG mills is of great relevance to guarantee their operational continuity due to their economic and environmental impact. Various authors have tried to use Machine Learning techniques to identify the relationship between the variables and the underlying overload phenomenon. Using a combination of techniques integrated into a framework, we seek to establish a model that learns and detects overloads, taking care of aspects such as selecting variables, the generation of an encode that maximizes the learning using a Gram's matrices approach, and that consider the imbalance of the classes to training a Convolutional Neural Network. Our proposed framework allowed us to establish a mechanism that statistically exceeds the metrics presented by other authors and opens an interesting space of exploration for the continuous improvement of predictive models.

Keywords: SAG Overload forecasting · Multivariate times series forecasting · Conditional Mutual Information (CMI) · Gramian Angular Difference Field (GADF) · Encoding time series as images · Convolutional neural networks · Snowball

1 Introduction

In mining, a Semi-Autogenous Grinding mill (SAG) is the equipment used at mineral processing plants in the size reduction process, making the ore suitable for the next stage of flotation. Its use lies in significant processing and reduction capability offered by this kind of mill, where maximize the operational continuity and production is essential.

Estimations reveal that for every 1% in mineral production increase, revenue increases between US\$80MM and US\$160MM per year for each SAG [9]. Furthermore, a 1% production increase, with the same electricity consumption,

This work was supported in part by Basal Project AFB 1800082.

© Springer Nature Switzerland AG 2021
J. M. R. S. Tavares et al. (Eds.): CIARP 2021, LNCS 12702, pp. 392–401, 2021.
https://doi.org/10.1007/978-3-030-93420-0_37

the emission avoid of between 200 and 400 tons of CO_2 per year in each SAG [3,8,9]. To achieve this growth, it is necessary to increase the mill's production under overload-free conditions, for example, through optimization and control techniques that are generally performed by expert operators and knowledge database. An overload is an instability condition of the mill that occurs under certain conditions, such as excessive ore volume, insufficient charge's movement, or an incorrect percentage of solids, providing an inefficient production. Identifying in advance an overload becomes especially relevant. For this reason, the identification of overburden is a central problem to optimize the grinding process.

In our research, we had found authors like McClure et al. [7] who explored the overload detection comparing methods as Kernel Principal Component Analysis (KPCA), Support Vector Machine (SVM), and Locally Linear Embedding (LLE); their results are significant to understand the relationship of overload and the variables associated with the mill. On the other hand, authors as Bardinas et al. [2] proposed a method using distance matrices encode to extract features from time series to detect mills' operating states, given us lights about a way that we will affront the feature transformations in our case. In both perspectives, besides the difficulty of obtaining a model to classify the overload state, the authors described the challenge to obtain a method to model the underlying process to the mill.

Using these investigations and inspired by the way in how specialists determine the overloads; observing the relationship in pairs of variables, our research will propose the use of Gram matrices to represent the angular difference of pairs of characteristics, all of them joined in images with several channels that will be applied to train a convolutional neural network to predict overloads.

However, the prediction of overloads is not only tricky because of the stochastic nature of the process. Exists at least two additional issues to resolve: the proper selection of the features and the classes' imbalance (less 2%). Given this, we proposed us create a framework that hopes to solve the overload forecast as an entire problem, using Conditional Mutual Information for feature selection, the generation of a convolutional model based on Gram's matrices for time series, and the treatment of the imbalanced classes using a modified snowball implementation.

In the next chapters, we will explain the overload problem in more detail and how we hope to resolve it. In Sect. 2, we will expand the definition of overload, and as this case has been attended by different researchers. In Sect. 3, we will detail our proposal and the way we approach the problem using a mix of technologies as a framework. In Sect. 4, we will present our experiments and results, and in Sect. 5, we will describe the conclusions and next steps in our investigation.

2   Prediction of Overloads

The main goal of a SAG mill is to maximize ore treatment. However, the objective of how many mass that the mill can process is limited. Some conditions, such as excessive ore volume, composition, particle size distribution, hardness, underestimated spin speed, incorrect percent solids, among others, can lead to overload.

Exists three types of overloads can occur in a SAG mill (Fig. 1). The first, called volumetric overload, occurs when there is an overfeed, and factors such as mineral composition, particle size, and the mineral's hardness are not correct. This type of overload usually prevents the effect *cataract* (fall of the mineral from the top of the mill to the inner ground), losing the mill efficiency. The second type of overload is when a pool in the lower part of the mill produces a decrease in solids, reducing the mineral's impact, limiting the size reduction mineral, accumulating, and producing an overload. Finally, the third type of overload occurs when the lifters (the plates that line the mill) are covered with pulp, losing their profile, preventing the mineral from rising, inhibiting the cascade necessary to produce the load's internal impact.

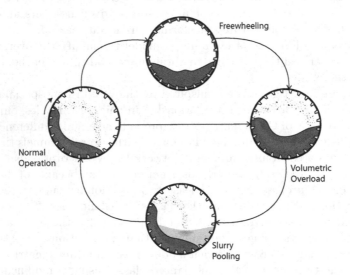

Fig. 1. Types of overloads in a SAG.

On the other hand, the filling, distribution, and composition of the load can influence an overload, which can be determined, for example, by confirming the decrease in the moment of inertia of the mill, because under this condition the torque required to turn decreases the power consumed, however, this behavior is observable only at the beginning of an overload state [1], without offering enough anticipation.

The complexity of the relationship between the various factors underlying an overload makes it difficult to obtain a physical model to determine whether an overload will occur within the next few minutes. A data-driven approach, however, allows us to establish models that learn these complex relationships. Thus, in this research, we will tackle predicting overload using a multivariate time series classification approach, where our model will seek to learn temporal relationships between the features selected and overload in the future.

Mathematically we define the vector $\mathbf{y} = [y_1, \ldots, y_{T'}]^T$ with $t \in [1..T']$ to the set of binary values representing the mill overload condition at different times t, where we will describe $y_t = 1$ as an overload and $y_t = 0$ the absence of it at instant t. We will assume, on the other hand, the $T' \times N$ matrix $\mathbf{X} = [\mathbf{x}_1| \ldots |\mathbf{x}_{T'}]^T$ describes the set of time vectors $\mathbf{x}_t$ composed by the N features, i.e. $\mathbf{x_t} = [x_t^{(1)}, \ldots, x_t^{(N)}]^T$. Thus, $x_t^{(j)}$ denotes the value of jth feature at time t. Our objective will be to predict the overload condition in y_{t+k}, hence in a future of k steps from each observation $\mathbf{x_t}$. Also, we define a T'-dimensional vector $\mathbf{ts}$ that contains the timestamps of the entire dataset.

2.1 Related Work

The problem of solving the overloads of a mill has two research approaches identified in our analysis. One seeks to obtain a method to achieve a functional space free of overloads, and the other seeks to establish a mechanism to anticipate these. Among the recent investigations leading to establishing a mechanism that guarantees a satisfactory operation, free of overloads, is the one exposed by Wang et al. [12], who analyze the method of discrete elements (DEM), of the speed of operation of a mill to avoid overload, describing the relationship between speed, quantity and size of the particles to define an optimal operating space.

On the other hand, research such as those by Salazar et al. [10] seeks to establish a mechanism for predicting and controlling the mill's behavior through a technique called Multiple Input-Multiple Output Model Predictive Control (MIMO MPC) to describe the effect of the operational variables on the total tonnage. Their work showed that it was possible to evaluate stable operating conditions through the developed simulator on an eventual overload condition, keeping the variables controlled without variation.

Perhaps one of the closest investigations to our development line is the one published by McClure et al. [7], who makes a comparison regarding the benefits and consequences of using techniques such as KPCA, SVM, and LLE to predict overloads. In this research, a comparison was made between the three techniques, obtaining a sensitivity and specificity greater than 90% with the LLE technique in conjunction with LDA. However, as we will see, our research was based on a set of characteristics and data different from those presented in said publication. Therefore, we had to run the model described above to convert it into a baseline to measure our performance.

3 Proposed Method

The overload forecast problem was not only a prediction issue; we faced selecting the proper features, resolving the unbalanced classes' obstacle, and achieving a robust model to obtain some promising results. Next, we describe some techniques that we used to overcome these difficulties, but first, we will define a conceptual frame.

Considering that our issue corresponds to a multivariate time-series problem, which represents a stochastic process, we will set our objective of constructing a model that generates a prediction for a moment $t + k$ for the class y, where k is the steps at the future. Given that we do not know the distribution associated with the underlying process, we will base our model on assuming that we use previously observed values for overload forecasts. We will be able to model the occurrence of the values $p(y_{t+k}|\mathbf{x}_t, \ldots, \mathbf{x}_{t-w-1})$ in w steps.

3.1 Feature Selection

As we see in detail in the next section, we had to use a convolutional network to train a set of matrices from Gram's encode, grouped by features pairs. Our first challenge was to select suitable features combined to maximize the relationship between these pairs and our classification target.

Usually, an attempt to describe the SAG mill's operation has been made through its operational variables, whether they are manipulated or resulting. We will define as non-controllable or manipulable variables those perturbations of the phenomenon to be solved, in this case, the distribution of particles that feed the mill. As the resulting variables, we will define the tonnage, the speed, and the solid ratio, leaving the pressure and power as manipulated variables.

Our first challenge is to select the variables that add more helpful information to our model. As will be seen in the next section, we need to select pairs of features that maximize our target information. We resolved it using a novel method based on Information Theory called Conditional Mutual Information (CMI) by [6], where the objective is to select a small number of features that can carry as much information as possible concerning a third (target). We can compute the CMI of a pair of feature with respect to the target as:

$$I(x^{(i)}, x^{(j)}|y) = E_y[D_{kl}(P_{(x^{(i)}, x^{(j)}|y)}||P_{(x^{(i)}|y)} \otimes P_{(x^{(j)}|y)})], \tag{1}$$

where $I(x^{(i)}, x^{(j)}|y)$ is the expected value with respect to y, of the *Kullback-Leibler divergence* or *relative entropy* D_{kl} from the conditional joint distribution $P_{(x^{(i)}, x^{(j)}|y)}$ to the product $\otimes$ of the conditional marginals $P_{(x^{(i)}|y)}$ and $P_{(x^{(j)}|y)}$. Then we define $\mathbf{z}_r = (z_r^{(1)}, z_r^{(2)})$ as the two-column matrix that contains the pair of features $x^{(i)}$ and $x^{(j)}$ with the rth largest CMI obtained using Eq. (1). For example, the vector $z_3 = (z_3^{(1)} = x^{(i)}, z_3^{(2)} = x^{(j)})$, where $x^{(i)}$ and $x^{(j)}$ are the features that obtains the 3rd highest CMI. And $z_{tr}^{(i)}$ denotes the value of ith feature at time t of rth pair selected. Also we define $\tilde{\mathbf{Z}}$ as a three-dimensional array with dimensions $w \times w \times R$, where w is the window length, and R is the number of selected pairs, composed of each $\tilde{z}_{tr}^{(i)}$.

3.2 Convolutional Neural Networks and Gram Matrices

Inspired in obtain overloads through the information contained in the relationship of the features, regarding this kind of events, we bet a method based on the right encoding, conserving, and highlighting the variables' temporal relation, and how these describe an overload when grouping them in pairs.

Convolutional networks are an essential way for pattern, and image recognition problems, however, they have also proven useful in time series problems [4]. We use an approach by converting variables to matrices representing the values' angular change (GADF), obtaining an encode useful to train our forecast model. Also, using GADF encoding, we can revert the conversion and obtain the original features' values [5,13] when as required.

Once we are selected our R better pairs of features from the time series $\mathbf{X}$ using the CMI technique, then we scale and translate each selected pairs $\mathbf{z}_1, \ldots, \mathbf{z}_R$ to the space $[-1, 1]$ using a *minmax scaler* technique showed as follow:

$$\tilde{z}_{tr}^{(i)} = \frac{z_{tr}^{(i)} - \min\left(\mathbf{z}_r^{(i)}\right)}{\max\left(\mathbf{z}_r^{(i)}\right) - \min\left(\mathbf{z}_r^{(i)}\right)}, i = 1, 2, \ t = 1, \ldots, T', \ r = 1, \ldots, R, \qquad (2)$$

where $\min\left(\mathbf{z}_r^{(i)}\right)$ and $\max\left(\mathbf{z}_r^{(i)}\right)$ are minimum and maximum values of the ith feature respectively, over all time instants.

Using $\tilde{z}_{tr}^{(i)}$, we transform it in polar coordinates encoding the value by the cosine of the angle of it and using the timestamp ts_t to obtain the radius ρ, given by

$$\theta_{rt}^{(i)} = \arccos\left(\tilde{z}_{tr}^{(i)}\right), -1 \leq \tilde{z}_{tr}^{(i)} \leq 1, \qquad (3)$$

$$\rho_t = \frac{ts_t}{\max\left(\mathbf{ts}\right)}, \ ts_t \in \mathbf{ts},$$

where $\mathbf{ts}_t$ is the value of tth timestamp and $\max\left(\mathbf{ts}\right)$ is the maximum value of all timestamps.

With the above equation is possible to preserve the temporal relationship and obtain a single representation of the time series in polar coordinates, and since the *cosine* function is bijective and monotonous when its argument $\theta \in [0, \pi]$, the conversion process has a single reverse map. The next step is to apply the Gramian Difference Angular Field (GAFD) over each pair, obtaining a three-dimensional array G with dimensions $w \times w \times R$. Here, w is the window length, and R is the number of selected pairs, and $g_{r(i,j)}$ represents the pixel of position (i, j) of the rth Gram's matrix of G, computed as follows:

$$g_{r(i,j)} = \cos\left(\theta_{ri}^{(1)} + \theta_{rj}^{(2)}\right), i, j = 0, \ldots, w - 1, \qquad (4)$$

where $\theta_{ri}^{(1)}$ and $\theta_{rj}^{(2)}$ represents the pairs selected and transformed previously. Hence, we can view G as a image with R channels, and each of this images represents the angular variation of the selected pairs. The resulting set of images G will be received by a deep convolutional network (Fig. 2), composed of two convolutional layers of n_1 filters of size $k_1 \times k_1$, a 2×2 max-pooling, followed by a layer of batch normalization. To these are added, others two convolutional layers of n_2 filters of size $k_2 \times k_2$, a max-pooling of 2×2, also followed by a batch normalization layer. The convolutional layers are zero-padding, and with

ReLU activation function, all of the above is sent to a full connected flatten layer, ending with a dense layer of n_3 nodes with *Dropout*, which connects to an output of a single neuron for binary classification with a *sigmoidal* activation function. The cost function used is *binary crossentropy*.

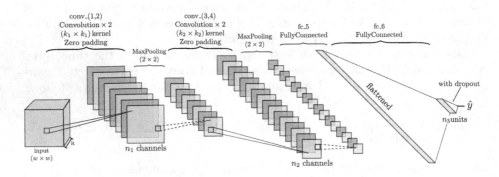

Fig. 2. Network architecture.

3.3   Unbalanced Classes and Snowball Method

Once we defined our model, we faced a significant issue associated with the binary class's unbalancing that describes the overloads. As expected, the overload condition is a rare situation that is sought that does not occur. Less than 1.5% of overload in all of the data is a usual scenario and a significant challenge to resolve in our training process. We define our class in terms of y_t, where $y_t = 0$ is a normal operation of the mill and $y_t = 1$ is a overload registered in the time t. To face this scenario, we adapted a technique called *snowball* proposed by Wang et al. [11]. The main idea of *snowball* is to separate the $\{0, 1\}$ classes into positive and negative subsets and performing training in cycles, starting with the complete subset of $y_t = 1$ class training and then adding elements of the $y_t = 0$ class, which are reinforced by new training of the $y_t = 1$ class in case of not achieving convergence in any of the cycles.

Formally, we will denote N_P to the amount of positive elements, and N_i to each of the disjoint subsets of negative elements of the experiment, with $i \in [0, q - 1]$. Initially, the algorithm suggests that random partitions form the N_i subsets and that a set of sweep constants be defined. However, we changed this method to use the benefits of current *early stopping* techniques, stopping the seek once we obtain a convergence.

3.4   Summarize as a Whole Framework

As we can see in Fig. 3, as well as in the previous sections, we based our proposal on implementing a framework that includes several steps or stages, each one in charge of solving one of the aspects that we define as intermediate objectives.

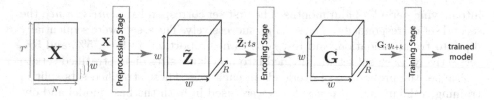

Fig. 3. Framework to train a overload forecast model.

Since we proposed a convolutional neural network training based on an encode representing the interaction between two variables regarding the target, the first step is to know which pairs of features add more information. We use the conditional mutual information for each possible combination of variables to measure it and choose the set of R pairs with the best results in terms of CMI. Then, we scale them to $[-1, 1]$ space for each pair to go to the right domain for each feature $\tilde{z}$. These steps we called *Preprocessing Stage*.

In the next phase, called *Encoding Stage*, the scaled features $\tilde{z}_{tr}^{(i)}$, and the time-stamps vector ts are transformed to polar coordinates. We use it in the generation of the Gram three-dimensional arrays G, that as we have seen, represents the angular difference between each pair of features previously selected. These matrices of each windows' time are stacked in R channels, obtaining a three-dimensional array for each time t, which we use to train in the final stage, called *Training Stage*. In this final stage, we will use G for each time t and its respective target in y_{t+k}. Given the overload class's unbalanced condition, we use a training process based on groups of elements called *Snowball*. The goal of *Snowball* is to obtain a model susceptible to identifying overloads, learning with the few samples of this type of state using the convolutional neural network described in Fig. 2.

4   Experiments and Results

Our goal is to obtain the overload forecast ten minutes in the future. With at time resolution of 30 s, we setted y_{t+k} with $k = 20$, and each window to the past with $w = 60$ steps, that is 30 minutes. The dataset has 693356 strongly unbalanced records over twelve months, with an unbalanced ratio of 1.46%. The data are composed of 72 features, but only 48 of these are linearly independents of each other.

As we have said, the first step has been to find the pairs of characteristics that maximize the overload information. We have chosen top 10, corresponding to pairs $\{(df, dg), (pw, mnt), (tn, mnt), (df, di), (dg, di), (pr, vl), (pw, vl), (pw, pr), (pr, mnt), (di, pr)\}$. Once we obtain the pairs of features, we generated the respective G matrices after scaling and converting each value to polar coordinates, resulting in 685450 matrices, which meant a loss of 1.14% respect original data due to missing data to create each window time-spaced. This condition degraded the unbalanced ratio to 1.38%. To train, we divided our data into moving periods of four months. The first set corresponded to $m_1, \ldots, m_4$, the second set corresponded to $m_2, \ldots, m_5$, successively. These sets were sequentially flip into train, validation, and test subsets in proportions of 90%, 5%, and 5%, respectively. Each subset was divided into non-same sizes eight stratified sets to guarantee the proportion of overloads in each one. These stratified sets split in training, validation, and test subsets were used in both the base model and our proposed model.

Network hyperparameters were set after a grid search process. After several tests, the channels of the convolutional networks were configured in $n_1 = 32$ and $n_2 = 128$, in addition to the number of nodes of the dense network of $n_3 = 32$ nodes. The kernel sizes were setted in $(k_1 \times k_1) = (k_2 \times k_2) = (3 \times 3)$. We used *Adam* optimizer with learning rate parameter setted in $lr = 0.003$.

Table 1. Average and standard deviation over S test sets. Best results are bolded.

Model	Sensitivity	Specificity	F1
LLE+LDA	0.555($\pm$0.425)	0.453($\pm$0.321)	0.440($\pm$0.298)
Proposed model	**0.643**($\pm$0.227)	**0.894**($\pm$0.106)	**0.609**($\pm$0.084)

As we mentioned before, the data used for this research have different periods and features from those mentioned by the reference model. To make an adequate base comparison, we limit ourselves to training both models with the same dataset. Also, the LLE + LDA model had training with the same features our model had access to, i.e., the variables present in pairs selected in the pre-processing stage using CMI.

Table 1 depicts average sensitivity, specificity, and F1-score with the respective standard deviation over our model and regard a model based on LLE+LDA as is suggest in [7]. To verify if the difference obtained was significant between the methods, we performed a paired student's t-test, getting a p-value = 0.009. This confirm that our proposal statistically outperform the LLE+LDA approach.

5   Conclusions and Future Works

We set out to investigate a way to forecast overloads in SAG mills. This challenge meant finding a complex scenario to obtain a model that adequately learns the underlying phenomenon. So we propose a framework that faces each phase of the process, splitting it into three stages. In the pre-processing stage, we proposed using CMI as a variable selection mechanism to address the creation of Gram matrices that will represent each pair's angular difference. We take advantage of the characteristics offered by CNNs to generate a model that allows us to learn the relationships of these variables concerning overloads. Finally, we include a block-based training mechanism using a technique called Snowball to face the target's strong unbalance. Experimental results show that our framework statistically outperforms a state-of-the-art approach based on KPCA, SVM, and LLE. This might be explained because the base model is not deep, but we select a non-deep model considering the similarity of the application.

In the future, we will investigate further other network architectures, to initiate a process of gradual improvement of its efficiency (evolutionary processes) and other methods such as the study of Robust Methods to reduce the impact of variability over the results. We are also interested in addressing additional elements that allow obtaining a coding that contributes more and provides better qualities. In particular, we are interested in the use of alpha channel to indicate the fall in time relation of each characteristic with respect to moment t.

References

1. Apelt, T., Asprey, S., Thornhill, N.: Inferential measurement of sag mill parameters. Miner. Eng. **14**(6), 575–591 (2001)
2. Bardinas, J., Aldrich, C., Napier, L.: Predicting the operating states of grinding circuits by use of recurrence texture analysis of time series data. Processes **6**(2), 17 (2018)
3. Bouchard, J., Desbiens, A., Poulin, E.: Reducing the energy footprint of grinding circuits: the process control paradigm. IFAC-PapersOnLine **50**(1), 1163–1168 (2017)
4. Gamboa, J.C.B.: Deep learning for time-series analysis. arXiv preprint arXiv:1701.01887 (2017)
5. Hatami, N., Gavet, Y., Debayle, J.: Classification of time-series images using deep convolutional neural networks. In: Tenth International Conference on Machine Vision (ICMV 2017) 10696, 106960Y (2018)
6. Liang, J., Hou, L., Luan, Z., Huang, W.: Feature selection with conditional mutual information considering feature interaction. Symmetry **11**(7), 858 (2019)
7. McClure, K., Gopaluni, R.: Overload detection in semi-autogenous grinding: a nonlinear process monitoring approach. IFAC-PapersOnLine **48**(8), 960–965 (2015)
8. Ortiz, J.M., Kracht, W., Pamparana, G., Haas, J.: Optimization of a sag mill energy system: integrating rock hardness, solar irradiation, climate change, and demand-side management. Mathematical Geosciences, pp. 1–25 (2019)
9. Pontt, J., Valderrama, W., Olivares, M., Rojas, F., Robles, H., L'Huissiers, S., Leiva, F.: Uso eficiente de la energia en procesos mineros. Centro de automatizacion para la industria minera, Chile (2012)
10. Salazar, J.L., Valdés-González, H., Vyhmesiter, E., Cubillos, F.: Model predictive control of semiautogenous mills (sag). Miner. Eng. **64**, 92–96 (2014)
11. Wang, J., Jean, J.: Resolving multifont character confusion with neural networks. Pattern Recogn. **26**(1), 175–187 (1993)
12. Wang, X., Yi, J., Zhou, Z., Yang, C.: Optimal speed control for a semi-autogenous mill based on discrete element method. Processes **8**(2), 233 (2020)
13. Wang, Z., Oates, T.: Imaging time-series to improve classification and imputation. In: Twenty-Fourth International Joint Conference on Artificial Intelligence (2015)

Novel Time-Frequency Based Scheme for Detecting Sound Events from Sound Background in Audio Segments

Vahid Hajihashemi[1]([✉]) [ID], Abdorreza Alavigharahbagh[1] [ID], Hugo S. Oliveira[1] [ID],
Pedro Miguel Cruz[2] [ID], and João Manuel R. S. Tavares[3] [ID]

[1] Faculdade de Engenharia, Universidade do Porto, Porto, Portugal
Hajihashemi.vahid@ieee.org
[2] Bosch Security Systems S.A., Ovar, Portugal
[3] Departamento de Engenharia Mecânica, Faculdade de Engenharia,
Universidade do Porto, Porto, Portugal

Abstract. Usually, Sound event detection systems that classify different events from sound data have two main blocks. In the first block, sound events are separated from sound background and in next block, different events are classified. In recent years, this research area has become increasingly popular in a wide range of applications, such as in surveillance and city patterns learning and recognition, mainly when combined with imaging sensors. However, it still poses challenging problems due to existent noise, complexity of the events, poor microphone(s) quality, bad microphone location(s), or events occurring simultaneously. This research aimed to compare accurate signal processing and classification methods to suggest a novel method for detecting sound events from sound background in urban scenes. Using wavelet and Mel-frequency cepstral coefficients, the analysis of the effect of classification methods and minimization of the number of train data are some of the advantages of the proposed method. The proposed methods' application to a standard sounds database led to an accuracy of about 99% in event detection.

Keywords: Signal processing · Wavelet transform · Machine learning · Event detection

1 Introduction

Information processing algorithms are a paramount step in artificial intelligence growth. However, the current modes of human-machine communication are geared more towards living with the limitations of computer input/output devices, mainly as to sound and image, rather than the convenience of humans. Sound is one of the primary modes of communication among humans or between environment and humans. On the other hand, it would be interesting if computers could listen to sound and understand meanings. Automatic speech recognition and sound event detection are processes of deriving the word sequence or sound reason, given the speech waveform. Speech understanding goes one step

© Springer Nature Switzerland AG 2021
J. M. R. S. Tavares et al. (Eds.): CIARP 2021, LNCS 12702, pp. 402–416, 2021.
https://doi.org/10.1007/978-3-030-93420-0_38

further, and describe the meaning of the signal in terms of human communication. Intelligent agents such as mobile phones, hearing aids or robots could also hear, but they cannot exactly interpret what is heard. Sound is often a supplement to content such as video and contains information about the environment. The difference is that often the sound can be collected and processed in easier ways. The information gathered from a meaningful sound analysis can be useful for other processes such as robot routing, user alerting, or analysis and understanding details of an event.

A sound event is a designation commonly used to describe a recognizable event in an audio segment. This designation usually enables a person to understand the meaning of an event and how it relates to other events. Sound events can be used to represent a scene symbolically; for example, a hearing scene on a busy street includes cars passing, cars crash and footsteps of pedestrians. Sound scenes can be described with different specific sound events to assign the main subject, for instance, a street Semantic and automatic event detection and understanding are fundamental requirements in modern urban surveillance systems towards smarter and safer cities. While such systems rely heavily on imaging data, other types of data, such as audio data, can be used to overcome the weaknesses of the visual-based systems and enhance the outcomes of the systems towards better decision making by the city authorities. Because of the fuzzy nature of sound events interpretation, artificial intelligence still has many weaknesses in comparison to the human system. Based on challenges in sound event detection in urban scenes, in this article, a comprehensive analysis of usual state of the art features in sound event detection is presented, and a novel time-frequency method is suggested to automatically detect sound events from sound background in audio segments acquired in urban scenes. This article is structured as follows: the next section gives an overview of state-of-the-art researches in sound event detection. The third section describes the mathematical and theoretical fundamentals of wavelet transform (WT), Mel-frequency cepstral coefficients (MFCCs), K-nearest neighbor (KNN) and support vector machine (SVM), which are used in the proposed method. Section 4 presents details of the used database and then describes the proposed method. Simulation results and conclusions are given in Sects. 5 and 6, respectively.

2 Literature Review

In recent years, efforts have been made to expand the issue of sound event recognition to a comprehensive set of events in environments. Most audible scenes are complex in terms of events, as they usually involve several simultaneously active overlapping sound events. There are two ways to automatically detect a sound event: 1) Finding the start and end time of an event in an audio segment, and then make a single-channel (Monophonic) sequence including events as output [1]. This method is called single-channel detection (Monophonic Detection). 2) Finding some events in a multi-channel (Polyphonic) sequence of events, which is called multi-channel recognition [2,3]. A lot of research has been done in sound event detection. An unrelated field approach was proposed in [4], which consists of two steps: automatic background detection and sound event detection. A method for modeling the previous probabilities of overlapping events was proposed using a probabilistic latent semantic analysis (PLSA) to calculate previous probabilities and learn the relationships between event sources [2].

There are two ways to detect events in multi-source environments that can detect multiple overlapping sound events. The first uses uniform single-channel recorded sounds (mixed signals), and in the detection stage, uses several limited Viterbi [5] transitions to record overlapping events [6]. The second uses Unsupervised Source Selection as a processing step to minimize the impact of overlapping events, and the detection step is performed separately for each system, [7]. In [3], two methods based on an iterative algorithm for Exception Maximization (EM) used to select the desired voice: one, based on the most probable current selection; and another, based on the gradual elimination of the most probable current from the training. The relationship between sound and label in a sound database was studied by evaluating the semantic similarity of sample labels with similar semantic sounds in [8]. On the other hand, a method for combining sound similarity and semantic similarity in a single similarity criterion was proposed in [9].

In some research, the audio signal is recognized as a single-channel signal with one event at a time [10,11]. LeCun et al. proposed a system for detecting an event in a real-life recorded file, using deep learning [12]. In [2], a Probabilistic Latent Semantic Analysis (PLSA), a method close to Non-negative Matrix Factorization (NMF), was proposed to detect overlapping sound events. Simultaneously with the occurrence of events, the degree of overlap of a polyphonic part is represented. Cotton and Ellis applied NMF to MFCCs and tested proposed method on the detection of heterogeneous sound events [10]. In speech recognition applications, a usual assumption is the existence of a dominant source that should be analyzed [13], but this assumption is not true in event detection. One strategy to manage multi-voiced signals is to separate sound resources and analyze each source separately [7,14]. In [14], a study on computational analysis of auditory scenes was performed to study human-robot interaction by recognizing auditory information. A review of the latest research in the category of sound event categorization is presented in [15], where various types of convolutional neural network (CNN) architectures used to categorize sound events are described.

Many researchers have focused on sound denoising to increase the sound event detection accuracy. According to our review, considerable research has been done on urban noise modeling and not so on noise removal. Usual noise management approaches are focused on the reduction of the noise energy [16]. In event detection, noise is usually defined as any unwanted normal environment sound that may decrease the accuracy of the abnormal sound detection. Two categories have been introduced for noise removing: energetic masking (EM) and informational masking (IM). EM uses similar time-frequency locations [17,18] and has weakness in high-energy [17,19]; therefore, EM alone is not a good choice [20]. IM is an indirect saliency-based method [21,22] of auditory attention [23,24].

A wide range of features in different domains has been used to detect sound events, namely: Spectrogram [25], patterns similarity in the time domain [26,27], and spectrum as suggested in [28]. Linear predictive coding was used in [29] for sound-based rare-event detection. Mel scale [30], Discrete Cosine Transform [31,32], Mel-Frequency Cepstral Coefficients [33,34], Wavelet decompression [35], Perceptual linear prediction (PLP) [36], Linear prediction cepstral coefficients (LPCCs) [37], and Line spectral frequencies (LSFs) [38] are other features that have been used for sound event detection in different researches.

Each of the aforementioned features has its weaknesses and advantages and none of the mentioned research works were able to specify a feature as the best in sound event detection. Based on our review, some challenges in sound event detection are the presence of different events in one soundtrack, unbalanced number of event data versus normal data in the training process, dynamic context-dependent form and different speed of occurrence. Recent researches usually use MFCCs and Wavelet based features as better features. Various classification methods can be used (based on differences between features) for sound event detection and understanding. Some of the most well-recognized classification methods that have been used in this topic are Logistic Regression, SVM, KNN, Fuzzy C-means clustering, Adaptive Neuro-Fuzzy Inference Systems, Naïve-Bayes, and Deep learning (mainly, Convolutional Neural Networks). In the proposed method, MFCCs and wavelet were selected as feature extractors, and a novel statistical scheme based on normalized histogram is used for feature processing. In the second step, SVM and KNN are used as detection methods.

3 Theoretical Framework

3.1 Wavelet Transform

One of the common feature extraction methods in sound processing is WT. In practice, audio signals are time-domain signals in their raw format. That is, whatever the signal is conveying, is a function of time. In many cases, the most distinguished information is hidden in the frequency content of the signal. The need for WT arises because in sound events, is necessary to have both the time and the frequency information at the same time depending on the particular application, and the nature of the signal in hand, since no frequency information is available in the time-domain signal, and no time information is available in the Fourier space [2,3].

The basic element of WT is known as the "mother wavelet" function, $\Psi(t)$. The Fourier transform of $\Psi(t)$, which is defined as $\Psi(\omega)$, must satisfy the following condition:

$$\int_{-\infty}^{+\infty} \frac{|\Psi(\omega)|^2}{\omega} d\omega = C_{\Psi} < +\infty. \tag{1}$$

Performing scaling and translation operations on $\Psi(t)$ creates a family of scaled and translated versions of the mother wavelet function:

$$\psi_{a,b}(t) = \frac{1}{\sqrt{|a|}} \psi \left(\frac{t-b}{a} \right), \tag{2}$$

where a is the scaling and b is the translation parameters, respectively. Given a mother wavelet function $\Psi(t)$, the continuous wavelet transform (CWT) of function $f(t)$ is:

$$CWT_f(a,b) = |a|^{-1/2} \int_{-\infty}^{\infty} f(t) \psi^* \left(\frac{t-b}{a} \right) dt \ a,b \in R, a \neq 0, \tag{3}$$

where $*$ denotes the complex conjugate. For discrete wavelets, scale-time parameters a and b are discretized as $a = a_0^m$ and $b = nb_0 a_0^m$.

This family of mother discretized wavelet functions $\{\Psi_{m,n}(t)\}$ is given as:

$$\Psi_{m,n}(t) = a_0^{-m/2} \Psi \left(a_0^{-m} t - nb_0 \right) \qquad m,n \in Z. \tag{4}$$

So, by using Eq. (4), Eq. (3) can be rewritten as:

$$(DWT_f)_{mn} = a_0^{-m/2} \int_{-\infty}^{\infty} f(t) \psi \left(a_0^{-m} t - nb_0 \right) dt. \tag{5}$$

The mother wavelet functions used in this work are:

$$\Psi(t) = \frac{2}{\sqrt{3}} \pi^{-\frac{1}{4}} \left(1 - t^2 \right) \exp \left(-\frac{t^2}{2} \right), \tag{6}$$

$$F_n(t) = \sum_{j=-n}^{n} \left(1 - \frac{|j|}{n+1} \right) e^{ijt} = \frac{1}{n+1} \left\{ \frac{\sin \frac{n+1}{2} t}{\sin t/2} \right\}^2, \tag{7}$$

$$\Psi(t) = \begin{cases} 1 & 0 \leq t < 0.5 \\ -1 & 0.5 \leq t < 1 \\ 0 & otherwise \end{cases}. \tag{8}$$

3.2   Mel-Frequency Cepstral Coefficients

MFCCs are based on the known variation of the human ear's critical bandwidths. In MFCC, some filters spaced linearly at low frequencies and logarithmically at high frequencies have been used to capture the phonetically important characteristics of sound signals. The Mel-frequency scale is a combination of linear frequency spacing 1000 Hz and a logarithmic spacing 1000 Hz. A block diagram of the structure of a MFCC scheme is given in Fig. 1. The audio signal is typically recorded at a sampling rate above 10 kHz. These sampled signals can capture all frequencies up to 5 kHz, which cover most energy of sounds that are heard by humans. The main purpose of the MFCCs is to mimic the behavior of the human ears. In addition, rather than the sound waveforms themselves, MFCC's are shown to be less susceptible to noise.

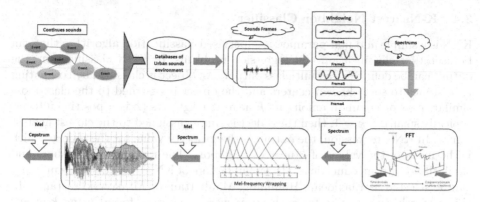

Fig. 1. Block diagram of the MFCC scheme.

3.3 Support Vector Machine

Since SVM classifiers are suitable for binary classification, in this study, a SVM is used for building a binary classifier between any sound event and sound background in audio segments. For the SVM classifier, different kernels were tested. SVM classifies the input data by building an imaginary hyperplane based on its kernel and tries to maximize the margin of that hyperplane to build a safe boundary for binary classification, as well as helping to find non-linear data pattern to classify input. Given a training set of N data points $\{y_k, x_k\}_{k=1}^N$ where $x_k \in R^n$ is the k-th input pattern and $y_k \in R$ is the k-th output pattern, the classifier can be constructed using the SVM method in the form:

$$y(x) = sign\left[\sum_{k=1}^N \alpha_k y_k K(x, x_k) + b\right], \tag{9}$$

where α_k is a non-negative Lagrange multiplier, b is a constant, and $K\left(\cdot, \cdot\right)$ is the kernel, which can be either $K\left(x, x_k\right) = x_k^T x$ - linear SVM, $K\left(x, x_k\right) = (x_k^T x + 1)^d$ - polynomial SVM of degree d, $K\left(x, x_k\right) = \tanh[\kappa\, x_k^T x + \theta\,]$ - multi-layer perceptron SVM, or $K\left(x, x_k\right) = \exp\{-\|x - x_k\|_2^2 / \sigma^2\}$ - RBF SVM, where κ, θ and σ are constants. First, a safety margin (Λ) is defined as:

$$
\begin{aligned}
if(x \in class \quad 1) &\Rightarrow \sum_{k=1}^N \alpha_k y_k K(x, x_k) + b \geq \Lambda, \\
if(x \in class \quad -1) &\Rightarrow \sum_{k=1}^N \alpha_k y_k K(x, x_k) + b \leq -\Lambda.
\end{aligned}
\tag{10}
$$

The SVM training step uses kernel function parameters, α_ks and b, to maximize Λ and the total accuracy. Many analytical, numerical and heuristic methods have been suggested for finding α_ks, b and the selected kernel parameters using training data.

3.4 K-Nearest Neighbor Classifier

KNN is a simple and non-parametric supervised classification algorithm that can be useful for classification and regression problems. In KNN classification, the output can be defined as a multiclass output. An object is classified by computing its distance to some known centers and the object is assigned to the class most similar, near or common among its k nearest neighbors (k is a positive integer, typically small). If $k = 1$, then the object is simply assigned to the closest nearest center. In KNN regression, the output is the property value for the object, which is the mean of the values of k nearest neighbors. The number of neighbors and similarity or distance metric are the main factors of KNN. The distance measure can be selected as Euclidean, Hamming, Manhattan, or Minkowski distance. In this research, the centers for each event were separately found using k-means clustering. The distance of new input were compared to all centers and each one assigned to background or event.

4 Proposed Method

The train step pseudo-code of the proposed system is given by Algorithm 1.

Algorithm 1: *Training procedure*

Input: Labelled recorded signal from urban scenes ($\mathbf{S_i}$)

    Split recordings into one-second non-overlapping sections ($\mathbf{S_{si}}$)

        Apply one-dimensional wavelet transform to $\mathbf{S_{si}}$ and make two output signals ($\mathbf{cA_{ssi}}$, $\mathbf{cD_{ssi}}$)

        Reshape the outputs to N × 8 matrices

        Calculate the 16-bin normalized histogram of two output signals (According to columns) and make 16×16 feature matrix

        Assign event and non-event label to each feature matrix

    Train classifier (**SVM**)

Output: Trained classifier

For sound event detection, many databases were collected and labelled by humans. In the evaluation of our method, the US-SED dataset [39,40] was used because, at first, it is easily accessed and converted to different software formats and, second, it covers several important urban sound events.

4.1 The US-SED Dataset

US-SED is a large dataset of 10,000 ten-second soundscapes and includes ten different sound classes: air conditioner, car horn, children playing, dog bark, drilling, engine idling, gunshot, jackhammer, siren and street music, which has been used for training and evaluating Sound Event Detection (SED) algorithms. All soundscapes were extracted from the UrbanSound8K dataset, approximately 1000 per each of ten urban sound sources (each clip contains one of the ten sources), as the soundbank. UrbanSound8K is pre-sorted into 10 stratified folds, and so can be used as folds 1–6 for generating 6000 training soundscapes, 7–8 for generating 2000 validation soundscapes, and 9–10 for generating 2000 test soundscapes. Soundscapes were generated using the following strategy: first, a background sound normalized to -50 LUFS (Loudness Units relative to Full Scale) was added. The same background sound file for all soundscapes was used in combination with a 10-second clip of Brownian noise, which resembles the typical "hum" often heard in urban environments. By using a purely synthesized background, the database maker was guaranteed that it does not contain any spurious sound events that would not be included in the annotation. Next, the label was chosen randomly from all 10 available sound classes, and the source file was chosen randomly from all clips matching the selected label [39,40].

4.2 Pre-processing

In the pre-processing, the steps of removing noise and dividing the audio signal into non-overlapping segments are performed. In the first step, signals with a frequency of less 20 Hz and above 20 kHz, whose range is outside the human hearing range, are removed. Hence, all noises outside the hearing frequency band, which may have been transmitted to the signal, are eliminated. In the second step, the event situation is assumed constant in each non-overlapping segment. To make this assumption correct, the audio signals are divided into segments of typically 100 ms without overlap. The classifier then classifies the features extracted from these segments. In other words, the event in one tenth second signal is supposed constant. Based on this assumption, each incoming sound is broken into non-overlapping segments with one tenth second length. After splitting the sound into non-overlapping segments, each audio segment is labeled related to including or non including a sound event. In this case, a binary vector is made, which was established based on event segments versus background segments in the database. The 0 (zero) is equivalent to the background segment, and one shows that a sound event has occurred. After denoising, splitting and labeling steps, the audio signal is ready to enter the feature extraction block.

4.3 Feature Extraction

At this block, WT is applied to the input audio signal in order to decompose it and obtain the approximation and detail coefficients. A total of 15 mother wavelet functions with different parameters were implemented in the feature extraction block and studied in terms of the accuracy and efficiency. The used wavelet functions are indicated in Table 1.

Due to the sampling frequency of the input audio signal, which was 44100 Hz, enough samples are available for wavelet. The coefficients of two approximations coefficients cA1 and detail coefficients cD1 resulting from the WT are arranged into two $8 \times N$ matrices and then merged in a $16 \times N$ matrix. For each row, the probability density function (PDF) of the amplitudes is calculated. The number of intervals for PDF is 16 and the length of all intervals are supposed the same. Finally, the output of all rows is added together and arranged as a vector with length equal to 256. Obviously, the length of the feature vector is equal for all types of wavelet functions. The second approach is MFCC. In MFCC,

Table 1. Designation and number of used wavelet functions.

Wavelet designation	Wavelet number
Haar	–
Daubechies	10, 20
Symlet	2, 10, 20
Coiflet	1
Discrete Meyer	–
Fejer-Korovkin	4, 8, 22
Reverse biorthogonal	1.1, 2.4
Biorthogonal	1.1, 2.4

firstly the short-term Fourier transform (STFT) is applied to the signal using the Hanning window. The selected Hanning window is considered periodic and its length is equal to 512. In addition, 128 overlapping samples are considered for each segment. The output of the short-time Fourier transform is applied as input to the MFCC feature extraction step, and according to the sampling frequency available in the database, the feature-length of the MFCC output is a 13×11 matrix. Each row of this matrix is analyzed using a PDF, according to 20 intervals, in order to have the closest adaptation to the wavelet features. The final feature vector of MFCC has 260 elements. The feature vectors are input of detection block.

4.4   Detection

In this study, due to the very high imbalance problem between the background sound segments and the ones with specified sound events, a KNN or a binary SVM is trained to separate background from event sound segments. In this case, the classifier does not need to know type of events occurred in one tenth second segment; therefore, an audio segment is tested in such a way that it can include any event or has only background sound. Simulations were performed to evaluate the accuracy of the different SVM kernels and KNN on different feature types, including mother wavelet functions and MFCCs. After training, the trained classifier should be tested. The pseudo code of the test step is given by Algorithm 2.

Algorithm 2: *Test procedure*

Input: Labelled recorded signal from urban scenes (S_i)
  Split recordings into one-second non-overlapping segments (S_{si})
   Apply one-dimensional wavelet transform to S_{si} and make
   two output signals (cA_{ssi}, cD_{ssi})
   Reshape the outputs to N × 8 matrix
   Calculate the 16-bin normalized histogram of two output
   signals (according to columns) and make a 16×16 feature
   matrix
   Apply feature matrix to the trained classifier
  Calculate the method accuracy for test data
Output: Accuracy

---

5 Simulation Results

In order to determine whether WT can be a good option for separating events from background sound in audio segments, the time domain and two channels of corresponding WT for four different classifiers were considered, Fig. 2.

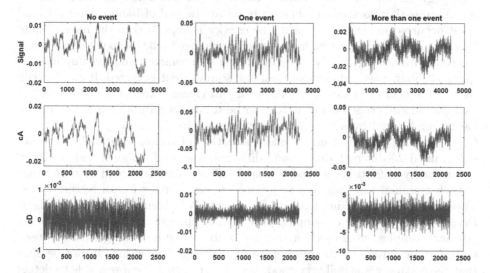

Fig. 2. Original audio signal and corresponding wavelet output for background and event sound segments.

As can be seen in Fig. 2, when the number of events increases, the signal pattern is compressed over time. Especially in cD, the amplitude and shape changes are obvious. These changes can be seen in time domain, but in WT, the changes are even clearer. It should be noted that all three audio segments in Fig. 2 were selected from the same environment and microphone, so the background noise and sound recording conditions are exactly similar for the three signals. To show the difference between event and background sound segments in MFCC, the short-time Fourier transform of the signals were extracted using Hanning window as a main block of MFCC, Fig. 3.

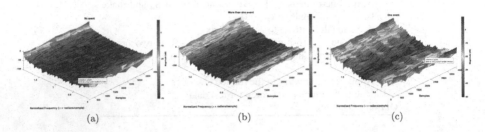

Fig. 3. (a) 3D STFT output for a background sound segment; (b) 3D STFT output for a segment including one sound event, and (c) 3D STFT output for a signal including more than one sound event.

Given the three-dimensionality of the analysis, it can be seen that in a no event segment, the amplitude range is between −50 and 5 dB, and the maximum signal strength is located in a narrow band at the end of bands. After a sound event occurred in the environment, the energy is expanded to the intermediate bands, and the signal pattern has completely changed relative to the signal without sound events (Fig. 3). Additionally, It is more compact and the amplitude is changed significantly when more than one sound event occurred. It can be seen that in the field of Mel coefficients, the signal changes are quite obvious in the background and event sounds, which can be used as a criterion for detecting the background signal from including events signal. In the training classifier step, for eliminating the effect of a random selection of training data, the classifier was trained 20 different times using different wavelet functions, and the average accuracy of 20 times is reported as the final value. It should be noted that in each training, the train and test samples were similar for all feature types. The results obtained using KNN and different SVM kernels are given in Fig. 4. Due to a large number of samples, only 10% of the existing 1 million segments (100,000 samples) were used for training, and all the remaining data was used as test data. Although the percentage of training data is low, it can be seen that the accuracy was still very good, and the trained system was able to detect background vs event sounds in the audio segments.

In Fig. 4, the accuracy for KNN and studied SVM kernels, including linear, RBF and Polynomial, is given separately for train and test sets. In RBF and Polynomial cases, Mel's coefficients, which have been cited as best feature in various researches, showed better accuracy than other methods, although, in linear kernel and KNN, Mel coefficients did not work well due to their non-linear properties.

Among each of the studied kernels, the best accuracy was obtained using MFCC with Polynomial Kernel (99.9%), which indicates the very good accuracy in separating the event signal from the background signal. Figure 4, shows that

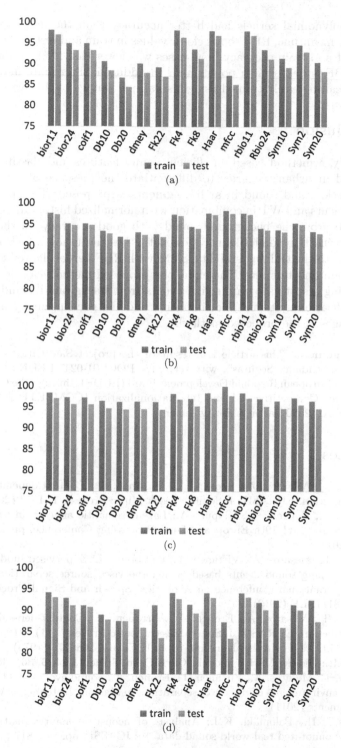

Fig. 4. Accuracy obtained by three different SVM kernels: a) linear, b) RBF, c) Polynomial, and d) KNN.

RBF and polynomial kernels had better accuracy than the linear kernel and KNN. In the meantime, RBF shows closer values in train and test sets (difference between test and train accuracy in all cases was lower than 1%), so for training stability, RBF is better than polynomial. In addition, the results indicate that the system can be easily trained using a small percentage of the available data.

6   Conclusion

In this study, a method based on efficient sound features and classifiers, which can be used in urban scenarios to differentiate the presence of sound events from the background sound in audio segments is proposed. The method uses Mel's coefficients and WT in combination with normalized histogram to separate sound events from the background sound with good accuracy. In the training step, SVM with Polynomial kernel showed the best accuracy, which was equal to 99.9%. As to the training stability, SVM with RBF kernel showed the closest values in train and test sets. Therefore, the proposed method can be used as a pre processing step to separate sound events from background sound in sound event classification systems. The simplicity and the good accuracy are among the advantages of the proposed method.

Acknowledgement. This article is a result of the project Safe Cities - "Inovação para Construir Cidades Seguras", with reference POCI-01-0247-FEDER-041435, co-funded by the European Regional Development Fund (ERDF), through the Operational Programme for Competitiveness and Internationalization (COMPETE 2020), under the PORTUGAL 2020 Partnership Agreement.

References

1. Heittola, T., Mesaros, A., Eronen, A., Virtanen, T.: Context-dependent sound event detection. EURASIP J. Sound Speech Music Process. **1**, 1–13 (2013)
2. Mesaros, A., Heittola, T., Klapuri, A.: Latent semantic analysis in sound event detection. In: 2011 19th European Signal Processing Conference, pp. 1307–1311. IEEE (2011)
3. Heittola, T., Mesaros, A., Virtanen, T., Gabbouj, M.: Supervised model training for overlapping sound events based on unsupervised source separation. In: 2013 IEEE International Conference on Acoustics, Speech and Signal Processing, pp. 8677–8681. IEEE (2013)
4. Heittola, T., Mesaros, A., Eronen, A., Virtanen, T.: Context-dependent sound event detection. EURASIP J. Sound Speech Music Process. **1**(1), 1 (2013)
5. Forney, G.D.: The viterbi algorithm. Proc. IEEE **61**, 268–278 (1973)
6. Eddy, S.R.: Hidden Markov models. Curr. Opin. Struct. Biol. **6**, 361–365 (1996)
7. Heittola, T., Mesaros, A., Virtanen, T., Eronen, A.: Sound event detection in multisource environments using source separation, in Machine Listening in Multisource Environments (2011)
8. Mesaros, T.H., Palomäki, K.J.: Analysis of acoustic-semantic relationship for diversely annotated real-world sound data. In: ICASSP, pp. 813–817 (2013)

9. Mesaros, T.H., Palomäki, K.: Query-by-example retrieval of sound events using an integrated similarity measure of content and label. In: 2013 14th International Workshop on Image Analysis for Multimedia Interactive Services (WIAMIS), pp. 1–4 (2013)

10. Cotton, V., Ellis, D.P.: Spectral vs. spectro-temporal features for acoustic event detection. In: Applications of Signal Processing to Sound and Acoustics (WAS-PAA), pp. 69–72 (2011)

11. Mesaros, A., Heittola, T., Virtanen, T.: Metrics for polyphonic sound event detection. Appl. Sci. **6**(6), 162 (2016)

12. LeCun, Y., Bengio, Y., Hinton, G.: Deep learning. Nature **521**, 436 (2015)

13. Barker, J.P., Cooke, M.P., Ellis, D.P.: Decoding speech in the presence of other sources. Speech Commun. **45**, 5–25 (2005)

14. Wang, D., Brown, G.J.: Computational Auditory Scene Analysis: Principles, Algorithms, and Applications. Wiley-IEEE Press, Hoboken (2006)

15. Hershey, S., Chaudhuri, S., Ellis, D.P., Gemmeke, J.F., Jansen, A., Moore, R.C., et al.: CNN architectures for large-scale sound classification. In: IEEE International Conference on Acoustics, Speech and Signal Processing (ICASSP), pp. 131–135 (2017)

16. Brown, A.L.: Soundscapes and environmental noise management. Noise Control Eng. **58**, 493 (2010)

17. Gelfand, S.A.: Hearing: An Introduction to Psychological and Physiological Acoustics, 6th edn. CRC Press, Boca Rato (2017)

18. Kidd, G.J., Mason, C.R., Richards, V.M., Gallun, F.J., Durlach, N.I.: Informational masking. Audit. Percept. Sound Sources, Springer Handb. Audit. Res. pp. 143–189 (2008)

19. Nilsson, M., Bengtsson, J., Klaeboe, R.: Environmental Methods for Transport Noise Reduction. CRC Press, Boca Raton (2014)

20. Westermann, A., Buchholz, J.M.: The influence of informational masking in reverberant, multi-talker environments. J. Acoust. Soc. **138**, 584–593 (2015)

21. Nilsson, M., et al.: Perceptual effects of noise mitigation. In: Environmental Methods for Transport Noise Reduction, pp. 195–220 (2014)

22. Oldoni, D., et al.: A computational model of auditory attention for use in soundscape research. J. Acoust. Soc. Am. **134**, 852–861 (2013)

23. Kaya, E.M., Elhilali, M.: Modelling auditory attention. Philos. Trans. R. Soc. B Biol., vol. 372 (2017)

24. Kaya, E.M., Elhilali, M.: Investigating bottom-up auditory attention. Front. Hum. Neurosci. **8**, 1–12 (2014)

25. Laffitte, P., Wang, Y., Sodoyer, D., Girin, L.: Assessing the performances of different neural network architectures for the detection of screams and shouts in public transportation. Expert Syst. Appl. **117**, 29–41 (2019)

26. Atrey, P.K., Maddage, N.C., Kankanhalli, M.S.: Sound based event detection for multimedia surveillance. In: 2006 IEEE International Conference on Acoustics Speech and Signal Processing Proceedings, Vol. 5, (2006)

27. Kong, Q., Xu, Y., Sobieraj, I., Wang, W., Plumbley, M.D.: Sound event detection and time-frequency segmentation from weakly labelled data. IEEE/ACM Trans. Sound Speech Lang. Process. **27**(4), 777–787 (2019)

28. Imoto, K., Ono, N.: Spatial cepstrum as a spatial feature using a distributed microphone array for acoustic scene analysis. IEEE/ACM Trans. Sound Speech Lang. Process. **25**(6), 1335–1343 (2017)

29. Janjua, Z.H., Vecchio, M., Antonini, M., Antonelli, F.: IRESE: an intelligent rare-event detection system using unsupervised learning on the IoT edge. Eng. Appl. Artif. Intell. **84**, 41–50 (2019)
30. Lim, M., et al.: Convolutional neural network based sound event classification. KSII Trans. Internet Inf. Syst. **12**(6) (2018)
31. Kürby, J., Grzeszick, R., Plinge, A., Fink, G.A.: Bag-of-features acoustic event detection for sensor networks. In: Proceedings of the Detection and Classification of Acoustic Scenes and Events Workshop (DCASE), pp. 55–59 (2016)
32. Vafeiadis, A., Votis, K., Giakoumis, D., Tzovaras, D., Chen, L., Hamzaoui, R.: Sound content analysis for unobtrusive event detection in smart homes. Eng. Appl. Artif. Intell. **89**, 103226 (2020)
33. Kumar, A., Raj, B.: Sound event detection using weakly labeled data. In: Proceedings of the 24th ACM international conference on Multimedia, pp. 1038–1047 (2016)
34. Derakhshan, M., Marvi, H.: Providing an adaptive model with two adjustable parameters for sound event detection and classification in environmental signals. Tabriz J. Electr. Eng. **49**(2), 565–576 (2019)
35. Crockett, B.G., Seefeldt, A.J.: U.S. Patent No. 10,523,169. Washington, DC: U.S. Patent and Trademark Office (2019)
36. Nasiri, A., Cui, Y., Liu, Z., Jin, J., Zhao, Y., Hu, J.: SoundMask: robust sound event detection using mask R-CNN and frame-level classifier. In: 2019 IEEE 31st International Conference on Tools with Artificial Intelligence (ICTAI), pp. 485–492. IEEE (2019)
37. Soni, S., Dey, S., Manikandan, M.S.: Automatic sound event recognition schemes for context-aware sound computing devices. In: 2019 Seventh International Conference on Digital Information Processing and Communications (ICDIPC), pp. 23–28. IEEE (2019)
38. Hadi, M., Pakravan, M.R., Razavi, M.M.: An efficient real-time voice activity detection algorithm using teager energy to energy ratio. In: 2019 27th Iranian Conference on Electrical Engineering (ICEE), pp. 1420–1424. IEEE (2019)
39. Salamon, J., Jacoby, C., Bello, J.P.: A dataset and taxonomy for urban sound research. In: Proceedings of the 22nd ACM International Conference on Multimedia, pp. 1041–1044 (2014)
40. Salamon, J., Bello, J.P.: Deep convolutional neural networks and data augmentation for environmental sound classification. IEEE Signal Process. Lett. **24**(3), 279–283 (2017)

Computer Vision

Generalized Conics with the Sharp Corners

Aysylu Gabdulkhakova[✉] and Walter G. Kropatsch

Pattern Recognition and Image Processing group, Technische Universität Wien,
Favoritenstrasse 9-11, Vienna, Austria
{aysylu,krw}@prip.tuwien.ac.at

Abstract. This paper analyses the properties of generalized conics that
are defined by N focal points with weights. The generated shapes can
be convex or concave. By varying the weights, it is possible to obtain
up to N corners associated with the focal points. From the shape analysis
perspective, the generalized conics extend the capabilities of the ellipse by
adding the single extra parameter. In general, they enrich the potential
of approaches to describe or represent the shape.

Keywords: Generalized conics · Corner · Shape representation

1 Introduction

A generalized conic is a locus of points satisfying the equidistance property of
the conic section (a parabola, a hyperbola, or an ellipse) that is extended to
accept infinitely many focal points. Originally, this subject raised interest in
the mathematical community. In particular, a multifocal ellipse (also called n-
ellipse or polyellipse) plays a crucial role in solving Fermat-Torricelli [10, 11] and
Weber [12] problems. The present paper aims to explore the representational
power of the generalized conics for shape analysis.

Our previous research studied the geometric properties of an ellipse and a
hyperbola. The developed framework efficiently computes the confocal elliptic
and hyperbolic distance fields. Later [3] it was extended to accept infinitely many
weighted focal points. The geometric nature of the focal points was reconsid-
ered for accepting not only the points but also the shapes. It enables applying
various weighting schemes for objects, their parts, and groups, and promotes
a hierarchical representation. The application scenarios include smoothing [6],
space tessellation [4], skeletonization [2,4,6], facility location problem [3], and
route planning [3]. This paper questions the potential for shape representation
when using the generalized conics. In particular, the interest lies in analysing
the conditions causing the corners in the level sets.

A. Gabdulkhakova—Supported by the Austrian Agency for International Cooperation in Education and Research (OeAD) within the OeAD Sonderstipendien program,
financed by the Vienna PhD School of Informatics.

© Springer Nature Switzerland AG 2021
J. M. R. S. Tavares et al. (Eds.): CIARP 2021, LNCS 12702, pp. 419–429, 2021.
https://doi.org/10.1007/978-3-030-93420-0_39

The remaining of the paper is organized as follows. Sect. 2 provides the main definitions and properties of the generalized conics. Section 3 describes a method for efficient computation of the corresponding distance fields by applying Distance Transform (DT) [1,3]. Section 4 and 6 analyse the configurations of weighted focal points producing convex and concave corners. The discussion continues in Sects. 5 and 7 by introducing the approaches to change the angle at the corresponding corner. Section 8 discusses the potential applications in the shape representation domain. Finally, Sect. 9 concludes the paper.

2   Generalized Conics

The properties of conic sections, such as a parabola, a hyperbola, and an ellipse, can be extended to accept infinitely many focal points. The resultant level sets are called generalized conics. Each level set depends on a number of the focal points, their corresponding weights, and a distance metric.

Consider the Euclidean distance, denoted as δ, between the two 2D points, $P = (x_P, y_P)$ and $Q = (x_Q, y_Q)$:

$$\delta(P, Q) = \sqrt{(x_P - x_Q)^2 + (y_P - y_Q)^2} \tag{1}$$

Definition 1. *A multifocal ellipse (also referred to as n-ellipse, or polyellipse), denoted as $ME(w_1 F_1, w_2 F_2, ..., w_N F_N)$, is a locus of points $P \in \mathcal{R}^2$ with a constant sum of the weighted distances to its N focal points:*

$$ME = \sum_{i=1}^{N} w_i \delta(P, F_i) = const \tag{2}$$

Remark 1. The weights of the focal points in $ME(w_1 F_1, w_2 F_2, ..., w_N F_N)$ are necessarily positive.

Definition 2. *Let $\mathcal{F}$ and $\mathcal{G}$ be the sets of focal points with M and N elements, respectively. A multifocal hyperbola, $MH(w_1 F_1, \cdots, w_M F_M | \nu_1 G_1, \cdots, \nu_N G_N)$, is a locus of points $P \in \mathcal{R}^2$ such that the following absolute difference of the distances remains constant:*

$$MH = |\sum_{i=1}^{M} w_i \delta(P, F_i) - \sum_{j=1}^{N} \nu_j \delta(P, G_j)| = const \tag{3}$$

Property 1. The level set $ME(w_1 F_1, w_2 F_2, ..., w_N F_N)$ is convex and compact [7].

Property 2. The Eq. 2 reaches a global minimum at exactly one point when the focal points are non-collinear or there is an odd number of collinear focal points [9].

Property 3. If there is an even number of collinear focal points, Eq. 2 has the minimum in all points of the line segment connecting a pair of middle focal points [9].

3   Generalized Conics from Distance Transform

For the efficient discrete computation of the distance fields showing the generalized conics, it is proposed to use the classical image processing approach called Distance Transform (DT) [1]. Now coordinates are integers, while Definitions 1 and 2 use continuous coordinates. Some properties may suffer from sampling, for instance, if the continuous minimum is between two sampling points.

Definition 3. *Let I_{binary} be a 2D binary image and $\mathcal{F} \subset I_{binary}$ - a non-empty set of feature elements: $I_{binary}(F) = 0$ and $I_{binary}(M) = 1$, $\forall F \in \mathcal{F}$, $\forall M \notin \mathcal{F}$. Distance Transform (DT) is an operator that converts a binary into a grayscale image, $\mathcal{D} : I_{binary} \mapsto \mathcal{R}$. It assigns to each pixel its distance to the nearest feature element regarding the selected metric d: $\mathcal{D}(M) = \min\{d(M, F) \mid F \in \mathcal{F}\}$, $\mathcal{D}(F) = 0$. DT generated from $\mathcal{F}$ is denoted by $\mathcal{D}_{\mathcal{F}}$.*

According to Eq. 2, the multifocal ellipses represent a sum of DTs of the focal points taken as feature elements $\mathcal{F} = \{F_1, F_2, ..., F_N\}$ with their weights:

$$CMEF_{\mathcal{F}} = \sum_{i=1}^{N} w_i \mathcal{D}_{F_i} \qquad (4)$$

In Eq. 4, $CMEF_{\mathcal{F}}$ denotes Confocal Multifocal Elliptic Field, defined on an image of the same size as I_{binary}.

Confocal Multifocal Hyperbolic Field, denoted as $CMHF_{\mathcal{F}|\mathcal{G}}$, is represented by the difference between the $CMEF$ fields generated from the sets $\mathcal{F}$ and $\mathcal{G}$:

$$CMHF_{\mathcal{F}|\mathcal{G}} = CMEF_{\mathcal{F}} - CMEF_{\mathcal{G}} \qquad (5)$$

As opposed to MH, some pixels in $CMHF_{\mathcal{F}|\mathcal{G}}$ are mapped to a negative distance value. It means that these pixels are closer to $CMEF_{\mathcal{F}}$ in terms of a minimum total distance. The positive sign defines the pixels that are closer to $CMEF_{\mathcal{G}}$. Zero values - the pixels equidistant from $CMEF_{\mathcal{F}}$ and $CMEF_{\mathcal{G}}$.

4   Multifocal Ellipse with Corners

This section discusses the configurations and properties of N focal points that generate corners passing through them. By corner, we define a curve point, where the left-hand tangent differs from the right-hand tangent. In the case of multifocal ellipses, the level sets passing through the focal points contain the corners. To simplify the upcoming discussion, let us introduce a normalization of N weights $w_1, w_2, \ldots, w_N$. Each weight is divided by the maximum value $max(w_1, w_2, \ldots, w_N)$ and ranges in the interval $(0, 1]$.

Property 4. For the given set of N focal points, the weighted multifocal ellipse may contain up to N corners.

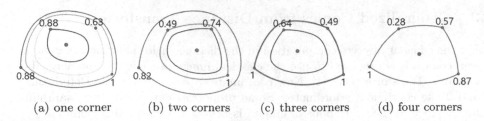

(a) one corner   (b) two corners   (c) three corners   (d) four corners

Fig. 1. The level sets containing one or several corners. The numbers indicate the weights of the focal points. (Color figure online)

First, consider N focal points forming a convex hull and their level sets. By taking the different values of weights, a corner is a part of either a level set passing through a single focal point or is a part of a closed sequence of arcs connecting multiple focal points. Figure 1 shows the example of a convex hull containing four points. Here, the red point highlights the global minimum of the distance field. Changing the weights generates the level sets containing one (Fig. 1a), two (Fig. 1b), three (Fig. 1c), and four (Fig. 1d) corners. In the latter, the corner with the largest angle has a focal point with a smallest weight.

Second, let a set of N focal points have at least one inside the convex hull. As stated in Property 1, the level sets are convex. So a level set contains maximally as many corners as there are focal points in a convex hull. Let $ME(\omega A, \nu B, \mu C)$ be the multifocal ellipse (Fig. 2a), where ω, ν, and μ are the weights. Let us add the points D and E with the corresponding weights η and ζ inside the convex hull (Figs. 2b and 2c). According to Property 1, there is no level set that contains A, B, C, D, and E. Instead, the added points can either be the global minimum (Fig. 2b) or the corner of another level set (Fig. 2c).

Property 5. Consider a level set passing through all the given focal points. The corresponding combination of the normalized weights is unique.

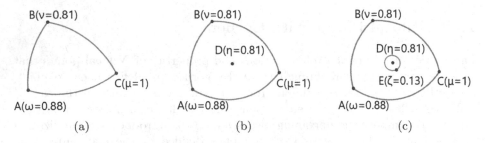

(a)   (b)   (c)

Fig. 2. The multifocal ellipses for (a) convex, (b)–(c) non-convex sets of focal points. The numbers indicate the weights of the focal points.

As follows from the axiom about a unique line passing through two points [5], there is a unique polygon that connects N focal points and, thus, a unique set of the respective angles. Similar reasoning applies to the angles at the corners. Property 5 stems from the fact that the angle formed at the corner of the level set depends only on the weight of the corresponding focal point. For example, if N focal points form an equiangular polygon, the level set passing through them requires all weights to be 1. In the general case, the largest weight corresponds to a point with the smallest angle.

5 Changing the Angle of the Egg-Shape Corner

In the previous sections, we discussed the multifocal ellipses regarding the level sets containing corners. One finding established a relation between the angles at the corners and the weights of the focal points. This correspondence is formalized for the level sets conforming to an oval, or an egg-shape, generated from a pair of focal points with non-equal weights. According to Definition 1, the confocal ellipses with the two weighted focal points, F_1 and F_2, are defined as:

$$ME(w_1F_1, w_2F_2) = \{P \in \mathcal{R}^2 | w_1\delta(P, F_1) + w_2\delta(P, F_2) = const\} \qquad (6)$$

Applying the normalization strategy to Eq. 6 results in having a single weight $0 < \mu = \frac{min(w_1, w_2)}{max(w_1, w_2)} \leq 1$. In the specific case, the distance field contains of confocal ellipses ($\mu = 1$). Otherwise, the level sets represent an egg-shape with various sharpness. For example, observe the distance field of multifocal ellipses for $\mu = 0.47$ (Fig. 3a). The particular interest lies in a level set passing through the focal point, thus having a sharp corner. Let us now define the angle that corresponds to it.

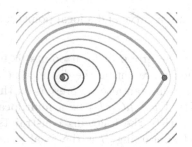
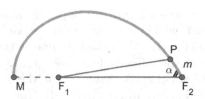

(a) the level sets for the pair of weighted focal points

(b) half of the level set that contains the corner at the focal point

Fig. 3. The egg-shape with the sharp corner, $\alpha = 62°$, $\mu = 0.47$.

Theorem 1. *Consider an egg-shape, $ME(F_1, \mu F_2)$, with a sharp corner at F_2. The cosine of the angle α between the major axis and the tangent passing through F_2 equals approximately its weight μ. (see Appendix A for the proof).*

The implication of Theorem 1 to shape representation is the enrichment of the geometric primitives to describe an object or its part. In contrast to ellipses, the generated egg-shapes can fit the corner of the polygonal shape by having one extra parameter for the weight. The angle ϕ at the corner formed by the two tangents at F_2 is $\phi = 2\alpha$. Then the level set satisfies:

$$\delta(F_1, P) + \cos\frac{\phi}{2}\delta(F_2, P) = \delta(F_1, F_2) \tag{7}$$

Given a shape satisfying Eq. 7, it is possible to find its parameters (Fig. 3b). The symmetry axis intersects the egg-shape at the two points, M and F_2, where the latter is at the corner. The tangent at F_2 forms an angle α with the symmetry axis. Assuming $P = M$ and $\frac{\phi}{2} = \alpha$ in Eq. 7 results in:

$$\delta(F_1, F_2) = \frac{\delta(M, F_2) \cdot (1 + cos\alpha)}{2} \tag{8}$$

Consequently, F_1 can be found by moving from F_2 to M by $\delta(F_1, F_2)$.

6   Multifocal Hyperbola with Corners

Generalized conics as multifocal ellipses produce only convex level sets. Concave level sets correspond to multifocal hyperbolas. Such level sets do not satisfy the properties regarding convexity and the global minimum (refer to Property 1–3).

Property 6. There exist no level set that passes through all the negatively and positively weighted focal points of the multifocal hyperbola.

Property 6 stems from the fact that a multifocal hyperbola tessellates the space based on proximity to one of the sets of focal points. As a result, the curve mapped to the zero distance values separates the two sets of focal points.

Property 7. A multifocal hyperbola might contain a curve and a focal point that does not belong to it.

The weights with the opposite signs cause one group of focal points to be minima, while the other group - maxima. When representing the distance field in 3D (Fig. 4a), the distance value at the focal point can also be present on the slope of the different focal point(s). Let $MH(\omega A, \nu B, \mu C | \eta D)$ be a multifocal hyperbola (Fig. 4). Assume that the weights satisfy the following constraints: $\eta \in [-1\ldots 0)$; $\mu, \omega, \nu \in (0 \ldots 1]$. The level set containing B is the focal point itself since it is the global minimum. The point A is the local minimum, hence, its distance value is also present on the slope of B. Similarly, the level set of C contains the focal point itself and the closed curve surrounding A and B. Finally, the point D is the local maximum, and the corresponding level set is the point itself and the closed curve surrounding all the focal points.

In general, focal points with a negative weight enable creating concavities in level sets. For instance, consider the negatively weighted focal point added at the global minimum of the multifocal ellipse with identical weights (Fig. 5). Varying the negative weight value changes the strength of concavity.

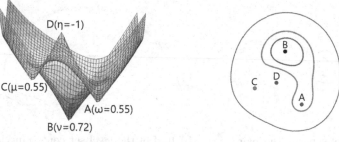

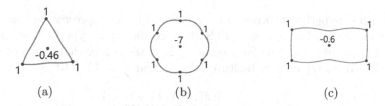

(a) a 3D representation of the distance values

(b) the level sets containing the focal points (top view)

Fig. 4. The level sets of the multifocal hyperbola $MH(\omega A, \nu B, \mu C | \eta D)$

(a)

(b)

(c)

Fig. 5. The level sets of the multifocal hyperbola. The numbers indicate the weights of the corresponding focal points.

7 Changing the Angle of the Hyperbolic Shape Corner

Similar to the egg-shape, it is possible to define the relation between the angle at the corner of the hyperbolic shape and the weight of the corresponding focal point. With respect to Definition 2, the weighted hyperbola generated from the two focal points, F_1 and F_2, can be formalized as:

$$MH(w_1 F_1 | w_2 F_2) = |w_1 \delta(P, F_1) - w_2 \delta(P, F_2)| = const \qquad (9)$$

After applying normalization to Eq. 9, there is only one weighting parameter $0 < \mu = \frac{min(w_1, w_2)}{max(w_1, w_2)} \leq 1$. In the specific case ($\mu = 1$), the level sets of the distance field contain confocal hyperbolas.

Theorem 2. *Consider the hyperbolic shape, $MH(F_1 | \mu F_2)$, with the sharp corner at the focal point F_2. The cosine of the angle β between the line segment $\overline{F_1 F_2}$ and the tangent passing through F_2 equals approximately its weight μ taken with the opposite sign. (see Appendix B for the proof).*

The angle ψ of the concave corner formed by the two tangents at F_2 is $\psi = 2\beta$, where β is larger than 90°. Then the level set satisfies:

$$\delta(F_1, P) + \cos \frac{\psi}{2} \delta(F_2, P) = \delta(F_1, F_2) \qquad (10)$$

 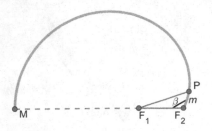

(a) the level sets for the pair of weighted focal points

(b) half of the level set containing the corner passing through the focal point

Fig. 6. The hyperbolic shape with the sharp corner, $\beta = 118°$, $\mu = 0.47$.

The described hyperbolic corners have the potential for representing concavities of the shape. By similar reasoning to Sect. 5, consider the hyperbolic shape that satisfies Eq. 10 (Fig. 6b). The focal point F_2 is at the corner that forms the angle β with the symmetry axis. Substitute $P = M$ and $\frac{\psi}{2} = \beta$ in (10):

$$\delta(F_1, F_2) = \frac{\delta(M, F_2) \cdot (1 + cos\beta)}{2} \tag{11}$$

Consequently, F_1 can be found by moving from F_2 to M by $\delta(F_1, F_2)$.

8   Shape Representation with the Generalized Conics

The paper introduced an approach to represent and reconstruct the shape that is generated from a pair of focal points with weights (Sects. 5 and 7). It aims to define the egg-shapes and hyperbolic shapes with the corners using the three parameters: the two endpoints from the symmetry axis and the angle at the corner. Such representation is invariant to translation, rotation, and scaling. This paper does not cover an extension to N focal points. Since the number of equations increases with the number of focal points, one possibility to represent a complex shape might be connected with machine learning. In this case, the vector of the weights of the focal points captures the shape structure.

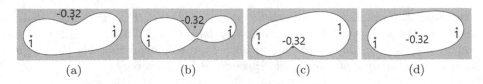

(a)   (b)   (c)   (d)

Fig. 7. The examples of the shapes generated from the three focal points. The numbers reflect the weights of the corresponding focal points.

9   Conclusion

This paper introduced the analysis of the generalized conics from the perspective of shape representation. Changing the weights of the focal points in multifocal ellipses enables generating the shapes with convex corners. In turn, the multifocal hyperbolas make it possible to create concave corners. The proposed findings broaden the view on shape representation and have the potential to generate a complex contour with a few focal points (Fig. 7).

Appendix A

Proof of Theorem 1

The pair of focal points, F_1 and F_2, generates the weighted multifocal ellipse, such that the weight, $0 < \mu \le 1$, corresponds to F_2 (Fig. 3b). Let us denote $\delta(F_1, F_2) = 2f$, $\delta(F_1, P) = n$, $\delta(F_2, P) = m$, and $\widehat{F_1 F_2 P} = \alpha$. By definition the level set contains a group of points mapped to the same distance value. So the distance value at F_2 is identical to the one at P. According to the normalized version of Eq. 6, it equals:

$$\delta(F_1, F_2) + \mu\delta(F_2, F_2) = \delta(F_1, F_2) = 2f \qquad (12)$$

As noted from Eq. 12, the distance value corresponding to the level set with the corner equals the length of the line segment $\overline{F_1 F_2}$. Let us substitute it in the analogical equation for P:

$$\delta(F_1, P) + \mu\delta(F_2, P) = 2f \qquad (13)$$

$$n + \mu m = 2f \qquad (14)$$

$$\implies n = 2f - \mu m \qquad (15)$$

To derive an alternative estimate of m, consider a triangle $\triangle F_1 P F_2$. According to the law of cosines [8]:

$$m^2 + 4f^2 - 4mf \cos\alpha = n^2 \qquad (16)$$

Substituting the value of n from Eq. 15 in Eq. 16 leads to:

$$m^2 + 4f^2 - 4mf \cos\alpha = 4f^2 - 4\mu m f + m^2\mu^2 \qquad (17)$$

$$\implies m = \frac{4f(\mu - \cos\alpha)}{\mu^2 - 1} \qquad (18)$$

The important assumption about the point P in continuous space states that it is infinitely close to F_2. This implies that the length of m converges to zero. Then Eq. 18 is further simplified:

$$m = \frac{4f(\mu - \cos\alpha)}{\mu^2 - 1} = 0 \qquad (19)$$

$$\implies \mu = \cos\alpha \qquad (20)$$

According to Eq. 20, the angle formed at the corner of the level set depends on the weight of the focal point and not on the distance between the focal points.

Appendix B

Proof of Theorem 2
The pair of focal points, F_1 and F_2, generates the weighted multifocal hyperbola (Fig. 6b). The level set passing through F_2 contains the sharp corner. Let us denote $\delta(F_1, F_2) = 2f$, $\delta(F_1, P) = n$, $\delta(F_2, P) = m$, and $\widehat{F_1 F_2 P} = \beta$. Assuming the normalized version of Eq. 9, the distance value at F_2 equals:

$$|\delta(F_1, F_2) - \mu\delta(F_2, F_2)| = \delta(F_1, F_2) = 2f \tag{21}$$

Similar to the proof for the egg-shape, consider the triangle $\triangle F_1 P F_2$ and derive the following relations:

$$n = 2f + \mu m \tag{22}$$

$$m = \frac{-4f(\mu + \cos\beta)}{\mu^2 - 1} \tag{23}$$

In continuous space, m is infinitely small. In discrete space, it can be assigned to zero, resulting in: $\mu = -\cos\beta$. So the angle at the corner of the hyperbolic shape depends only on the weight of the respective focal point.

References

1. Borgefors, G.: Distance transformations in digital images. Comput. Vis. Graph. Image Process. **34**(3), 344–371 (1986)
2. Gabdulkhakova, A., Kropatsch, W.G.: Confocal ellipse-based distance and confocal elliptical field for polygonal shapes. In: Proceedings of the International Conference on Pattern Recognition, pp. 3025–3030 (2018)
3. Gabdulkhakova, A., Kropatsch, W.G.: Generalized conics: properties and applications. In: Proceedings of the International Conference on Pattern Recognition, pp. 10728–10735 (2020)
4. Gabdulkhakova, A., Langer, M., Langer, B.W., Kropatsch, W.G.: Line Voronoi Diagrams using elliptical distances. In: Bai, X., Hancock, E.R., Ho, T.K., Wilson, R.C., Biggio, B., Robles-Kelly, A. (eds.) S+SSPR 2018. LNCS, vol. 11004, pp. 258–267. Springer, Cham (2018). https://doi.org/10.1007/978-3-319-97785-0_25
5. Hilbert, D.: The Foundations of Geometry. Open Court Publishing Company, Chicago (1902)
6. Langer, M., Gabdulkhakova, A., Kropatsch, W.G.: Non-centered Voronoi Skeletons. In: Couprie, M., Cousty, J., Kenmochi, Y., Mustafa, N. (eds.) DGCI 2019. LNCS, vol. 11414, pp. 355–366. Springer, Cham (2019). https://doi.org/10.1007/978-3-030-14085-4_28
7. Ostresh, L.M., Jr.: On the convergence of a class of iterative methods for solving the Weber location problem. Oper. Res. **26**(4), 597–609 (1978)
8. Pickover, C.A.: The Math Book: From Pythagoras to the 57th Dimension, 250 Milestones in the History of Mathematics. Sterling Publishing Company, New York (2009)

9. Sekino, J.: n-ellipses and the minimum distance sum problem. Am. Math. Mon. **106**(3), 193–202 (1999)

10. Torricelli, E.: Opere di Evangelista Torricelli. In: Vassura, G., Loria, G., Montanari, G. (eds.) vol. I/2, pp. 90–97 (1919)

11. Torricelli, E.: Opere di Evangelista Torricelli. In: Vassura, G., Loria, G., Montanari, G. (eds.) vol. III, pp. 426–431 (1919)

12. Weber, A., Pick, G.: Über den Standort der Industrien. Mohr, J.C.B (1922)

Automatic Face Mask Detection Using a Hide and Seek Algorithm

Pratyaksh Bhalla[1], Soumya Snigdha Kundu[1]($\boxtimes$), S. Deepanjali[1], G. Vadivu[1], and Sapdo Utomo[2]

[1] SRM Institute of Science and Technology, Kattankulathur, India
{pb3739,sk7610,deepanjs,hod.it.ktr}@srmist.edu.in
[2] National Chung Cheng University, Chiayi County, Taiwan

Abstract. The coronavirus has affected millions around the world and has inevitably brought about a necessity to wear face masks in official and public places to take the first step in keeping one's self safe. To monitor personnel and public areas and prevent the spread of the disease we present a scalable and deployable face mask detection system in a real time setting using a novel hide and seek algorithm. Our model, based on openCV library and dlib environment utilizes the facial landmarks where in the algorithm detects face masks through the presence and absence of facial markers. We call this process as seeking and hiding. We overcome present issues of high computational cost of deep learning models and low inference speeds of general detection paradigms. We also validate our algorithm on several aspects which affect the accuracy of other models such as image and face orientation, type of face masks and more. As our model requires no data for model training, we eliminate the highly sensitive issue of acquiring facial data and bias. Our model achieves 98.79% precision and 94.81% recall.

Keywords: Detection · Face masks · Facial-landmarks

1 Introduction

The pandemic caused due to covid-19 [1] brought about the absolute necessity to wear face masks to prevent a person from catching and spreading the virus as it was the sole reason to affect 2.7 million people world-wide and crush an entire country's economy stand-point. Along with that there has always been the presence of severe acute respiratory syndrome (SARS) and the Middle East respiratory syndrome (MERS) disease. This has led to the creation of a new task of detection of these face masks on people's faces. This was of utmost importance as people gathering in public or official spaces had to be monitored for the presence of masks on every single individual. Companies and business had to install systems to function within the safety regulations and avoid any potential outbreaks (Fig. 1).

P. Bhalla and S. S. Kundu—Equal contribution.

© Springer Nature Switzerland AG 2021
J. M. R. S. Tavares et al. (Eds.): CIARP 2021, LNCS 12702, pp. 430–439, 2021.
https://doi.org/10.1007/978-3-030-93420-0_40

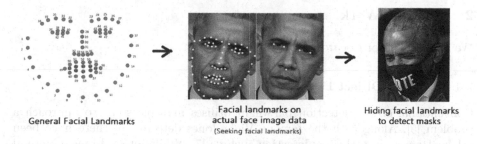

General Facial Landmarks   Facial landmarks on actual face image data (Seeking facial landmarks)   Hiding facial landmarks to detect masks

Fig. 1. Proposed working pipeline

This inevitably generated a entire new level of difficulty in detection tasks [2] as the masks not only come in different shapes, sizes and colors but also have to be detected in mass amounts as they are to be worn by everyone. Few other issues in tandem to the problem was data collection of people specific to a region as facial data landmarks are highly sensitive in nature and the problem associated with scalability and deployment perspective due to high computational costs of previous models. The demand of face mask detectors is quite high as it is virtually required everywhere. This prevents deep-learning models [3,4] from being efficiently and cost-effectively deployed in rural areas. Current research utilises NVIDIA GeForce RTX 2080 Ti which proves to be quite expensive in mass deployment. It is also recorded that training models on high-end GPUs generates the carbon equivalent of a flight across the United States [20]. Mass deployment of such models will be quite dangerous in the long run. For the above reasons face mask detection has become a crucial computer vision task to help the global society, but research related to face mask detection is limited as it is recently aroused problem.

To tackle the existing problems we came with a non-deep learning solution so we can accustom to all the variability present in the problem and the other challenges present along with it. Inability to access facial data due to security reasons becomes detrimental to the model therefore we present a data-less solution. Our contributions are as follows -

- We present a face mask detector algorithm called Hide and Seek which demands no previous data or manual pre-processing steps prior to its usage.
- Our solution runs on cpu driven backend which is ideal for deployment and scalability due to its high performance and fast processing speeds in tandem to low computational power demand. It also shows our approach to lower carbon-footprint computing.
- Our system also has virtually no time lag between different iterations of detection and different types of masks and, is unhampered by various other external factors such as type of face masks, person in frame, face orientation and more. This makes the model highly generizable.
- We attach a small explanation in the end as to validate the exact points we have used in our algorithm.

2   Related Work

We focus our prior research on two broad aspects:

2.1   General Object Detection

Conventional object detection paradigm utilises multiple steps to approach a problem [5]. Along with the famous Viola-Jones detector [6] there have been other attempts in real time detection such as [7]. While integral image method and Haar feature descriptors are the focus of [6] to process the algorithm it is still quite computationally intensive. To counter this [8] proposed an effective feature extractor called HOG which accounts for the directions and magnitudes of oriented gradients over image cells. [19] has used the OpenCV software to detect faces and emotions involved in them which aligns with our objective of facial-point analysis.

Instead of using manually curated features, deep learning based detector demonstrated outstanding performance recently, due to its robustness and high feature extraction capability [21]. Majorly, there are two broad divisions of detectors -

- One-stage object detectors [21]
- Two-stage object detectors [23]

Two-stage detector generates region proposals in the first stage and then fine-tune these proposals in the second stage. While the two-stage detector can provide high detection performance it generally falls prey to its lower speeds [22].

2.2   Convolutional Neural Networks

State of the art vision problems are generally approached and solved through the utilization of Convolutional Neural Networks [9,26]. This is due to the spatial feature extraction and understanding ability of the architecture. CNN's use convolution kernels to convolve with the original images or feature maps and extract higher-level features with the same. The following variations of the base model are utilised for the task of object detection.

The work proposed in [10] utilizes selective search to propose some important regions which may contain the desired object. These values are then fed into a CNN model to extract features, and a support vector machine (SVM) is used to recognize classes of objects. This is known as RCNN. To overcome the computationally expensive second-stage [11] proposes Faster-RCNN model which by introduces the a region of interest (ROI) pooling layer so that all the input proposal regions are interpreted at once. There has also been the research attempt to limit the speeds of detectors through selective search in the form of region proposal networks [12].

3 Proposed Approach

Our proposed approach is based on a novel Hide and Seek algorithm which is the developed on general purpose cross-platform software library Dlib [24] and the facial landmark localization through the OpenCV library [25]. A few of such landmarks [27] are the:

1. Eyes   2. Eyebrows   3. Jawline   4. Mouth   5. Nose

The landmarks help in localizing the face and letting the algorithm be aware of the face present even in various environments. The awareness of the algorithm does not require any form of previous training and is adaptable to any type of face or face mask. The model first begins the process by detecting whether the face present in the frame by looking or "seeking" the facial landmarks. If the landmarks are present or "seeked", then that indicates the fact that there is no mask present. This parallelly increments the *countnomask* value. If the landmarks are "hidden", then the program interprets that there is a mask and parallelly increments the *countmask* value (Fig. 2).

Table 1. Explanation of confusion matrix values accounted for

Metric	True positive	False positive	True negative	False negative
Mask	Detected and present	Detected but not present	Not detected and not present	Not detected but present
No-mask	Detected and not present	Mask detected but present	Mask present and not detected	Mask detected but not present

---

Input: Facial images with and without face mask
Result: Face Mask Detection

Set var countmask and var countnomask $\longrightarrow$ 0

FOR every iteration of face data:

Generate $\longrightarrow$ selective facial data-points to **"seek"** the face in the frame.

IF (Points are **"hidden"**):
// Face mask is detected
countmask = countmask + 1
ELSE:
// Face mask is absent
countnomask = countnomask + 1
[end IF and FOR loop]

Algorithm 1: Hide and Seek Algorithm

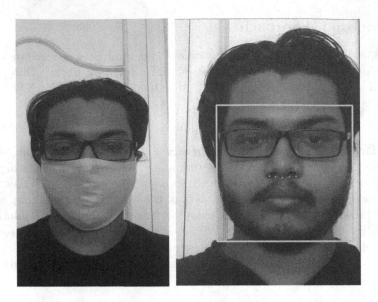

Fig. 2. Example of unmasked "seeked marked" faces (right) and masked "hidden marked" face masks (left) by the algorithm

Table 2. Comparison of proposed model (Hide and Seek) with previous approaches

Method	Face		Mask	
	Precision	Recall	Precision	Recall
Baseline [13]	89.6%	85.3%	91.9%	88.6%
RetinaFaceMask+ResNet [3]	91.9%	96.3%	93.4%	94.5%
RetinaFaceMask+MobileNet [3]	83.0%	95.6%	82.3%	89.1%
Hide and Seek	98.79%	94.81%	98.23%	95.18%

4   Experimentation

Here we discuss about the dataset collection and the extra set of metrics which have been utilised to further provide evidence towards the benefits of a non-deep learning approach.

4.1   Dataset

We sampled images from several online sources such as Face Mask Dataset [13], Wider Face [14] and MAsked FAces dataset (MAFA) [15]. We also manually scraped random celebrity images both from google images and various open-sourced github repositories as well to extend the overall dataset. Our focus was on single face detection which is generally applied in a security setting for offices and public gathering spots. The samples include a variety of situations for confusing the algorithm such as:

- Faces with masks,
- Faces without masks
- Confusing images without masks
- Different colors of mask
- Patterned masks
- Slight shift in orientation of faces with and without masks

Different datasets were utilised to show that our model is generizable to any form of data. The problem scope is world wide and testing the data on a certain region or area will not help the standing cause.

4.2   Evaluation Metrics

We measure our algorithm in major 2 division of metrics.

Quantitative Metrics - Here we quantify the model through metrics such as accuracy, precision and recall. They are defined as -

$$accuracy = \frac{\text{No. of correct predictions}}{\text{No. of incorrect Predictions}} \tag{1}$$

$$precision = \frac{TP}{TP + FP} \tag{2}$$

$$recall = \frac{TP}{TP + FN} \tag{3}$$

where, TP, FP and FN denotes the true positive, false positive and false negative, respectively. We also evaluate model through measured training time to account for the scalability and deployability of the model.

Qualitative Metrics - We measure our model on external parameters such as data requirement, presence of bias and computational power requirement. These are accounted for as data is quite hard to attain in such sensitive matters and bias is a longstanding problem in deep-learning models. The computational aspect is considered as their is an expanding research avenue towards "greener" artificial intelligence [16].

5   Result Analysis

5.1   Visual Intuition Development

There is no learning involved in our model unlike AI models. Hence our approach requires 0 data points to train on. This makes it highly beneficial when the model is transferred over different regions because any form of human data, especially facial data is extremely sensitive in nature and can be quite difficult to obtain. We strongly avoided the training phases solely to avoid the problem of collecting data and generating a model which can avoid any variability or bias.

5.2   Bias Elimination

[18] explains how bias is a growing problem for AI related achievements as they are very heavily dependent on data. Through our pure algorithmic approach we eliminate any of the following biases as depicted in Fig. 3 -

1. Gender   2. Region   3. Skin Color

Fig. 3. Depiction of invariancy of bias (region, skin color, gender)

Table 3. Confusion matrix for the hide and seek algorithm

		Prediction outcome		
		2981	19	
Real value	p′	TN	FP	P′
	n′	FN	TP	N′
		229	1558	

5.3   Confusion Matrix and It's Analysis

From Table 2 and Confusion Matrix [Table 3] we see that our algorithm has a extensive increase in precision by nearly 7% in the case of face images and slightly over-performs in the case of mask images. While it has a lesser recall for face images it again has a higher recall for masks which is the important aspect of detection. We also attain a calculated 94.81% accuracy value. All the values were accounted on the basis of the *countmask* and *countnomask* values specified in Algorithm 1 and seen in Table 1. We were not able to directly compare this with other methods as such values for other methods have not been published.

5.4   Computational Power Comparison

Lower Carbon Footprint - Carbon footprint emission of machine learning models is of utmost concern due to the vast amount models which are upcoming and the energy they require to run through to obtain the results [17]. Here we use a CPU instead of GPU which greatly eliminates the carbon footprint and energy consumption of the model.

Lesser Training Time - Lesser training time is used wisely as a metric in many fields of computer vision. As the model is not monitored or susceptible to variability in data native training time becomes null. This makes our model very feasible to scalability and achieve easy deployment procedures in many scenarios. Face mask detection is a necessity everywhere and not all vendors and users can afford GPU Cloud services.

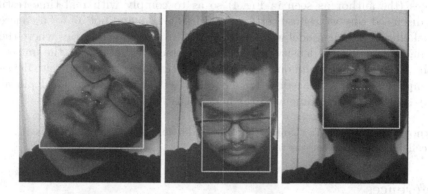

Fig. 4. Data-points are intact irrespective of image and facial orientations.

5.5   Invariant to Orientation of Face

Our model accounts for various orientations of faces both through facial orientation and image orientations as seen in Fig. 4. We took personal photos outside our mentioned dataset and made sure our methods generalises well. We see that irrespective of the tilt or orientation of the face or the image, our model still gives the right result. Our algorithm can also take in various size of image dimension. It is not bound by any specific dimension.

6   Discussion

6.1   Facial Point Choice

As mentioned in the algorithm and seen in various places, we have carefully considered as to which points to consider for our algorithm. This helped us gain faster inference speeds for algorithm 1 as greater number of points will lead to longer calculation time. If we used all the points then the algorithm would've demanded the mask to cover everyone's chin. That is always not the case. For number of points also lead to more coverage and more variance to tilt. We also excluded any points directly linking the mouth because the downward tilt at times failed to be considered. Any lesser number of points can not be considered because they again fail towards the downward tilt.

7   Conclusion and Future Works

We present an efficient, scalable and low powered face mask detection algorithm based on facial landmarks using the openCV environment and dlib library called Hide and Seek. We compare our method qualitatively and quantitatively to the previous state of the art approaches through precision and recall values and see that we majorly perform better. We also compared our method with individual images (the author as seen in Fig. 4) so as to comply with real time testing and discussed why we choose certain facial-landmarks and not all for a concrete model efficiency. Our model was made in mind to suit the countries where there is a higher population density and account for more variability in the unknown setting. We hope to extend our work to work with videos as well so as to develop a complete screening procedure for any office setting and present a deployable ready solution for the same with the help of cost-effective hardware.

Acknowledgement. We thank the Taiwan-India Joint Collaboration for facilitating the collaboration between the universities.

References

1. Wu, F., Zhao, S., Yu, B., et al.: A new coronavirus associated with human respiratory disease in China. Nature **579**, 265–269 (2020). https://doi.org/10.1038/s41586-020-2008-3
2. Zhao, Z., Zheng, P., Xu, S., Wu, X.: Object detection with deep learning: a review. IEEE Trans. Neural Netw. Learn. Syst. **30**(11), 3212–3232 (2019). https://doi.org/10.1109/TNNLS.2018.2876865
3. Jiang, M., Fan, X., Yan, H.: RetinaMask: a face mask detector. arXiv:2005.03950, [cs.CV] (2020)
4. Meenpal, T., Balakrishnan, A., Verma, A.: Facial mask detection using semantic segmentation. In: 2019 4th International Conference on Computing, Communications and Security (ICCCS), Rome, Italy, pp. 1–5 (2019). https://doi.org/10.1109/CCCS.2019.8888092
5. Zou, Z., Shi, Z., Guo, Y., Ye, J.: Object detection in 20 years: a survey. arXiv preprint arXiv:1905.05055 (2019)
6. Viola, P., Jones, M.: Rapid object detection using a boosted cascade of simple features. In: Proceedings of the 2001 IEEE Computer Society Conference on Computer Vision and Pattern Recognition. CVPR 2001, vol. 1, p. I. IEEE (2001)
7. Fleuret, F., Geman, D.: Coarse-to-fine face detection. Int. J. Comput. Vis. **41**, 85–107 (2001). https://doi.org/10.1023/A:1011113216584
8. Dalal, N., Triggs, B.: Histograms of oriented gradients for human detection. In: IEEE Computer Society Conference on Computer Vision and Pattern Recognition (CVPR 2005), vol. 1, pp. 886–893. IEEE (2005)
9. Gu, J., et al.: Recent advances in convolutional neural networks. arXiv:1512.07108 [cs.CV] (2017)
10. Girshick, R., Donahue, J., Darrell, T., Malik, J.: Rich feature hierarchies for accurate object detection and semantic segmentation. In: Proceedings of the IEEE Conference on Computer Vision and Pattern Recognition, pp. 580–587 (2014)

11. Girshick, R.: Fast R-CNN. In: Proceedings of the IEEE International Conference on Computer Vision, pp. 1440–1448 (2015)
12. Ren, S., He, K., Girshick, R., Sun, J.: Faster R-CNN: towards real-time object detection with region proposal networks. In: Advances in Neural Information Processing Systems, pp. 91–99 (2015)
13. Chiang, D.: Detect faces and determine whether people are wearing mask (2020). https://github.com/AIZOOTech/FaceMaskDetection
14. Yang, S., Luo, P., Loy, C.-C., Tang, X.: Wider face: a face detection benchmark. In: Proceedings of the IEEE Conference on Computer Vision and Pattern Recognition, pp. 5525–5533 (2016)
15. Ge, S., Li, J., Ye, Q., Luo, Z.: Detecting masked faces in the wild with LLE-CNNs. In: Proceedings of the IEEE Conference on Computer Vision and Pattern Recognition, pp. 2682–2690 (2017)
16. Lacoste, A., Luccioni, A., Schmidt, V., Dandres, T.: Quantifying the carbon emissions of machine learning. arXiv:1910.09700 [cs.CY] (2019). https://mlco2.github.io/impact/
17. Henderson, P., Hu, J., Romoff, J., Brunskill, E., Jurafsky, D., Pineau, J.: Towards the systematic reporting of the energy and carbon footprints of machine learning. arXiv:2002.05651 [cs.CY] (2020)
18. Verma, S., Gao, R., Shah, C.: Facets of fairness in search and recommendation. In: Boratto, L., Faralli, S., Marras, M., Stilo, G. (eds.) BIAS 2020. CCIS, vol. 1245, pp. 1–11. Springer, Cham (2020). https://doi.org/10.1007/978-3-030-52485-2_1
19. Lee, C.-H.J., Wetzel, J., Selker, T.: Enhancing interface design using attentive interaction design toolkit. In: ACM SIGGRAPH 2006 Educators Program, Massachusetts, Boston, 30 July–03 August 2006 (2006)
20. Gault, M.: Yes, AI Has a Carbon Footprint (2019). VICE
21. Chen, K., et al.: Towards accurate one-stage object detection with AP-loss. In: 2019 IEEE/CVF Conference on Computer Vision and Pattern Recognition (CVPR), Long Beach, CA, USA, pp. 5114–5122 (2019). https://doi.org/10.1109/CVPR.2019.00526
22. Soviany, P., Ionescu, R.T.: Optimizing the trade-off between single-stage and two-stage deep object detectors using image difficulty prediction. In: 20th International Symposium on Symbolic and Numeric Algorithms for Scientific Computing (SYNASC), Timisoara, Romania, pp. 209–214 (2018). https://doi.org/10.1109/SYNASC.2018.00041
23. Shih, K.-H., Chiu, C.-T., Lin, J.-A., Bu, Y.-Y.: Real-time object detection with reduced region proposal network via multi-feature concatenation. IEEE Trans. Neural Netw. Learn. Syst. 31(6), 2164–2173 (2020). https://doi.org/10.1109/TNNLS.2019.2929059
24. King, D.E.: Dlib-ml: a machine learning toolkit. J. Mach. Learn. Res. 10, 1755–1758 (2009)
25. OpenCV. Open Source Computer Vision Library (2015)
26. Guo, G., Wang, H., Yan, Y., Zheng, J., Li, B.: A fast face detection method via convolutional neural network. Neurocomputing 395(28), 128–137 (2020). https://doi.org/10.1016/j.neucom.2018.02.110
27. Johnston, B., Chazal, P.: A review of image-based automatic facial landmark identification techniques. J. Image Video Proc. 2018, 86 (2018). https://doi.org/10.1186/s13640-018-0324-4

A Feature Extraction Approach Based on LBP Operator and Complex Networks for Face Recognition

João Gilberto de Souza Piotto🆔 and Fabrício Martins Lopes(✉)🆔

Departamento Acadêmico de Computação (DACOM), Universidade Tecnológica
Federal do Paraná (UTFPR) Câmpus Cornélio Procópio, Av. Alberto Carazzai,
1640 - CEP, Cornélio Procópio, PR 86300-000, Brazil
`fabricio@utfpr.edu.br`

Abstract. Face recognition is one of the most important tasks in computer vision research. Despite the advances achieved, novel approaches are needed to extract meaningful information and produce interpretable results. This work presents a new approach to feature extraction, called LBP-CN, based on Local Binary Pattern (LBP) and Complex Network (CN) measurements for face recognition. Initially, the face region is subdivided into smaller regions and the network nodes are defined from the pixels of these sub-regions. The LBP operator is applied to the pixels of the subregions and the edges are defined by the distances between all pixels. Therefore, measurements of complex networks are extracted from the network and thresholds are applied to remove edges with higher values. Five face data sets and four competing methods were adopted to evaluate the proposed approach. The results show the adequacy and robustness of the proposed approach to facial recognition, getting better accuracy and smaller standard deviation than the competing methods in a relatively simple way, using only two parameters. In addition, the LBP-CN results for the HPI and FEI data sets also showed the adequacy of the proposed approach to angle variations.

Keywords: LBP operator · Complex Networks · Face recognition · Texture description · Feature extraction

1 Introduction

Since the first techniques of face recognition and detection emerged, many problems have already been solved. High-impact algorithms such as Eigenfaces [36] and Viola-Jones [37] have been proposed, but they do not work very well in detection at different scales, rotations or lighting variations. Over the years, these methods have undergone updates and inspired the development of new techniques, with more accurate and reliable results and a diverse set of applications.

Currently, there are almost invariable feature extraction methods at rotation and scale, which are robust to lighting variation such as the *Scale Invariant*

© Springer Nature Switzerland AG 2021
J. M. R. S. Tavares et al. (Eds.): CIARP 2021, LNCS 12702, pp. 440–450, 2021.
https://doi.org/10.1007/978-3-030-93420-0_41

Feature Transform (SIFT) [22], the *Speeded Up Robust Features* (SURF) [7] and Local Binary Pattern (LBP) [27].

Regarding face recognition, many efforts have been made over the years, from classical algorithms such as Eigenfaces [36], FisherFaces [8] and LBPHFaces [2] to approaches based on deep convolutional networks such as DeepFace [34], FaceNet [31] and ArcFace [11]. However, most methods suffer from variations in facial angles and rarely consider local facial features, which are notably important for facial recognition [3, 25].

In particular, the LBP technique is very effective in describing texture features [2]. In addition, LBP allows very fast feature extraction because of its simplicity, leading to properties such as high-speed computing and rotation invariance, being successfully applied in image recovery, texture description and face recognition [1, 4, 18, 24, 32, 33].

On the other hand, Complex Networks (CN) have a great attention and many applications have been developed since the works of Watts and Strogatz [38] and Barabasi and Albert [6]. In particular, some methods that adopt complex networks have been proposed in the literature, showing their suitability for image analysis [5, 19, 20, 24, 30].

In this context, a new feature extraction method, called LBP-CN, based on LBP and CN measurements, is proposed for face recognition. The method starts by dividing the image of the face into sub-regions from which the LBP features are obtained, each sub-region is then mapped to a CN. Then, some CN measurements [10] are extracted iteratively. Therefore, the extracted measurements make up a feature vector that is applied to the sub-region classification. Five face datasets and four of the most important methods of face recognition were adopted to evaluate LBP-CN.

2   Complex Networks

A network is a model that represents the relationships between pairs of objects (nodes). The most classic example is the social networks, which connect people for specific relationships such as kinship, friendship or work relationship. The network topological structures represent their behavior and can model many real-world problems [6, 9, 17, 20, 21, 26, 38].

A simple network is usually represented by a graph with a set of nodes, which are connected by edges representing the connection between such nodes. However, a problem represented by a graph can cause a model composed of hundreds of nodes, where the connections between them can make it difficult to analyze and visualize the network [26]. Complex network theory is the branch of graph theory that analyzes the topological behavior of the network to understand the relationships between nodes and edges. A network can be represented by its measurements, so that its topological structure can be used in the analysis of its properties, including representation, characterization, classification, modeling and dynamics, i.e. the change of topological structure over some parameter (time, threshold, perturbation, etc.) [10].

This work adopts some complex network measurements commonly used to explore them in the proposed approach, such as average degree, betweenness, eigenvector, assortativity, clustering coefficient and average shortest path length, which were successfully applied to characterize networks.

3   Methodology

The proposed approach is defined initially by adopting a pre-processing step in order to define the face region of the image. The Viola-Jones method [37] was adopted. Thus, selected face region is divided into smaller sub-regions by adopting square windows of size m, which is an input parameter.

Then, a third step is applied on each subregion for feature extraction and composition of a feature vector. Figure 1 presents the overview of the proposed approach.

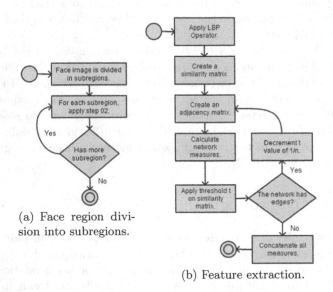

(a) Face region division into subregions.

(b) Feature extraction.

Fig. 1. Overview of the feature extraction approach for face recognition.

More specifically, each square subregion of face image is considered to creating a complex network by using its pixels as nodes of the network. For this, the Local Binary Pattern (LBP) [27] was adopted and applied to the pixels of each subregion. Thus, to select only the most promising operators, all non-uniform LBP operators were discarded [28].

Using the LBP representation, the distances between all pixels are calculated, thus forming a weighted similarity matrix. Then, the matrix is scaled so that its values adjust to the [0, 1] range, which makes it more suitable for the next step: removing the edges. A threshold is applied to the rescaled similarity matrix to create a dynamic from weaker to stronger edges. Initially, all nodes in the network are connected. Then, some complex network measurements are obtained. These measurements are stored and a threshold t is applied to remove edges with values greater than t. Considering the edge removal process, a change in network topology is expected. Therefore, the network topology measurements are obtained for each t value, i.e. different scales. The initial value of t is set to $1 - \frac{1}{n}$, where n is the total number of iterations performed by the method. The value n is an input parameter, which defines the number of iterations, adjusting the sensitivity in the extraction of the network measurements (features). Then, the value of t is decreased at the rate of $\frac{1}{n}$ with each iteration of the algorithm. In this way, topological measurements are obtained at different network resolutions (scales), creating a dynamic for these measurements, i.e. how they are affected by the edge removal process. As a result, a single feature vector is created for each sub-region, considering all the topological measurements extracted and the respective threshold t applied in the network.

Regarding the parameter m, each face is represented by sub-regions and their respective feature vectors. Therefore, it is necessary to classify each feature vector individually, a feature vector for each face. For this reason, facial recognition methods may have their performance diminished by the occurrence of occlusion, rotation, irregular illumination, among others [3,25]. In the proposed approach, each sub-region has its own feature vector, and it is classified individually. In order to avoid a new parameter, i.e. how many sub-regions are needed to correspond to a face and deal with partial occlusion, the frequency of the classes achieved in the classification of the sub-regions was adopted. Thus, the class with the highest frequency considering all classified subregions is selected for the face classification. Figure 2 shows an overview of the feature extraction process.

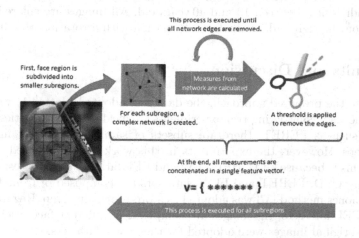

Fig. 2. Feature extraction process performed individually for each sub-region.

In doing so, the proposed feature extraction approach, called LBP-CN, combines some important properties, such as the robustness against different facial expressions, illumination and aging of subjects from LBP [2], the rotation invariance of Complex Networks [12] and its classification based on the frequency that allows dealing with partial occlusion.

3.1 Materials

In order to validate the proposed method, Caltech Face Dataset (CFD) [39], Color FERET Database [29], Head Pose Image Database (HPI) [13], Labeled Faces in the Wild (LFW) [16] and FEI Face Database [35] were adopted.

The samples from Caltech's face data set were taken from the front position of the individual, with differences in lighting and especially in the background's composition of each image. The background is composed of several objects to ensure the robustness of the method. The Caltech data set comprises 450 faces of 27 different individuals. However, some individuals contributed with only one sample. The FERET data set contains 11,338 images, collected from 994 people from various angles and different situations. It has become very popular because it is used to compare performance and accuracy of different face recognition methods.

The HPI data set comprises two series of 93 images with variations in skin tone, hair and accessories (glasses). The face position was determined by two different angles, ranging from -90 to $+90°$.

The LFW data set contains over 13,000 images of faces labeled with the person's name in the image. In addition, 1680 samples have two or more distinct images in the data set.

Finally, the FEI data set contains 2,800 images, with 14 images for each of the 200 individuals between 19 and 40 years old. All images are colored with a homogeneous background, in the frontal position with a rotation of $180°$.

4 Results and Discussion

To evaluate the proposed approach, the data sets adopted (Sect. 3.1) were used in the same way, considering two experiments. Regarding the properties of each data set, such as FERET, there are subsets of samples for the training and testing steps. However, the experiments in this work were not based on these subsets. This is because CFD, HPI, LFW and FEI datasets have no test subsets. In addition, CFD, FERET and LFW data sets are composed of front faces, so the Viola-Jones method [37] was adopted as a pre-processing step. Regarding the HPI and FEI data sets, the images are composed of different face orientations. Thus, the original images were adopted for these two data sets.

Initially, an experiment was performed to evaluate the input parameters m (number of face divisions) and n (number of algorithm interactions for removing edges from the network) defined in Sect. 3. The best results were achieved with $m = 8$ and $n = 30$, which were adopted for all experiments.

The first round of experiments was performed using the classical classification algorithms from the WEKA machine learning software [15], adopting its default parameters, i.e. no tuning of the classification algorithms has been performed. The classification results were validated using the 10-fold cross-validation method. Table 1 summarizes the results considering some classical classification methods, indicating the suitability of the proposed feature extraction (LBP-CN) for facial recognition.

Table 1. Average accuracy of LPB-CN considering different classification algorithms.

Classifier	Dataset	Average accuracy
KNN	CFD	99.10%
	FERET	99.91%
	HPI	98.99%
	LFW	99.20%
	FEI	98.01%
Naive Bayes	CFD	97.38%
	FERET	98.61%
	HPI	97.91%
	LFW	97.78%
	FEI	96.91%
SVM	CFD	92.56%
	FERET	90.66%
	HPI	92.80%
	LFW	93.80%
	FEI	91.47%
Random Forest	CFD	**100.0%**
	FERET	**100.0%**
	HPI	**99.97%**
	LFW	**98.78%**
	FEI	**99.89%**
Neural network	CFD	99.33%
	FERET	98.88%
	HPI	96.34%
	LFW	98.21%
	FEI	97.97%

It can be noted that all classifiers have consistent results and higher accuracy rates, even considering that SVM results depend on the adopted kernel and KNN depends too much on the k parameter, i.e. the number of adopted neighbors. Thus, indicating the robustness of LBP-CN considering different classification algorithms. Thus, the Random Forest classifier was adopted as the classifier for the second round of experiments.

The second round of experiments was performed to compare LPB-CN with competing methods, all data sets were considered in the same way as previous experiments and also adopted 10-fold cross validation. Classic algorithms such as Eigenfaces [36], FisherFaces [8] and LBPHFaces [2] available in the OpenCV (Open Source Computer Vision Library) were adopted. In addition, FaceNet [31] based on deep convolutional network was also adopted. Table 2 shows the results achieved by the methods and data sets adopted.

Table 2. Average accuracy of LPB-CN compared to competing methods.

Method	CFD	FERET	HPI	LFW	FEI
LBP-CN	**100.0%**	**100.0%**	**99.97%**	98.78%	**99.89%**
EigenFaces	90.15%	92.49%	81.34%	83.21%	78.78%
FisherFaces	91.63%	93.98%	88.57%	86.39%	76.87%
LBPHFaces	92.62%	97.28%	94.49%	95.82%	93.75%
FaceNet	**100.0%**	**100.0%**	98.81%	**99.94%**	98.69%

For the CFD dataset, the results indicate that the LBP-CN obtained better accuracy when compared to other methods. This behavior can be explained by the composition of the samples of this data set, as the faces are always accompanied by other background elements. Even considering that a pre-processing step by the Viola-Jones method was adopted for all methods, some background content always appears on the face. These background elements may have produced some noise during the execution of the Eigenfaces, FisherFaces and LBPHFaces methods, which would explain the average accuracy rates obtained by these methods. Since LBP-CN considers the higher frequency in classification of the face sub-regions, the interference caused by noise was lower.

Regarding the results achieved for the Color FERET data set, all methods have a higher average accuracy than the CFD data set. In particular, LBPHFaces shows a significant improvement. This is because the FERET data set has a clean background, indicating consistent results.

Experiments with the HPI data set have revealed that the Eigenfaces and FisherFaces algorithms are more sensitive to angle variations, since it is the main characteristic of these data sets. On the other hand, the LBP-CN had better accuracy rates than all methods, indicating the adequacy of the proposed approach to angle variations.

The LFW data set presents a wide variety of people with objects in the background. The results reinforce the suitability of the LBP-CN, which surpasses the classical methods and presents similar performance to FaceNet, which is based on deep convolutional network. However, the LBP-CN produces explainable and interpretable results, so that experts in the field can be convinced how the method produced the results relatively simply, using only two parameters. Methods based on deep neural networks can produce insightful interpretations, focus on specific architectures, and can be difficult to generalize [14,23]. The FEI dataset also contains angle-varying faces. The LBP-CN again outperforms

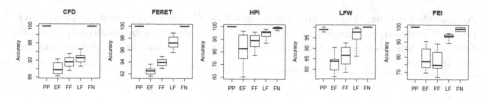

Fig. 3. Box plots of the accuracy distribution on each data set. PP (LBP-CN), EF (EigenFaces), FF (FisherFaces), LF (LBPHFaces) and FN (FaceNet) obtained from 10-fold cross validation.

all competing methods, which reinforces its suitability for face recognition and its robustness in the presence of angle variations.

To verify the robustness, Fig. 3 shows the box plots of the accuracy considering the results of the 10-fold cross validation. Regarding the CFD dataset, the classical methods have relatively higher distribution, showing that they were more affected by the cross-validation method. As explained previously, noise causes this variation, because of the elements present in the image's background. The Color FERET data set experiment has already shown variation in the accuracy rate of all classical methods. The results of HPI and FEI show the sensitivity of Eigenfaces and FisherFaces to angular variations and the robustness of the LBP-CN, which presents higher accuracy and lower distribution of its results.

5   Conclusions

The LBP is a suitable texture description method widely used in computer vision, including facial recognition. LBP features have important properties such as their robustness in relation to facial expression, aging, lighting and alignment [2]. Complex network measurements were shown as a method invariant to rotation [12]. This paper proposes LBP-CN, a novel approach to feature extraction based on LBP and complex network measurements. In addition, the classification of each face is based on the frequency of the sub-region classification that allows dealing with partial occlusion.

Two experiments were performed to evaluate the proposed approach. The Caltech Face Dataset (CFD), the Color FERET Database, the Head Pose Image Database (HPI), the Labeled Faces in the Wild (LFW) and the FEI Face Database were adopted. The accuracy obtained by LBP-CN were also compared with important facial recognition methods such as EigenFaces, Fisher Faces, LBPHFaces and FaceNet. In particular, LBP-CN results for the HPI and FEI data sets also indicated the adequacy of the proposed approach to angle variations. In general, LBP-CN achieved higher accuracy rates and lower variations in its results when compared to competitive methods.

Furthermore, LBP-CN produces explainable and interpretable results, so that experts in the field can be convinced of how the method produced the results relatively simply, using only two parameters.

Acknowledgment. This study was financed by Conselho Nacional de Desenvolvimento Científico e Tecnológico (CNPq) (Grant number 406099/2016-2) and the Fundação Araucária e do Governo do Estado do Paraná/SETI (Grant number 035/2019).

References

1. Ahonen, T., Hadid, A., Pietikäainen, M.: Face description with local binary patterns: application to face recognition. IEEE TPAMI **28**(12), 2037–2041 (2007)
2. Ahonen, T., Hadid, A., Pietikäinen, M.: Face recognition with local binary patterns. In: Pajdla, T., Matas, J. (eds.) ECCV 2004. LNCS, vol. 3021, pp. 469–481. Springer, Heidelberg (2004). https://doi.org/10.1007/978-3-540-24670-1_36
3. Al-Waisy, A.S., Qahwaji, R., Ipson, S., Al-Fahdawi, S.: A multimodal deep learning framework using local feature representations for face recognition. Mach. Vis. Appl. **29**(1), 35–54 (2017). https://doi.org/10.1007/s00138-017-0870-2
4. Alahmadi, A., Hussain, M., Aboalsamh, H.A., Zuair, M.: PCAPooL: unsupervised feature learning for face recognition using PCA, LBP, and pyramid pooling. Pattern Anal. Appl. **23**(2), 673–682 (2019). https://doi.org/10.1007/s10044-019-00818-y
5. Backes, A.R., Casanova, D., Bruno, O.M.: Texture analysis and classification: a complex network-based approach. Inf. Sci. **219**, 168–180 (2013)
6. Barabási, A.L., Albert, R.: Emergence of scaling in random networks. Science **286**(5439), 509–512 (1999)
7. Bay, H., Tuytelaars, T., Van Gool, L.: SURF: speeded up robust features. In: Leonardis, A., Bischof, H., Pinz, A. (eds.) ECCV 2006. LNCS, vol. 3951, pp. 404–417. Springer, Heidelberg (2006). https://doi.org/10.1007/11744023_32
8. Belhumeur, P., Hespanha, J., Kriegman, D.: Eigenfaces vs. fisherfaces: recognition using class specific linear projection. IEEE TPAMI **19**(7), 711–720 (1997)
9. Breve, M.M., Lopes, F.M.: A simplified complex network-based approach to mRNA and ncRNA transcript classification. In: BSB 2020. LNCS, vol. 12558, pp. 192–203. Springer, Cham (2020). https://doi.org/10.1007/978-3-030-65775-8_18
10. Costa, L.d.F., Rodrigues, F.A., Travieso, G., Villas-Boas, P.R.: Characterization of complex networks: a survey of measurements. Adv. Phys. **56**(1), 167–242 (2007)
11. Deng, J., Guo, J., Xue, N., Zafeiriou, S.: ArcFace: additive angular margin loss for deep face recognition. In: The IEEE Conference on Computer Vision and Pattern Recognition (CVPR), June 2019
12. Gonçalves, W.N., de Andrade Silva, J., Bruno, O.M.: A rotation invariant face recognition method based on complex network. In: Bloch, I., Cesar, R.M. (eds.) CIARP 2010. LNCS, vol. 6419, pp. 426–433. Springer, Heidelberg (2010). https://doi.org/10.1007/978-3-642-16687-7_57
13. Gourier, N., Hall, D., Crowley, J.L.: Estimating face orientation from robust detection of salient facial structures. In: FG Net Workshop on Visual Observation of Deictic Gestures, pp. 1–9. FGnet Cambridge, UK (2004)
14. Grisci, B.I., Krause, M.J., Dorn, M.: Relevance aggregation for neural networks interpretability and knowledge discovery on tabular data. Inf. Sci. **559**, 111–129 (2021)
15. Hall, M., Frank, E., Holmes, G., Pfahringer, B., Reutemann, P., Witten, I.H.: The WEKA data mining software: an update. SIGKDD Exp. Newsl. **11**(1), 10–18 (2009)

16. Huang, G.B., Ramesh, M., Berg, T., Learned-Miller, E.: Labeled faces in the wild: a database for studying face recognition in unconstrained environments. Technical report 07–49, University of Massachusetts, Amherst (2007)
17. Ito, E.A., Vicente, F.F., Katahira, I., Lopes, F.M., Pereira, L.P.: BASiNET - BiologicAl Sequences NETwork: a case study on coding and non-coding RNAs identification. Nucleic Acids Res. **46**(16), e96–e96 (2018)
18. Kumar, A., Kaur, A., Kumar, M.: Face detection techniques: a review. Artif. Intell. Rev. **52**(2), 927–948 (2018). https://doi.org/10.1007/s10462-018-9650-2
19. de Lima, G.V.L., Castilho, T.R., Bugatti, P.H., Saito, P.T.M., Lopes, F.M.: A complex network-based approach to the analysis and classification of images. In: CIARP 2015. LNCS, vol. 9423, pp. 322–330. Springer, Cham (2015). https://doi.org/10.1007/978-3-319-25751-8_39
20. de Lima, G.V., Saito, P.T., Lopes, F.M., Bugatti, P.H.: Classification of texture based on bag-of-visual-words through complex networks. Expert Syst. Appl. **133**, 215–224 (2019)
21. Lopes, F.M., Martins, D.C., Jr., Barrera, J., Jr., Cesar, R.M., Jr.: A feature selection technique for inference of graphs from their known topological properties: revealing scale-free gene regulatory networks. Inf. Sci. **272**, 1–15 (2014)
22. Lowe, D.: Object recognition from local scale-invariant features. In: Proceedings of the 7 IEEE International Conference on Computer Vision, vol. 2, pp. 1150–1157 (1999)
23. Lu, Y., Fan, Y., Lv, J., Noble, W.S.: DeepPINK: reproducible feature selection in deep neural networks. In: Advances in Neural Information Processing Systems, pp. 8676–8686 (2018)
24. de Mesquita Sá Junior, J.J., Backes, A.R., Cortez, P.C.: Texture analysis and classification using shortest paths in graphs. PRL **34**(11), 1314–1319 (2013)
25. Min, R., Hadid, A., Dugelay, J.L.: Improving the recognition of faces occluded by facial accessories. In: Face and Gesture 2011, pp. 442–447. IEEE (2011)
26. Newman, M.E.J.: The structure and function of complex networks. SIAM Rev. **45**(2), 167–256 (2003). https://doi.org/10.1137/S003614450342480
27. Ojala, T., Pietikäainen, M., Mäenpää, T.: Multiresolution gray-scale and rotation invariant texture classification with local binary patterns. IEEE TPAMI **24**, 971–987 (2002)
28. Ojala, T., Pietikäainen, M., Harwood, D.: A comparative study of texture measures with classification based on featured distributions. Pattern Recogn. **29**(1), 51–59 (1996)
29. Phillips, P., Moon, H., Rizvi, S., Rauss, P.: The FERET evaluation methodology for face-recognition algorithms. IEEE TPAMI **22**(10), 1090–1104 (2000)
30. Piotto, J.G.S., Lopes, F.M.: Combining surf descriptor and complex networks for face recognition. In: CISP-BMEI, pp. 275–279 (2016)
31. Schroff, F., Kalenichenko, D., Philbin, J.: FaceNet: a unified embedding for face recognition and clustering. In: IEEE CVPR (2015)
32. Subash Kumar, T.G., Nagarajan, V.: Local curve pattern for content-based image retrieval. Pattern Anal. Appl. **22**(3), 1233–1242 (2018). https://doi.org/10.1007/s10044-018-0724-1
33. Tabatabaei, S.M., Chalechale, A.: Noise-tolerant texture feature extraction through directional thresholded local binary pattern. Vis. Comput. **36**(5), 967–987 (2019). https://doi.org/10.1007/s00371-019-01704-8
34. Taigman, Y., Yang, M., Ranzato, M., Wolf, L.: DeepFace: closing the gap to human-level performance in face verification. In: IEEE CVPR (2014)

35. Thomaz, C.E., Giraldi, G.A.: A new ranking method for principal components analysis and its application to face image analysis. Image Vis. Comput. **28**(6), 902–913 (2010)
36. Turk, M., Pentland, A.: Face recognition using eigenfaces. In: IEEE CVPR, pp. 586–591 (1991)
37. Viola, P., Jones, M.: Rapid object detection using a boosted cascade of simple features. In: IEEE CVPR 2001, vol. 1, pp. I-511–I-518 (2001)
38. Watts, D.J., Strogatz, S.H.: Collective dynamics of small-world networks. Nature **393**, 440–442 (1998)
39. Weber, M.: Frontal face dataset. California Institute of Technology (1999). http:// www.vision.caltech.edu/html-files/archive.html

End-to-End Deep Sketch-to-Photo Matching Enforcing Realistic Photo Generation

Leonardo Capozzi[1,2]($\boxtimes$) ⓘ, João Ribeiro Pinto[1,2] ⓘ, Jaime S. Cardoso[1,2] ⓘ, and Ana Rebelo[1] ⓘ

[1] INESC TEC, Porto, Portugal
{leonardo.g.capozzi,joao.t.pinto,jaime.cardoso,arebelo}@inesctec.pt
[2] Faculdade de Engenharia da Universidade do Porto, Porto, Portugal

Abstract. The traditional task of locating suspects using forensic sketches posted on public spaces, news, and social media can be a difficult task. Recent methods that use computer vision to improve this process present limitations, as they either do not use end-to-end networks for sketch recognition in police databases (which generally improve performance) or/and do not offer a photo-realistic representation of the sketch that could be used as alternative if the automatic matching process fails. This paper proposes a method that combines these two properties, using a conditional generative adversarial network (cGAN) and a pre-trained face recognition network that are jointly optimised as an end-to-end model. While the model can identify a short list of potential suspects in a given database, the cGAN offers an intermediate realistic face representation to support an alternative manual matching process. Evaluation on sketch-photo pairs from the CUFS, CUFSF and CelebA databases reveal the proposed method outperforms the state-of-the-art in most tasks, and that forcing an intermediate photo-realistic representation only results in a small performance decrease.

Keywords: Digital forensics · Sketches · Generation

1 Introduction

Over the years, convolutional neural networks (CNN) have been very successful for several pattern recognition and computer vision tasks, including that of face recognition. Recent publications in this topic often report significantly improved accuracy in the matching process when using deep learning methodologies [2,11,18]. However, face recognition based on photos or video is obviously an easier task than face recognition based forensic sketches, since a sketch might not be the most accurate representation of an individual, as it was drawn based on descriptions from eye witnesses [14]. On the problem of sketch-to-face matching, several recent state-of-the-art methods have also used CNNs to match a sketch to the corresponding identity [1,4,6,10]. However, several of these do not take full advantage of the potential of deep learning as they are not end-to-end deep

© Springer Nature Switzerland AG 2021
J. M. R. S. Tavares et al. (Eds.): CIARP 2021, LNCS 12702, pp. 451–460, 2021.
https://doi.org/10.1007/978-3-030-93420-0_42

approaches. The inclusion of blocks manually tuned or separately optimised can cause dissonance between different processes and limit achievable performance. One of the examples of separate processes in sketch-to-face recognition is the prior transformation of a sketch to resemble a real face photo. Several literature approaches apply such transformations including, sophisticated adversarial methods based on CycleGANs [6] and cGANs [1,10]. Having a photo-realistic representation of the suspect's face is a great advantage for manual identification, when the automatic matching process fails to deliver useful results. Nevertheless, the separate optimisation of these processes will, as emphasized, induce performance limitations. Hence, this work tackles these two important aspects of forensic sketch-to-photo matching, that have not yet been addressed in the literature. The first aspect is avoiding the combination of separate processes that are individually optimised, which often limits achievable performance. The second aspect is enforcing the end-to-end model to offer an intermediate representation that is photo-realistic, so the authorities have access to a realistic face rendering that will help manual identification of the suspect. To achieve this, we propose an end-to-end model composed of a cGAN and a matching CNN that are jointly optimised. When trained, the model receives a sketch and returns a template that can be used for matching using simple distance metrics. Although the approach is end-to-end, the training strategy induces the cGAN to generate intermediate latent representations that look realistic and are similar to the corresponding real photographs. Hence, we avoid the performance limits often linked to non-end-to-end approaches, while retaining the intermediate realistic images that could help an alternative manual identification process.

2   Proposed Methodology

The proposed method is an end-to-end model that, although integrated and optimizable as a whole, is composed of two main parts: a sketch-to-render generator and a matching network (see Fig. 1). The sketch-to-render generator will transform the input sketch into a face rendering that is photo-realistic and similar to the real face that corresponds to the sketch. The matching network will receive the realistic rendering and output a template that can be used for matching through a simple distance measure. In the next subsections, the architecture of both parts, the loss function, and the training process of the model are described in higher detail.

2.1   Network Architecture

Sketch-to-Render Generator. The sketch-to-render process is performed by an image-to-image model that receives a sketch and transforms it into a realistic face rendering. For this model, we use the generator of a cGAN [9], that follows the typical structure of a U-Net, with an encoder, that reduces data resolution at each level, followed by a decoder, that processes data up to the original resolution (see Fig. 2). Skip connections enable the transmission of information between corresponding levels from the encoder to the decoder. The U-Net encoder mimicked

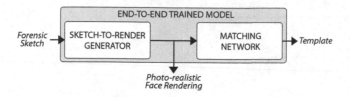

Fig. 1. Overview schema of the proposed method.

the architecture of VGG-16 [17] and it is composed of 13 convolutional layers and 4 max-pooling layers distributed over 5 resolution levels. Convolutional layers have 64–512 filters, with size of 3×3 and stride of 1×1. Max-pooling layers use window size and stride of 2×2. The decoder mirrors the structure of the encoder with convolutional and deconvolution layers. Convolutional layers have 64–512 filters with size and stride similar to their respective encoder counterparts. Deconvolution layers have 64–512 filters with size and stride of 2×2. The discriminator used during training is a CNN adapted from the cGAN of pix2pix [5] (see Fig. 2), which receives as input the photo-realistic representation outputted by the generator and the corresponding sketch, and outputs a prediction on whether it is real or generated. This discriminator is composed of 3 convolutional layers, 1 fully-connected layer, and batch normalization. The convolutional layers have 64–256 filters with size of 4×4 and stride of 2×2.

Matching Network. After the sketch is transformed into a photo-realistic rendering of the face, the matching part of the model transforms this rendering into a template that can easily be used for matching. In this work, the matching network follows the structure of the VGG-16 [17] and uses pretrained VGG-Face weights [11]. Given a sketch, the output of the matching network (and of the method as a whole) is a template that can be used for matching. This template (or face descriptor) is a numerical uni-dimensional vector of 2622 features that describe the face represented on the input sketch. The sketch is thus matched with face photos from a database by computing the cosine distance between the respective templates.

2.2 Loss

The loss function used for training is a composition of several loss components relative to each part of the proposed model. The first component of the loss corresponds to the sketch-to-render, and comes from the typical training methodology of a cGAN. This part can be described as following:

$$\mathcal{L}_{cGAN}(G, D) = \mathbb{E}_{x,y}[\log D(x, y)] + \mathbb{E}_x[\log(1 - D(x, G(x)))], \tag{1}$$

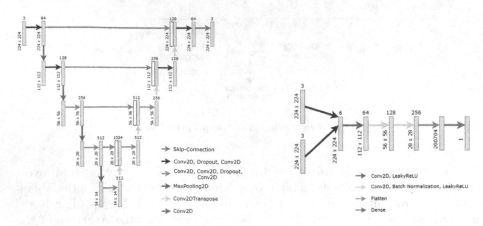

Fig. 2. Network architecture of the generator (on the left) and discriminator (on the right).

where x is a sketch and y is the corresponding ground-truth photo. The generator (G) tries to minimize this loss and the discriminator (D) competes to maximize it. This competition, if adequately balanced, will lead the generator to output increasingly realistic face images. Nevertheless, the generator should not only transform the sketch into a photo-realistic image, but also should closely mimic the respective ground-truth face photograph. In order to achieve this, a second loss term is defined as:

$$\mathcal{L}_{L1}(G) = \mathbb{E}_{x,y}[||y - G(x)||_1], \qquad (2)$$

which minimizes the $L1$ norm of the difference between the real image (y) and the generated image ($G(x)$). Moreover, we need to ensure the identity information in the sketch is preserved and refined throughout the network, and matches that of the respective ground-truth image.

Hence, a third component is added to the loss function, corresponding to the matching part, such as:

$$\mathcal{L}_{match}(G) = \mathbb{E}_{x,y}[||V(y) - V(G(x))||_2]. \qquad (3)$$

Notice that the weights of the VGG-Face network (V) are frozen, since we are using the pretrained weights. The only weights that are adjusted in the matching loss are the weights of the generator. Combining the loss components mentioned above, the final loss function becomes as:

$$\mathcal{L}(G, D) = \min_G \max_D \mathcal{L}_{cGAN}(G, D) + \lambda_1 \mathcal{L}_{L1}(G) + \lambda_2 \mathcal{L}_{match}(G). \qquad (4)$$

3   Experimental Settings

3.1   Data

The proposed model has been trained using pairs of sketches and corresponding face images from the CUHK Face Sketch database (CUFS) [19], the CUHK Face Sketch FERET dataset (CUFSF) [19,21] and the CelebA Dataset [8]. The CUFS dataset contains 606 sketch-photo pairs. The sketches correspond to face images acquired from students of the Chinese University of Hong Kong (CUHK) and from the AR and XM2GTS databases. Due to the current unavailability of the latter two databases, this work only used the 188 sketch-photo pairs relative to CUHK students. The original data split was used, which contained 88 pairs for model training and 100 pairs for testing. The CUFSF contains 1189 sketches that correspond to photos in the FERET database [12,13]. Of those, 1089 pairs were used for training and 100 pairs were used for testing. The CelebA dataset is a large scale dataset, which contains 202599 face images of 10177 identities. We randomly select 2000 images for testing and the remaining for training. The selected images for testing contain 400 different identities and 5 images per identity. This dataset does not contain sketches, therefore we generated our own sketches using an edge detection algorithm [10].

3.2   Pre-processing

To uniformize the CUFS and CUFSF databases, the DeOldify API[1] was used to colorize the photos of CUFSF from the FERET database. The use of this API allows the proposed model to generate color images. After visually inspecting the colorized images, we noticed that the colorization process did not negatively affect their quality.

In style transfer problems using conditional GANs, image spatial consistency is paramount. When the locations of image landmarks are consistent between the input and the expected output, the network is able to offer realistic results. However, in the case of sketch-to-photo transformation, the shape, size, and location of facial landmarks in the sketch and the ground truth can vary considerably. These situations reflect on the generator loss values and generally cause distortions that damage the realism of the rendering. To solve this problem, the sketches and photos are transformed so that the faces are aligned and the position of the eyes are consistent in each sketch-photo pair. This enabled higher photo-realism in the generated renderings. Additionally, the accuracy of the matching process is also improved [16].

3.3   Training

The model was trained on two experimental settings. The first one followed the aforementioned loss function to promote both an intermediate realistic face

---

[1] DeOldify API. Available on: https://github.com/jantic/DeOldify.

rendering and high matching accuracy. Here, the model was trained during 780, 1500 and 2 epochs, for CUFSF, CUFS and CelebA respectively, with batch size 8, using the Adam optimizer with a learning rate of 2×10^{-4} and parameter $beta_1 = 0.5$. The loss parameters λ_1 and λ_2 were experimentally set to 100 and 1, respectively. This allowed for a good balance of all the losses and increased the quality of the generated images. The pix2pix paper [5] also recommended a value of 100 for the λ_1 parameter. The second setting studied the effect on performance if the realistic face rendering generation was disposable. To achieve this, the loss terms $\mathcal{L}_{cGAN}$ and $\mathcal{L}_{L1}$, that promotes the realistic intermediate representation, were removed. In this setting, the model was trained during 1320, 2800 and 2 epochs, for CUFSF, CUFS and CelebA respectively, with batch size 12, using the Adam optimizer with a learning rate of 1×10^{-3} and parameter $beta_1 = 0.9$. To improve the robustness of the model and to avoid overfitting, data augmentation was applied to each pair of images. These were randomly cropped and horizontally mirrored before each epoch.

4   Results and Discussion

To evaluate the matching performance of the proposed model, its rank-N accuracy was computed on the test sets of CUFS, CUFSF and CelebA. Rank-N accuracy measures the fraction of test instances where the true correspondence is successfully found among the N strongest predictions offered by the model. In this case, we consider $N \in \{1, 5, 10\}$. After training the model with the CUFSF dataset (1089 train pairs), 54% rank-1 accuracy (100 test pairs) was attained. Over all considered ranks, the matching accuracy of the proposed method is superior or aligned to the alternative methods, as presented in Table 1. However, when dismissing photo-realistic rendering generation, the matching accuracy increases significantly, showing a trade-off between intermediate representation realism and matching performance that could be tuned to fit the objectives of specific application scenarios.

Table 1. Matching accuracy on the CUFSF and CUFS datasets, using different methods to enhance the sketch (r.r.g.: realistic rendering generation).

Method	CUFSF			CUFS		
	Rank-1	Rank-5	Rank-10	Rank-1	Rank-5	Rank-10
Sketch	49	77	88	47	80	90
pix2pix	53	83	92	52	79	86
IPMFSPS [7]	**74**	94	97	**80**	**95**	**97**
HFFS2PS [1]	–	–	–	36	69	–
Proposed (with r.r.g)	54	87	96	44	77	86
Proposed (without r.r.g)	**74**	**98**	**99**	59	85	91

On the CUFS dataset, using 88 sketch-photo pairs for training and 100 pairs for testing, the proposed method attained 44% rank-1 matching accuracy (see Table 1). These performance results are below, but nevertheless aligned with the alternative methods evaluated on the same settings. When removing the rendering realism constraints in the loss, the accuracy at all ranks increases considerably, achieving much better performance.

Considering the significantly smaller size of the CUFS training set *vs.* the size of the training set of the CUFSF dataset, these results may denote that the proposed method is more sensitive to scarce data than the alternatives, failing to offer more general solutions. Furthermore, knowing the images in the CUFS dataset are much less diverse than those of CUFSF (regarding subject ethnicity, skin color, hair color, background), one can argue that the proposed method offers the greatest advantages over the alternatives in more challenging scenarios.

On the CelebA dataset, using 200599 sketch-photo pairs for training and 2000 test pairs, a rank-1 accuracy of 67% was attained (see Table 2). In order to calculate the accuracy we select one sketch-photo pair per identity and add it to the query set. We add the rest of the sketch-photo pairs to the gallery. Then we generate a realistic representation of every sketch in the query set, and order the identities in the gallery by their average distance to the generated image. We use the ordered list of identities to calculate the rank-N accuracy. Since the process of selecting the query sketch-photo pairs is random we repeat this process 10 times and take the average.

4.1   Realistic Generation Performance

After the evaluation of matching performance, the realism of the intermediate representations generated by the proposed method were evaluated. Examples of these photo-realistic images and the corresponding sketches and ground-truth photos from the CUFSF, CUFS and CelebA datasets can be seen in Fig. 3.

Measuring realism is a difficult task, since the concept is highly subjective. However, we can assume that, if a generated image is sufficiently similar to the corresponding ground-truth, it is realistic. Hence, we use similarity/dissimilarity

Table 2. Matching accuracy on the CelebA dataset, using different methods to enhance the sketch (r.r.g.: realistic rendering generation).

Method	Accuracy (%)		
	Rank-1	Rank-5	Rank-10
Sketch	18	33	43
QGS2PIS [10]	66	80	**92**
FAGDS2PS [6]	66	–	–
Proposed (with r.r.g.)	63	81	86
Proposed (without r.r.g)	**67**	**83**	88

metrics that are common in related literature works for an objective evaluation, along with a visual subjective inspection of the test results. These metrics were the Fréchet Inception Distance (FID) [3], the Inception Score (IS) [15], and the Structural Similarity Index (SSIM) [20].

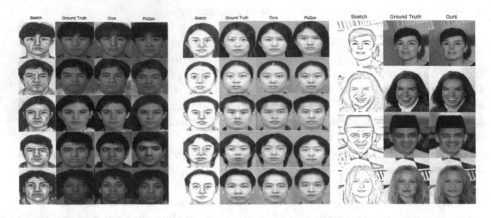

Fig. 3. Images generated by our method using the CUFSF dataset (on the left); Images generated by our method using the CUFS dataset (on the middle); Images generated by our method using the CelebA dataset (sketches were generated using an edge detection algorithm) (on the right).

As expected, the results through the similarity/dissimilarity metrics are better when the proposed method includes realistic rendering generation (see Table 3), as the method has learnt to generate images that are similar to the respective ground-truth photos. With CUFSF data, the results with the proposed method are slightly inferior but comparable to the alternatives, once again showing that the proposed method finds advantages in more challenging data.

Table 3. Comparison between different methods to enhance the sketch, using the CUFSF and CUFS datasets (r.r.g.: realistic rendering generation).

Method	CUFSF			CUFS		
	FID ↓	IS ↑	SSIM ↑	FID ↓	IS ↑	SSIM ↑
Ground truth	0.0	1.71	1.0	0.0	1.39	1.0
pix2pix	**70.54**	**1.69**	0.59	**41.46**	1.26	0.62
HFFS2PS [1]	–	–	–	58.50	1.43	**0.70**
Proposed (with r.r.g)	83.51	1.47	**0.60**	74.74	**1.54**	0.61
Proposed (without r.r.g)	330.41	1.48	0.32	321.96	1.35	0.45

Fig. 4. Intermediate representation of the sketches using the CUFSF dataset (on the left); Intermediate representation of the sketches using the CUFS dataset (on the right).

Visually, we can confirm that the results of the proposed method are, indeed, very similar to the alternative method (see Fig. 3). Considering the improvement in performance, maintaining the degree of photo-realism is a positive aspect of the proposed method. Furthermore, the proposed method delivers, in most cases, images that retain most information needed to identify the person represented in the sketch. Nevertheless, the method presents a worrying lack of diversity in specific facial features (such as hair, skin, or eye color) that should be addressed. In fact, when trained without photo-realistic generation, the method avoids these details by using unrealistic color schemes (see Fig. 4). To improve this shortcoming, an approach similar to the one proposed in [4], which consists in giving facial attribute information along with the sketch, could be an important addition to the proposed model.

5   Conclusion

This paper proposes an end-to-end deep method for sketch-to-photo matching that promotes the generation of a photo-realistic intermediate representation of the face depicted on the input sketch. As an end-to-end model, jointly trainable, it aims to eliminate performance limitations associated with separately optimized processes. Upon evaluation, the matching process showed performance improvements over the state-of-the-art and the generation of face renderings offered realistic results. When disregarding realistic rendering generation, the performance results improved. Despite the promising results, further efforts should be devoted to improve the face rendering generation component. Namely, the limited diversity of face characteristics like hair, eyes, and skin color on generated images should be addressed, in order to improve both the realism of the results and the matching performance.

Acknowledgements. This work is financed by National Funds through the Portuguese funding agency, FCT - Fundação para a Ciência e a Tecnologia, within project UIDB/50014/2020, and within the PhD grant "SFRH/BD/137720/2018". Portions of the research in this paper use the FERET database of facial images collected under the FERET program, sponsored by the DOD Counterdrug Technology Development Program Office.

References

1. Chao, W., Chang, L., Wang, X., Cheng, J., Deng, X., Duan, F.: High-fidelity face sketch-to-photo synthesis using generative adversarial network. In: ICIP, pp. 4699–4703 (2019)
2. Deng, J., Guo, J., Xue, N., Zafeiriou, S.: Arcface: additive angular margin loss for deep face recognition. In: The IEEE Conference on Computer Vision and Pattern Recognition (CVPR) (2019)
3. Heusel, M., Ramsauer, H., Unterthiner, T., Nessler, B., Hochreiter, S.: Gans trained by a two time-scale update rule converge to a local nash equilibrium. In: NeurIPS, pp. 6626–6637 (2017)
4. Iranmanesh, S.M., Kazemi, H., Soleymani, S., Dabouei, A., Nasrabadi, N.M.: Deep sketch-photo face recognition assisted by facial attributes. In: IEEE BTAS, pp. 1–10 (2018)
5. Isola, P., Zhu, J.Y., Zhou, T., Efros, A.A.: Image-to-image translation with conditional adversarial networks. In: CVPR (2017)
6. Kazemi, H., Iranmanesh, M., Dabouei, A., Soleymani, S.M. Nasrabadi, N.: Facial attributes guided deep sketch-to-photo synthesis. In: WACVW, pp. 1–8 (2018)
7. Lin, Y., Ling, S., Fu, K., Cheng, P.: An identity-preserved model for face sketch-photo synthesis. IEEE Signal Process. Lett. **27**, 1095–1099 (2020)
8. Liu, Z., Luo, P., Wang, X., Tang, X.: Deep learning face attributes in the wild. In: ICCV (2015)
9. Mirza, M., Osindero, S.: Conditional generative adversarial nets. arXiv (2014)
10. Osahor, U., Kazemi, H., Dabouei, A., Nasrabadi, N.: Quality guided sketch-to-photo image synthesis. arXiv 2005.02133 (2020)
11. Parkhi, O.M., Vedaldi, A., Zisserman, A.: Deep face recognition. In: British Machine Vision Conference (2015)
12. Phillips, P.J., Moon, H., Rizvi, S.A., Rauss, P.J.: The FERET evaluation methodology for face recognition algorithms. IEEE Trans. Pattern Anal. Mach. Intell. **22**, 1090–1104 (2000)
13. Phillips, P.J., Wechsler, H., Huang, J., Rauss, P.: The FERET database and evaluation procedure for face recognition algorithms. Image Vision Comput. J. **16**(5), 295–306 (1998)
14. Pramanik, S., Bhattacharjee, D.D.: An approach: modality reduction and face-sketch recognition. arXiv (2013)
15. Salimans, T., et al.: Improved techniques for training gans. In: NeurIPS, pp. 2234–2242 (2016)
16. Schroff, F., Kalenichenko, D., Philbin, J.: Facenet: a unified embedding for face recognition and clustering. In: CVPR (2015)
17. Simonyan, K., Zisserman, A.: Very deep convolutional networks for large-scale image recognition. In: International Conference on Learning Representations (2015)
18. Wang, M., Deng, W.: Deep face recognition: a survey. arXiv (2018)
19. Wang, X., Tang, X.: Face photo-sketch synthesis and recognition. IEEE Trans. Pattern Anal. Mach. Intell. (PAMI) **31**, 1955–1967 (2009)
20. Wang, Z., Bovik, A.C., Sheikh, H.R., Simoncelli, E.P.: Image quality assessment: from error visibility to structural similarity. IEEE Trans. Image Process. **13**(4), 600–612 (2004)
21. Zhang, W., Wang, X., Tang, X.: Coupled information-theoretic encoding for face photo-sketch recognition. In: CVPR (2011)

Forensic Analysis of Tampered Digital Photos

Sara Ferreira[1(✉)], Mário Antunes[2,3], and Manuel E. Correia[1,3]

[1] Faculty of Science, University of Porto, Porto, Portugal
up201606726@up.pt, mcc@dcc.fc.up.pt
[2] CIIC, School of Technology and Management, Polytechnic of Leiria,
Porto, Portugal
mario.antunes@ipleiria.pt
[3] INESC-TEC, CRACS, University of Porto, Porto, Portugal

Abstract. Deepfake in multimedia content is being increasingly used in a plethora of cybercrimes, namely those related to digital kidnap, and ransomware. Criminal investigation has been challenged in detecting manipulated multimedia material, by applying machine learning techniques to distinguish between fake and genuine photos and videos. This paper aims to present a Support Vector Machines (SVM) based method to detect tampered photos. The method was implemented in Python and integrated as a new module in the widely used digital forensics application Autopsy. The method processes a set of features resulting from the application of a Discrete Fourier Transform (DFT) in each photo. The experiments were made in a new and large dataset of classified photos containing both legitimate and manipulated photos, and composed of objects and faces. The results obtained were promising and reveal the appropriateness of using this method embedded in Autopsy, to help in criminal investigation activities and digital forensics.

Keywords: Digital forensics · Deepfake · Photo tampering · Support Vector Machines · Discrete Fourier Transform

1 Introduction

Cybercrime assumes different shapes. By having a computer connected to the Internet, cybercriminals are able to carry on a widespread of illegitimate and malicious activities against companies and individuals. In the last five years, there has been an increase of 67% in the incidence of security breaches worldwide [7], being malicious activities like phishing, ransomware, and crypto-jacking, some of the most popular threats to cybersecurity [14].

Intrinsically related, and in some way more silent, defacing and deepfake take advantage of multimedia contents manipulation techniques to modify digital photos and videos. In this type of crime, attackers are interested in defacing individuals' digital identity, by spreading malicious multimedia content and exposing

© Springer Nature Switzerland AG 2021
J. M. R. S. Tavares et al. (Eds.): CIARP 2021, LNCS 12702, pp. 461–470, 2021.
https://doi.org/10.1007/978-3-030-93420-0_43

individuals in an odd context. Broadly speaking, the motivations for deepfake are digital kidnap, usually involving under-aged and vulnerable victims [4].

Digital forensics analysis integrates techniques and procedures for the collection, preservation, and analysis of evidence in electronic equipment. It is an imperative tool for criminal investigation teams, namely in the analysis and identification of artifacts and digital evidence. Criminal investigation has recently encountered several challenges in detecting manipulated multimedia content, being even more affordable as cybercriminals massively use digital equipment connected to the Internet. Once seized, these equipment have to be analyzed to identify digital evidence and artifacts of suspicious activity.

Machine Learning (ML) has boosted the automated detection and classification of artifacts in a digital forensics investigation. Existing techniques to detect manipulated photos [1] are not yet properly integrated into forensic applications and therefore a module to automate this type of detection is relevant. The enhancements observed in the reported ML methods were not yet been translated into substantial improvements for cybercrime investigation, as those are not yet massively incorporated in state-of-the-art digital forensics tools.

Autopsy (https://www.autopsy.com/) is an open-source digital forensics application, widely used by criminal investigation police, dedicated to analyse and identify digital evidence and artifacts of suspicious and anomalous activities. It incorporates a wide set of native modules to process digital images (e.g. disks) and also allows the community to develop others more specific.

This paper describes the deployment and development of a module for Autopsy, that incorporates an SVM based method [6] to detect manipulated photos. The method was developed as a Python based module for Autopsy and is able to detect distinct anomalies in photos, like splicing and copy-move. The features were calculated by the Discrete Fourier Transform (DFT) and extracted for further processing by a SVM-based method. The module was tested with a classified dataset of about 40,000 photos, composed of both faces and objects, where it is possible to find examples of splicing and copy-move manipulations. The results proved the precision of the SVM-based method that achieved an averaged precision and recall of 99,4%, when detecting manipulated photos.

The remaining of the paper is organized as follows. Section 2 describes the fundamentals behind the topics covered in this paper, namely digital forensics, detection techniques, and deepfake. Section 3 depicts the overall architecture and pipeline delineated to process the input photos. Section 4 presents the experimental setup, the datasets, and the performance metrics used. Section 5 describes the results obtained. Finally, Sect. 6 describes the main conclusions and delineates the future work.

2   State of the Art

This Section describes the fundamentals of digital forensics, video manipulation techniques, and the most relevant ML techniques to deal with the detection of fake multimedia content.

2.1 Digital Forensics

Digital forensics embodies techniques and procedures to collect, preserve and analyze digital evidence in electronic equipment, namely disks, smartphones, and other devices with storage capacity. The underpinning main goal is to produce a sustained reconstruction of events, that may help digital forensics investigators to build a list of evidence that may dictate about suspect's innocence or guilt.

Cybersecurity professionals understand the value of digital forensics information and the importance of maintaining it protected. For this reason, it is essential to establish strict guidelines and procedures, namely detailed instructions about authorized rights to retrieve digital evidence, how to properly prepare systems for evidence retrieval, where to store any recovered evidence, and how to document these activities to guarantee data authenticity and integrity. Among the digital forensics tools to extract, collect, and analyze digital artifacts, Autopsy, EnCase, FTK, XRY are the most relevant and widely used ones.

2.2 Multimedia Manipulation Techniques

Photos manipulation is appealing, mostly in the context of spreading fake news, defacing, deepfake, and digital kidnap. There are three main types of photo manipulation, that are described below.

Copy-move (Fig. 1(a)) consists of copying or moving part of a photo to another place in the same photo. The goal is to give the illusion of having more elements in the photo than those that are there.

(a) Copy-move   (b) Splicing

Fig. 1. Photos manipulation types.

Splicing (Fig. 1(b)) consists of superimposing different regions of two photos, being deepfake the most relevant consequence. It is an artificial and automated manipulation of media, usually by means of artificial intelligence techniques, in which a person's face in an existing photo or video is swiped by someone else's face [2].

While deepfake of photos and videos is not new and can be seen in numerous digital contents, it has leveraged powerful ML and artificial intelligence techniques to improve contents manipulation. The most common ML methods used to improve deepfake are based on deep learning and involve training generative neural network architectures, such as auto encoders or Generative adversarial networks (GANs) [3]. Deepfake has garnered widespread attention, as it has been used in digital campaigns of spreading fake news. This manipulation technique is also responsible for digital kidnap, revenge porn, and financial fraud [16].

Finally, Resampling consists of changing the scale or even the position of an element in a photo.

2.3   Techniques Used to Detect Photos Manipulation

Bearing in mind that the use of deepfake and copy-move in digital crimes is a growing problem and has a great impact on today's society, there are already documented algorithms to tackle with this type of manipulation.

The difference between Gaussian (DoG) and Oriented Rotated Brief (ORB) are techniques used to detect copy-move in manipulated photos. The method suggested by Niyishaka et al. [8] comprises three steps: corners detection with Sobel algorithm; features extraction with DoG and ORB; and finally, features correspondence.

Unmasking deepfake with DFT and ML is a method described in [6]. It is based on a classical frequency domain analysis with DFT, followed by a classification based on ML techniques. The frequency characteristics of a photo can be analyzed in a space defined by a Fourier transform, namely a spectral decomposition of the input data indicating how the signal's energy is distributed over a range of frequencies. In this method it is used a DFT, which is a mathematical technique to decompose a discrete signal into sinusoidal components of various frequencies ranging from 0 (constant frequency, corresponding to the image mean value) up to the maximum of the admissible frequency, given by the spatial resolution. The frequency-domain representation of a signal carries information about the signal's amplitude and phase at each frequency, and can be computed as (1):

$$X_{k,l} = \sum_{n=0}^{N-1} \sum_{0}^{M-1} x_{n,m} \cdot e^{(-\frac{i2\pi}{N}k_n)} \cdot e^{(-\frac{-i2\pi}{M}l_m)} \tag{1}$$

After applying a Fourier Transform to a photo, the returned values are represented in a new domain but within the same dimensionality. Therefore, given that we work with photos, the output still contains 2D information. We then apply azimuthal average to compute a robust 1D representation of the DFT power spectrum. At this point, each frequency component is the radial average from the 2D spectrum. Support Vector Machines (SVM) is then used to create a model based on a training dataset with manipulated and genuine photos. The model will then applied to a test dataset, to identify an optimal separating hyperplane that maximizes the margin between both classes.

Image splicing detection with artificial blurred boundary is a method described in [13]. It is based on image edge analysis and blur detection and has two steps: image edges analysis along with feature extraction and detection algorithm. For the first step, it is performed a Non-Subsampled Contourlet Transform (NSCT) in order to get more details of the high-frequency components with the size of all the directions the same as that of the original image. The last step is to classify all features with SVM, similarly to the method described above. By using NSCT the authors claim to achieve a 95.12% true positive rate on detecting image splicing. The architecture developed in this paper, which is described in Sect. 3, applies the DFT method, having in mind the promising results previously obtained [6].

3   Architecture

This Section describes the architecture that was deployed to process input photos and to classify them as being genuine or manipulated. It also describes the Autopsy module developed to classify photos in a digital forensics context.

3.1   General Architecture

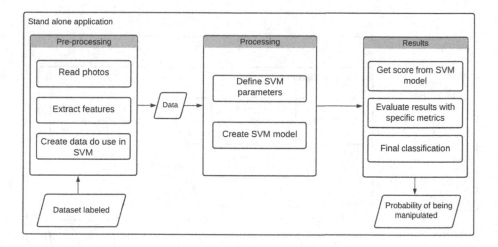

Fig. 2. Overall architecture.

The overall architecture of the standalone application developed to classify photos is depicted in Fig. 2. It has three main building blocks: pre-processing, processing, and results.

Pre-processing consists of extracting the features from the photos, by applying DFT method described in Sect. 2.3 [6] to produce the labeled input datasets for both training and testing. The pre-processing phase reads the photos through the `OpenCV` library and further extracts their features [6]. Using this method, exactly fifty features were obtained for each photo, that were then loaded into a new file with the corresponding label (0 for fake photos and 1 for the genuine ones). A training file with the features previously extracted, was created being then used to feed the SVM model.

The processing phase corresponds to the SVM processing. SVM is included in a set of kernel-based learning methods, in which the problem is addressed by mapping the data to a larger dimension space. This mapping may not be linear, as the function that allows this mapping is called a kernel [18]. The RBF (Radial basis function) kernel and a regularization parameter of 3 were chosen based on the experiments. The implementation of SVM processing was made through `scikit-learn` library for Python 3.9. The tests were carried on with a split of the whole dataset in two parts: 67% for training and 33% for testing. Both datasets (training and testing) are balanced regarding the amount of fake and genuine photos.

The results obtained in each processing are the following: SVM score; confusion matrix, precision, and recall; and the calculated prediction that allow us to deduce the probability of an image being manipulated.

3.2   Autopsy Module Architecture

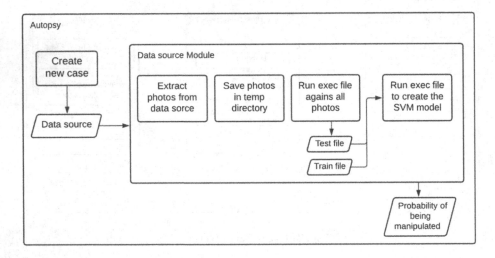

Fig. 3. Autopsy module architecture.

As stated before, Autopsy is among the most used digital forensics applications and is open to the integration of third-parties modules. Autopsy processes the input data and shows the user the results by means of report modules.

Autopsy uses Jython in new modules development, to enable Python scripting. Jython is converted into Java byte code and runs on the JVM. However, it is limited to Python 2.7. To overcome this limitation and the fact that some libraries used by the SVM classification method do not work in python 2.7, two Python executables were created to: one to process the photos; and another to create the SVM model and to classify the photos.

The data source ingest module, that runs against a data source added to the Autopsy, was developed and its architecture is depicted in Fig. 3. To start this analysis, it is needed to create a new "case" inside Autopsy and add one data source to it. An example of a data source is a disk image. Then, the module starts by extracting each photo within the data source added to the Autopsy case and saves them in a temporary directory. Next, the first Python executable extracts the features of the photos stored in the temporary directory, being all of them labeled with 0 (we assume that all photos are fake until proven otherwise). With the output obtained and the training file already created and distributed with the module, the second Python executable comes to action and the SVM classification model is created. Finally, the artifacts with the classification results are calculated and post in the Autopsy blackboard, which are further displayed to the user.

It is possible to note that the standalone application architecture corresponds also to the method used in Autopsy data source ingest module (Fig. 3). The standalone application was developed before the Autopsy module, which gave the possibility to develop and test the method while disregarding the needed compatibility with the Python libraries and with the strict format that is required by Autopsy to develop new modules.

4   Experimental Setup

This Section describes the dataset used for the experiments and the evaluation metrics that were applied to evaluate the SVM classification.

4.1   Datasets

A dataset containing both people's faces and objects was created to train the classification model. The dataset described in [6], which was used in the tests, is a compilation of photos available in CelebA-HQ dataset [9], Flickr-Faces-HQ dataset [10], "100K Facesproject" ("https://generated.photos/") and "this person does not exist" project (https://thispersondoesnotexist.com/).

Some complexity was added to the dataset, by including objects and other people's faces being possible to detect other types of manipulations aside deepfake. COVERAGE dataset [11] was included to add photos with objects and other people's faces, as well as a copy-move forgery database with similar but genuine objects that contains 97 legitimate photos and 97 manipulated ones. Columbia Uncompressed Image Splicing Detection Evaluation Dataset [12] was also added, which consists of high-resolution images, 183 authentic (taken using just one camera and not manipulated), and 180 spliced photos. Finally, 14 legitimate and 14 fake ad-hoc photos were also added, containing splicing and copy-move manipulations created by us. Putting it all together, the new dataset used in this paper is balanced and has 40,629 photos divided into two classes: 20,335 genuine (or real) photos and 20,294 that were manipulated.

4.2   Evaluation Metrics

To have a correct evaluation of the classification method, it is necessary to discuss the evaluation metrics. The metrics used to evaluate the results were Precision (P), Recall (R), and F1-score, which can be calculated through the well-known and documented confusion matrix [17].

In the confusion matrix, each row represents the instances in a predicted class, while each column represents the instances in an actual class. The positive class refers to the manipulated photos, while the negative class refers to the genuine ones. TP represents the events where the model has correctly predicted the positive class, that is a manipulated photo. TN calculates the events that were correctly predicted as negative, that is genuine photos. FP and FN evaluate the events that were incorrectly predicted by the model, namely those that legitimate photos classified as manipulated and those manipulated that were classified as genuine, respectively.

Precision and Recall correlate the metrics described above. Precision measures the percentage of examples identified as true that are really true. That is, those photos that are manipulated, from those that were classified as manipulated. Precision is calculated by (2):

$$P = \frac{TP}{(TP + FP)} \tag{2}$$

Recall is the percentage of manipulated images that we could find of the total number of manipulated images. Recall corresponds to the following (3):

$$R = \frac{TP}{(TP + FN)} \tag{3}$$

F1 is an harmonic mean between Precision and Recall. The range for the F1-score is between [0, 1] and measures the preciseness and robustness of the classifier. That is, the number of instances that were correctly classified and those that were misclassified, respectively. F1 measure is calculated by (4):

$$F1 = 2 * \frac{P * R}{(P + R)} \tag{4}$$

5   Results Analysis

This Section describes the results obtained from the experiments and the corresponding analysis. Ten experiments were made and, for each one, the dataset was randomly divided into 33% for testing and 67% for training. As can be seen in Table 1, the results obtained were very satisfactory, with a high number of correctly classified photos and a residual number of FP and FN.

It is possible to observe that we managed to achieve in these tests a precision, recall, and F1-score of approximately 100%. Comparing with the results documented in [6], even enriching the dataset with photos containing objects and other types of manipulation, it was possible to achieve the mean P, R, and F1

Table 1. Results obtained with 10 different runs.

	TP	TN	FP	FN	Precision	Recall	F1-score
Run 1	6629	6646	14	43	0.99789	0.99355	0.99571
Run 2	6580	6698	19	35	0.99712	0.99470	0.99591
Run 3	6636	6633	23	40	0.99654	0.99400	0.99527
Run 4	6644	6618	25	45	0.99625	0.99327	0.99476
Run 5	6621	6648	15	48	0.99774	0.99280	0.99526
Run 6	6713	6554	14	51	0.99792	0.99246	0.99518
Run 7	6674	6608	21	29	0.99686	0.99567	0.99627
Run 8	6584	6683	21	44	0.99682	0.99336	0.99509
Run 9	6600	6680	10	42	0.99849	0.99368	0.99608
Run 10	6640	6635	19	38	0.99715	0.99431	0.99573
Average	6632	6640	18	41	0.99728	0.99378	0.99553

above 99.3% (P = 99.73%, R = 99.38% and F1 = 99.55%). The mean values for
FP and FN are residual (FP = 0.13%, FN = 0.3%) and, by analysing the misclassified photos, it is possible to infer that were related to the resolution of the
photos. A richer dataset with heterogeneous examples regarding the resolution
of the photos would benefit the overall results obtained.

6   Conclusion

This paper described the development of an SVM based method [6] to tackle
the detection of manipulated photos. An Autopsy module that incorporates the
proposed standalone SVM-based method, was also developed, giving a helping
hand to digital forensics investigators and leveraging the use of ML techniques
to fight cybercrime activities.

The overall architecture and development make use of two well-known and
documented techniques: Discrete Fourier Transform (DFT) technique to extract
features from photos; SVM-based method to create a learning model. Both techniques were incorporated in the proposed SVM-based learning standalone application, which was further integrated as an Autopsy module in a digital forensics
context. The dataset proposed in [6] was extended with different sources, mainly
to accommodate objects and other manipulation types, besides faces and splicing respectively. The final dataset has about 40,000 photos, composed of both
faces and objects, where it is possible to find examples of splicing and copy-move
manipulations. The results obtained were promising and in line with previous
ones documented in the literature. It was possible to achieve a mean F1-score of
99.55% on the detection of manipulated photos.

Future work has two major topics: to enhance the dataset present in this
paper, by adding more genuine and manipulated photos; to apply the presented
methodology and architecture to detect these types of manipulations in videos.

References

1. Tolosana, R., Vera-Rodriguez, R., Fierrez, J., Morales, A., Ortega-Garcia, J.: Deepfakes and beyond: A survey of face manipulation and fake detection. arXiv preprint arXiv:2001.00179 (2020)
2. Kietzmann, J., Lee, L., McCarthy, I., Kietzmann, T.: Deepfakes: trick or treat? Bus. Horiz. **63**(2), 135–146 (2020)
3. Nguyen, T., Nguyen, C., Nguyen, D.T., Nguyen, D.T., Nahavandi, S.: Deep learning for deepfakes creation and detection 1 (111573) (2019)
4. Spivak, R.: Deepfakes": the newest way to commit one of the oldest crimes. Georget. Law Technol. Rev. **3**(2), 339–400 (2019)
5. Harris, D.: Deepfakes: false pornography is here and the law cannot protect you. Duke Law Technol. Rev. **17**, 99 (2018)
6. Durall, R., Keuper, M., Pfreundt, F.J., Keuper, J.: Unmasking deepfakes with simple features. arXiv preprint arXiv:1911.00686 (2019)
7. Bissell, K., LaSalle, R.M., Dal Cin, P.: The cost of cybercrime-Ninth annual cost of cybercrime study. Technical report, Accenture, 2019. Independently conducted by Ponemon Institute LLC and jointly developed by Accenture (2019). https://www.accenture.com/_acnmedia/PDF-96/Accenture-2019-Cost-of-Cybercrime-Study-Final.pdf. Accessed 9 Jan 2021
8. Niyishaka, P., Bhagvati, C.: Digital image forensics technique for copy-move forgery detection using DoG and ORB. In: Chmielewski, L.J., Kozera, R., Orłowski, A., Wojciechowski, K., Bruckstein, A.M., Petkov, N. (eds.) ICCVG 2018. LNCS, vol. 11114, pp. 472–483. Springer, Cham (2018). https://doi.org/10.1007/978-3-030-00692-1_41
9. Karras, T., Aila, T., Laine, S., Lehtinen, J.: Progressive growing of gans for improved quality, stability, and variation. arXiv preprint arXiv:1710.10196 (2017)
10. Karras, T., Laine, S., Aila, T.: A style-based generator architecture for generative adversarial networks. In Proceedings of the IEEE Conference on Computer Vision and Pattern Recognition, pp. 4401–4410 (2019)
11. Wen, B., Zhu, Y., Subramanian, R., Ng, T.T., Shen, X., Winkler, S.: COVERAGE - a novel database for copy-move forgery detection. In: IEEE International Conference on Image processing (ICIP), pp. 161–165 (2016)
12. Hsu, Y.F., Chang, S.F.: Detecting image splicing using geometry invariants and camera characteristics consistency. In: 2006 IEEE International Conference on Multimedia and Expo, pp. 549–552 (2006)
13. Liu, G., Wang, J., Lian, S., Dai, Y.: Detect image splicing with artificial blurred boundary. Math. Comput. Model. **57**(11–12), 2647–2659 (2013)
14. Moore, M.: Top Cybersecurity Threats in 2020. University of Sandiego. https://onlinedegrees.sandiego.edu/top-cyber-security-threats/. Accessed 12 Jan 2021
15. Christian, J.: Experts fear face swapping tech could start an international showdown. The Outline. https://theoutline.com/post/3179/deepfake-videos-are-freaking-experts-out?zd=1&zi=hchawpks. Accessed 13 Jan 2021
16. Roose, K.: Here Come the Fake Videos, Too. The New York Times. https://www.nytimes.com/2018/03/04/technology/fake-videos-deepfakes.html. Accessed 13 Jan 2021
17. Shung, K.P.: Accuracy, Precision, Recall or F1?. Towards Data Science.https://towardsdatascience.com/accuracy-precision-recall-or-f1-331fb37c5cb9. Accessed 6 Jan 2021
18. Hearst, M.A., Dumais, S.T., Osuna, E., Platt, J., Scholkopf, B.: Support vector machines. IEEE Intell. Syst. Appl. **13**(4), 18–28 (1998)

COVID-19 Lung CT Images Recognition: A Feature-Based Approach

Chiara Losquadro[✉], Luca Pallotta, and Gaetano Giunta

Department of Engineering, Roma Tre University, Rome, Italy
{chiara.losquadro,luca.pallotta,gaetano.giunta}@uniroma3.it

Abstract. The SARS-CoV-2 is quickly spreading worldwide resulting in millions of infection and death cases. As a consequence, it is increasingly important to diagnose the presence of COVID-19 infection regardless of the technique applied. To this end, this work deals with the problem of COVID-19 classification using Computed Tomography (CT) images. Precisely, a new feature-based approach is proposed by exploiting axial CT lung acquisitions in order to differentiate COVID-19 versus healthy Computed Tomography (CT) images. In particular, first-order statistical measures as well as numerical quantities extracted from the autocorrelation function are investigated with the aim to provide an efficient classification process ensuring satisfactory performance results.

Keywords: COVID-19 · Classification · Feature extraction · Lung Computed Tomography (CT) images

1 Introduction

On December 31st 2019, the Health Commission of Wuhan in China, informs the World Health Organization (WHO) about a cluster of unknown pneumonia cases in the province, later identified as the SARS-CoV-2. The initially unknown SARS-CoV-2 is a member of the beta-coronavirus family, currently named as COVID-19, and it can lead to mild to severe infections of respiratory tract and the traditional clinical symptoms are pneumonia, cold and fever, headache, painful throat and weakness as traditional clinical symptoms. In the worst scenario, the infected patients could also present dyspnoea or hypoxaemia signs but some asymptomatic cases were also observed where any acute COVID-19 symptoms are exhibited [6]. However, for both asymptomatic or acute situation, an early and accurate diagnosis represents a crucial step in order to reduce the COVID-19 mortality and to improve the treatment of this insidious disease. Nowadays,several methods are employed to provide a definitive diagnosis of COVID-19, including reverse transcriptase-polymerase chain reaction (RT-PCR) [11], serology tests, and medical imaging techniques. In particular, Computed Tomography (CT) scans and chest X-Ray have been proved to be a great supplement to assess the infection severity [2]. Artificial intelligence technologies

© Springer Nature Switzerland AG 2021
J. M. R. S. Tavares et al. (Eds.): CIARP 2021, LNCS 12702, pp. 471–478, 2021.
https://doi.org/10.1007/978-3-030-93420-0_44

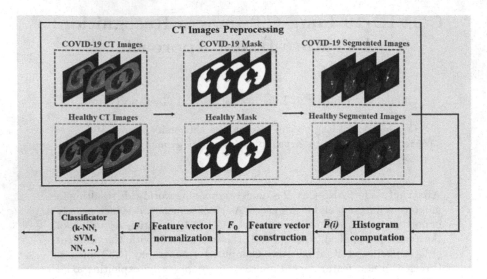

Fig. 1. Block scheme of the proposed CT images features extraction algorithm.

are achieving remarkable progress with respect to medical image analysis, with the aim to assist the clinicians classifying different diseases, due to the possibility to extract quantitative features from the acquired images [18]. Recently, several studies have focused on these methodologies in order to provide a secure and automatic way to diagnose also the COVID-19 infection applying several machine learning methods [3,4,8–10,14,16]. Following the line of reason of these works, this study proposes and investigates a new feature-based methodology in order to classify COVID-19 versus normal CT images. The analyses have been conducted on the challenging CT axial scans data deeply described in [13]. Interestingly, the obtained results have shown the effectiveness of the proposed features in distinguish among COVID-19 and healthy patients.

2   Proposed Description and Proposed Classification Algorithm

The proposed solution focuses on the exploitation of the intrinsic discriminative properties of textural features extracted from lung CT images. In particular, a joint statistical and numerical approach to texture analysis is used starting from the gray-level distribution within the images in order to differentiate them. To this end, the texture information is derived from the histogram computed from the lung CT images, through which both first-order statistic measures and numerical quantities associated with the autocorrelation function are evaluated. The proposed classification algorithm is synthetically represented in Fig. 1, and it can be seen as composed by few steps that are described in the following lines.

2.1 Lung Segmentation

In order to achieve accurate classification results, the segmentation process represents a preliminary step of paramount importance [1] when dealing within the framework of medical images. Precisely, lung segmentation aims to basically separate the pixels corresponding to the lung cavity from the surrounding chest anatomy in order to avoid erroneous classification due to the overall anatomy of the patients under observation. For such a purpose, in the proposed procedure, a segmentation technique has been implemented.

2.2 Histogram Computation

In the attempt to differentiate COVID-19 from healthy patients CT scans, a first-order statistical approach is investigated by exploiting gray-level distribution within the CT images [5]; in particular, as a result of first-order statistical techniques, texture information is derived from the histogram, that is computed for each individual image to classify. A histogram example of a COVID-19 CT scan versus a healthy patient CT scan is shown in Fig. 2. It is worth to observe that, through a preliminary visual inspection of Fig. 2, a histogram study may reveal useful to discriminate a COVID-19 image from a healthy one.

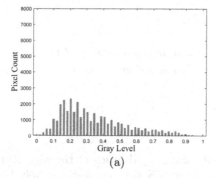 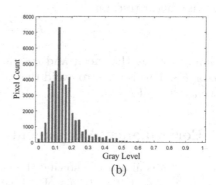

(a)   (b)

Fig. 2. Histogram computation. Subplots refer to (a) COVID-19 and (b) healthy patients.

2.3 Feature Extraction

Starting from the considerations made in Subsect. 2.2, a histogram-based feature extraction method is presented to classify COVID-19 and healthy patients CT axial scans. Therefore, the initial point is the histogram normalized to have unit area, say $\overline{P}(i)$, derived at the previous step of the proposed algorithm, from which several quantities are extracted to construct the so-called feature vector indicated with F_0 in Fig. 1. More precisely, F_0 comprises four first-order statistical measures, viz. mean, standard deviation, skewness and kurtosis, lined-up to an additional measure, the median, and three other indices, namely the peak sidelobe level (PSL) ratio and two different definitions of the integrated sidelobe level (ISL) ratio of the autocorrelation function [15] directly computed from $\overline{P}(i)$.

As to the PSL and the two different definitions of the ISL, they are computed from the normalized autocorrelation of $\overline{P}(i)$, indicated as $C_P(k)$, $k = 0, ..., K-1$, as follows:

$$\text{PSL} = \max_k \frac{|C_P(k)|}{|C_P(0)|}, \tag{1}$$

$$\text{ISL}_1 = \frac{\sum_{k=1}^{K-1} |C_P(k)|}{|C_P(0)|}, \tag{2}$$

and

$$\text{ISL}_2 = \frac{\sum_{k=1}^{K-1} |C_P(k)|^2}{|C_P(0)|}, \tag{3}$$

having indicated with $|\cdot|$ the modulus of its argument. The last step of the devised algorithm consists in normalizing the resulting feature vector, $\boldsymbol{F}_0$, thanks to the following linear rescaling:

$$\boldsymbol{F} = \frac{\boldsymbol{F}_0 - \mu_{\boldsymbol{F}_0}}{\sigma_{\boldsymbol{F}_0}} \tag{4}$$

meaning $\mu_{\boldsymbol{F}_0}$ as the mean and $\sigma_{\boldsymbol{F}_0}$ as the standard deviation of the feature vector $\boldsymbol{F}_0$. This is done to avoid that a very strong feature value could polarize the classifier's decision.

3   Performance Assessment

This section is aimed at showing the classification capabilities of the algorithm described in Sect. 2 in terms of CT scans discrimination exploiting their histograms. Therefore, a deep description of the patients dataset is first provided; then, the classification procedure is detailed together with the discussion of the obtained results. is discussed in the following subsections.

3.1   Patients Dataset

The analyses are conducted on patients with either COVID-19 condition and healthy condition, which are included in the dataset herein exploited, namely the Cov19-Healthy Dataset. The latter is selected from the original COVID-CTset, that has been gathered from Negin Medical Radiology Center located at Sari (Iran) between March and April, 2020 [13]. In particular, COVID-CTset contains the CT axial scans of both COVID-19 and healthy people, acquired using a CT scanner (SOMATOM Scope Power, Siemens Healthcare) and stored in a 512×512 DICOM format.

For the purpose of this study, 180 CT scans (i.e., Cov19-Healthy Dataset) are extracted from the original dataset including 80 COVID-19 infected images and 100 uninfected images, respectively. Table 1 summarizes the patients population details in terms of their sex and age (expressed as mean with its confidence interval in terms of standard deviation (SD), namely, ±SD).

Table 1. Sex and age details of patients enrolled in the study.

		COVID-19	Healthy
Sex	*Male*	48	48
	Female	32	52
Age y (±*SD*)		49.8 ± 14.8	39.8 ± 10.6

3.2 Classification Procedure and Results

This subsection is aimed at showing the effectiveness of the devised methodology in discriminating the COVID-19 and healthy patients described in the previous subsection.

The classification is performed dividing the dataset into two non-overlapped groups: the training set, composed by 70% of the available data and the test set, composed by the remainder 30% (i.e., excluding all the data of training phase). Moreover, since the aim of this work is to show the effectiveness of the proposed feature-extraction based method, we use as classifier the k-nearest neighbour (k-NN) because of its low computational burden, with the parameter k set equal to 5 [7]. Then, to prove the effectiveness of the proposed algorithm, the classification accuracy, say A_{cc}, is used as figure of merit, whose analytic expression is given by

$$A_{cc} = \frac{\text{TP} + \text{TN}}{\text{TP} + \text{FP} + \text{TN} + \text{FN}}, \tag{5}$$

where TP is the total number of true positives, FP is the total number of false positives, TN is the total number of true negatives, and finally FN represents the total number of false negatives.

Moreover, in order to provide a statistical characterization of the entire classification method, the average classification accuracy is estimated by means of a standard Monte Carlo approach. More precisely, for each independent Monte Carlo trial a different selection of the training and test sets is randomly chosen for each class (i.e., COVID-19 CT scan and healthy CT scan). This process reiterates N number of times (note that in this study N is set equal to 100) the training and test procedures by independent random extractions. The results in terms of average A_{cc} are shown also in comparison with the feature extraction algorithm based on the exploitation of the four statistical measures (i.e., mean, standard deviation, skewness and kurtosis) exclusively, indicated as first order based approach (FOA).

Table 2 shows the classification results in terms of classification accuracy for the above defined two classes. From the table it is evident that the proposed feature extraction method of Sect. 2.3 is capable of ensuring a satisfactory performance reaching the 84.63% of average correct classification. Conversely, the FOA provides a lower correct classification percentage. In particular, these results are also confirmed by the fact that the maximum achieved classification accuracy of the proposed algorithm is equal to 92.60% still overcoming that reached by the FOA.

Table 2. Classification accuracy (expressed in percentage) for COVID-19 versus healthy patients.

	Average A_{cc}	Maximum A_{cc}
FOA	76.85%	83.33%
Proposed method	84.63%	92.60%

To give further insights about the classification capabilities of the proposed algorithm, the average confusion matrices and the one associated with the maximum A_{cc} for the quoted algorithms are reported in Fig. 3.

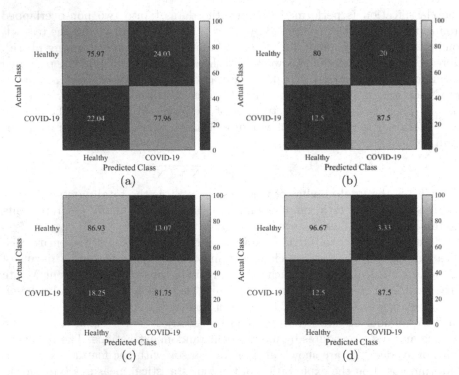

Fig. 3. Confusion matrices, actual class versus predicted class (%). Subplots refer to (a) average confusion matrix of FOA, (b) confusion matrix of the best case of FOA, (c) average confusion matrix of the proposed method, and (d) confusion matrix of the best case of the proposed method.

As can be clearly observed, the confusion matrices corroborate the results obtained in terms of overall classification capabilities. As a matter of fact, the use of all 8 features extracted through the proposed procedure allows to identify COVID-19 patients in spite of the very low computational cost of this method. It is finally worth to highlight that the promising results of the considered feature extraction method could be surely improved considering for instance more sophisticated classifiers such as a support vector machine (SVM) [17], neural networks [12], and so on.

4   Conclusions

In this paper a new feature extraction procedure has been proposed aiming to discriminate COVID-19 patients from healthy ones. The presented method has dealt with the exploitation of existing intrinsic discriminative textural properties from lung CT images, starting from their histograms computation. In particular, the histogram-based approach has allowed to investigate several first-order statistical measures and numerical quantities derived from the histogram auto-correlation which have been used as input to a k-NN classifier. The performance of the proposed extracted features has been assessed on CT axial scans of both COVID-19 affected and not affected people, ensuring a satisfactory average correct classification result of 84.63% and reaching, in the best case, the high performance of 92.60% of classification accuracy.

Possible future research works might consider the use of other features to be line-up to those proposed in this paper to improve its recognition capabilities as well as the test of the proposed features by means of neural networks.

Aknowledgments. The authors would like to thank the authors of [13] for providing the CT axial data. All data supporting this paper are available at https://github.com/mr7495/COVID-CTset.

References

1. Aggarwal, P., Renu, V., Bhadoria, S., Dethe, C.: Role of segmentation in medical imaging: a comparative study. Int. J. Comput. Appl. **29**(1), 54–61 (2011). https://doi.org/10.1109/36.739146
2. Ai, T., et al.: Correlation of chest CT and RT-PCR testing for coronavirus disease 2019 (COVID-19) in China: a report of 1014 cases. Radiology **296**, 32–40 (2020). https://doi.org/10.1148/radiol.2020200642
3. Apostolopoulos, I.D., Mpesiana, T.A.: COVID-19: automatic detection from X-ray images utilizing transfer learning with convolutional neural networks. Phys. Eng. Sci. Med. **43**(2), 635–640 (2020). https://doi.org/10.1007/s13246-020-00865-4
4. Barstugan, M., Ozkaya, U., Ozturk, S.: Coronavirus (COVID-19) Classification using CT Images by Machine Learning Methods (2020)
5. Chen, C.H., et al.: Radiomic features analysis in computed tomography images of lung nodule classification. PloS one **13**(2), e0192002 (2018)

6. Das, S., Das, S., Ghangrekar, M.M.: The COVID-19 pandemic: biological evolution, treatment options and consequences. Innov. Infrastructu. Solut. **5**(3), 1–12 (2020). https://doi.org/10.1007/s41062-020-00325-8

7. Duda, R.O., Hart, P.E.H., et al.: Pattern classification and scene analysis, vol. 3. Wiley, New York (1973)

8. Gomes, J.C., et al.: IKONOS: An intelligent tool to support diagnosis of Covid-19 by texture analysis of x-ray images. medRxiv (2020). https://doi.org/10.1101/2020.05.05.20092346

9. Ito, R., Iwano, S., Naganawa, S.: A review on the use of artificial intelligence for medical imaging of the lungs of patients with coronavirus disease 2019. Diagn. Interv. Radiol. **26**(5), 443 (2020)

10. Latif, S., et al.: Leveraging data science to combat COVID-19: a comprehensive review. IEEE Trans. Artif. Intell. **1**(1), 85–103 (2020). https://doi.org/10.1109/TAI.2020.3020521

11. Li, Y., et al.: Stability issues of RT-PCR testing of SARS-CoV-2 for hospitalized patients clinically diagnosed with COVID-19. J. Med. Virol. **92**, 903–908 (2020). https://doi.org/10.1002/jmv.25786

12. Meng, L., et al.: A deep learning prognosis model help alert for COVID-19 patients at high-risk of death: a multi-center study. IEEE J. Biomed. Health Inf. **24**(12), 3576–3584 (2020). https://doi.org/10.1109/JBHI.2020.3034296

13. Mohammad, R., Abolfazl, A., Seyed, S.: A Fully Automated Deep Learning-based Network for Detecting COVID-19 from a New and Large Lung CT Scan Dataset (2020). https://doi.org/10.20944/preprints202006.0031.v2

14. Ozsahin, I., Sekeroglu, B., Musa, M.S., Mustapha, M.T., Uzun Ozsahin, D.: Review on diagnosis of covid-19 from chest CT images using artificial intelligence. Comput. Math. Methods Med. **2020** (2020)

15. Persico, A.R., et al.: On model, algorithms, and experiment for micro-doppler-based recognition of ballistic targets. IEEE Trans. Aerosp. Electron. Syst. **53**(3), 1088–1108 (2017). https://doi.org/10.1109/TAES.2017.2665258

16. Qian, X., et al.: M^{3}Lung-Sys: a deep learning system for multi-class lung pneumonia screening from CT imaging. IEEE J. Biomed. Health Inf. **24**(12), 3539–3550 (2020)

17. Sharma, S., Khanna, P.: Computer-aided diagnosis of malignant mammograms using Zernike moments and SVM. J. Digit. Imaging **28**(1), 77–90 (2014). https://doi.org/10.1007/s10278-014-9719-7

18. Zhang, J., Xia, Y., Xie, Y., Fulham, M., Feng, D.D.: Classification of medical images in the biomedical literature by jointly using deep and handcrafted visual features. IEEE J. Biomed Health Inf. **22**(5), 1521–1530 (2018). https://doi.org/10.1109/JBHI.2017.2775662

A Topologically Consistent Color Digital Image Representation by a Single Tree

Pablo Sánchez-Cuevas[1], Fernando Díaz-del-Río[1], Helena Molina-Abril[2(✉)], and Pedro Real[2]

[1] Department of Computer Architecture and Technology, University of Seville, Sevilla, Spain
[2] Department of Applied Mathematics I, University of Seville, Sevilla, Spain
habril@us.es

Abstract. A novel, flexible (non-unique) and topologically consistent representation called CRIT (Contour-Region incidence Tree) for a color 2D digital image I is defined here. The CRIT is a tree containing all the inter and intra connectivity information of the constant-color regions. Considering I as an abstract cell complex (ACC), its topological information can be packed as a smaller (in terms of cells) ACC, whose 2-cells are the different constant-color regions of I. This modus operandi overcomes the classical connectivity paradoxes of digital images by working with lower-dimensional cells such as 0-cells, 1-cells, and 2-cells. The CRIT structure allows to describe this smaller ACC in a non-redundant way. The proposed technique is based on the previous construction of the Homological Spanning Forest (HSF) structures for encoding homological information of the ACCs canonically associated to I, in terms of rooted trees connecting digital object elements without redundancy.

Keywords: Color 2D digital image · Abstract cell complex · Homological Spanning Forest · Contour-region incident tree

1 Introduction

Representing nD digital images simply using a rectangular array of pixel values, has several drawbacks. One of the most important is the "deficiencies" that this representation has with regards to getting consistent local and global spatial topological (region-contour, interior-boundary,...) information. Digital Topology has dealt with these problems of connectedness and continuity in this discrete context, proposing two kind of solutions in terms of enriched representations: (a) those based on neighborhood graphs [2,10]; (b) those based on abstract cell complexes (ACC) of dimension n [3]. The first ones propose for binary images, different neighborhood adjacency between pixels of the same color. They potentially suffer from connectivity paradoxes and incompleteness in adequately gathering all the information of topological interest. Nevertheless, there has been a lot of successful contributions of this type or image representation in the design

© Springer Nature Switzerland AG 2021
J. M. R. S. Tavares et al. (Eds.): CIARP 2021, LNCS 12702, pp. 479–488, 2021.
https://doi.org/10.1007/978-3-030-93420-0_45

of algorithms for modelling and analysing images [11]. ACC-based solutions work at inter-pixel level and with cellular structures in which connectivity is initially measured by the notion of incidence between cells of different dimension. Some important progress in this sense has also been published [4].

In this paper, working with a color 2D digital image I as an abstract cell complex(ACC) of dimension two, a new topological representation suitably containing all its intra and inter connectivity information of constant-color regions is defined by using a single tree. We call this representation the Contour Region incidence Tree (CRIT for short). CRITs promise to be simple, easy to manipulate and fast to compute in an almost fully parallel manner, due to the fact that the method is based on the HSF (Homological Spanning Forest) framework for topological parallel computing of 2D digital objects [7–9].

2   Related Works

Among all the topological data structures representing digital images that are available in the literature, there are numerous works dealing with the idea of representing images as neighborhood graphs whose nodes are the pixels of the image and whose edges are connecting two adjacent pixels (with regard, for example, 4-adjacency relationship, for square physical pixels), such that its topological information is condensed in a smaller graph, where 4-connected regions represent nodes in the graph and an edge between two nodes exists if the corresponding regions share a 4-connected frontier.

A classical example of such compact representations is the Region Adjacency Graph. RAGs are simple graphs (no parallel edges, no self-loops), which represent the neighboring relationships of whole regions. A region is a collection of connected pixels, which share some properties. RAGs effectively encode the image into a different representation, that stores the layout of the image, and describes which regions are next to each other and which are not. Unfortunately, some important topological information is missing in semantically correct RAGs of digital images, such as the number of frontiers between two regions (also called degree of contact) or, how these frontiers are connected to each other. RAGs are simple graphs, which implies a disadvantage if we want to encode more complex configurations of regions (i.e. region being included in another region). In this sense, we can affirm that RAG representations are not enough for segmentation and other image processing purposes. In [1], a consistent topological representation is the one that "shows the interactions between regions and then realize the topology of the image". Despite using this not very rigorous definition, we can venture to assert that RAGs are not topologically consistent representations, even in 2D. There are topologically different sets of inter-connections between regions in a continuous or digital image that are locally represented by the same graph pattern in its RAG. For example, Fig. 1 shows two topologically different color images and their RAG, from which they both can not be distinguished.

More complex representations have been proposed to overcome these problems. Dual graphs [5], for instance, can be considered as an extension of the

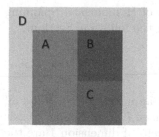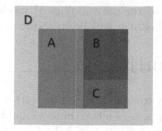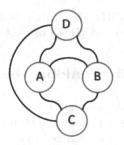

Fig. 1. Left: Two topologically different color images. Right: The RAG representation of these images (same for both)

RAG, were images are represented by a pair of graphs that are finite, planar, connected, not simple in general and dual to each other. These graphs have been used to build a hierarchy of image partitions called Dual Graph Pyramids. Although dual graphs are able to encode inclusion relationships and multiple borders, some ambiguities still remain unsolved, were for instance two images in which two regions are interchanged will have exactly the same dual graph representation. Furthermore, processing these two graphs is not time efficient, and its practical application to real images is quite limited.

Other structures such as combinatorial maps [6] overcome these ambiguities by representing the image as a set of half-edges called darts, and two permutation functions. Advancing in complexity, generalized maps have advantage over the classical combinatorial maps that they can also represent non-orientable objects. The problem with these more complicated structures (such as combinatorial and generalized maps) is their complexity and difficult manipulation. Summing up, from the viewpoint of considering a color 2D digital image I as a ACC object whose 2-dimensional cells are the physical pixels of I, RAG-based representations mainly represent the topology of the image using the connectivity graph involving all the incidence relations between 2-cells and 1-cells of I and combinatorial maps from the graph involving all the incidence relations between 0-cells and 1-cells of I. In 2D, using exclusively one of these graphs and topological duality properties, it is possible to extract topological information of I in terms of number of color-constant connected components and 1-dimensional homological holes, but it seems difficult to solve using these models, for instance, the problem of *homological hole's classification*: to determine the homological equivalence class to which a simple closed curve inside the object belongs to. This shortcoming and other more complex problems involving homotopy information (that is, in which the boundary information of any cell is treated as a loop structure) prevents these models from being fully "topologically consistent".

A CRIT representation of I is a subgraph of the connectivity graph of I, from which it is automatic to get a more compact and topologically informative ACC, whose 2-cells are the different constant-color regions of I. In this way, the previous difficulties encountered for preventing an enriched RAG description to be a topologically consistent representation are overcome, due to the fact that

CRIT description is based on the topological model of Homological Spanning Forest (HSF) of an ACC, which will be introduced in the following Section.

3   Building the CRIT

An abstract cell complex (or ACC, for short, [4]) can appropriately describe all the regions and their relations of a two dimensional image I. In this paper, pixels are considered to be cells of dimension 2. Interpixel cells of dimension 1 are the edges between pixels and the corners between pixels are defined as cells having dimension 0 (see Fig. 2). The keypoint in this ACC is that its cells can hold all the topological information of a 2-dimensional image composed of several constant-color regions, that is, a puzzle of objects (represented by two-dimensional cells) and their adjacency relations (represented by one and zero-dimensional cells).

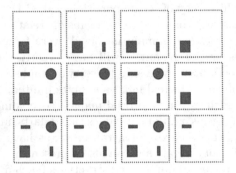

Fig. 2. ACC for a 2-dimensional image. Squares are 2-cells (image pixels), dotted lines are 1-cells (contours) and circles are 0-cells (crosses).

In Fig. 3 a 3×5 image with three regions is represented by its ACC. Not only regions are labelled with their colors A, B, C, but their contours and crosses are congruently labelled with those colors of their adjacent cells. Contours are a mix of two colors, and crosses are defined by the colors of their four surrounding pixels. Those 0 and 1 cells which are in the interior of a region, can be simply labelled by one color.

Once the different 0, 1, 2-dimensional objects are represented by a set of cells, their HSFs can be straightforwardly obtained for them. Roughly speaking, an HSF of a digital image I is a set of trees living at interpixel level within an abstract cell complex version of I (or ACC, for short, [4]) and appropriately connecting all cells without redundancy. This notion is in principle independent of the pixel's value within the image. Using this concept, any digital 2D image I can be seen as a cell complex formed exclusively by two trees (see [7,8]). One of them, is formed by all the cells of dimension 0 and some of dimension 1 and the other one, is formed by the rest of cells of dimension 1 and all the cells of dimension 2 of I. More concretely, crosses are only represented by a unique

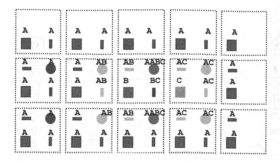
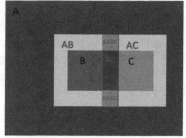

Fig. 3. Left: A 2-dimensional color image seen as an ACC. Regions are represented by 2-cells, contours by 1-cells and crosses by 0-cells. Each cell is labeled with its representative color. Right: The same image seen as a puzzle of 0, 1 and 2 dimensional objects.

tree (just one node, that is, a 0-cell). Contours are a set of 0 and 1-cells, whose "homology" information can be represented by one 1-cell and a set (which can be empty) of 0-cells representing the existence of holes. And finally, regions are a set of 0, 1 and 2-cells, whose homology information can be represented by one 2-cell and a set (which can be empty) of 1-cells representing the existence of holes. To fix ideas, the set of HSFs of Fig. 3 is depicted only by the cells representing their homology information in Fig. 4 (center), and a possible set of HSFs on the left of this image. The construction of HSFs is based on the concept of primal and dual vectors. Given two incidence cells c and c' of dimensions k and $k + 1$ respectively, a primal (resp. dual) $(k, k + 1)$ vector (c, c') (resp. (c', c)) connects the tail c (resp. c') with the head c' (resp. c). In this case, HSFs have been built following a boundary criterion for the primal vectors: from each k-cell one of its boundary $(k - 1)$-cells (of the same object) has been chosen (represented by an arrow). Dual vectors are represented by a line, and they link each $(k - 1)$-cell with all of its coboundary (k)-cells (of the same object).

In this case, the homology of region A can be condensed to one 2-cell and one 1-cell, whereas regions B and C can be described only by one 2-cell each, thus conforming the table on the right of the Figure. The three contours have no holes, that is, each of them is represented by one 1-cell, called AB, AC and BC. Each one of the two crosses are given by one 0-cell (named $AABC$). Finally, note that this process is fully generic for any 2-dimensional image with the 4-adjacency criterion

To this point, a color image can be condensed into a set of representative homology cells, each one being the root of a certain HSF tree in the case of two-dimensional objects. Additionally, these cells can be labelled with the colors that embody the relations among objects (see Fig. 4, right). The next step is taken here for efficiency purposes. Instead of managing relations among these cells or linking them in a graph (like in the case of RAG), it is preferable to condense all the homology information into a unique tree, called CRIT (Contour Region incidence Tree). CRIT combines the information of the flat regions, contours and

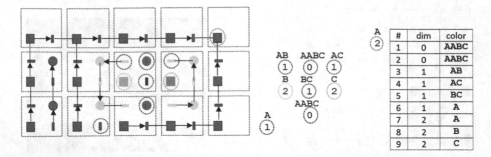

Fig. 4. Left: A possible set of HSFs for the image on Fig. 3. Cells representing homology information are surrounded by a dashed circle. Center: The set of homology information of the image indicating their dimension and representative color. Right: A table with the homology information of the image.

crosses of the original color image, plus the relations between them. According to Fig. 4, this supposes the conversion of the table of Fig. 4 (right) into a unique tree. To do this, some linking rules must be defined so as to prevent graphs. This can be achieved for instance by:

- For those cells of the same object, the same boundary criterion of the HSF building can be obeyed, that is, linking each representative (k)-cell with only one $(k-1)$-cell, until a 0-cell is reached. This comprises a new set of primal vectors.
- Inserting additional primal vectors by linking the rest of k-cells with one $(k-1)$-cell (of other object). Although, this usually supposes that 0-cells would remain disconnected to the rest of cells, for a simple image, the CRIT building may finish here. This is the case of Fig. 5, which contains a simple image of 4 pixels of different colors, so that a cross of 4 regions exists in the middle.
- Introducing additional links from the 0-cells to only one of their coboundary 1-cells, that is, a new set of dual vectors. This step must forbid forming graphs, for instance, by defining a direction for the links and preventing that the dual vectors go backwards along the selected direction.

In order to systematically build a CRIT in the most convenient way, a color incidence criterion can be followed. One possible set of rules is the following:

a) Firstly, "closing" each object by linking each k-cell with one of its boundary $(k-1)$-cells (of the same object, that is, the same color). These are represented by continuous arrows in this paper. For instance, in Fig. 6, only one primal vector exists (region A).
b) Secondly, inserting the rest of additional primal vectors by linking the rest of k-cells with one $(k-1)$-cell (of other object) having the most similar color to the k-cell. These are represented by a dashed arrows in Fig. 6. No matter that two or more $(k-1)$-cells match in the similar color criterion. For example,

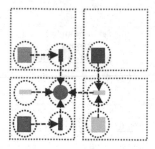

Fig. 5. A simple image of 4 pixels having different colors, so that a cross of 4 regions exist in the middle. In addition, this image can built a CRIT just by inserting additional primal vectors from the homology representative k-cells to one $(k-1)$-cell (of other object).

contour BC can be linked to the upper or the inferior cross. Similarly region B may link with contour BC or with the AB, and so on. For this image, five of these additional primal vectors must be inserted.

c) Finally, dual vectors are represented by double arrows in the same Figure. Note that the criterion has been "going outwards", that is, from the center to the image border. Thus, each 0-cell goes to the 1-cell (hole of region A) that surrounds the composite region B, C.

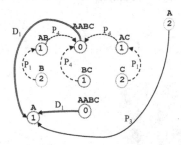

Fig. 6. Building the CRIT by linking the homology group cells of the image in Fig. 3

This criterion is followed in all the figures of this paper. Previous rules can be straightforwardly determined for bidimensional color images. More concretely, objects can be classified into five cases (Fig. 7, top), and therefore, only seven primal vectors and two dual vectors can exist (Fig. 7, bottom). Besides the link cases that appear in Fig. 6, the rest of possible links are drawn for a different color image in Fig. 8.

To sum up, note that a CRIT representation is consistent, so that any topological representation can be extracted from it. For instance, RAGs can be straightforwardly calculated by simply selecting the 1-dimensional objects (they

#	Object description	Set of homology groups (representative cells)
O_2	A region without holes	2
$O_{2\,1}$	A region with one or more holes	2, 1, ..., 1
O_1	A contour without holes	1
$O_{1\,0}$	A contour with a hole	1, 0
O_0	A cross (of 3 or 4 regions)	0

#	Additional Primal Vectors (description)	Example
P_1	2 → 1: from object O_2 to O_1	Fig. 6 from B to AB, from C to AC
P_2	2 → 1: from object O_2 to $O_{1\,0}$	Fig. 8 from 2-cell of E to 1-cell of ED
P_3	2 → 1: from object $O_{2\,1}$ to $O_{2\,1}$	Fig. 6 from 2-cell of A to 1-cell of A
P_4	1 → 0: from object O_1 to O_0	Fig. 6 from AB to AABC, from AC to AABC, from BC to AABC
P_5	1 → 0: from object $O_{1\,0}$ to $O_{1\,0}$	Fig. 8 from 1-cell of ED to 0-cell of ED
P_6	1 → 0: from object $O_{2\,1}$ to $O_{1\,0}$	Fig. 8 from 1-cell of E to 0-cell of F
P_7	1 → 0: from object $O_{2\,1}$ to O_0	Fig. 8 from 1-cell of F to FGH

#	Additional Dual Vectors (description)	Example
D_1	0 → 1: from object O_0 to $O_{2\,1}$	Fig. 6 from AABC to 1-cell of A (two vectors)
D_2	0 → 1: from object $O_{1\,0}$ to $O_{2\,1}$	Fig. 8 from 0-cell of ED to 1-cell of E

Fig. 7. Top: The five possible objects for bidimensional color images from a topological point of view. Bottom: Possible edges among topological objects for bidimensional color images.

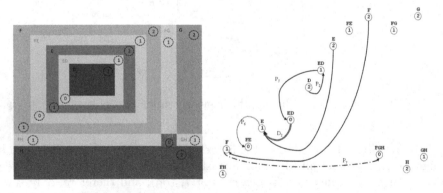

Fig. 8. Left: A 2-dimensional color image that contains additional links that are not present in Fig. 6. Each homology representative cell is drawn with a dashed circle and its dimension. Right: Homology representative cells and the additional links (P_2, P_6, P_7, D_2). Each cell is labeled with its representative color and its dimension. The rest of additional links are not drawn for clarity purposes.

are the edges of the RAG) and the 2-dimensional ones (they compose the RAG nodes). Each RAG edge directly associates the two nodes of its representative color. However, an interesting circumstance is when two regions share two (or more) contours (Fig. 9). In this case, regions K and L share two common contours, namely $KL1$, $KL2$. Usual RAG representation is shown in Fig. 9, right, where the edge between K and L is sometimes weighted by a factor of 2 to indicate this double contact. Nonetheless, common RAGs do not represent the hole of M, nor the fact that K and L are contained in it.

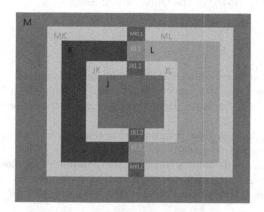

Fig. 9. Left: An image with two regions K and L having two common boundaries, namely $KL1$, $KL2$. Because of external M and internal J regions, there are also 4 crosses that can be represented by the same colors in two couples $(MKL1, MKL2, JKL1, JKL1)$. Right: Classical RAG representation for this image.

4 Conclusions and Future Work

A CRIT model of a color 2D digital image I (seen as an ACC) is a new topologically consistent tree-representation having as nodes the different 0-cells, 1-cells and 2-cells of I and having as edges some of the incidence relationships between 0-cells and 1-cells or 1-cells and 2-cells. The set of the rest of incidence relationships for any node in a CRIT that are not expressed as edges of it is suitably stored as node information. From a CRIT, it is automatic to construct the smaller ACC that encodes the homological information of I, having as 2-cells the different constant-color connected regions of I, as 1-cells, the different connected cracks between two regions and as 0-cells the cross points of I. Moreover, although this model is not unique for one image and its construction depends on some fusion criterion for the regions, it is also straightforward to change one CRIT into another one. In a near future, we plan to advance in the following issues:

- to design and implement an almost-full parallel algorithm for computing a CRIT of a digital image of dimension $n \times m$ of complexity near $\log(n + m)$.
- to design fast algorithms for transforming one CRIT into another one that meets specific incidence requirements;
- to define new geometric and topological features or characteristics associated to a CRIT.

Acknowledgments. This work has been supported by the Ministerio de Ciencia e Innovacion of Spain and the AEI/FEDER (EU) through the research project PID2019-110455GB-I00 (Par-HoT) and the IV-PP of the University of Seville.

References

1. Fiorio, C.: A topologically consistent representation for image analysis: the frontiers topological graph. In: Miguet, S., Montanvert, A., Ubéda, S. (eds.) DGCI 1996. LNCS, vol. 1176, pp. 151–162. Springer, Heidelberg (1996). https://doi.org/10.1007/3-540-62005-2_13
2. Klette, R.: Cell complexes through time. Communication and Information Technology Research Technical report 60 (2000)
3. Kovalevsky, V.: Algorithms and data structures for computer topology. In: Bertrand, G., Imiya, A., Klette, R. (eds.) Digital and Image Geometry. LNCS, vol. 2243, pp. 38–58. Springer, Heidelberg (2001). https://doi.org/10.1007/3-540-45576-0_3
4. Kovalevsky, V.: Geometry of Locally Finite Spaces. Baerbel Kovalevski, Berlin (2008)
5. Kropatsch, W.: Building irregular pyramids by dual-graph contraction. In: IEEE Proceedings-Vision, Image and Signal Processing, pp. 366–374 (1995)
6. Lienhardt, P.: Topological models for boundary representation: a comparison with n-dimensional generalized maps. Comput.-Aided Des. **23**(1), 59–82 (1991)
7. Molina-Abril, H., Real, P.: Homological spanning forest framework for 2D image analysis. Ann. Math. Artif. Intell. **64**(4), 385–409 (2012)
8. Diaz-del Rio, F., Real, P., Onchis, D.: A parallel homological spanning forest framework for 2d topological image analysis. Pattern Recognit. Lett. **83**, 49–58 (2016)
9. Diaz-del Rio, F., Sanchez-Cuevas, P., Molina-Abril, H., Real, P.: Parallel connected-component-labeling based on homotopy trees. Pattern Recognit. Lett. **131**, 71–78 (2020)
10. Rosenfeld, A.: Adjacency in digital pictures. Inf. Control **26**, 24–33 (1974)
11. Saha, P., Strand, R., Borgefors, G.: Digital topology and geometry in medical imaging: a survey. IEEE. Trans. Med. Imaging **34**(9), 1940–1964 (2015)

Author Index

Printed in the United States
by Baker & Taylor Publisher Services

Printed in the United States
by Baker & Taylor Publisher Services